ADVANCES IN CHEMICAL PHYSICS

VOLUME XXX

ADVANCES IN CHEMICAL PHYSICS—VOLUME XXX

I. Prigogine and Stuart A. Rice—Editors

Molecular Scattering: Physical and Chemical Applications

Edited by K. P. LAWLEY

Department of Chemistry, Edinburgh University

AN INTERSCIENCE® PUBLICATION

JOHN WILEY AND SONS

LONDON · NEW YORK · SYDNEY · TORONTO

Library of Congress Cataloging in Publication Data:

Lawley, K. P.
Molecular scattering: physical and chemical applications.

(Advances in chemical physics; v. 30)
'A Wiley-Interscience publication.'
1. Excited state chemistry—Addresses, essays, lectures.
2. Chemical reaction, Rate of—Addresses, essays, lectures.
I. Title. II. Series.

QD453.A27 vol. 30 [QD461.5] 541'.08s [539'.6] 74-23667

ISBN 0 471 51900 6

Printed in Great Britain by J. W. Arrowsmith Ltd., Bristol

INTRODUCTION

In the last decades chemical physics has attracted an ever-increasing amount of interest. The variety of problems, such as those of chemical kinetics, molecular physics, molecular spectroscopy, transport processes, thermodynamics, the study of the state of matter and the variety of experimental methods used, makes the great development of this field understandable. But the consequence of this breadth of subject matter has been the scattering of the relevant literature in a great number of publications.

Despite this variety and the implicit difficulty of exactly defining the topic of chemical physics, there are a certain number of basic problems that concern the properties of individual molecules and atoms as well as the behaviour of statistical ensembles of molecules and atoms. This new series is devoted to this group of problems which are characteristic of modern chemical physics.

As a consequence of the enormous growth in the amount of information to be transmitted, the original papers, as published in the leading scientific journals, have of necessity been made as short as is compatible with a minimum of scientific clarity. They have, therefore, become increasingly difficult to follow for anyone who is not an expert in this specific field. In order to alleviate this situation, numerous publications have recently appeared which are devoted to review articles and which contain a more or less critical survey of the literature in a specific field.

An alternative way to improve the situation, however, is to ask an expert to write a comprehensive article in which he explains his view on a subject freely and without limitation of space. The emphasis in this case would be on the personal ideas of the author. This is the approach that has been attempted in this new series. We hope that as a consequence of this approach, the series may become especially stimulating for new research.

Finally, we hope that the style of this series will develop into something more personal and less academic than what has become the standard scientific style. Such a hope, however, is not likely to be completely realized until a certain degree of maturity has been attained—a process which normally requires a few years.

At present, we intend to publish one volume a year, and occasionally several volumes, but this schedule may be revised in the future.

In order to proceed to a more effective coverage of the different aspects of chemical physics, it has seemed appropriate to form an editorial board. I want to express to them my thanks for their cooperation.

I. Prigogine

CONTRIBUTORS TO VOLUME XXX

A. P. M. BAEDE, Dordogne 3, Leusden-C, Holland

G. G. BALINT-KURTI, School of Chemistry, Bristol University, Cantock's Close, Bristol BS8 1TS

UDO BUCK, Max-Planck-Institut fur Strömungforschung, 34 Gottingen, Böttingerstrasse 6–8, West Germany

R. GRICE, Department of Chemistry, Cambridge University, Lensfield Road, Cambridge CB2 1EW

VOLKER KEMPTER, Fakultät für Physik, Universitat Freiburg, Hermann-Herder-Strasse 3, D-7800 Freiburg i. Br., West Germany

WALTER S. KOSKI, Department of Chemistry, The Johns Hopkins University, Baltimore, Maryland 21218, U.S.A.

K. P. LAWLEY, Department of Chemistry, West Mains Road, Edinburgh EH9 3JJ

D. A. MICHA, Department of Chemistry, University of Florida, Gainesville, Florida 32601, U.S.A.

W. MILLER, Department of Chemistry, University of California, Berkeley, California 94720, U.S.A.

J. REUSS, Fysich Laboratorium, Kathiolieke Universiteit, Dreihuizerweg 200, Nijmegen, Holland

CONTENTS

INTRODUCTION: NEW DIRECTIONS IN MOLECULAR BEAMS   1
    By K. P. Lawley

QUANTUM THEORY OF REACTIVE MOLECULAR COLLISIONS   7
    By D. A. Micha

CLASSICAL S-MATRIX IN MOLECULAR COLLISIONS   77
    By W. H. Miller

POTENTIAL ENERGY SURFACES FOR CHEMICAL REACTION   137
    By G. G. Balint-Kurti

SCATTERING OF POSITIVE IONS BY MOLECULES   185
    By W. Koski

REACTIVE SCATTERING   247
    By R. Grice

ELASTIC SCATTERING   313
    By U. Buck

SCATTERING FROM ORIENTED MOLECULES   389
    By J. Reuss

ELECTRONIC EXCITATION IN COLLISIONS BETWEEN NEUTRALS   417
    By V. Kempter

CHARGE TRANSFER BETWEEN NEUTRALS AT HYPERTHERMAL ENERGIES   463
    By A. P. M. Baede

SUBJECT INDEX   537

INTRODUCTION: NEW DIRECTIONS IN MOLECULAR BEAMS

K. P. LAWLEY

Department of Chemistry, West Mains Road, Edinburgh EH9 3JJ

The first areas of chemical interest to be explored by beam techniques in the 1960's were simple bimolecular reactions and elastic scattering, both involving alkali metals. The early success in the first of these fields was the recognition of different limiting models for direct reactions and their link with the product angular scattering pattern and the energy partitioning among the product degrees of freedom. In elastic scattering the observation of the rainbow phenomenon and, later, supernumerary rainbows led to the first unambiguous measurement of intermolecular potentials over fairly well-defined ranges of the separation. Those models of chemical reactions have become part of our way of thinking about molecular dynamics and the intermolecular potentials derived from elastic scattering now set a standard for other, less direct methods. These two areas continue to attract much interest. In the field of reactive scattering, studies have been extended from the A + BC type of reaction to those in which four atoms or groups play a part. Several non-alkali systems have also been investigated, a very important step in that it helps to correct any bias that may have entered into our overall view of chemical reactions that comes from studying only the alkali metals. Within the reactive scattering field the tendency is towards more precise energy analysis, with the internal state now being probed spectroscopically in favourable cases, either by chemiluminescence or laser induced fluorescence. These trends are surveyed by R. Grice in Article 6 and there are hints of more exciting things to come as the beam technique is extended to higher kinetic energies so that reactions involving appreciable activation energy can be studied and the overlap with conventional chemical kinetics is thereby completed.

The use of accelerated beams, however, raises the old question in chemical kinetics of the relative efficiencies of vibrational and translational energy in supplying the activation energy of a reaction. While vibrational population inversion in a beam can be achieved in selected cases by optical pumping, any beam method in this area will have to compete with chemical laser techniques. In these the decay of emission from the upper vibrational states is monitored in the presence of a quenching gas (i.e. the reaction partner) in the optical cavity itself.

New directions in elastic scattering are covered by U. Buck in Article 7. Here the recent emphasis has been on non-alkali systems and on measurements of the highest possible angular (and, equally necessary, energy) resolution. The observation of high-frequency interference between the positive and negative branches of the deflection function is often the additional piece of information that makes possible the satisfactory unfolding of the differential cross section to give the deflection function; the inversion of this to the potential is then fairly standard. Atomic scattering over the range of energies below the threshold for electronic excitation is now in a very advanced state. Above the excitation threshold, elastic scattering is depleted by loss to the new channels and an optical potential analysis may be attempted. Much more information is contained in the scattering pattern if a localized curve crossing is involved and the probability of crossing is not too close to zero or unity. An entirely new interference structure between trajectories following the two potential surfaces then appears, and in favourable cases a portion of both potentials can be mapped.

Elastic scattering studies are increasingly being extended to excited atoms where they overlap the field of vacuum ultraviolet spectroscopy. Molecular line spectra are only observed if both the ground and excited diatom states have potential minima; the beam technique is free from this limitation and also from the operation of Franck–Condon factors, which limit the range of separations effectively scanned in a vibronic band. Nevertheless, except under the highest angular resolution, a differential cross section does not contain as much information as a line spectrum converging to a band head.

Ion/neutral reactive scattering studies are in some ways easier experimentally than neutral/neutral work. The better energy selection and analysis provided the first detailed energy disposal information on simple reactions. The use of ions also extended the types of potential energy surface encountered. Recent developments are surveyed by W. Koski (Article 5), and include experiments in translation–vibration energy transfer by the energy loss technique that are made possible by the high energy resolution of ion beams. The subject of inter-molecular energy transfer had its beginnings in vibrational energy relaxation in shock heated gases and the dispersion of the velocity of sound. Now, besides beam scattering techniques, crossed beam I.R. fluorescence can be used to measure changes in vibrational populations if the spectrum lies in a convenient region. From a molecular point of view, translational–internal energy transfer is governed by the cross terms in the interparticle Hamiltonian that involve both internal and relative coordinates. To probe these terms using scattering techniques, we can either measure inelastic cross sections or, for the angle dependent terms in the potential, the orientation dependence of the elastic cross section. From a quantum point of view, molecular orientation must be achieved by selecting the rotational state and this is the subject of Article 8 by J. Reuss. The technique is difficult, but the non-central part of the potential has otherwise proved particularly elusive. In the absence of rotational state selection, these angle dependent terms tend to quench the structure in the differential and total cross sections. For diatomics the effect is not very pronounced (which is an advantage in deriving the central or angle averaged part of the potential) and it has not proved a useful route to measuring these aniso-tropic terms.

As the wealth of experimental detail concerning simple chemical reactions grows, so does the need for a theoretical basis to correlate them. Theory can intervene at several points. The scattering pattern may suggest a simple phenomenological model such as direct or complex, stripping or recoil, but these usually only apply in a particular energy range. More fundamental is the underlying potential energy surface and here the question of uniqueness enters. Detailed though our experimental knowledge of some reactions may be, an unambiguous many-body potential cannot be deduced and the *ab initio* or semi-empirical calculation of some features of the surface then becomes an important part of the fitting. These calculations are discussed by G. G. Balint-Kurti in Article 4.

Classical mechanics has well-known limitations, indeed they were expected to show up in conventional chemical kinetics where the tunnelling of protons was much sought after. In particular, there is the question of classical behaviour when the amount of initial and final quantum state averaging is small, and of the threshold energy dependence of the cross section of a new channel—something that is uniquely well probed by beam techniques. These uncertainties must be borne in mind as the number of classical calculations grow and it is important to lay proper foundations for a general theory of reactive scattering. This is done by D. A. Micha in Article 2, and W. H. Miller in Article 3. Theory in its present phase of applica-tion is more concerned with the limits of validity of approximations such as transition state theory, statistical and impulse models, the optical potential and one-dimensional calculations rather than with full *a priori* calcula-tions of the cross sections of simple reactions. The bridge between classical and quantum mechanics lies in the semi-classical approximation which associates a phase with every classical path through the action integral and thus permits the interference effects which are prominent in detailed cross sections. The development of this in the field of reactive scattering is discussed in Article 3, where particular emphasis is placed on the validity of Monte Carlo methods.

Beam studies have until recently been largely confined to systems in which the dynamics are governed by a single potential surface. The use of classical trajectory studies and adiabatic correlation diagrams in predicting the reaction path are both implicitly founded on the Born–Oppenheimer approximation which allows us to deal with only one electronic state during the collision. Electronically excited states are, however, not far away in many reactions of ground state species as the observation of chemiluminescence (Article 6) from strongly exothermic reactions such as $M_2 + Cl$ reminds us. More generally, when scattering measurements are extended beyond thermal energies or the incident species are in an excited electronic or vibrational state (e.g. $N_2^{\dagger} + Na$), new parts of the potential energy surface are explored in which electronic states come much closer and may even cross. To be sure, in almost the first reactive system to be studied by crossed beams, that of the alkali metals with the halogens, an avoided curve crossing was involved but it transpired that at thermal speeds the motion over the crossing was essentially adiabatic and the system remained on the lower surface.

Some of the earliest evidence for non-adiabatic effects in scattering came from ion/atom systems, e.g. He^+/Ne. There are both experimental and theoretical reasons why such effects might be more prominent in ionic than in neutral systems. In the former there is greater likelihood of close lying charge transfer states of the same symmetry and their interaction may even be strong enough to be studied by the perturbation of the elastic scattering. Non-adiabatic effects in neutral systems can be studied by beam techniques in basically two ways, either by the observation of excited products from the scattering of ground state species or through the reverse process of quenching. The field of quenching, usually of resonance radiation, has until recently been the preserve of kinetic spectroscopy where many of the excited species studied are too short-lived for conventional beam techniques. However, beam experiments involving the more energetic metastable atoms (and with lifetimes $\gtrsim 10^{-3}$ s) are beginning to yield results, principally with the ^3P states of the rare gases. The He $(1s2s)$ ^1S and ^3S states are also widely studied and together with the $2s$ state of H are convenient in that M substate selection is not necessary. Indeed, the angular resolution in current work on He*/He is so high that perturbing effect of a third He_2 state can be seen.

The three non-adiabatic effects most easily studied by beam techniques are (1) collisional excitation followed by the emission of radiation for the upper state, (2) collisional ionization to $A^+ + B^-$, (3) Penning ionization. Collision induced fluorescence is discussed by V. Kempter in Article 9; the technique is a relatively sensitive one and cross sections down to 10^{-21} cm^2 can be detected, though this still falls short of the sensitivity of bulb spectroscopic methods. The technique is also powerful in that separate cross sections for excitation of members of a multiplet can be obtained although, as always with a beam technique, absolute values are difficult to obtain.

If the ground state of the colliding pair should come close to an ionic one, the ion pair becomes a possible exit channel. A. P. M. Baede in Article 10 describes the measurement of interpretation of the energy dependence of the total and differential cross section for ion production in the 1–10 eV region where the process reaches its peak efficiency. A longer established technique is Penning electron spectroscopy in which the key quantities are the autoionizing width, $\Gamma(R)$, the neutral and the ionic potentials. The detailed information contained in the energy distribution of the Penning electrons enables a rough fit for $\Gamma(R)$ to be obtained, together with information about one of the potentials, thus making the phenomenon a rather well understood one from an essentially classical point of view. In systems in which quenching is principally through Penning ionization, $\Gamma(R)$ also plays a role in the total scattering cross section and the elastic differential cross section so that these areas should all link up. Many new problems in interpretation are raised by extending scattering measurements to excited states. Besides the diagonal matrix elements of Hamiltonian of the composite system (i.e. the adiabatic potential energy functions) for at least two electronic states, the off-diagonal elements that lead to coupling between these states now play a part. Fortunately the Landau–Zener formula continues to serve the two state situation well, in spite of much critical attention. Together with Coriolis coupling it forms the basis of interpretation of the bulk of the results in Articles 9 and 10. When augmented by the phase factors introduced by Stueckelberg, the Landau–Zener formula accounts well for differential inelastic scattering involving a well-defined curve crossing.

Interesting developmens that can be expected in the field of excited state scattering are exploration of vibronic coupling, increased atomic state selection for atoms with $J > \frac{1}{2}$ and scattering from increasingly short-lived species (possibly optically pumped) so that beam and kinetic spectroscopy finally link up. In these more complicated situations, the relative kinetic energy dependence of excitation or quenching cross sections obtained by beam methods will play a vital role in identifying the mechanism of energy transfer in processes of relatively large cross sections.

QUANTUM THEORY OF REACTIVE MOLECULAR COLLISIONS

D. A. MICHA

*Quantum Theory Project, Departments of Chemistry and Physics,
University of Florida, Gainesville, Florida 32611*

CONTENTS

I. Introduction 7
  A. Scope of this Review. 7
  B. Organization of this Review 9
II. Computational Modelling 11
  A. General Aspects 11
  B. Collinear Motion 12
  C. Coplanar and Spatial Motion 31
III. Many-Channel Approaches 37
  A. Statistical Models 37
  B. Optical-Potential Models 48
  C. Coupled-Channels Method 52
  D. Electronic Transitions 59
IV. Many-Body Approaches 60
  A. Impulsive Models 60
  B. Faddeev–Watson Equations 62
V. Discussion 66
  A. Summary of Results 66
  B. Expectations and Future Needs 68
Acknowledgements 69
References 69

I. INTRODUCTION

A. Scope of this Review

A great wealth of information on reactive molecular collisions has continued to accumulate during the last few years, thanks to experimental developments in the fields of molecular beams and of spectroscopy of reacting species. This in turn has motivated a number of theoretical contributions, some of which are designed to interpret experimental results, others to isolate concepts and methods of general validity and of predictive value.

The wider field of molecular collisions has periodically been reviewed in the past, to keep pace with the literature. By comparison, there have been few reviews in the last years concentrating on reactive molecular collisions.

One of these was written by Nikitin (1968); another one by Light (1971), emphasizing quantum calculations with coupled equations and related approximations. An overview of the quantum theory has recently been presented by Kouri (1973) who concentrated on exact quantum mechanical methods. In the present review we wish to include computational and analytical developments of the quantum theory, and both approximate and formally exact treatments. We have covered the literature since 1970 and have referred to previous articles only occasionally, when they have been a basic source of information. The coverage extends to articles that had appeared or were scheduled to appear by the end of 1973.

The more general reviews on molecular collisions contain a good deal of information relevant to our aim, which we would like to single out. The formalism of the quantum theory of reactive collisions, as it relates to molecular systems, was covered by Levine (1969), who also described a number of approximate models. Several important contributions appeared in the volume edited by Schlier (1970), with work by Cross on inelastic processes, Ross and Greene on elastic processes in reactive systems, Ross on formal reactive scattering theory, Henglein on ion–neutral reactions, Karplus on distorted-wave methods and Bunker on unimolecular decomposition. Fluendy (1970) has reviewed molecular scattering. A volume edited by Alder and others (1971) contains several computational treatments by Gordon, Light, Lester and Secrest on integration of sets of coupled scattering equations. Unimolecular reactions are treated in detail in books by Robinson and Holbrook (1972), Weston and Schwarz (1972) and Forst (1973). An extensive list of references may be found in a review by Levine (1972). Long-lived states in atom–molecule collisions, both reactive and non-reactive, have been briefly discussed by Micha (1973). The most recent reviews are by George and Ross (1973) who also covered many aspects of reactive scattering, by Secrest (1973) on rotational and vibrational energy transfer, and by Connor (1973) on both reactive and non-reactive collisions, with emphasis on trajectory studies. A forth-coming review (Micha, 1974) presents in detail effective hamiltonian methods for molecular collisions. Recent theoretical and experimental advances have been reported in conferences, with Proceeding published in the *Discussions of the Faraday Society* (vol. 55, 1973), which contains reviews by Marcus and by Polanyi, and in *Berichte der Bunsengesellschaft fur physikalische Chimie* (vol. 77, ner. 8, 1973).

A thorough understanding of reactive collisions is not possible without reference to potential energy surfaces. Expositions of interest to dynamical studies have been written by Karplus (see Ch. Schlier, ed., 1970), by Krauss (1970) and by Certain and Bruch (1972), see also Article 4, this volume. Several groups of quantum chemists are presently producing potential surfaces, for light atoms, of useful chemical accuracy. Among these are McLean,

Wahl and their collaborators (Lester, ed., 1971) and Schaefer and collaborators (e.g. Schaeffer, 1973). Recent work of significance for dynamical calculations has been that on diatomics-in-molecules (Ellison, 1963) as recently employed by Kuntz and Roach (1972) and by Tully (1973a, b).

Reviews of experimental work relevant to the theory of elementary reactions include those of Steinfeld and Kinsey (1970), Polanyi (1971) on non-equilibrium processes, Lee (1971) on reactive scattering of atoms and molecules, several articles in the volume edited by Polanyi (1972), written by J. C. Kinsey on molecular–beam reactions, Dubrin and Henchman on ion–molecule reactions, and Carrington and Polanyi on chemiluminescence reactions. Dubrin (1973) has recently covered reactions of high kinetic energy species. J. P. Toennies (1974) surveys molecular beam experiments and discusses elastic, inelastic and reactive collisions. Both experimental and theoretical aspects of ion–molecule reactions are described by McDaniel and others (1970) and by Friedman and Reuben (1971), and in contributions by Henchman and by Herman and Wolfgang in two volumes on *Ion-molecule Reactions* (Franklin, ed., 1972). An introductory monograph by Fluendy and Lawley (1973) deals with chemical applications of molecular–beam scattering, and another by Levine and Bernstein (1974) outlines many of the concepts and phenomena of molecular reaction dynamics.

Work in several related areas usually contain relevant material. One such area is chemical lasers (Dzhidzhoev and colleagues, 1970; Moore, 1971; Moore and Zittel, 1973; Kompa, 1973); another area is gaseous chemical kinetics (Spicer and Rabinovich, 1970; Troe and Wagner, 1972; Westenberg, 1973). Bersohn (1972) has discussed chemical applications of optical pumping. The field of molecular reactions owes a great deal to nuclear and high-energy physics, where several of the models have been developed (see e.g. Hodgson, 1971). A suggestive description of potential practical applications of basic research in molecular dynamics has been given by Bernstein (1971). The activated-complex approach to chemical kinetics has continued its development during the last years; and has been recently reviewed by Daudel (1973) and by Christov (1972, 1974). This approach has several points in common with the statistical theories we shall describe, and may provide a link between collision theory and chemical kinetics in liquids.

B. Organization of this Review

There is a great variety of approaches to the calculation of cross sections for reactive molecular collisions. This variety reflects the different interests and backgrounds of investigators in the field, which has attracted both chemists and physicists. Although some approaches will eventually prove more general and efficient than others, no single approach is likely to provide the answer for each of the many collision problems that arise when going from some chemical species to others or when changing physical parameters.

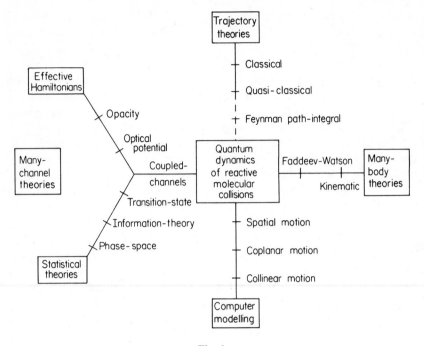

Fig. 1.

This review is organized after methods of approach, as shown in Fig. 1. We can distinguish four main categories: computer modelling, many-channel theories, many-body theories and trajectory theories. Since the present study focuses on quantum theories we shall not consider methods based on classical trajectories, but we refer to previous reviews. Trajectory methods originated with very successful classical calculations using model potential surfaces with variable parameters (see reviews by Bunker, 1971; Polanyi and Schreiber, 1973), and have been significantly extended to account for quantal interference and tunnelling, particularly by Miller and collaborators (1971) and by Marcus and collaborators (1971). These latter quasi-classical methods may be considered to be approximations to the exact formulation of quantum mechanics in terms of Feynman path-integrals.

Computer modelling has provided a wealth of numerical results. Many accurate cross sections are now available for collinear collisions of atoms with diatoms, where motion is constrained to a line at all times. These models are clearly not for immediate application to physical situations, but they are useful in other respects. They provide exact results with which approximations may be tested. This testing may prevent waste of effort while developing approximations for physical (spatial motion) situations. Results on coplanar motion are available and others are likely soon on spatial motion, in both cases making use of the computational techniques developed for collinear motion.

Many-channel theories in their simplest forms employ statistical or effective Hamiltonian concepts. Statistical theories, listed in order of increasing reference to the conformation of the collision complex, are the phase-space model, the information-theory (or apparent temperature) model and the transition-state models. Effective Hamiltonians provide justification for introducing opacity models and optical-potential models. The coupled-channels method may be used to introduce all these models and, on the other hand, it may be made as accurate as desired, although at great cost. Many-body theories, where the bodies are atoms or ions, are relatively new in the field of molecular reactions. We have included here impulsive models and the Faddeev–Watson method. Impulsive models make extensive use of kinematic conservation rules and very little use of potential surfaces. They have been very successful in interpreting experimental results and are simple to apply. The Faddeev–Watson equations are themselves relatively new and offer the most rigorous description of three interacting bodies. Some kinematic models may be derived from them, and recent computational developments make it now possible to obtain numerical results with these equations.

The classification we are following is not strict. For example, coupled-channel equations appear also in computational modelling although other aspects are more important here. Furthermore, as the different approaches are refined, some of them should become equivalent.

Most of the reviewed applications deal with atom–diatom rearrangement collisions. This is not a big restriction, however, because the dynamics of three atoms includes most of the challenges of larger systems. Also, most of the discussed reactions involve only ground electronic states, i.e. they are adiabatic. The amount of work done so far on non-adiabatic transitions in chemical reactions is small, and we shall only briefly refer to it.

II. COMPUTATIONAL MODELLING

A. General Aspects

A large part of the computational work has been influenced by the introduction of curvilinear coordinates, designed to take advantage of the topography of potential surfaces. These coordinates allow for a smooth change from reactant to product conformations and in effect transform the rearrangement problem into the much simpler one of inelastic collisions. The various treatments have employed: reaction-path (or natural collision) coordinates; less restricted reaction coordinates; atom-transfer coordinates, somewhat analogous to those used for electron-transfer; and, for planar and spatial motion, bifurcation coordinates.

To illustrate the general procedure let us consider collinear reactions of type $A + BC \rightarrow AB + C$. This means that the three atoms move on a line, e.g. the x-axis, and furthermore that velocities are also along the same line. Indicating relative atomic distances by x_{AB}, x_{BC} and x_{CA}, we introduce centre-of-mass coordinates, which for reactants are $x = x_{BC}$ and X, the distance from A to the centre of mass of BC. Similar coordinates could be defined for products. Because of restrictions in the type of motion it is not possible to simultaneously account for the $B + CA$ rearrangement. To do this one must proceed to planar or spatial motion and, for example, introduce bifurcation coordinates.

Curvilinear coordinates may be defined in terms of (x, X) and potential parameters. We sketch their use here, leaving detailed definitions and references for the following sections. Reaction path coordinates are usually designated s, along the path, and ρ, perpendicular to it. The Schrodinger equation becomes

$$\left\{ \frac{1}{\eta}\left(\frac{\partial}{\partial s} \frac{1}{\eta} \frac{\partial}{\partial s} + \frac{\partial}{\partial \rho} \eta \frac{\partial}{\partial \rho} \right) + \frac{2\mu}{\hbar^2}[E - V(s, \rho)] \right\} \Psi(s, \rho) = 0$$

where μ is the reduced mass of reactants, and $\eta = 1 + \varkappa(s)\rho$, with $\varkappa(s)$ the path curvature. Expanding Ψ in a basis of N internal states $\varphi_n(\rho; s)$,

$$\Psi_m = \sum_{n=1}^{N} \psi_{nm}(s)\varphi_n(\rho; s) = \boldsymbol{\varphi}\boldsymbol{\psi}_m$$

where $\boldsymbol{\varphi}$ is a row matrix and $\boldsymbol{\psi}_m$ a column matrix for reactants initially in state φ_m. The $N \times N$ matrix $\boldsymbol{\psi}$, whose rows are the $\boldsymbol{\psi}_m$, satisfies

$$\left[\mathbf{1} \frac{d^2}{ds^2} + \mathbf{P} \frac{d}{ds} + \mathbf{Q}(E) \right] \boldsymbol{\psi}(s) = 0$$

The physical boundary conditions for scattering are,

$$\psi_{nm} \underset{s \to -\infty}{\sim} \sqrt{\frac{\mu}{k_m^{(-)}}} e^{ik_m^{(-)}s} \delta_{nm} + \sqrt{\frac{\mu}{k_n^{(-)}}} e^{-ik_n^{(-)}s} \mathbf{S}_{nm}^{(-)}$$

$$\underset{s \to +\infty}{\sim} \sqrt{\frac{\mu}{k_n^{(+)}}} e^{ik_n^{(+)}s} \mathbf{S}_{nm}^{(+)}$$

where reactants correspond to $s \to -\infty$ and products to $s \to +\infty$, and $\mathbf{S}^{(\pm)}$ is the scattering amplitude $N \times N$ matrix. Reaction probabilities are given by $\mathbf{P}_{nm}^{(+)} = |\mathbf{S}_{nm}^{(+)}|^2$, also variedly called $\mathbf{P}_{nm}^T$ or $\mathbf{P}_{nm}^R$ in the literature, and direct probabilities by $\mathbf{P}_{nm}^{(-)} = |\mathbf{S}_{nm}^{(-)}|^2$, also called $\mathbf{P}_{nm}^R$ (for reflection) or $\mathbf{P}_{nm}^V$. Wavenumbers are $k_m^{(\pm)} = [(2\mu/\hbar^2)(E - W_m^{(\pm)})]^{1/2}$, with $W_m^{(\pm)}$ the internal plus dissociation energies.

The procedure has usuaily been to choose a parametrized potential surface, to vary some of its features such as position and shape of barriers, or isotopic masses, and then to extract physical insight from the computed output. This program has been carried out for a variety of surfaces, integration algorithms and expansion basis sets. Converged numerical results have been compared among themselves and with approximations such as the classical, several quasi-classical and distorted-wave ones.

Efforts to interpret numerical, and also experimental, results have led to analytical models for describing internal excitation of products. These have also made use of curvilinear coordinates and will be reviewed in the following subsection.

A different approach which has recently provided many results for collinear collisions is the direct integration of the partial differential equation. This approach by-passes the use of curvilinear coordinates and does not require expansions in terms of internal (vibrational) states.

In the following paragraphs we review these and other developments in the last few years, including the more recent treatments that extend to coplanar and spatial motions.

B. Collinear Motion

Reaction-path coordinates were first described in detail by Marcus (1966). Choosing a curve $\mathscr{C}$ in the two-dimensional configuration space (x, X) for the reaction $AB + C \rightarrow A + BC$, he introduced two new variables: the distance s along $\mathscr{C}$, and r, the shortest distance of nearby points in the plane to $\mathscr{C}$. He then proposed an adiabatic-separable method that included curvilinear motion effects. Writing for the potential V, without loss of generality,

$$V(s, r) = V_1(s) + V_2(r, s)$$

he defined the reaction path as the curve $r_0(s)$ on which the classical local vibrational and internal centrifugal forces balance. The Hamiltonian operator was as given before, with $\rho = r - r_0$. He proposed an adiabatic-separable method based on trial wavefunctions

$$\Psi(s, \rho) = \Phi^{(1)}(s, \alpha)\Psi^{(2)}(\rho, \alpha)$$

with parameters α denoting constants and quasi-constants of motion, and described non-adiabatic corrections and expressions for transmission coefficients.

A set of coordinates closely related to the previous one was introduced by Light and coworkers (Rankin and Light, 1969). These authors defined a new arbitrary set of reaction coordinates (u, v) chosen by convenience in the plane (x, X). Letting

$$u = u_0 - \gamma/x + \beta x$$

and taking v to be the perpendicular distance of a point to the curve, one can express x and X in terms of u and v and obtain a kinetic energy operator similar to the previous one except that now

$$\eta = [1 - K(u)v]\, ds/du$$

where K is a positive curvature.

It is then possible to introduce a basis set for the local vibrational motion with effective mass and vibrational constant dependent on u. For the harmonic oscillator set $\{\varphi_{iu}(v)\}$ the total wavefunction was expanded in the form

$$\Psi(u, v) = \eta^{-1/2} \sum_i f_i(u)\varphi_{iu}(v)$$

to obtain the set of coupled equations

$$\frac{d^2}{du^2} f_n(u) + \sum_m V'_{nm}\frac{df_m}{du} + V_{nm}f_m = 0$$

It has been pointed out that, since the operator in this equation is not Hermitian, an S matrix calculated with a truncated set would not be unitary. But in practice numerical deviations from unitarity could be kept small.

This set of coupled equations was solved by a modified exponential method (Chan *et al.*, 1968). For an N-dimensional basis set, a $2N$-dimensional column matrix $\mathscr{F}$ was defined with the first N elements given by the f_i and the second N elements by df_i/du. This matrix satisfies the differential equation

$$d\mathscr{F}/du = \begin{pmatrix} \mathbf{0} & \mathbf{1} \\ -\mathbf{V} & -\mathbf{V'} \end{pmatrix}\mathscr{F}$$

whose solution may be expressed in terms of a translational matrix $\mathbf{U}$ such that

$$\mathscr{F}(u) = \mathbf{U}(u, -\infty)\mathscr{F}(-\infty)$$

and

$$\mathbf{U}(u_0, u_0) = \mathbf{1}$$

$$\mathbf{U}(u_n, u_0) = \sum_{i=0}^{n-1} \mathbf{U}(u_{i+1}, u_i)$$

$$\mathbf{U}(u_{i+1}, u_i) = \exp\left\{\begin{bmatrix} \mathbf{0} & \mathbf{1} \\ -\mathbf{V}(\bar{u}_i) & -\mathbf{V'}(\bar{u}_i) \end{bmatrix}(u_{i+1} - u_i)\right\}$$

with $2\bar{u}_i = u_{i+1} + u_i$. The S matrix could then be obtained from the collection of solutions $\mathscr{F}^{(j)}(u)$ for $u \to \pm\infty$, constructed such that the incoming particles were in the jth internal state for $u \to -\infty$.

This approach was applied to several reactions using realistic potential surfaces, equal to sums of a potential along u plus a Morse potential in v with u-dependent parameters, and centred at the local minima $v_0(u)$.

One of these reactions was $H + Cl_2$ (Miller and Light, 1971a, b; Light, 1971a, b). In this case the number of open channels is different for reactants and products and a certain number of closed channels must be used with the exponential method in the intermediate region. Four aspects of the surface were varied: the path's curvature; steepness of the potential; position of the barrier; and vibrational force constant around the saddle point or col. Vibrational distributions of products were determined, besides the dependence of probabilities on collision energy.

Another work (Russell and Light, 1971) was on tunnelling in isotopically substituted reaction of $H + H_2$. It showed that tunnelling is determined, not only by the reaction-path height and width, but also by the curvature $K(u)$ and the perpendicular vibrational frequency, which change from one isotopic reaction to another.

C. G. Miller (1973) has compared $H + Cl_2 \rightarrow HCl + Cl$ with the isotopic variation $D + Cl_2 \rightarrow DCl + Cl$. The path of the second reaction has smaller curvature, which would favour low excited products, but this is in part compensated by the smaller vibrational spacing of DCl, which favours high excitation. These two reactions may be considered of type L + HH, with L for light and H for heavy atom. The reaction $Cl + HI \rightarrow HCl + I$ was also taken up as an example of H + LH. The fraction of product vibrational excitation in these reactions was investigated as a function of the position of the col and of the vibrational constant there.

A computational approach of a very different nature, usually referred to as the 'boundary value method', introduces a multidimensional finite-difference mesh to directly solve the partial differential equations of reactive scattering. This approach has been taken by Diestler and McKoy (1968) in work related to a previous one by Mortensen and Pitzer (1962). While the second authors used an iterative procedure to impose the physical boundary conditions of scattering, the more recent work constructs the wavefunction Ψ as a linear combination of independent functions χ_j which satisfy the scattering equation for arbitrarily chosen boundary conditions.

For a collinear, electronically adiabatic reaction in the variables x_{AB}, x_{BC}, the procedure is to digitate these variables and number the finite-difference mesh so that the differential operator becomes a banded matrix. Then, the Schrodinger equation becomes

$$\mathbf{B}\chi_j = \mathbf{b}$$

where $\mathbf{B}$ is a banded matrix of order equal to the number of interior mesh points, χ is a column matrix with values of the χ_j at the mesh points and $\mathbf{b}$ is a column matrix of constants. The χ_j functions were chosen to asymptotically contain both trigonometric and exponential functions for open and closed channels, respectively.

This procedure does not require expansions in internal states. Hence, internal states need not be specified (except asymptotically), nor does one need to calculate matrix elements with respect to them. However, the approach requires solving a set of many (about 2000) coupled linear equations. The formalism was applied to three model potentials for the exchange of identical particles.

Extensive computational studies of collinear collisions have been carried out by Kuppermann and collaborators. Their work has made use of the potential surface of Wall and Porter, with the parameters of Shavitt (barrier height of 0·424 eV) and otherwise adjusted to *a priori* results (Shavitt *et al.*, 1968). The first results (Truhlar and Kuppermann, 1970, 1971; Kuppermann, 1971) were based on the boundary-value computational method of Diestler and McKoy, with some modifications that improved accuracy. Calculations were done at several grid sizes and extrapolated to infinitely fine grids. The results were subject to an R-matrix analysis to obtain transition probabilities to about 1%.

Calculations for $H + H_2$ at energies below threshold for excitation of the first vibrational level were compared with two approximations which assumed separability of motions along the reaction path and transversal to it. One approximation assumed conservation of transversal vibrational energy, and the second one conservation of transversal vibrational quantum number, i.e. adiabatic motion. The second assumption was closer to the exact results both in magnitude and in the shape of the reaction threshold region. However, as collision energies were increased and other vibrational states of products became excitable, the behaviour became more nearly statistical, in that probabilities of reactions tended to decrease and become comparable.

It was also found that the elastic probability showed an increase as the threshold for the first vibrational state of products was passed, rather than a decrease, as classical arguments about competition for flux would suggest. This is a quantum interference effect which was pointed out as a manifestation of the quantum mechanical optical theorem. Figure 2 shows the energy dependence of calculated probabilities.

These accurate results were used (Truhlar and Kuppermann, 1971) to test the transition-state approximation (TSA) as it would apply to a collinear world. For the mentioned surface the activation energy is 0·277 eV. A calculation of the exact rate coefficient $k(T)$ could be fitted by an Arrhenius plot in the range 800°K to 1300°K with a pre-exponential factor $A = 2·20 \times 10^5$ cm molec/sec and $E_{act} = 0·299$ eV. However, at low temperatures deviations were evident, as large as 40% at 450°K. Plots of $k_{TSA}(T)/k(T)$ for different classical and quantal treatments of tunnelling show values close to unity at high temperatures, but again large deviations (sometimes of several orders of magnitude) at low temperature.

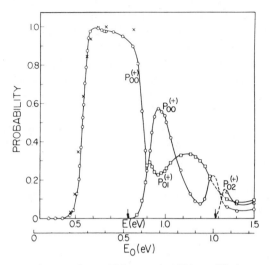

Fig. 2. Reactive collision probabilities $P_{0n}^{(+)}$, from ground state of the reactants, calculated by Truhlar and Kuppermann (1972), as functions of total energy E and relative kinetic energy E_0. Arrows indicate vibrational thresholds. Crosses are results of Mortensen and Gucwa (1962) for $P_{00}^{(+)}$, shifted to the left by 0·057 eV.

A comparison with classical results was also carried out by Bowman and Kuppermann (1971). Relative collision energies were varied from 0·20 to 1·28 eV and from 0·07 to 0·70 eV for ground and for singly excited states of reagents. Classical thresholds were 0·07 eV and 0·09 eV above the quantal ones, respectively. Classical results decreased beyond threshold with quantal results oscillating around them, but classical oscillations were also found near threshold. A comparison of classical and quantal rate coefficients showed good agreement (larger than 75 %) at high temperatures, with activation energies of 0·300 eV and 0·299 eV respectively, but not at temperatures below about 700°K.

A complete coverage of most of the foregoing results has been presented by Truhlar and Kuppermann (1972) together with a detailed list of references. This work includes a description of the computational procedure and adds to previous comparisons another one relating to the statistical phase-space theory as developed by Lin and Light (1966). It was concluded that at total energies above 0·86 eV, for which singly excited products are possible, the phase-space prediction that reaction probability equals 0·50 for each of the two open reaction channels is verified on the average. Some of these conclusions are evident in Fig. 2.

An instructive analysis of quantum effects has been made by Schatz *et al.*
(1973) for $F + H_2 \rightarrow FH + H$. They employed Muckerman's surface (1971)
with barrier height 0.0471 eV and exoergicity 1.3767 eV, and used the foregoing computational scheme. For collision energies up to 0.4 eV, four vibrational states of HF were open. Results for the total probability $P_0^{(+)}$ of
reaction from ground state of H were 2.5 times smaller than classical ones,
for collision energies below 0.075 eV. Only $P_{02}^{(+)}$ and $P_{03}^{(+)}$ contributed appreciably to $P_0^{(+)}$. The authors concluded that the extent of agreement between
quantum and classical calculations depends also on the nature of the potential
surface. Figure 3 shows results and a comparison.

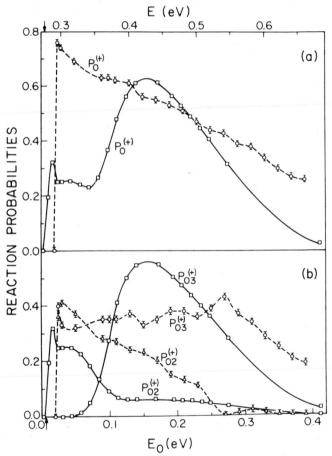

Fig. 3. Reaction probabilities for collinear $F + H_2$: (*a*) Total
reaction probabilities; (*b*) Contributions to (*a*). Solid lines are
quantum calculations and dashed ones classical calculations. The
vertical arrow indicates the $n = 3$ threshold of HF (from Schatz
et al., 1973).

A calculation of delay times $\tau_{ij}^{(+)}$ and phase shifts $\delta_{ij}^{(+)}$ was the basis for a discussion of interference of direct and resonance processes in reactions of $H + H_2$ (Schatz and Kuppermann, 1973). With their previous potential, these authors found resonances for $i = 0$, $j = 0$ at total energies 0·90 and 1·276 eV with widths 0·05 and 0·008 eV, respectively. The maximum delay time $\tau_{00}^{(+)}$ at the 1·276 eV resonance is about an order of magnitude greater than the period of symmetric stretch vibration of H_3 at the saddle point. An Argand plot was also given for these resonances (see Fig. 4).

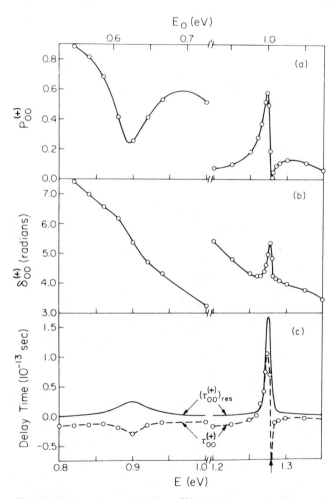

Fig. 4. (a) Reaction probability $P_{00}^{(+)}$; (b) phase-shift $\delta_{00}^{(+)}$; and (c) time delay $\tau_{00}^{(+)}$ and its resonance component $(\tau_{00}^{(+)})_{res}$ for collinear $H + H_2$. The arrow indicates the $n = 2$ threshold of H_2, at 1·280 eV (from Schatz and Kuppermann, 1973).

The same surface and computational method were applied by Truhlar *et al.* (1973) to reactions with isotopes: (1) $H + H_2$; (2) $D + D_2$; (3) $D + H_2$; and (4) $H + D_2$. Plots of $\ln P_0^{(+)}$ versus energy in the threshold regions (collision energies between 0·05 and 0·25 eV) gave straight lines. It was also found that barrier heights $E(VAZC)$, from the vibrational-adiabatic zero path curvature approximation, agreed within 0·01 eV with threshold energies $E^{0·5}$ corresponding to $P_0^{(+)} = 0·5$, and within 0·02 eV with activation energies E^a for Arrhenius fits within 750–1250°K.

A direct test of the vibrationally adiabatic approximation for $H + H_2$ has also been made (Bowman *et al.*, 1973). This test was done by projecting accurate wavefunctions on the vibrationally adiabatic functions for zero curvature, and measuring deviations of the resulting probability weight from unity. The symmetric stretch motion was found to be adiabatic to within 10% for total energies between 0·51 and 0·72 eV, but adiabaticity was lost at lower and higher energies.

Very recently Duff and Truhlar (1973a) compared converged results of $P_{00}^{(+)}$ for $H + H_2$ in a Porter–Karplus surface. Values from the finite-difference boundary value method at six threshold energies agreed within 2% with Diestler's (1971) and Wu *et al.* (1973a), but there was disagreement with those of Wu and Levine (1971). Further comparison with several approximations showed that distorted-wave calculations gave a good description of threshold energy dependence.

The same authors (1973b) compared quantal and semi-classical results for $H + H_2$, to ascertain whether disagreements, when present, should be ascribed to the semiclassical assumptions or to neglect of complex-valued trajectories. They found disagreement at high energies, where complex trajectories do not play a role. Hence these could not be blamed. In fact, they concluded that the classical S-matrix approximation is most useful and most accurate when complex trajectories govern the processes. Other comparisons of semi-classical, quasiclassical and exact quantum results for this system were made by Bowman and Kupperman (1973).

The computational approach of Kuppermann (1971) has recently been applied by Baer (1974) to $H + Cl_2$ and $D + Cl_2$ reactions, partly to improve upon previous calculations that neglected some closed channels. He used a LEPS surface with a barrier of 0·108 eV in the entrance valley. Reaction probabilities for $H + Cl_2$ from $v_i = 0, 1, 2$ to $v_f \leq 7$ showed that the $0 \rightarrow 4$ transition dominated at low energies, while $0 \rightarrow 5$ and then $0 \rightarrow 6$ dominated as energies increased. The general trend was the same for $v_i = 1$ or 2, but, in detail, the distributions of v_f depended on v_i. These dependencies were discussed in terms of a model of vertical non-adiabatic transitions between two displaced vibrational wells. Results with $v_i = 0$ for $D + Cl_2$ showed that $0 \rightarrow 5$ dominated at low energies. Total transition probabilities were weakly dependent on both v_i and isotopic masses.

Persky and Baer (1973) also did a study of $Cl + H_2, D_2, T_2$, with $v_i = 0$, 1 in each case. Their aim was to compare the isotope ratios of rate coefficients. k_{Cl+H_2}/k_{Cl+D_2} and k_{Cl+H_2}/k_{Cl+T_2}, with experiment in the temperature range $230°K–400°K$. They found good agreement (to 10%) with differences of activation energies but magnitudes were too large by about 30%.

A study of reactive $H + H_2$ collisions has also been carried out by Wu and Levine (1971) using the close-coupling method and natural collision coordinates. They employed a Porter–Karplus (1964) surface and investigated the collision energy range between 9 and 35 kcal/mole. For the interaction they used

$$V(s, \rho) = V_1(s) + \mu\omega(s)^2\rho^2/2$$

with V_1 the potential along the reaction path and $\omega(s)$ the local vibrational frequency. They expanded the total wavefunction in an adiabatic set of local harmonic vibration states, and solved the coupled equations using up to 9 channels below 20 kcal/mole and 11 channels above that. Non-adiabatic coupling terms were classified as static (SC), resulting from the dependence of ω or s, and dynamic (DC), brought about by the varying curvature along the path. The first (SC) widens the valley in the col region, which makes more translational energy available there, and increases reaction probability. The second (DC) introduces a dynamical centrifugal barrier which decreases probabilities and dominates at high energies. The difference found between classical and quantal results at threshold was ascribed not to tunnelling but to different degrees of translation–vibration energy transfer in the two cases. More translational energy remains available in the quantum case, to surmount the barrier. Converged results were presented of reflection and transmission probabilities for transitions $0 \rightarrow 0$, $0 \rightarrow 1$, $1 \rightarrow 0$ and $1 \rightarrow 1$. The $0 \rightarrow 0$ results differed somewhat from those of Diestler (1971) for the same potential.

The same authors investigated resonances in reactive collisions (1971). For the particular case of curvature $K = 0$, they found one at a total energy $E = 12.3$ kcal/mole with time-delay $\tau = 3 \times 10^{-14}$ sec and another at $E = 28.7$ kcal/mole. Observing that $V_1(s) + (n + \frac{1}{2})\hbar\omega(s)$ shows a well capable of sustaining bound states of positive energy only for $n > 0$, one concludes that the first resonance is analogous to a shape resonance, while the second corresponds to a compound-state resonance. A shape resonance is shown in Fig. 5.

Isotopic variations of $H + H_2$, obtained by replacing hydrogens by deuteriums, were considered (Wu et al., 1973a) within a different computational scheme (Johnson, 1972) and, in connection with threshold behaviours and resonances for varying isotopic combinations. No significant reaction probabilities were found for total energies below the static barrier potential $V_1(s)$. One-dimensional barriers provided reasonable probabilities only when constructed in accordance with the potential $V_1(s) + \epsilon_0(s)$, which includes zero-point vibration and DC. Resonances were again investigated, with the proper curvature, and were delineated by means of Argand plots of $\mathrm{Re}\,(S^{(+)})$ versus $\mathrm{Im}\,(S^{(+)})$, with energy as a parameter. Resonances for reactive $H + H_2$ were found at $E = 10\cdot3$, $21\cdot7$ and $30\cdot7$ kcal/mole. A similar low energy resonance was found for $D + D_2$, but shifted and broadened due to mass effects.

A series of 'diagnostic' (London–Eyring–Polanyi–Sato) surfaces were introduced (Wu *et al.*, 1973b) to study effects of: (1) location of barriers and magnitude of exoergicity, (2) change of vibrational frequency along the reaction path and (3) resonances. It was found that, at given total energies, reactant translational energy was more effective than vibrational energy in promoting reaction for a surface with a barrier in the entrance channel and no exoergicity, and vice versa for a barrier in the exit channel and no exoergicity. Strong vibrational excitation of products was found in an 'early-downhill' surface with barrier in the entrance valley, while strong de-excitation was the rule for a 'late-downhill' surface with barrier in the exit

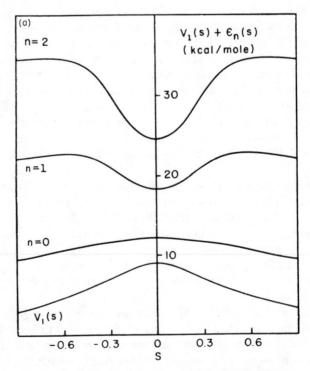

Fig. 5. (*a*) Adiabatic channel potentials for the Porter–Karplus potential of $H + H_2$ along the reaction coordinate s.

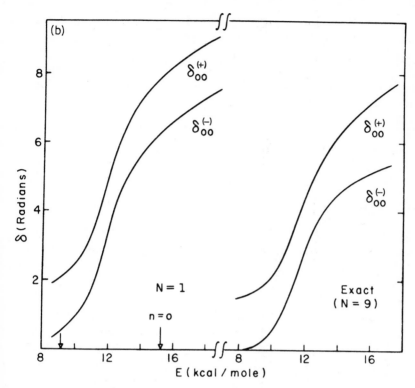

Fig. 5. (*b*) Reactive and non-reactive phase-shifts in the threshold region, in the neighbourhood of a shape resonance (from Levine and Wu, 1971).

channels. All these conclusions agree with previous ones from classical results. Resonances were found whenever the $V_1(s)$ potential showed a well. The $F + H_2 \rightarrow H + HF$ reaction probabilities were calculated to illustrate the general conclusions.

A comparison of exact results for $H + H_2$ with semiclassical ones (Wu and Levine, 1973c) was done for reactive $0 \rightarrow 0$ and $0 \rightarrow 1$ transitions between $E = 12{\cdot}5$ and 30 kcal/mole. Comparisons were made with the primitive semiclassical approximation, a simplified version of the uniform semiclassical approximation, and a near-classical approximation (Levine and Johnson, 1970). These approximations were only qualitatively correct.

Variational techniques have been used by Mortensen and Gucwa (1969) for $H + H_2$. They based their treatment on the Kohn variational principle, whereby the integral

$$I_i = \int d\tau \chi_i (H - E)\chi_i$$

must be made stationary for variations of the trial χ_i functions subject to scattering boundary conditions. These were taken to be

$$\chi_i(x_{AB}, 0) = \chi_i(0, x_{BC}) = \chi_i(\infty, \infty) = 0$$

i.e. they excluded dissociation. At energies below threshold for vibrational excitation the chosen form was

$$\chi_i = B_{1i}\varphi_v(x_{BC})[\varphi_{ts}(X_1) + \tan \epsilon_i \cdot \varphi_{tc}(X_1)]$$
$$+ B_{3i}\varphi_v(x_{AB})[\varphi_{ts}(X_3) + \tan \epsilon_i \cdot \varphi_{tc}(X_3)]$$
$$+ \sum_{jk} c_{ijk}\varphi_{vj}(x_{AB})\varphi_{vk}(x_{BC})$$

where φ_v is a vibrational eigenfunction of the Morse oscillator, while φ_{ts} and φ_{tc} are relative motion functions going asymptotically as $\sin kX$ and $\cos kX$ respectively. These last are in turn a superposition of two distorted waves, solutions of distortion potentials with variationally chosen parameters. Numerical results were actually obtained minimizing either I_i^2 or the norm of $(H - E)\chi_i$. Once the χ_i were determined, they were combined to form scattering functions ψ_1 with the physical asymptotic behaviour. Transmission probabilities obtained with up to 25 φ_v functions were stable except at the total energy $E = 17.5$ kcal/mole, for which the phase angle ϵ_1 in the χ_i was too close to $-\pi/2$. Results for the Sato potential with Weston's parameters agreed with results of Mortensen and Pitzer (1962) to within 2%. It was pointed out that the instability could be avoided by choosing a different asymptotic form for the χ_i functions.

Another approach to the $A + BC \rightarrow AB + C$ reaction has been developed by Diestler (1971) making use of two symmetric coordinate systems: (x_{AB}, x_{BC}) which is appropriate for reactants, and (x_{AC}, x_{AB}) for products. The common coordinate x_{AB} is initially large, decreases during collision and increases again whether direct or atom-exchange events take place. The total wavefunction is written in the form

$$\Psi = \sum_m f_m^\alpha(x_{AC})\varphi_m^\alpha(x_{BC}) + \sum_n f_n^\beta(x_{AC})\varphi_n^\beta(x_{AB})$$

with α and β reactant and product channels, $0 \leq m \leq N_\alpha$ and $0 \leq n \leq N_\beta$, and with Morse oscillator functions used for internal states.

Schrodinger's equation leads then to $N_\alpha + N_\beta$ second-order coupled differential equations with first derivatives. These are solved transforming the set into $2(N_\alpha + N_\beta)$ first-order coupled equations, and generating $N_\alpha + N_\beta$ linearly independent solutions by choosing suitable boundary conditions at small r_{AC}. A Bashforth–Moulton fourth-order predictor-corrector algorithm was used in the integration.

This procedure does not require an *a priori* choice of curvilinear coordinates. However, the expression for Ψ is not convenient for introducing scattering boundary conditions, which requires reexpansion of Ψ in centre-of-mass coordinates once each for reactants and products. This in turn requires calculation of new integrals of type

$$I_{mm'} = \int dx_{BC}\varphi_m^\alpha(x_{BC})\exp(ik_m^\alpha\Omega_\alpha x_{BC})\varphi_{m'}^\alpha(x_{BC})$$

All quadratures were done by means of the Gauss–Hermite algorithm.

Results using the Porter–Karplus surface show that reaction from $m = 0$ does not occur until the total energy (measured from H + H + H) is greater than the barrier height value of -4.351 eV, and becomes almost certain between -4.2 eV and -4.0 eV. Reaction probabilities from either $m = 0$ or 1 into $n = 1$ are larger than 0.5 between -3.85 and -3.55 eV.

These quantum mechanical (QM) results have been compared (Diestler and Karplus, 1971) with classical mechanical (CM) ones obtained from Monte Carlo averages of classical trajectories with the same potential. The QM threshold is about 0.07 eV lower than the CM one, and QM total probabilities of reactions are consistently larger. The CM results did not show the sharp QM dip present just above threshold for vibrational excitation of products.

The same quantum mechanical computational approach has been used (Diestler, 1972) to study the influence of variations of a potential surface on reaction probabilities. The employed potential was an eight-parameter Wall–Porter surface. For H + H_2 this is symmetric in reactant and product coordinates. The entrance and product valleys correspond to a Morse potential (3 parameters) and the intermediate region to a rotated Morse potential (5 parameters). Attention was concentrated on variation of (1) barrier height, (2) curvature of reaction path and (3) slope and barrier width along the reaction path. Some of the salient observations are: (1) the difference between threshold energy and barrier height is inversely related to the height; (2) the degree of vibrational excitation of products, just above threshold, decreases with increasing barrier height and increases both with increasing path curvature and increasing path slope; but (3) at energies well above this threshold the mentioned trends are reversed. These conclusions were compared with the corresponding ones of CM (Fong and Diestler, 1972). Trends deduced from CM and QM were found to agree on the average, with the extent of agreement dependent on the shape of the surfaces. In all these studies it is well to remember that variation of potential parameters leads to changes that, while more noticeable in certain regions, are of global rather than local nature.

The above-mentioned computational technique has been compared with the finite difference boundary-value method as modified by Truhlar and Kupperman (Diestler et al., 1972) for a Wall–Porter fit to the Shavitt et al. (1968) surface. Results over the total energy range 0.014 a.u. to 0.400 a.u. agree typically to 1%, except for some very small probabilities.

An instructive description of the $H + H_2$ reaction was provided by McCullough and Wyatt (1971a, b). They constructed a wavepacket and followed its time development on the Porter–Karplus surface. They introduced centre-of-mass coordinates appropriate for reactants and used these for all times. The wavefunction Ψ at time $t = n\,\Delta t$ was constructed, from the time-evolution operator U, in the form

$$\Psi(\mathbf{r}, t) = U(\Delta t)^n \Psi_0(\mathbf{r})$$

$$U(\Delta t) = [1 - e^{i\Delta t H/(2\hbar)}]/[1 + e^{i\Delta t H/(2\hbar)}]$$

with an initial wavepacket

$$\Psi_0(\mathbf{r}) = \varphi(X)\chi(x_{BC})$$

where $\chi(x)$ is a Morse oscillator and $\varphi(X)$ a plane wave times a normalized Gaussian amplitude. A two-dimensional finite-difference mesh was employed for the spatial integration, which transformed the equation for $\Psi(t)$ into a matrix problem.

Time-evolution could conveniently be described in terms of the probability density $\rho(\mathbf{r}, t) = |\Psi(\mathbf{r}, t)|^2$ and flux

$$\mathbf{j}(\mathbf{r}, t) = (\hbar/\mu)\,\mathrm{Im}\,[\Psi^*(\mathbf{r}, t)\nabla\Psi(\mathbf{r}, t)]$$

The most noticeable dynamical features were the tendency of the wavepacket to move to the outside of the reaction path near the col, climbing the wall as it moved into the products valley, and the appearance of vortices in the flux stream around the col. A very strong resemblance was found between wavepacket motion and hydrodynamical flow. A comparison with classical flux lines showed very good agreement in cases where classical motion was allowed.

Defining the reaction probability at time t as the fraction of probability density on the products side of the symmetric stretch line at time t, it was possible to plot reaction probabilities versus time for the quantal and classical cases. Comparing results at several energies, classical mechanics underestimated probabilities at large t and reaction times, but overall it gave the correct behaviours and magnitudes.

A broken-path model considered by Middleton and Wyatt (1972) consists of dividing the surface into several regions and introducing local Cartesian coordinates s, ρ in each of them. These coordinates are chosen to lie close to the curvilinear coordinates and rotate from region to region. In each region $V(s, \rho)$ is assumed separable in $V_1(s) + V_2(\rho)$, with V_1 a linear potential and V_2 a Morse potential. Hence, the wavefunction is locally of the form

$$\Psi(s, \rho) = \sum_{i=1}^{N} \chi_i(s)\varphi_i(\rho)$$

where the φ_i are Morse oscillator functions and the χ_i are combinations of regular and irregular Airy functions. Integration of the scattering equation is accomplished in terms of propagation matrices, which differ from those for step-translation in that they include matching of functions and derivatives at region boundaries. Calculations were carried out for the Wall–Porter surface with Shavitt's parameters (1968). They studied the influence of closed-channels, non-adiabaticity and the effect of changing N from region to region. Probability conservation could be kept within 5%, although deviations from reversibility were larger. Agreement with results of Truhlar and Kuppermann was only qualitative.

Walker and Wyatt (1972) have also performed a distorted-wave calculation for $H + H_2$, based on the Porter–Karplus surface. They constructed reactant and product distortion potentials assuming adiabatic vibrational motion in each case, and obtained numerical solutions for the relative motions. Their results show that by choosing adequate potential parameters it is possible to reproduce the threshold behaviour, but that probabilities grow above unity soon after the threshold energy.

An approach somewhat related to the broken-path method, but more accurate, has been employed by Johnson (1972). It also makes use of curvilinear coordinates (s, ρ) chosen with constant curvature, and divides the intermediate region into several sectors. In each of these the variables s, ρ are separable. Calculations were done with the amplitude density technique, matching functions and derivatives at each sector's boundary. The potential was that of Porter and Karplus, and the local vibrational motion was assumed to be harmonic. Good agreement was found with Diestler's $P_{00}^{(+)}$ at low energies.

Shipsey has employed his own set of coordinates (1969, 1972) in a calculation (1973) of $P_{0 \to v_f}^{(+)}$ for $O + HBr \to OH + Br$, with $v_f = 0, 1, 2$. This reaction has an exoergicity of about 1 eV. The potential surface was a LEPS one with a barrier of 0.00264 eV in the entrance valley. The probability $0 \to 1$ was the largest at all collision energies from 0.0 to 0.315 eV; the $0 \to 2$ one was not appreciable until after about 0.12 eV. Good agreement was found (Koeppl, 1973) between these numerical results and absolute reaction rate theory at threshold energies.

Two other attempts have been made by Crawford. In the first one (1971a) an R-matrix approach was used, following the Wigner–Eisenbud procedure. It was concluded that the R-matrix parameterization did not provide a simple solution. The second one (1971b) employed the Kohn variational principle and a basis of Gaussian functions. Results from a normalized variational S matrix were compared to exact ones and showed some promise.

Several distorted wave approximations were explored by Gilbert and George (1973). Their best results were obtained by fitting the Porter–Karplus surface separately for reactants and products, which leads to two quasi-adiabatic surfaces. Each of these were given, in relative coordinates, by a fit which assumed a Morse potential with variable parameters, for vibrational motion. Their unitarized results, obtained using an exponential form, gave reasonable values near threshold.

Idealized potentials have been useful for checking computational procedures. A detailed study with idealized potentials was carried out by Tang, Kleinman and Karplus (1969) with a procedure applicable to particles of arbitrary mass. They divided the potential surface into entrance valley, exit valley and intermediate region, using centre-of-mass coordinates in the first two and suitable polar coordinates in the third. For piecewise surfaces constructed from constant potentials, they obtained probabilities in terms of integrals calculated by Gauss–Legendre quadratures. The discussion included quantum-mechanical tunnelling and resonance scattering.

Descriptions of scattering by an L-shaped potential in terms of Cartesian coordinates are affected by instabilities at certain energies. It was pointed out (Robinson, 1970) that when collision energies coincide with eigenvalues of intermediate homogeneous boundary-value problems, it is necessary to replace basis functions of type $\sin ax \sin by$ by others such as $x \cos \alpha x \sin \beta y - y \cos \alpha x \sin \beta y$, in order to obtain convergent results. The modified basis set was employed by Dion, Milleur and Hirschfelder (1970) to bring older results into agreement with those of Tang et al. at the critical energies mentioned above.

Idealized potentials of a different type have been used by James and North (1972). These authors employed the Faddeev equations, to which we return later on, to obtain numerical results for three particles interacting through a sum of pairs of delta-function potentials. This interaction was chosen for computational convenience, in order to solve easily for the two-body off-energy-shell transition operators required in the method and to compare the numerical results with exact ones. They considered three identical particles, each pair with a single bound state of energy $-11\cdot2$ eV. Their transmission probabilities agreed with the exact ones and were given between zero kinetic energy and that for the break-up threshold.

The great majority of reactive collinear collisions have been studied computationally, with physical insight being extracted from the final numerical results. An alternative to this approach is the development of models that may be interpreted in terms of approximate analytical solutions. This has been done by Hofacker and collaborators. A non-adiabatic model was introduced (Hofacker and Levine, 1971) to describe population inversion in reactions. It makes use of reaction path coordinates and assumes harmonic vibrational motion perpendicular to the path.

Introducing vibrational creation and annihilation operators b, b^+, such that $[b, b^+] = 1$, and expanding the Hamiltonian in an adiabatic basis, the total Hamiltonian is written as a sum of the three terms

$$H_0 = \sum_\alpha [E_\alpha + \omega_{\alpha\alpha} b^+ b] |\alpha\rangle\langle\alpha|$$

$$V_d = \sum_\alpha g_{\alpha\alpha} |\alpha\rangle\langle\alpha| (b^+ + b)$$

$$V_{nd} = \sum_{\alpha \neq \beta} [g_{\alpha\beta}(b^+ + b) + \omega_{\alpha\beta} b^+ b] |\alpha\rangle\langle\beta|,$$

where $|\alpha\rangle$ is the relative motion wavefunction along the reaction path for relative energy E_α, $\omega_{\alpha\beta}$ is the matrix in the $|\alpha\rangle$ set of the local vibrational frequency and $g_{\alpha\beta}$ is a matrix, in the same set, which measures the change of the relative kinetic energy operator $T^s(s, \rho)$ with vibrational coordinate ρ. Of these terms, H_0 represents the unperturbed motion, V_d the diagonal part of the vibration–translation coupling, and V_{nd} the rest. The term V_d may be eliminated by the unitary transformation

$$U = \exp\left[\sum_\alpha (g_{\alpha\alpha}/\omega_{\alpha\alpha})(b^+ - b) |\alpha\rangle\langle\alpha| \right]$$

Of the two terms $\tilde{H}_0$ and $\tilde{V}_{nd}$ in the transformed Hamiltonian $\tilde{H}$, the first one corresponds to adiabatic oscillatory motion along a dynamically displaced path. The second one, $\tilde{V}_{nd}$, provides the coupling leading to vibrational transitions. Depending on whether the vertical displacement of vibrational wells is larger or smaller than a vibrational quantum, it is predicted that high or low vibrational states will be found in the products, respectively. The foregoing approach was used (Hofacker and Levine, 1972) to modify the transition-state theory so as to account for product vibrational populations in direct reactions. With the previous mechanism in mind, the phase space available to the transition complex was limited by requiring that the probability distribution of the complex's states be a maximum under the constraints of fixed total energy and average vibrational energy $\langle E_V \rangle$ of products. This leads to definitions of effective vibrational entropy and temperature T_V (see Section III.A). For negative temperatures the most likely vibrational quantum number should be $v_{max} > 0$. This would happen whenever $\langle E_V(T_V) \rangle > \langle E_V(T_V \to \infty) \rangle$.

Working with a similar model Fischer and Ratner (1972) developed a theory applicable to polyatomics, but simplified in practice by introduction of reaction path variables, s, along the path and ρ_i for the other degrees of freedom (neglecting rotations). Rather than focusing on cross sections, they developed expressions for rate coefficients to predict the effect on vibrational excitation of products of relative energy, curvature of the reaction path, changes in normal-mode frequencies and position of the saddle point of the surface.

Another analytic approach has been developed by Basilevsky (1972). He discusses bimolecular reactions $A + B \rightarrow C + D$ in terms of curvilinear coordinates (s, ρ_i), after separating the six (or five in the linear case) degrees of freedom of centre-of-mass and overall orientation. He defines a zeroth order Hamiltonian neglecting curvature along the reaction path and assumes a potential equal to a sum of an Eckart potential along s and Morse potentials along the ρ_i. The remaining terms in the Hamiltonian are considered a perturbation and a Born expansion is proposed for the scattering amplitudes.

A more recent paper by the same author (1973) evaluates coupling terms of the Hamiltonian between adiabatic vibrational states in a semiclassical fashion, which leads to equations similar to the Landau–Zener one for electronic excitation. The treatment applies to slow motion along the path and fast vibrational motion in $A + BC \rightarrow AB + C$, a situation complementary to that treated by Hofacker and Levine (1971). The author indicates the presence of a spurious asymptotic coupling in this other work.

Parameters required in the model of Hofacker have been approximated (Hofacker and Rösch, 1973) for the $F + H_2$ surface recently calculated by Schaefer *et al.* Values of the curvature $\varkappa$, the effective reduced mass along the path, and a semiclassical approximation to $g(s)$ were given. From the maxima of g for $F + H_2$ and $F + D_2$, the quotient $g_{DF}/g_{HF} = 1\cdot24$ was found to be in good agreement with the experimentally extracted value of $1\cdot19$. The magnitudes of g were highly sensitive to the shape of the potential surface.

We can reach a number of conclusions from all the foregoing work, at least for the restricted world of collinear collisions:

(a) There are now accurate scattering probabilities for $H + H_2$, for two different potential surfaces (the Porter–Karplus and Shavitt ones) and from a variety of computational methods. Results include energy dependences, vibrational distribution of products, isotope effects and threshold behaviour.

(b) Different opinions exist as to the meaning and role of tunnelling. This is not an essential problem insofar tunnelling is only meant to measure deviations of classical calculations. It relates, however, to the permeabilities of transition-state theory, but using these does not seem necessarily to improve agreements between transition-state and accurate rate coefficients.

(c) The existence of long-lived or compound states in reactions has been computationally established, and resonance structures have been found in the probabilities both near threshold (shape resonances) and above (compound-state resonances).

(d) Classical and quasi-classical results appear to be qualitatively correct, and to agree on the average with quantal ones, but the extent of agreement depends on the surface's shape and the range of energies.

(e) Adiabatic behaviour of vibrational modes depends on the choice of co-ordinates. It is restricted to regions immediately above threshold but it disappears at higher energies, and in fact is replaced by a statistical behaviour whenever several product channels are open. Analytical models appear promising for the interpretation of non-adiabatic transitions.

(f) Although seldom used so far, variational methods appear capable of providing good numerical results, and an avenue to the study of more complex processes. Distorted-wave calculations have given good results at threshold but there is not, so far, a satisfactory way of choosing distortion potentials.

C. Coplanar and Spatial Motion

Quantum mechanical calculations of reactions for three atoms moving on a plane, or even more generally in space, have been for some time the aim of researchers in the field. In this section we consider efforts that are directly linked to use of curvilinear coordinates, or that are an extension of the approaches for collinear motion. Approaches from different angles are left for the other main sections.

Early attempts to extend collinear calculations have allowed for rotation of the axis on which the three atoms are positioned, so that velocities may point in other directions and impact parameters do not need to be zero. More recently, the first computed results on coplanar motion have appeared. Studies on spatial motion are being developed at present, but equations have only been solved for schematic potentials.

Collinear models may be extended to include the centrifugal forces that arise from rotation of the three-atom axis. This has been done by Wyatt (1969), who studied $H + H_2$ reactive collisions with a vibrational adiabatic model. The required coordinates are (x_{AB}, x_{BC}), and new ones (θ, φ) which specify the axial orientation. Transforming (x_{AB}, x_{BC}) to curvilinear coordinates (s, r) he expressed the total wavefunction as

$$\Psi(r, s, \theta, \varphi) = \sum_{l=0}^{\infty} R^{-1}\Psi_l(r, s)Y_l^0(\theta, \varphi)$$

$$\Psi_l(r, s) = \psi_l(s)\Phi_l(r, s)$$

where $\psi_l(s)$ is the wavefunction for motion along s, and then neglected non-adiabatic coupling terms in the determination of Φ_l. This procedure led to effective potentials

$$V_l(r, s) = V(r, s) + [l(l + 1) + 1]\hbar^2/(2\mu R^2)$$

where R is a radius of giration. The second term is a Coriolis potential which raises and widens the barrier along the reaction path. Wyatt obtained numerical reaction probabilities as functions of collision energy E, $p_l(E)$, for three models corresponding to the perturbed stationary-state (PSS) approximation, and to reaction-path (RP) and reaction-coordinate (RC) models. Plots of p_l versus E were found to move to the right with increasing l. Total (i.e. sum of partial) cross sections $S(E)$ were calculated by means of

$$S(E) = [\pi \hbar^2/(2\mu E)] \sum_{l=0}^{\infty} (2l + 1)p_l(E)$$

and threshold behaviours indicated that $S(\text{PSS}) > S(\text{RP}) > S(\text{RC})$ at low energies. All these results were obtained with the Porter–Karplus surface for H_3.

Another modified collinear model was developed by Connor and Child (1970). They constrained atoms A, B and C to a line, but allowed this to rotate. In this way impact parameters could be different from zero, although velocities remained always in the same plane. The two coordinates (X, x) in centre of mass were replaced by natural ones, (s, ρ). The wavefunction Ψ was taken as a product of an adiabatic vibrational state $\xi_n(s, \rho)$ times a translational function $\psi_n(s, \theta)$, with θ the axial angle. Then

$$\Psi = (s^2 + \rho^2)^{1/2}\psi_n(s, \theta)\xi_n(s, \rho)$$

$$\psi_n(s, \theta) = \sum_{l=0}^{\infty} A_l S_{nl}(s)p_l(\cos \theta)$$

and, for a potential $V(s, \rho)$, neglecting the curvature of the reaction path, they obtained

$$\left(\frac{d^2}{ds^2} + \frac{2}{\hbar^2}\left\{E - V_n(s) - \frac{\hbar^2[l(l + 1) + 1]}{2(s^2 + \rho_e^2)}\right\}\right)S_{nl}(s) = 0$$

with ρ_e the equilibrium value of ρ. The potential $V_n(s)$ was chosen as (1) a Lorentzian shape; (2) a downhill one; and (3) a double well with a barrier. The one-dimensional problem was solved semiclassically, and sums over l were carried out by means of the stationary phase method. In all cases the reactive differential cross sections were forward peaked. An estimate was made of reactive probabilities in the spirit of the absorption model, by replacing the short-range portion of the double well by a repulsive potential and comparing the large angle elastic differential cross sections. Results were correct within a factor of 2.

The first preliminary calculations for coplanar $H + H_2$ were published by Saxon and Light (1971). This system is favourable for exploratory calculations because of the large rotational spacing of H_2. Total energies were varied from 13 to 17·5 kcal/mole and total angular momenta from 0 to 12 $\hbar$. Calculations were made for an analytical fit to the Shavitt et al. (1968) surface, within a manifold of 19 rotational states (from $j = -9$ to $j = 9$)

for $14 \cdot 5 \leq E \leq 17 \cdot 5$ kcal/mole, whereby open channels have $|j| \leq 7$, and of 17 rotational states ($j = -8$ to 8) for $13 \cdot 0 \leq E \leq 14 \cdot 5$ kcal/mole, with open channels for $|j| \leq 6$.

Cross sections (of dimension length) for fixed j and varying collision energy were nearly superimposable, with those for larger j growing more rapidly. Small values of $|j_i - j_f|$ were favoured and, for $j_i = 0, 2$, the reaction probabilities descreased monotonically with increasing orbital angular momentum l.

In more detailed publications, Saxon and Light (1972a, b) pointed out that no computationally satisfactory set of curvilinear coordinates could be found going smoothly from reactants to products. They chose to make a transformation from reactants to product coordinates at an intermediate region. For reactants, the internal polar coordinates (r, θ) and relative polar coordinates (ρ, φ) were transformed into (u, v, θ, φ) with the meaning of (u, v) similar to that for collinear collisions. For products, they transformed (r', θ') and (ρ', φ') into (u, v, θ, φ'), and related (θ', φ') to (θ, φ).

Indicating with $\Theta_j(u, v, \theta)$ the internal vibration–rotation states and with $Y_l(\varphi)$ the orbital states, scattering boundary conditions for direct and reactive collisions are, respectively

$$\Psi_i \underset{u \to -\infty}{\sim} e^{ik_i u \cos \varphi} + \sum_{jl} g_{ji}^{(-)}(l)(k_j u)^{-1/2} e^{-ik_j u} \Theta_j Y_l$$

$$\underset{u \to +\infty}{\sim} \sum_{jl} g_{ji}^{(+)}(l)(k_j u)^{-1/2} e^{ik_j u} \Theta_j Y_l$$

and the differential cross section is

$$d\sigma_{ji}^{(\pm)}/d\varphi = k_i^{-1} \left| \sum_l g_{ji}^{(\pm)}(l) Y_l \right|^2$$

The calculation was set up in a manner analogous to that of Miller and Light (1971a, b). It involved expanding Ψ in the set $\Theta_j Y_l$ for reactants, to obtain a set of coupled equations, and solving this set by the exponential method. Taking advantage of symmetry, the integration had to be done only on one side of the col in the (u, v) plane with a potential $V(u, v, \theta - \varphi)$. The change of basis from reactants to products was carried out in two parts, firstly by determining a surface on which to match the wavefunctions and their normal derivatives, and secondly by expanding the products basis set in the reactants basis set.

The most serious shortcoming of the foregoing approach is within the procedure for transforming the basis. It is valid only for a symmetric reaction and has been carried out setting v equal to its value at the reaction path, in order to define a matching surface. Furthermore the computed transformation matrix was only approximately orthogonal.

In detail, the potential was parametrized in the form

$$V(u, v, \chi) = [D_{H_2} - 0.89f(u)](\exp\{-\alpha(u, v)[v - v_0(u)]\} - 1)$$
$$+ 0.89f(u)[1 + h(\chi)]$$

where $\chi = \theta - \varphi$, $D_{H_2} = 109$ kcal/mole, $f(u)$ is the potential along the reaction path with six parameters, $v_0(u)$ is the equation of the reaction path, and $\alpha(u, v)$ (with two parameters) and $h(\chi)$ (with three parameters) were fitted to give symmetric stretching and bending force constants at the col. The employed vibrational states were solutions of the quartic oscillator.

Reaction probabilities versus impact parameter b show that rearrangement occurs only at small b, $<2a_0$, and are nearly symmetric around $b = 0$. This indicates that reaction is nearly the same whether orbital and rotational angular momenta are parallel or antiparallel. Angular distributions must be obtained between $0°$ and $360°$, because once the rotation sense is fixed one must distinguish between upper and lower half-planes. Impact parameter dependencies are shown in Fig. 6.

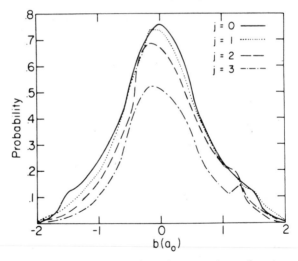

Fig. 6. Total probability of reactive scattering as function of impact parameter b and for several rotational quantum numbers j, at total energy of 17 kcal/mole (from Saxon and Light, 1972b).

A comparison with classical results (Saxon and Light, 1972c) shows that classical cross sections are lower near threshold, which was attributed to tunnelling. At higher energies classical results are larger, unlike the relation found in collinear studies. Classical mechanics provides a reasonable description of impact parameter dependence.

Further comparisons with semiclassical probabilities were made by Tyson et al. (1973). The classical S matrix was obtained as described by Miller (1970), by a sum over paths α, of the form

$$S_{ji}^{(\pm)} = \sum_{\alpha\pm} [\sqrt{D_{ji}^{(\alpha)}}/(2\pi i\hbar)\,e^{i\Phi_{ji}^{(\alpha)}/\hbar}$$

where $\Phi^{(\alpha)}$ is the classical action and $D^{(\alpha)}$ is a determinant of derivatives of internal action variables with respect to angle variables. Internal angle-action variables are (θj) for rotational motion and (q, n) for vibrational motion. Choosing initial $(n_1 j_1)$ values, acceptable paths are those for which the final $n_2(\theta_1 q_1)$ and $j_2(\theta_1, q_1)$ have the required integer values. In the course of the path search for fixed (n_2, j_2) it was found that the stationary-phase approximation leading to S_{ji} was not justifiable at some values of (q_1, θ_1).

During 1974, both Light et al. and Kuppermann et al. have obtained new results on coplanar motion, which are in substantial agreement with each other. These new results correct some of the conclusions in the foregoing publications: quantum mechanical angular distributions show only back-ward peaks, and a correlation between the parities of the total angular momentum J and of $|j_f - j_i|$ has disappeared.

Very recently there has been a distorted-wave calculation (Walker and Wyatt, 1973) for planar reactive $H + H_2$ using two different distortion potentials. Results were used to approximate three-dimensional total and differential cross sections.

Marcus generalized his previous reaction-path coordinates to three-dimensional motion (1967, 1968) by introducing natural bifurcation co-ordinates. Two different curves $\mathscr{C}$ and $\mathscr{G}$ were defined in a space-fixed reference frame, and for the six-dimensional configuration space given by (x, y, z, X, Y, Z), the set of internal and relative coordinates of reactants. The first curve was introduced to describe $AB + C \to A + BC$ and the second one to describe $AB + C \to AC + B$. This allows for interference of different product channels, a feature that has not been included in collinear treatments. A transformation to a body-fixed frame was specified by Euler angles (θ, φ, χ) such that A, B and C were in the new (y, z) plane. The new (z, Z) variables were replaced by the distance s along the path $\mathscr{C}$ (defined in terms of z and Z) and the perpendicular distance n, while polar coordinates m (a distance) and ζ (an angle) replaced the new (y, Y). Finally (n, m) were re-expressed in terms of a radial distance r and an angle γ, and ζ was chosen as a given function of s. In this fashion a natural coordinate set $(s, \rho = r_0 - r, \theta, \varphi, \chi, \gamma)$ was defined, with r_0 a function of s. The two curves $\mathscr{C}$ and $\mathscr{G}$ in the plane (z, Z) were distinguished by taking $z > 0$ for $\mathscr{C}$, and $z < 0$ for $\mathscr{G}$. The interaction potential, which does not depend on the Euler angles, may be written without loss of generality as

$$V = V_1(s) + V_2(s, \rho, \gamma).$$

A different set of coordinates for three atoms in 3-dimensional space has been introduced by Shipsey (1969, 1972). This set is intended to treat in an optimum manner reaction paths with small radius of curvature and seems useful for systems with a bent intermediate configuration. Euler angles are introduced to relate space-fixed and body-fixed frames, and parabolic co-ordinates are defined in this last frame. The coordinates are illustrated for the reaction $^{16}O + {}^{14}N\,{}^{18}O \rightarrow {}^{18}O + {}^{14}N\,{}^{16}O$, and it is shown that the orientation coordinate for relative motion becomes the bending normal co-ordinate of NO_2 in the interaction region.

The natural collision coordinates for spatial motion introduced by Marcus were made the starting point for a development (Wyatt, 1972) of reactive collision equations for $AB + C \rightarrow A + BC$. The treatment may be regarded as an extension of previous work (Curtis and Adler, 1952) for inelastic collisions. The kinetic energy operator was simplified by introducing two approximations appropriate for linear intermediates, and the interaction was chosen of form

$$V(s, \rho, \gamma) = V_1(s) + V_2(\rho, s) + V_3(\gamma, s)$$

The total wavefunction Ψ was written as a linear combination of states

$$\Psi_{jln}^{JM} = \sum_K D_{MK}^J(\mathcal{R})\chi_{jln}^{JK}(s, \rho, \gamma)$$

where D is Wigner's rotation matrix for a rotation $\mathcal{R}$ of Euler angles $(\theta\varphi\chi)$. Furthermore $\chi(s, \rho, \gamma)$ was expanded in vibration–rotation states

$$\Lambda_{jln}^{JK}(\gamma, \rho, s) = R_{jl}^{JK}(\gamma, s)U_n(\rho, s)$$

with U_n the local vibrational states and $R(\gamma, s)$ rotation bending states. These last were discussed in the cases of: (A) large $\pm s$, i.e. free rotation; (B) moderate $\pm s$, i.e. slightly hindered rotation; and (C) $s = 0$, where rotation is strongly hindered and K may be assumed a good quantum number.

These developments have been extended to include the two possibilities $AB + C \rightarrow A + BC$ and $\rightarrow AC + B$, by employing bifurcation coordinates (Harms and Wyatt, 1972). This means introducing in the plane of the atoms the two sets (s, ρ, γ) and (s', ρ', γ') for the two possibilities, respectively. Furthermore, in order to relate these two sets and to allow the system to 'wobble' between the two paths $\mathscr{C}$ and $\mathscr{C}'$, γ and γ' were replaced by a common variable $0 \leq u \leq 1$. The kinetic energy operator in the variables (s, ρ, u) was approximately formed so that it reduces to previous ones for no bifurcation, and for planar motion. The potential was chosen of form

$$V(s, \rho, u) = V_t(s) + V_v(\rho, s) + V_b(u, s)$$

with $V_t(s)$ the potential along path $\mathscr{C}$.

For planar motion the wavefunction in bifurcation coordinates was expanded as

$$\Psi(s, u, \rho, \theta) = \sum_{J=-\infty}^{+\infty} X^J(s, u, \rho)\, e^{iJ\theta}/\sqrt{2\pi}$$

and the torsional potential V_b was chosen as

$$V_b(u; s) = \tfrac{1}{2}V_0(s)(1 - \cos 4\pi u)$$

where now $2\pi u$ is equivalent to γ in the free rotor region. Eigenfunctions $M_j^J(u, s)$ for torsional motion have been studied by Walker and Wyatt (1972b). As s changes from $-\infty$ to 0 the corresponding eigenenergies $E_j^J(s)$ raise and cross for different torsional quantum numbers j, and the density $|M_j^J|^2$ changes from a free-rotor one into a density with maxima corresponding to bifurcation.

The use of bifurcation coordinates has been computationally illustrated (Middleton and Wyatt, 1973) with a schematic model based on a two-variable potential $V(s, x)$. At low energies, flux goes smoothly into one or the other final paths, but at high energies the motion is 'turbulent'.

It is at present too soon to judge the usefulness of curvilinear coordinates for coplanar and spatial motions. From the numerical results and analytical developments published so far, it is clear that generalizations to a plane or space have been possible only through a large increase of effort over collinear calculations, due to the increased complexity of the equations to be solved. However, this is a usual situation in the preliminary stages of most work and may change in the future. Another open question, that will have to be answered, is the extent to which curvilinear coordinates will be useful in situations where surface 'hopping', i.e. electronic transitions, are involved.

III. MANY-CHANNEL APPROACHES

A. Statistical Models

The number of scattering channels that partake in molecular reactions may increase very rapidly when chemical species or kinematical parameters are changed. For example rotational channels multiply as the moments of inertia of reactants or products decrease; vibrational modes become numerous in polyatomic molecules; and generally the number of accessible channels increases with increasing collision energies. In these cases it may not be possible, and sometimes it is not needed, to develop a detailed treatment of cross sections. Rather, one may want to know only averages of cross sections, for example over a distribution of collision energies. Let us indicate the scattering channels for one rearrangement by a, for another one by b, and so on. If $\sigma_{ba}(k_a)$ is the cross section for $a \to b$ reaction at collision energy $E_t = \hbar^2 k_a^2/(2\mu_a)$, and the number of E_t values within dE_t is given by $\rho(E_t)\,dE_t$, which we assume is peaked at a particular k_a, then

$$\langle \sigma_{ba}(k_a) \rangle_{\mathrm{Av}} = \int_0^\infty dE'_t\, \rho(E'_t)\sigma_{ba}(k'_a) \Big/ \int_0^\infty dE'_t\, \rho(E'_t)$$

This cross section may be expressed in terms of S-matrix elements, which are in turn obtained from the asymptotic form of total wavefunctions Ψ. Let $\Psi_a^{(c)}(\mathbf{R}_c, \mathbf{r}_c)$ be the component of Ψ for incoming reactants in channel a and outgoing products in channel c. If j_c and l_c are rotational (or internal) and orbital (or relative) angular momenta respectively, it is convenient to introduce the total angular momentum coupling $(j_c l_c JM)$ and to expand $\Psi_a^{(c)}$ in the corresponding eigenstates $\mathscr{Y}_{l_c j_c}^{JM}$. Expansion coefficients $\psi_{\gamma\alpha}^J(R_c)$, where α contains l_a, must satisfy

$$\psi_{\gamma\alpha}^J \underset{R_c \to \infty}{\sim} -\sqrt{\frac{\mu_c}{\hbar k_c}}\big[\delta_{\gamma\alpha}\, e^{-i(k_a R_a - (l_a \pi/2)} - S_{\gamma\alpha}\, e^{+i(k_a R_a - (l_a \pi/2)}\big]$$

which defines scattering amplitude elements $S_{\gamma\alpha}^J$. In terms of these one gets

$$\sigma_{ba}(k_a) = \frac{\pi}{k_a^2}\frac{1}{2j_a + 1}\sum_{Jl_a l_b}(2J + 1)|S_{\beta\alpha}^J - \delta_{\beta\alpha}|^2$$

where an average and a sum over rotational orientation of reactants and products, respectively, have been carried out.

Performing next the energy average, the slowly varying k_a^{-2} factor may usually be taken outside the integral, and one is left with averages of $|S_{\beta\alpha}^J|^2$ and $\mathrm{Re}\,(S_{\beta\alpha}^J)$. Statistical models may then be developed by making assumptions about these averages, which must however take into consideration the unitarity (from flux conservation) and symmetry (from time reversal invariance) of $\mathbf{S}^J$, which are

$$\sum_\gamma S_{\beta\gamma}^J (S_{\alpha\gamma}^J)^* = \delta_{\beta\alpha}$$

$$S_{\beta\alpha}^J = S_{\tilde\beta\tilde\alpha}^J$$

with $\tilde\alpha$ the scattering-channel quantum numbers that follow from α by time-reversal.

A simple and popular approximation has been the phase-space model (see Light, 1967, for a review and previous references; also Nikitin, 1968, op. cit.). Within the foregoing scheme, this model may be obtained by assuming

$$\langle \mathrm{Re}\, S_{\beta\alpha}^J \rangle_{\mathrm{Av}} = 0$$

$$\langle |S_{\beta\alpha}^J|^2 \rangle_{\mathrm{Av}} = w_{\beta I}^J w_{I\alpha}^J$$

i.e. a random phase in $S_{\beta\alpha}^J$, and a decoupling of initial and final channels because of the large number of intermediate states involved. The properties of $\mathbf{S}^J$ imply that

$$w_{\beta I}^J = w_{I\alpha}^J = (N^J)^{-1/2}$$

where N^J is the total number of asymptotic channels accessible to the intermediate states, and compatible with conservation of total energy and any other fixed quantities. Then, with $b \neq a$,

$$\langle\sigma_{ba}\rangle_{\text{Av}} = \frac{\pi}{k_a^2} \frac{1}{2j_a + 1} \sum_J (2J + 1)\frac{n_a^J n_b^J}{N^J}$$

$$n_a^J = \sum_{|J - j_a| \leq l_a \leq l_{a,\text{max}}} 1$$

and similarly for n_b^J. The quotient n_b^J/N^J may be interpreted as a probability of decomposition of the complex, and the remaining factor as the cross section for its formation from channel a.

The criterion for choosing $l_{a,\text{max}}$ is crucial in this model, because $l_{a,\text{max}}$ is the only parameter that depends on the physical nature of the intermediate complex. It has usually been taken as

$$l_{a,\text{max}} = \text{Min}\ \{J + j_a, b_{\text{max}}k_a\}$$

with b_{max} a maximum impact parameter defined differently for channels without or with activation energies. In the first case b_{max} is determined by long-range forces, in the second case by the effective activation barriers.

Considering the assumptions in the phase-space model, this appears most appropriate for calculation of magnitudes of cross sections, and particularly of branching ratios. But it may be expected to fail with regard to product energy distributions, because these are fixed a priori by the process of state counting.

The phase-space model oversimplifies matters when it assumes randomness and decoupling in $\mathbf{S}^J$ irrespective of the values of J. In actuality, one expects different coupling regimes for different ranges of J (Lester and Bernstein, 1970). The strong coupling required by the phase-space model is not likely to exist for very small or very large J.

To compensate the previous drawbacks, the phase-space model offers a simple way of calculating cross sections. It has been applied in the last years to several neutral–neutral and ion–neutral reactions. Truhlar (1971) has applied it to K + HCl, to compare with disagreeing experiments. His results were inconclusive, but provided an overall description of the reaction. Assuming a reaction threshold smaller than 1 kcal/mole, he obtained large cross sections for rotational excitation. Because of the difference in reduced masses, products HCl are favoured over KCl. Calculated reaction cross sections were larger, by about a factor 2, than experimental ones. Truhlar

(1972) also discussed enhancement of the reaction cross section of He + $H_2^+ \rightarrow HeH^+ + H$ by vibrational excitation of H_2^+. Qualitative agreement was found with experiment, although the calculated enhancement was appreciably smaller than detected. The qualitative agreement remained after Truhlar and Wagner (1972) corrected some degeneracy factors, following comments on treatment of nuclear spin in the phase-space model.

Fullerton and Moran (1971) considered the role of dispersion and short range forces in reactions of $He^+ + N_2$, in the context of the phase-space model, and also (1972) reactions of C^+ with O_2 and N_2. The same authors (Moran and Fullerton, 1971) discussed collision-induced dissociation of excited O_2^+ and NO^+ ions, and in another paper (1972) rotational excitation of N_2^{*+} produced in thermal reactions $A^{*+} + N_2 \rightarrow N_2^{*+} + A$, with A a noble-gas atom. Phase-space results in the last paper compared favourably with experiment, with some discrepancies observed for the moderately exoergic reactions.

An application of the phase-space model to four bodies has been made by Wolf and Haller (1970). They treated reactions of type $A_2^+ + A_2$, i.e. cases where exchange resonance may arise. Introduction of a resonance potential to describe charge transfer in the reactant channel led to a relative increase of the non-reactive cross seection, compared with that expected from using only ion-induced dipole potentials.

The phase-space model has been extended by Miller (1970) in order to incorporate the effect of closed channels. He made use of a parametrized form of the S matrix previously developed for compound-state resonances in atom–molecule collisions (Micha, 1967). Indicating with $\mathbf{S}^d$ the S-matrix for direct scattering, i.e. in the absence of coupling to closed channels, one can write (omitting the index J),

$$\mathbf{S} = \mathbf{S}^d - i \sum_s (\mathbf{S}^d)^{1/2} \mathbf{c}_{os} \frac{\Gamma_s}{E - E_s + i\Gamma_s/2} \mathbf{c}_{os*}^\dagger (\mathbf{S}^d)^{1/2}$$

where E_s and Γ_s are the energy and width of resonance s and $\mathbf{c}_{os}$ is a column matrix relating the s resonance to open channels. Molecular resonances are in many cases sharp and non-overlapping, or may be assumed independent on the average. Concentrating on only the $s = r$ resonance, one finds

$$\langle |S_{\beta\alpha}|^2 \rangle_{Av} = P_{\beta\alpha}^{(0)} + p_{\beta r} p_{r\alpha} / \sum_{\text{all } \gamma} p_{\gamma r}$$

$$p_{\beta r} = p_{r\beta} = 2\pi |[(\mathbf{S}^d)^{1/2} \mathbf{c}_{or}]_\beta|^2 \rho(E_r)$$

where $\rho(E)$ is the spread of total energies used in the average, and $P_{\beta\alpha}^{(0)}$ and $p_{\beta r}$ may be considered respectively as the probabilities for the direct $\alpha \rightarrow \beta$ transition and for the $r \rightarrow \beta$ decomposition. From probability conservation, for a single resonance,

$$p_{\beta r} + \sum_\gamma P_{\beta\gamma}^{(0)} = 1$$

Different choices may be made for these probabilities, but some caution is required. One finds that if, within the $(ljJM)$ representation, one has

$$\langle |S^J_{j_b l_b, j_a l_a}|^2 \rangle_{\text{Av}} = p_r(j_b l_b) p_r(j_a l_a)/N^J$$

then a similar decomposition applies to other representations. In the helicity representation, with μ_a and μ_b the projections of $\mathbf{j}_a$ and $\mathbf{j}_b$ on $\mathbf{k}_a$ and $\mathbf{k}_b$ respectively, one finds

$$\langle |S^J_{j_b \mu_b, j_a \mu_a}|^2 \rangle_{\text{Av}} = p_r(j_b \mu_b) p_r(j_a \mu_a)/N^J$$

$$p_r(j\mu) = \sum_l p_r(jl)(\langle lj0m_j|Jm_j\rangle)^2$$

which means that decomposition probabilities in one representation determine those in another, and that consistency must be preserved.

Choices of $p_r(jl)$ corresponding to phase-space models have been considered in the paper by Miller, while the $p_r(j\mu)$ have been parametrized by Herschbach and collaborators (see e.g. Miller *et al.* 1967) in accordance with a fission model, which leads to

$$p_r(jl) = \text{cst. } \exp(\pm|c|\mu_j^2)$$

with $(+)$ for oblate complex structures and $(-)$ for prolate ones.

Expressions for reaction angular distributions have also been developed within the phase-space model by White and Light (1971). The equations show that the assumption of statistical averaging for all J leads always to angular distributions symmetric around $\theta_{\text{c.m.}} = 90°$. They compared their calculations on $\text{Rb} + \text{CsCl} \rightarrow \text{RbCl} + \text{Cs}$ with experimental measurements. Theory predicts too little internal excitation even though formation of a long-lived complex is likely. It has already been pointed out that long-lived complexes are not necessary to validate the phase-space model. The result above shows that they are not sufficient, either.

The phase-space model has been applied to triple collisions by F. T. Smith (1969) in a detailed study of termolecular reaction rates. He classified 3-body entry or exit channels into two classes, of pure and indirect triple collisions, and introduced kinematic variables appropriate to each class. These variables were then used to develop a statistical theory of break-up cross-sections. A recent contribution (Rebick and Levine, 1973) has dealt with collision induced dissociation (CID) along similar lines. Two mechanisms were distinguished in the process $A + BC \rightarrow A + B + C$. Direct CID, where the three particles are unbound in the final state, and indirect CID, where two of the particles emerge in a quasi-bound state. Furthermore, a distinction was made in indirect CID, depending on whether the quasi-bound pair is the initial BC or not. Enumeration of the product (three-body) states was made in terms of quantum numbers appropriate to three free bodies (see e.g. Delves and Phillips, 1969); the vibrational quantum number of a product pair is replaced by a generalized angular momentum λ. The phase-space summation over states was carried out for direct and indirect CID, and led in both cases to post-threshold energy dependencies of type $(E - E_0)^n/E_{tr}$, with $n(\text{direct}) > n(\text{indirect})$. Comparison with experimental results for $He + He_2^+(v)$ showed disagreements in the translational energy and total energy dependencies, suggesting that this process is not entirely statistical, but that a dynamical bias must be introduced in its description.

Time-reversal invariance, or microscopic reversibility, which we have already encountered in the phase-space model, has been the subject of several investigations. Anlauf $et\ al.$ (1969) considered the influence of reactant excitation upon the rate of endoergic reactions by relating this rate to the inverse one for production of excited products in exoergic reactions. Kinsey (1971) thoroughly investigated microscopic reversibility in reactions with partial resolution of product and reactant states. He considered experiments where transitions are determined from a set A of reactant channels a to a set B of product channels b. Indicating with $\bar{\varkappa}(A \to B)$ the measured rate for unit particle-density per unit volume, and with $\rho(A)\, dE$ the number of states in A per unit volume and for total energy within dE, he defined

$$\bar{\Omega}(A, B) = \bar{\varkappa}(A \to B)/\rho(B) = \bar{\varkappa}(B \to A)/\rho(A),$$

an averaged transition rate per reactant pair, satisfying $\bar{\Omega}(A \cdot B) = \bar{\Omega}(B, A)$ on account of microscopic reversibility. This $\bar{\Omega}$ reflects the intrinsic probability of the process, since it does not include the statistical bias of experiments. He furthermore gave the $\rho(A)$ densities for several cases: only temperature or total energy determined; intermediate resolution, such as that providing E_V (vibrational energy) and E_T (translational energy) of products; and complete resolution, only in ideal experiments. Microreversibility has been employed by Levine and Bernstein (1972) to discuss post-threshold energy dependence of cross sections for endoergic processes, on the basis of information on the inverse, exoergic processes. They considered vibrational excitation for diatomics whose relaxation, e.g. in a shock tube, is small, and reactions whose inverse involve negligible activation barriers.

Some of the previous articles have inspired several investigations on the relation between entropy and chemical change. In work by Bernstein and Levine (1972) optimal means of characterizing the distribution of product energies were discussed in terms of an information measure I (see e.g. Jaynes, 1963, Katz, 1967; for uses of I in thermodynamical problems). Global, detailed experiments provide average transition probabilities $\omega(A, B)$, from which the 'surprisal' $I_p(A, B)$ is defined to be

$$I_p(A, B) = -\lg \omega(A, B)$$

with lg the logarithm in basis 2. For less thorough experiments, only sums of the $\omega(A, B)$ are available, and the corresponding surprisals may be defined and compared among themselves. Entropy deficiencies may also be defined in terms of differences between averages of the surprisals for actual and for equilibrium (microcanonical) probability distributions.

These concepts have been applied by Ben-Shaul *et al.*, 1972a, b to analysis of product energy distributions and to the definition of apparent temperatures for non-equilibrium product distributions. Their procedure may be illustrated with a treatment of vibrational distribution of products, as measured in infrared chemiluminescence studies. The quantities measurable in this case are E and E_V. It is convenient to introduce fractional energies $f_x = E_x/E$, where $f_T + f_V + f_R = 1$. The probabilities actually determined are conditional probabilities $P(f_V|E)$ for fixed E. The surprisal is

$$I(f_V|E) = -\ln\left[P(f_V|E)/P^0(f_V|E)\right]$$

where P^0 is the *a priori* or equilibrium probability, derivable from quantal degeneracies. The entropy deficiency becomes, using $S_{eq} = 0$,

$$\Delta S(f_V) = S_{eq}(f_V) - S(f_V) = k_B \sum_{v=0}^{v^*} P(f_V|E)I(f_V|E)$$

with v^* the highest allowed vibrational quantum number for given E. The surprisal is then expanded around $f_V = 0$, and the first derivative

$$\lambda_V = [dI(f_V|E)/df_V]_0 = E/(k_B T_V)$$

is assumed (and often found to be) constant at fixed E, which defines an apparent temperature T_V. It follows that, provided higher derivatives may be neglected

$$P(f_V|E) = Q(E, \lambda_V)^{-1} P^0(f_V|E) e^{-\lambda_V f_V}$$

where Q is a normalization factor. This equation and similar ones for f_R, f_T have been applied to twelve reactions for which distributions are available from experiment or classical trajectory calculations. Parametrization in terms of λ_V proved very successful and, e.g. from $F + H_2, D_2$, it showed that λ_V is invariant under isotopic substitution. Fitting of $P(f_T)$ for $K + I_2 \rightarrow KI + I$ required two values of λ_T, which appears to correspond to the formation of $I(P_{3/2})$ or $I(P_{1/2})$. $P(f_R)$ on the other hand could not be fitted in this way at small f_R. The T_V were found to be negative in all cases, and the T_T and T_R positive in the acceptable ranges of constant λ. Figure 7 shows results for $F + D_2$.

Further analysis by Levine *et al.* (1973) of rotational distributions in $Cl + HI/DI$, $F + H_2/D_2$ and $H/D + Cl_2$ showed that useful parameters could be defined in these cases too, provided rotational distributions were defined for fixed E and fixed f_V. Within a manifold characterized by f_V, the available fraction of rotational energy is

$$E_R/(E - E_V) = f_R/(1 - f_V)$$

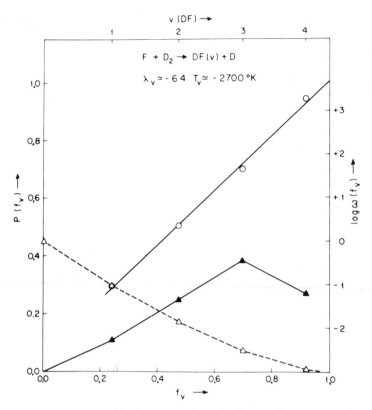

Fig. 7. Distribution of final vibrational energies for $F + D_2 \rightarrow FD(v) + D$. The parameter λ_v is essentially invariant under isotopic substitution. Filled and open triangles are measured and *a priori* distributions (scale to left); open circles are the surprisals (scale to right) (from Ben-Shaul *et al.*, 1972).

and the surprisal is

$$I(f_R|f_V, E) = -\ln\left[P(f_R|f_V, E)/P^0(f_R|f_V, E)\right]$$

The modified parameter

$$\theta_R = dI(f_R|f_V, E)/d[f_R/(1 - f_V)]$$

was found to be constant and isotopically invariant. This leads then to the parametrization

$$P(f_R, f_V|E) = \frac{P^0(f_R, f_V|E)}{Q_V Q_R(v)} \exp\left(-\lambda_V f_V - \theta_R \frac{f_R}{1 - f_V}\right)$$

An alternative parametrization has been proposed by Ben Shaul (1973).

More recent work (Levine and Bernstein, 1973) proposes a new measure of the influence of reactant energy distribution on product energy distribution. It defines the 'relevance' R as a measure of the deviation of the pair probability $P(f_T, f_{T'})$, with f_T for reactants and $f_{T'}$ for products, from the product $P(f_T) \cdot P(f_{T'})$.

The foregoing developments in terms of the 'surprisal' provide, at least, a successful parametrization for distributions of direct reactions. Non-existence of long-lived states in these reactions would make unjustifiable the introduction of temperatures in the sense of equilibrium thermodynamics. At best, parametrization may lead to dynamical models with predictive value. These are not available at the present time, though progress is being made in this area by Hofacker and coworkers (op. cit), using model Hamiltonians that include vibrationally diabatic effects, and introducing dynamical constraints in statistical models.

A third statistical approach, frequently applied during the last years to molecular beam experiments, is closely related to the Rice–Ramsperger–Kassel–Marcus (RRKM) theory of unimolecular decomposition (Marcus, 1968; for an introductory treatment see, e.g., Rice, 1967). The assumptions of the RRKM theory have been verified, with Monte-Carlo calculations, by Bunker and Pattengill (1968). They also observed that the critical configuration in the theory may not be arbitrary, but must be chosen in the region of high density of states of the complex. Another analysis of the model, relating to centrifugal effects in rate theory, has been made by Waage and Rabinovitch (1970).

Recent extensions to bimolecular reactions include a study by Marcus (1970) on the relation between state-selected cross sections of endothermic reactions and rate constants of exothermic reactions. This study was prompted by the investigation of Anlauf et al. (1969), and applied a micro-canonical activated-complex theory to bimolecular reactions.

A transition-state theory (Safron et al., 1972) has been developed within the context of scattering theory, to provide suitable models for crossed molecular beam processes. As in the case of RRKM theory, it is based on the premise that the probability of complex decomposition is a product of a probability of break-up and a probability of departure from the collision region. But it adds restrictions peculiar to bimolecular reactions, such as a limit on the maximum angular momentum that allows formation of the complex from reactants. Let $p(E_t')$ indicate the probability density for finding a product pair with kinetic energy E_t'. This may be written as

$$p(E_t') = N_{vr}^{+}(E' - E_t')A(E_t')$$

where $N_{vr}^{+}(E' - E_t')$ is the density of active energy levels of the complex (excluding translation along the reaction path) with energy $E' - E_t'$ which measures the probability of break-up at total energy E'), and $A(E_t')$ is the probability of departure, which on account of conservation of angular momentum depends on both reactant and product interaction constants. Classically, one finds

$$N_{vr}^{+}(\epsilon) = \epsilon^{s+r/2-2}$$

with s and r the number of active vibrational and rotational degrees of freedom of the complex, respectively, and

$$A(\epsilon) = (\epsilon/B'_m)^{(n-2)/n}, \qquad \epsilon < B'_m$$
$$= 1, \qquad\qquad \epsilon > B'_m$$

where B'_m is the maximum exit centrifugal barrier for a long-range interaction potential of form R^{-n}. The transition-state theory is closely related to the phase-space model, and in fact this last should be recovered from the first in the limit of a very loosely quasi-bound complex.

This transition-state theory has been applied to exchange reactions of Li atoms with alkali halides (Kwei et al., 1971; Lees and Kwei, 1973) where examples of short-lived, or osculating, complexes may be found, and to beam studies of unimolecular decay (Lee et al., 1972), in particular on the extent of internal equilibration.

More recent applications have dealt with diatom–diatom and with atom–polyatom reactions. W. B. Miller et al. (1972) considered four-atom collision complexes in exchange reactions of CsCl with KCl and KI. The statistical theory gives good agreement with the observed translational energy and angular distributions. But it overestimates reactive to nonreactive branching ratios, perhaps due to the presence of geometrical isomerism of the complex. Riley and Herschbach (1973) considered long-lived collision complexes of K, Rb and Cs with $SnCl_4$ and SF_6. The observed product angular and velocity distributions were consistent with theory. For the $SnCl_4$ reaction there was evidence of two decay modes, one for formation of alkali chloride and the other for formation of a heavier alkali compound. Fig. 8 compares experimental and theoretical product velocity distributions for Cs + SF_6.

Phase-space and RRKM theories have also been applied to a series of experiments designed to explore whether or not randomization in a reaction complex is fast relative to decomposition of the complex. Analysis of reactions of fluorine with ethylene (Parson and Lee, 1972) and with four butene isomers (Parson et al., 1973) show qualitative agreement between experiment and RRKM predictions of branching ratios. Product energy distributions, on the other hand, do not agree. This suggests that the assumption of randomization of the complex internal energy breaks down at the transition state. A similar analysis (Shobatake et al., 1973) of reactions of fluorine with aromatic and heterocyclic molecules indicate that: (1) when light groups are emitted, randomization is not faster than decomposition; and that (2) when heavy groups are emitted, all vibrational modes of the

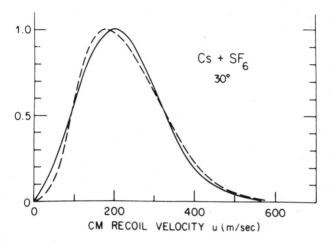

Fig. 8. Flux density distribution in arbitrary units for Cs + SF$_6$ in c.m. system. Solid line is from experimental results and dashed line from a fit to the functional form of the transition-state theory (from Riley and Herschbach, 1973).

complex participate in the reaction. All these conclusions are, however, affected by uncertainties on the energetics of the reactions and on the magnitude of the decomposition barriers.

M. Shapiro (1972) has carried out a computational investigation of unimolecular decomposition. He chose the Van der Waals complex Xe–D$_2$ as an example and described interference effects due to overlapping resonances of the complex within the context of scattering theory.

Quasi-statistical complexes in chemical reactions have been discussed by George and Ross (1972). Depending on the extent of statistical averaging, they proposed a classification of experimental results into the categories: direct interaction, doorway states (a concept borrowed from nuclear physics), partial statistical complex and complete statistical complex. They stressed that symmetric product angular distributions are indicative of complex formation, but these are neither necessary nor sufficient evidence; rather, they indicated that the observations necessary for a complete statistical complex are: symmetric angular distributions; velocity spectrum peaked near the centre-of-mass velocity; and statistical isotope distribution.

A study of long-lived states in atom–molecule collisions (Micha, 1973) has made use of theoretical methods developed for Van der Waals complexes, to discuss a number of reactive atom–molecule pairs where formation of long-lived states appears established. It employs the statistical approximation to the S-matrix for resonance processes, and points out the importance of both compound-state and shape resonances in reactions. It suggests that probabilities of departure from the collision region could be determined from the shortest-lived exit-channel resonances, which would agree with the Patengill–Baker (1968) criterion for choosing the critical configuration of complexes.

Orbiting, or shape, resonances also play a role in a theory of termolecular recombination developed by Roberts et al. (1969). Their model applies to reactions $A + B + M \rightarrow AB + M$, and assumes that A and B form an excited quasi-bound state, which is de-excited by the third body to form stable AB. Results on $H_2 + M$ were compared to those for $D_2 + M$ by Roberts and Bernstein (1970). More recently Dickinson et al. (1972) applied the model to ion–atom association in rare gases.

Resonance states, and in particular their corresponding time-delays, have been used by Van Santen (1972) to discuss association and dissociation reactions in the presence of third bodies. A density matrix formalism was employed to obtain expressions for rate coefficients valid for all pressures.

B. Optical-Potential Models

Optical potentials have been used, so far, mostly to extract semi-empirical information from molecular beam experiments. In cases where detection devices are sensitive to only one type of species (or a single state), i.e. to only one reaction (or scattering) channel, the effect of competing reactions is to apparently decrease the detected scattering flux over the incoming one. This effect may be described in terms of optical or complex potentials

$$V_{opt}(R) = V(R) - iW(R)$$

where both V and W are real. It is found that for this potential

$$\mathbf{V} \cdot \mathbf{J} = -\frac{2}{\hbar} W\rho$$

where $\mathbf{J}$ and ρ are the local flux and density of the scattering wave-function $\psi(\mathbf{R})$. Hence a positive $W(R)$ leads to flux absorption at R. By analogy with elastic scattering one can solve for ψ and (in this case) extract from it a complex phase shift $\delta_l = \xi_l + i\zeta_l$. Elastic and absorption integral cross sections are then, using $S_l = \exp(2i\delta_l)$, given by

$$\sigma_{el}^{(l)}(k) = (\pi/k^2)|1 - S_l|^2$$
$$\sigma_{abs}^{(l)}(k) = (\pi/k^2)(2l + 1)p_l$$
$$p_l(k) = 1 - |S_l|^2 = 1 - e^{-4\zeta_l}$$

with p_l the opacity function. A semiclassical analysis of this problem for reactive scattering has been developed by Ross and collaborators, and reviewed by Ross and Greene (see Ch. Schlier, ed., 1970; also McDaniel et *al.*, 1970). Indicating with $b = (l + \frac{1}{2})/k$ the impact parameter, with $\theta = \theta(b)$ the deflection function and with $d\sigma_{el}$ and $d\sigma_{el}^{(0)}$ the semiclassical differentials of elastic cross sections for given V_{opt} and V, respectively, one finds

$$p(k, b) = 1 - d\bar{\sigma}_{el}(k, \theta)/d\sigma_{el}^{(0)}(k, \theta)$$

for the semiclassical opacity. Replacing $d\sigma_{el}$ and $d\sigma_{el}^{(0)}$ by measured and calculated values, respectively, it is possible to estimate absorption cross section. This was done (see Ross and Greene *op. cit.*) for a series of reactions of potassium with diatomics and polyatomics, to estimate reaction cross sections (assumed equal to σ_{abs}).

Opacity functions are usually assumed to be fairly constant functions of b up to a cut-off point, where they rapidly fall. This behaviour, a rounded-off step function, was found in exact numerical calculations performed by Marriott and Micha (1969) with $W(R)$ proportional to R^{-12}, and fitted to reproduce quenching of glory undulations in non-reactive Li + HBr. Real and imaginary phase shifts and the opacity function, calculated from the parametrized potential, are shown in Fig. 9. Their results were compared by Roberts and Ross (1970) to semiclassical calculations, which turned out to be quite accurate. The rounded-off step functions were adopted in a detailed optical model analysis (Harris and Wilson, 1971) of non-reactive scattering of K, Rb and Cs from CCl_4, CH_3I and $SnCl_4$. Uniform semi-classical results were found to be in good agreement with quantum ones, and were employed in most calculations of elastic cross sections. A Lennard-Jones (12, 6) potential was used for CCl_4, an anisotropy was added to this for CH_3I and an electron-jump model was used for $SnCl_4$. It was found that the velocity dependence of the opacity was virtually removed by expressing p_l in terms of the distances of closest approach y. It was also shown that large anisotropies may produce sudden drops of intensity at large angles, an effect previously ascribed to reaction flux loss.

Numerical calculations have also been performed by Eu (1970) with the variable-phase method. He investigated changes of the opacity versus angular momentum curve for various collision energies and imaginary potentials, and for elastic collisions of K + HBr at energy $E = 1.49$ kcal/mole. Tables of results were presented in terms of reduced variables.

Paulsen *et al.* (1972) developed an optical model for vibrational relaxation in reactive systems. Only collinear atom–diatom collisions were considered, i.e. impact parameter dependencies were omitted. The model was applied to vibrational relaxation of electronically excited I_2 in inert gases, in which case dissociation of I_2 is responsible for flux loss. Olson (1972) used an absorbing-sphere model for calculating integral cross sections of ion–ion recombination processes $A^+ + B^- \rightarrow A^* + B + \Delta E$, with A or B atoms or molecules. He employed the Landau–Zener formula to obtain a critical crossing distance R_c, and assumed the opacity to be unity for distances

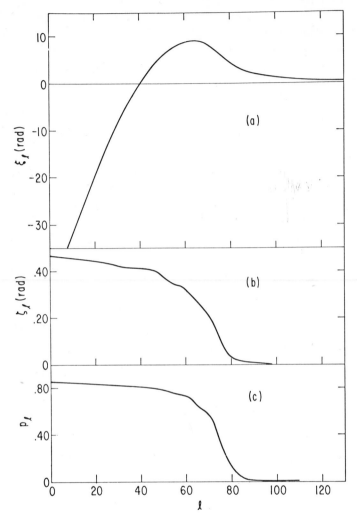

Fig. 9. (a) Real phase shifts, (b) imaginary phase shifts and (c) opacity function for Li + HBr at relative velocity $v = 1.5 \times 10^4$ cm/sec and for a potential $V_{opt} = (1 - i0.44)C_{12}R^{-12} - C_6R^{-6}$ (from Marriott and Micha, 1969).

within R_c. Optical models have also beeen applied by Rusinek and Roberts (1973, 1974) to studies of the effect of the vibrational state of reactants.

These and other results contain some warnings about semi-empirical procedures to extract opacities. Since opacities account for all non-elastic processes, i.e. both inelastic and reactive ones, the $p(k, b)$ may be identified with probabilities of reactions only for those impact parameters at which reaction predominates. Also, the expression presented for $p(k, b)$ is valid only when the deflection angle is a single-valued function of b, i.e. only at large b for most potentials.

A generalized opacity function P_l may simply be defined by rearranging the expression for σ_{ba} in terms of the S matrix, to obtain

$$\sigma_{ba} = (\pi/k_a^2) \sum_{l_a} (2l_a + 1) P_{l_a}(b, a)$$

Averaging this over a distribution of energies, one is led to a formal expression for $\langle P_{l_a} \rangle_{Av}$ which may be parametrized on physical grounds, to describe reaction processes. Parametrization may best be done by separating regions of strong channel coupling (intermediate values of l_a) from regions of weak coupling (large values of l_a). The expression above is convenient in processes dominated by reactant interaction; an equivalent expression for products follows changing l_a by l_b.

Opacity analysis based on P_l has been the subject of an investigation of inelastic collisions by Levine and Bernstein (1970). The same authors applied opacity concepts (1971a) to collision-induced dissociation at near post-threshold energies. They employed an expression valid for products, in this case three bodies. The weight $(2l_b + 1)$ must then be replaced by a new one $G(\lambda)$, with λ the generalized angular momentum for three bodies. Assuming that the opacity function $P_\lambda(CID)$ was constant up to a maximum λ_m, and then zero, they found a cross section of form $\sigma_{CID} = A \cdot (E_{tot} - E_0)^{5/2}/E_{tr}$, where E_{tot}, E_0 and E_{tr} are total, threshold and relative translational energies, respectively. A value of $E_0 = 1.9 \, \text{eV}$ was obtained from experiments with $He + He_2^+$; this is smaller than expected theoretically ($E_0 = 2.65 \, \text{eV}$), probably due to the simplying assumptions of the model. On the other hand, the qualitative behaviour with E_{tot}, a positive curvature of $\sigma(CID)$ vs E_{tot}, was correct.

Another opacity analysis has been presented (Levine et al., 1971) for reactions of $H^- + O_2$. It is based on kinematic restrictions in the final channel, and gives distributions of final kinetic energy that grow logarithmically with E_{tr} of products.

Molecular optical potentials for non-reactive processes may be rigorously defined by means of partitioning techniques (see e.g. Feshbach, 1962), which are based on the classification of scattering channels in two groups: the first one includes states which are asymptotically selected or detected, and is characterized by a projection operator P; the second one includes all other states (in practice those to which flux is lost) and is characterized by the projector Q. An optical potential operator $\mathscr{V}_P$ may then be constructed as

$$\mathscr{V}_P = P(H - H_0)P + PHQ(E^{(+)} - QHQ)^{-1}QHP$$

where H and H_0 are the total and asymptotic Hamiltonians, respectively.

The second term corresponds (from the right) to transitions from the set of states in P to those in Q, dynamical interaction among states in Q and finally return to those in P. An imaginary negative part is contained in the second term, where $E^{(+)}$ stands for $E + i\epsilon$, $\epsilon > 0$. The optical potential is clearly non-local and energy dependent.

The relevance of optical potentials to direct molecular reactions was considered by Micha (1969). Numerical results were presented for real and imaginary parts of the optical potential for $H + H_2$, in an adiabatic approximation that included vibrational and rotational motion of H_2. Distortion, adiabatic and sudden approximations to optical potentials, and their validity, have recently been described (Micha, 1974). This work also presents procedures for calculating upper and lower bounds to the second term of $\mathscr{V}_P$, in certain ranges of energies. The various approaches are developed in detail for atom–diatom collisions.

Although physical arguments may be presented for using potentials such as $\mathscr{V}_P$ in reactive collisions, rigorous theoretical justification for using them is not yet available. Practical procedures for constructing optical potentials for reactive processes do not exist at present. However, several authors (Chen, 1966; Hahn, 1966; George and Miller, 1972) have proposed schemes for constructing projection operators for reaction problems.

Computational procedures have been developed by Allison (1972) for single-channel optical potentials, and by Wolken (1972) for multi-channel (but a single rearrangement channel) optical potentials. White et al. (1973) have discussed these potentials within time-dependent quantum theory.

C. Coupled-Channels Method

The coupled-channels method may be developed within the language of wave-mechanics, or more formally (and more compactly) by means of operator equations. The common feature of both approaches is that the total scattering state is expanded in internal states of reactants and products. The nature of the colliding particles and the quantum numbers of the internal states define the reaction channel index $c = a, b, \dots$. We begin with the wave-mechanical approach, some of whose features have been presented in the section on statistical theories. For the total wavefunction ψ_a of reactants in channel a, with relative wave vector $\mathbf{k}_a$, we can write

$$\Psi_a = \sum_c \Psi_a^{(c)}(\mathbf{R}_c, \mathbf{r}_c)$$

where $\mathbf{R}_c$ and $\mathbf{r}_c$ are relative and internal position vectors in channel c. Each of the terms in the sum may be expanded in internal radial states $u_c(r_c)$, although care must be taken to avoid expanding in overcomplete sets. For example, in a three-atom problem, where $A + B + C$ is one of the channels, the set of internal states of the other channels should not include dissociation states (already in A + B + C). The expansion coefficients $\psi_{ca}(\mathbf{R}_c, \omega_c)$ may again be expanded in total angular momentum states $\mathscr{Y}^{JM}_{l_c j_c}(\Omega_c, \omega_c)$ and in spherical harmonics of the orientation angles of $\mathbf{k}_a$, to obtain relative radial functions $\psi^J_{\gamma\alpha}(R_c)$, with $\alpha = (a'v_a j_a l_a)$.

In practice, calculations must be carried out for small basis sets that would only produce approximations $\tilde{\Psi}^{(c)}_a$ to $\Psi^{(c)}_a$. To compensate for this, a possible approach is to write

$$\Psi_a = \sum_c \tilde{\Psi}^{(c)}_a(\mathbf{R}_c, \mathbf{r}_c) + \sum_n A_n \Phi_n(\mathbf{R}, \mathbf{r})$$

where the Φ_n are conveniently chosen normalizable states. This approach was for example taken in the collinear case by Mortensen and Gucwa, who determined the wavefunction parameters by the Kohn variational principle. A similar procedure was followed by Miller (1969) in the general (three-dimensional) case to obtain equations for A + BC. Extremizing the variation functional with respect to both the functions $\psi^J_{\gamma\alpha}$ and the coefficients A_n, one gets a set of coupled integrodifferential equations. To obtain scattering amplitudes one can take the following steps: (1) solve the equations with $A_n = 0$, to obtain $(\psi^J_{\gamma\alpha})_0$; (2) write the asymptotic form of these functions in terms of sines and cosines, to extract an $\mathbf{R}_0$ matrix; (3) in terms of this, the total $\mathbf{R}$ matrix (and hence $\mathbf{S}$), follows from

$$\mathbf{R} = \mathbf{R}_0 - \sum_{mn} \mathbf{C}_n (\mathbf{M}^{-1})_{nm} \mathbf{C}_m$$

where $\mathbf{M}$ is the matrix of $H - E$ in the Φ_n basis, and the $\mathbf{C}_n$ are column matrices relating the state Φ_n to each of the rearrangement indices γ. In actual applications it is not necessary to invert M, but only to solve a set of linear equations with this matrix of coefficients. This prevents difficulties associated with the roots of det $(\mathbf{M})$. The last mentioned article also contains a discussion of the simplifications that follow when two or three of the atoms are identical. A general view of this problem will be presented in connection with the Faddeev–Watson equations.

The above procedure has been followed by Wolken and Karplus (1974) for reactive H + H$_2$ collisions. Normalizable functions were not included in the expansion. Even so, this work constitutes a very significant step towards the solution of the three-dimensional problem. The coupled equations to be solved for the matrix ψ^J of states $\psi^J_{\gamma\alpha}$ is, omitting the J index,

$$[\mathbf{D}(R_a, E) + \mathbf{V}(R_a)]\psi(R_a) + \int dR_b\, \mathbf{U}(R_a, R_b)\psi(R_b) = 0$$

where $\mathbf{D}$ is the diagonal matrix of differential operators that remains when the direct coupling $\mathbf{V}$ and the exchange coupling $\mathbf{U}$ are dropped. The number of coupled equations may be significantly reduced for three identical atoms

by properly combining the equations. Scattering amplitudes are then super-
positions of the direct amplitude for $H + H'H'' \rightarrow H + H'H''$ and of the
exchange amplitudes for $\rightarrow H' + H''H$ and $\rightarrow H'' + HH'$.

Calculations were performed recasting the differential equation into
integral form, and replacing the kernel of the integral equation by a separable
approximation that resulted from using an N-point quadrature formula.
The channels $v = 0$, $j = 0$ to 6 were kept for $J = 0$, and $v = 0$, $j = 0$ to 3
for $J > 0$. Attention was focussed on $H + p - H_2(j = 0) \rightarrow H + o -$
$H_2(j = 1)$, for relative kinetic energies from 0·20 eV to 0·50 eV. Cross
sections for the Porter–Karplus surface showed significantly lower reaction
threshold than three-dimensional classical or distorted-wave calculations
(see Fig. 10). Angular distributions at $E_{rel} = 0.50$ eV were backward peaked

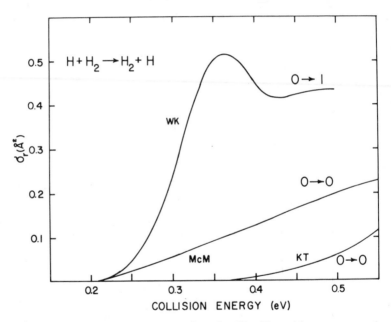

Fig. 10. Integral reaction cross sections for $H + H_2$, with reactants and
products in their ground vibrational states. Results are from coupled-channel
(WK) calculations (Wolken and Karplus, 1973) and $j_a = 0 \rightarrow j_b = 1$; from
two-state (McM) calculations (McGuire and Micha, 1973); and from distorted-
wave (TK) calculations (Tang and Karplus, 1971). The last two for
$$j_a = 0 \rightarrow j_b = 0.$$

in centre-of-mass, and did not go to zero at small angles. However, the
small-angle results appear somewhat doubtful.

The distorted-wave calculations referred to above have been performed
by Tang and Karplus (1971). For the reaction $a \rightarrow b$, this approximation

requires calculation of $\psi_{\alpha\alpha}^J$ and $\psi_{\beta\alpha}^J$, for given a, b. These states are approximated by solutions of uncoupled radial equations in R_a and R_b, respectively, that contain distortion potentials chosen on physical grounds. Letting $V_a(\mathbf{R}_a, \mathbf{r}_a)$ indicate the interaction potential in channel a, one writes

$$V_a = V_a^{(0)}(R_a) + V_a'(\mathbf{R}_a, \mathbf{r}_a)$$

where $V_a^{(0)}$ is the distortion potential, and similarly for channel b. From the radial solutions one obtains distorted waves $\chi_a^{(\pm)}(\mathbf{R}_a)$, in terms of which the transition amplitudes T_{ba} are

$$T_{ba}(\text{DWA}) = \langle \chi_b^{(-)} | V_b' | \chi_a^{(+)} \rangle = \langle \chi_b^{(-)} | V_a' | \chi_a^{(+)} \rangle.$$

The mentioned authors considered two distortion potentials: (1) $V_{a,\text{un}}^{(0)}$, the one produced by an unperturbed diatomic; and (2) $V_{a,\text{ad}}^{(0)}$, another one produced by an adiabatically stretched molecule. They carried out the integral in T_{ba} within a linear approximation, to obtain $T_{ba}(\text{DWLA})$. Results for $(v_a j_a) = (0, 0)$ to $(v_b, j_b) = (0, 0)$ showed that adiabatic potentials give larger and more reliable cross sections. Angular distributions with these potentials were backward peaked at $E_{\text{rel}} = 0.05\,\text{eV}$ and moved to larger angles with increasing E_{rel}. Integral cross sections for $j_a = 0 \rightarrow j_b = 1$ were largest, followed by $0 \rightarrow 2$ and $0 \rightarrow 0$, but all of them were nearly superimposable at threshold. Cross sections were multiplied by 2 to account (statistically) for the two final reaction channels $H' + H''H$ and $H'' + HH'$.

Distorted waves and a linear approximation were also used in earlier (and more approximate) work by Micha (1965). His results for $D + H_2 \rightarrow DH + H$ also gave backward cross sections, in agreement with recent experimental results (Geddes et al., 1972). Distorted-wave calculations for the same system, done by Tang (1972) also agreed with experiment.

The important role of distortion in rearrangement collisions has been emphasized in work by Gelb and Suplinskas (1970). These authors treated the internal motions within the perturbed-stationary state approximation, i.e. in an adiabatic fashion, and relative motions within both the Born (no distortion) and the distortion approximations. Calculations performed for $Ar^+ + HD$ with a model switching potential showed that distortion of the final relative motion was essential to obtain the correct product translational energies, because these are determined by the potential drop in the product valley. Isotope effect were, however, incorrect in both cases.

A procedure intermediate between extensive coupled-channel and distorted-wave calculations is that where coupling is allowed only between one reactant channel and one product channel. This approach was taken by McGuire and Mueller (1971), who performed calculations for $H + H_2$ with an assumed coupling between the reactant and product channels. Their work was appreciably extended by McGuire and Micha (1973). These authors made use of atom-transfer coordinates, i.e. the pair $(\mathbf{r}_{AC}, \mathbf{r}_{BC})$ for

A + BC and (r_{AC}, r_{AB}) for AB + C. As a result, coupling potentials became energy-dependent and were found to decrease with increasing relative energy. A second feature of their work was consideration at the outset of both electronic and nuclear motions. They expanded the total wavefunction, for both electrons and nuclei, in valence-bond electronic states, and determined the coefficients by means of Kohn's variational principle. The resulting scattering equations were differential, rather than integro-differential ones, because of the use of atom-transfer coordinates. They calculated integral and differential reaction cross sections of $H + H_2$, $D + H_2$ and $D + D_2$, for $(v_a j_a) = (0, 0) \rightarrow (v_b j_b) = (0, 0)$. Integral cross sections showed reaction thresholds (which do not change much with j_b) much closer to the accurate values than distorted-wave results, as is seen in Fig. 10. Angular distributions were found to be backward peaked at $E_{coll} = 0.50$ eV, and are shown in Fig. 11 for $D + H_2$. These results, as those for collinear and distorted-wave calculations, do not account for permutational exchange symmetry for identical nuclei, but simply include a statistical factor of two for homonuclear reactants.

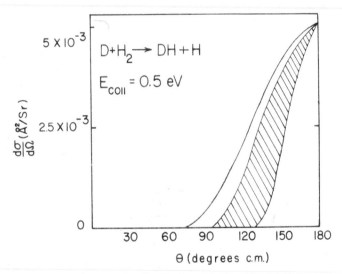

Fig. 11. Reaction angular distribution calculated by McGuire and Micha (1973) for $D + H_2$, and measured (cross-hatched region) by Geddes et al. (1973).

Elementary substitution reactions of type $R_1 + R_2R_3 \rightarrow R_1R_2 + R_3$, with R_k a molecular group, have been described in the context of the coupled-channel method by Brodsky and Levich (1973). These authors introduced distortion potentials for reactants and products and a parametrized, isotropic potential coupling. In practice, transition amplitudes were calculated within the distorted-wave approximation with hard-sphere distortion potentials. Estimated activation energies of hydrogen reactions were compared with experiment.

We return now to the coupled-channels approach based on operator equations. The formalism is adequately covered in several books (see e.g. Goldberger and Watson, 1964; Newton, 1966; Levine, 1969) and we shall only present the main equations. Assume for simplicity that only two reaction channels a (for A + BC) and b (AB + C) exist. The total Hamiltonian H may be split into two terms, a channel Hamiltonian H_c for free motion and a channel interaction V_c, with $c = a, b$. If Φ_a is the initial free state and we want the scattering states in channel a, i.e. those in the absence of rearrangement, then the Lippmann–Schwinger equation gives

$$\Psi_{aa} = \Phi_a + (E + i\epsilon - H_a)^{-1}V_a\Psi_{aa}$$

for the exact scattering wavefunction $\Psi_{aa}(\mathbf{R}_a, \mathbf{r}_a)$. In this equation one may let $\epsilon \to 0+$ only after scalar products are evaluated. If however, we want scattering states in channel b, then the wavefunction is $\Psi_{ba}(\mathbf{R}_b, \mathbf{r}_b) = \Psi_{aa}(\mathbf{R}_a, \mathbf{r}_a)$ and satisfies

$$\Psi_{ba} = \frac{i\epsilon}{E + i\epsilon - H_b}\Phi_a + (E + i\epsilon - H_b)^{-1}V_b\Psi_{aa}$$

The first term does not contribute to Ψ_{ba} when BC is bound, but it would contribute if it were unbound. Again, one may not put ϵ (or the first term) equal to zero until after scalar products are evaluated. It is in principle possible to solve the above equation for Ψ_{ba}, by expanding both sides in the internal states of AB. But the expansion must include continuum (dissociation) states of AB, which has discouraged researchers from attempting this approach so far. However, some efforts are being made in this direction.

A different approach, also based on operator equations, makes use of an expression for the rearrangement transition operator τ_{ca}, with $c = a, b$ given by

$$\tau_{ca} = V_a + V_c(E^{(+)} - H_{c'})^{-1}\tau_{c'a}$$

where $c' = a, b$ again. Baer and Kouri (1971a) investigated reactions on a constant piece-wise potential in two variables, using the previous equations in two forms. Firstly they put $c' = c$ and examined the analytical solution in the limit that no dissociative continuum is present, to conclude that this procedure would not conserve scattering flux if continuum contributions were only approximated. Secondly, they worked with the pair of coupled equations

$$\tau_{aa} = V_a + V_a(E^{(+)} - H_b)^{-1}\tau_{ba}$$

$$\tau_{ba} = V_a + V_b(E^{(+)} - H_a)^{-1}\tau_{aa}$$

in which case they obtained the correct answer even though the continuum contributions were again approximated. Baer and Kouri (1972a) obtained numerical results with these last equations for the Hulburt–Hirschfelder model, previously mentioned within collinear studies. The same authors considered three-dimensional models with piece-wise potentials and an infinitely heavy atom B, both for angular-independent (1971b, 1972b) and angular dependent (1972c) potentials. Further contributions along these lines were made by Hayes and Kouri (1971a, b), Kouri (1973) and Evers and Kouri (1973). Selection rules that determine the rotational state distribution of products have been analysed with these equations by Baer (1973) for a two-channel system with an infinitely heavy atom.

Operator equations have been employed by George and Ross (1971) to analyse symmetry in chemical reactions. In order to preserve the identity of electronic states of reactants and products, these authors worked within a quasi-adiabatic representation of electronic motions. By introducing a chain of approximations, going from separate conservation of total electronic spin to complete neglect of dynamics, they discussed the Wigner–Witmer angular momentum correlation rules, Shuler's rules for linear molecular conformations and the Woodward–Hoffmann rules.

Besides the described development on the numerical solution of coupled-channel equations, there have been several studies of the semiclassical limit of the coupled equations for rearrangement collisions. These studies are intended to provide simple analytical transition probabilities, or to take advantage of the short wavelengths involved in heavy-particle collisions, to simplify calculations. Marcus (1971) expressed the total complex wavefunction for rearrangement in polar form, with the phase satisfying a Hamilton–Jacobi equation and the amplitude satisfying an equation for flux conservation. He developed the theory in action-angle variables and obtained semiclassical scattering matrix elements by means of the stationary-phase approximation. Eu (1973) has developed a uniform semiclassical approximation. He expressed the non-local exchange coupling $U(R_a, R_b)$ in a separable form and used a non-iterative method for solving the integro-differential equations. Wavefunctions were given by linear combination of regular and irregular Airy functions, and S matrix elements by quadratures over these functions. An analogous treatment developed by Shipsey (1973) is based on a set of adiabatic reference functions (Dubrovskii, 1970) and is suitable for chemical reactions described in terms of curvilinear coordinates. Ritchie (1974) has used atom-transfer coordinates for $H + H_2$, separating the interaction potential into a central part plus a coupling part. A classical trajectory was calculated for the central part, and the coupling was introduced within a Born-type approximation, taking into consideration the identity of the nuclei. A somewhat related, but more general, perturbed elastic trajectory method has been briefly described by Cross (1973). He indicated how the sudden approximation may be applied to some of the degrees of freedom in inelastic and reactive scattering.

D. Electronic Transitions

We would like to complete this section by briefly describing some of the recent developments on electronically non-adiabatic reactions. From the standpoint of the coupled-channels method, there is in principle no added difficulty in treating more than one electronic state of the reactive system. This may be done, for example, by keeping electronic degrees of freedom in the Hamiltonian and expanding the total scattering wavefunction in the electronic states of reactants and products. In practice, however, some new difficulties may arise, such as non-orthogonality of vibrational states on different electronic potential surfaces. There is at present a lack of quantum mechanical results on this problem.

Some useful conceptual models have emerged from work (Krenos *et al.*, 1971) comparing experiments on $H^+ + D_2$ with classical trajectory calculations. In this case one finds two electronic potential surfaces with an avoided crossing. Reactions occurring in the lowest surface lead to formation of D^+. However, in certain instances D_2 or HD, receding after collision, may become vibrationally excited, reach the avoided crossing seam and lead to formation of HD^+ or D_2^+. A detailed study of this reaction was done by Tully and Preston (1971; see also Tully, 1973) in terms of 'surface-hopping' trajectories constrained to conserve total energy and angular momenta. Preston and Tully (1971) considered the H_3^+ system. The same conceptual model was taken up by Haas *et al.* (1972) in a discussion of electronic excitation induced by reactive collisions. This work employed curvilinear coordinates and stressed the role of dynamical displacement of the reaction path in producing vibrational excitation, which in turn increases Franck–Condon factors for non-adiabatic transitions. Nakamura (1973) has presented a quantum mechanical formulation for collinear reactions along the same lines. It assumes that the regions for vibrational excitation and electronic transitions are well separated, and that vibrational coupling is strong, while electronic coupling may be handled in a distorted-wave approximation. An extension of the classical 'hopping' model has been developed by Miller and George (1972) within the context of the classical S-matrix approximation and in terms of complex-valued trajectories. This work has been continued recently by Lin *et al.* (1973).

A different model, also for ion-reactions, has been proposed by Gislason (1972). It applies to $(ArH_2)^+$, for which the surfaces of $Ar^+(^2P_{1/2,3/2}) + H_2$ and of $Ar + H_2^+(v)$ are assumed to cross at large distances, where they are given by ion-induced dipole potentials. Cross sections at various energies are expressed in terms of the crossing radius and the radii of Langevin's theory of ion-reactions.

Electron-jump in reactions of alkali atoms is another example of non-adiabatic transitions. Several aspects of this mechanism have been explored in connection with experimental measurements (Herschbach, 1966; Kinsey, 1971). The role of vibrational motion in the electron-jump model has been investigated (Kendall and Grice, 1972) for alkali–dihalide reactions. It was assumed that the transition is sudden, and that reaction probabilities are proportional to the overlap (Franck–Condon) integral between vibrational wavefunctions of the dihalide X_2 and vibrational or continuum wave-functions of the negative ion X_2^-. Related calculations have been carried out by Grice and Herschbach (1973). Further developments on the electron-jump mechanism may be expected from analytical extensions of the Landau–Zener–Stueckelberg formula (Nikitin and Ovchinnikova, 1972; Delos and Thorson, 1972), and from computational studies with realistic atom–atom potentials (Evans and Lane, 1973; Redmon and Micha, 1974).

IV. MANY-BODY APPROACHES

A. Impulsive Models

In their simplest form, impulsive models make use of conservation of total mass, momentum, angular momentum and energy and assume mechanisms whereby reactant and product quantities may be related, with a minimum of information on the interaction potentials. Several useful models have arisen from interpretation of experiments, particularly those involving ions, and have been reviewed by Mahan (1970) and by Henglein (1974).

The spectator-stripping model (Henglein et al., 1965; Minturn et al., 1966) has been frequently used for reactions of type $A + BC \rightarrow AB + C$. It is most easily described in a laboratory frame, for BC initially unexcited and at rest. Assuming that a sudden collision leaves C at rest, conservation of momentum in the centre of mass frame leads to

$$W' = D' - D + \epsilon \left[1 - \frac{m_A m_C}{(m_A + m_B)(m_B + m_C)} \right]$$

where D, W and ϵ are dissociation, internal and relative-motion energies, respectively, and primes indicate product quantities. The model also predicts forward scattering in centre-of-mass, and a single value of ϵ'. These predictions are followed by a variety of high-energy ion collisions but with deviations, such as spread of scattering angles and of ϵ' values, that must be attributed to the role of the interaction. A modified spectator-stripping model was introduced by Wolfgang and coworkers (Herman et al., 1967), which incorporated attractive ion-induced dipole interactions for reactants and products. Light and Chan (1969) calculated isotopic distributions in $Ar/Kr + HD^+$ reactions with a similar model. They considered the effect of separating centre of charge from centre of mass of the reactant diatomic, and obtained good agreement with experiment. Another related model

(Chang and Light, 1970) assumed an intermediate impulsive event in which the relative radial velocities of reactants are reversed on impact, and was applied to angular distributions in $Ar^+ + D_2$. A different type of modification was considered by Connolly and Gislason (1972), to account for deviation of measured exoergicities from predicted ones. They noted that internal motion of BC would give to C a certain velocity before collision, even though BC is stationary, and calculated averages over all orientations of the C atom velocities. The same arguments were incorporated by Kuntz (1973) into a study of the effect of BC vibrational energy upon product angular and translational energy distributions.

Suplinskas (1968) developed an impulsive model for atom–diatom reactions in which the three atoms are represented by hard spheres. The reactant relative motion is treated classically. Following contact, the velocities of the three atoms are calculated and from them internal energies are found for each pair. If this internal energy is smaller than a chosen activation value, the pair will be bound, while if it is larger the pair will not form a bond. Reaction probabilities are derived by integrating over trajectory parameters and averaging over reactant molecular orientations. Suplinskas applied the model to $H + H_2$ and its isotopic variations, for collision energies above reaction threshold. George and Suplinskas (1969) extended the model to ion-reactions by including attractive reactant forces, and applied it to $Ar^+ + H_2/D_2$. The same authors (1971a) further included attractive product forces, and then (1971b) the torque effect due to the difference between centres of mass and of charge in HD, for which they calculated isotope effects. D. J. Malcolm-Lawes (1972a, b, c) developed a model for reaction of energetic hydrogen atoms using a hard sphere model and computer simulation. The effect of multiple hard-sphere collisions ('chattering') was commented upon by Kendall (1973).

A number of direct interaction models have originated in an analysis (Kuntz et al., 1969) of three-dimensional trajectories for reactions of alkali atoms M with halogen molecules XC, where an ionic bond is formed. These authors found that, in reactions proceeding by electron-jump, the magnitude of the repulsion energy between X and C in XC^- plays an essential role. The rejected atom C is not, therefore, a 'spectator'. A model of direct interaction with product repulsion (DIPR) was shown to account quantitatively for product angular and energy distributions, and to provide insight into transition from backward to sideways to forward scattering as product repulsion is decreased. Indicating with $\mathbf{P}$ a relative momentum, and with $\mathbf{I}$ the impulse transmitted to C along XC, application of momentum conservation to atom C in centre-of-mass leads to

$$\frac{m_C}{m_B + m_C}\mathbf{P} = \mathbf{P}' + \mathbf{I}$$

$$\cos\theta_{c.m.} = \mathbf{P}.\mathbf{P}'/PP'$$

where $\theta_{c.m.}$ is the centre-of-mass scattering angle. These equations reduce to the ones for spectator stripping in the limit $I \rightarrow 0$. A subsequent article by Kuntz (1970) developed the model in detail, to obtain product angular and energy distributions in terms of a model parameter q_V, equal to the speed change of C divided by its initial speed. This author (1972) extended the model, by allowing q_V to depend on the orientation angles of XC, to describe the $K + ICH_3$ reaction. It was found that the product energy distribution was quite sensitive to the newly introduced dependence, while the angular distribution was insensitive to it. The DIPR model has been recast by Marron (1973) in terms of a different choice of coordinates that results in considerably simpler equations.

Two other contributions (Grice, 1970; Grice and Hardin, 1971) have also incorporated orientation effects in an impulsive model, in this case involving only two hard spheres, which was applied to the alkali–iodine molecules $M + RI$ rebound reactions. Experimental results on the $O^+ + H_2/D_2/HD$ reaction have been compared (Gillen et al. 1973) to predictions of a sequential encounter model that also represents the atoms as hard spheres. Product angular distributions and their isotopic dependences are well represented by the model, which however, is less useful in predicting collision energy behaviours.

B. Faddeev–Watson Equations

An atom-transfer reaction is, in its simplest form, a three-body process. Collisions involving three bodies may in principle be described by means of the Lippmann–Schwinger equation. However, use of this equation is mathematically disadvantageous, when all reaction channels are included, because its integral kernel is not compact. Faddeev (1961, 1963) remedied this deficiency by introducing a set of coupled integral equations, closely related to previous ones by Watson (see, e.g., Watson and Nutall, 1967), such that after one iteration the kernel is compact. This mathematical development encouraged a number of studies on high-energy, nuclear and atomic three-body processes (Gillespie and Nuttall, 1968), and the development of methods of solution of the coupled equations (Sitenko, 1971; Delves, 1972; Chen, 1974).

The Faddeev–Watson equations have also provided new physical understanding in those cases where the interaction potential is given by a sum of pair-potentials. In molecular reactions however, the particles are atoms or ions and their internal structure play an essential role. As a result, the interaction potential is dominated by a three-body term whenever it is expressed only as a function of the positions of the three bodies, which would seem to invalidate the conceptual advantages of the equations. A solution to this problem has recently been presented (Micha, 1971, 1972), which may be simply described for three hydrogen atoms. A pair of ground-state hydrogens interact through $^1\Sigma$ or $^3\Sigma$ potentials, or equivalently through a spin-dependent potential $V_1(r_1)$ for, e.g., B and C at distance r_1, given by

$$V_1(r_1) = {}^1E_{BC}(r_1){}^1\mathcal{O}_1 + {}^3E_{BC}(r_1){}^3\mathcal{O}_1$$

$$^1\mathcal{O}_1 = \tfrac{1}{4} - s_B \cdot s_C, \qquad {}^3\mathcal{O}_1 = \tfrac{3}{4} + s_B \cdot s_C$$

where $^{1,3}\mathcal{O}$ are projection operators, and s_i is the atomic spin of atom i. Similar potentials V_2 and V_3 may be defined for AC and AB, and the total interaction V may be written as

$$V = V_1 + V_2 + V_3$$

This is a spin-dependent operator which, diagonalized in the two-dimensional space of doublet states (formed from three spin 1/2 spinors), leads to the London–Eyring–Polanyi potential surface, expressed in terms of singlet and triplet potentials. This provides a very accurate description (typically within 1 kcal/mole) of the H_3 surface. This procedure may be justified *a priori* within the Born–Oppenheimer approximation and using valence-bond electronic states, or, for a quite large class of atoms, within the diatomic-in-molecules approach. Alternatively, V may be interpreted as an interpolation expression that would lead to a large class of potentials, e.g. early- or late-downhill, by varying the strength of attractive and repulsive pair-potentials.

Returning to the Faddeev–Watson equations, we introduce channel indices 0, 1, 2, 3 and 4 for rearrangements A + B + C, A + (B, C), B + (C, A) and (A, B, C), respectively, where A + (B, C) means that A moves freely while B and C interact. The equations for the T operator are

$$T = V + VG_0 T = \sum_{j=1}^{3} T^{(j)}$$

$$T^{(j)} = T_j + \sum_{k=1}^{3} \bar{\delta}_{jk} T_j G_0 T^{(k)}$$

where G_0 is the propagator for three free atoms, T_j is the T-operator for V_j alone (i.e. requires the solution of only a two-atom problem), and $\bar{\delta}_{jk} = 1 - \delta_{jk}$. The operators T_j depend also on spins, but this dependance may readily be handled in terms of spin-recoupling coefficients. By so doing, a physical picture emerges in which a reaction is a succession of pair-collisions in which spins recouple to reflect temporary formation or breakup of bonds.

Quantitative results for high-energy processes may be obtained from a multiple-collision expansion. Indicating with Φ_k the free total state in channel k (a product of relative, internal and atomic spin states), the transition matrix element M_{31} for A + BC → AB + C is given by

$$M_{31} = \langle \Phi_3 | V_1 | \Phi_1 \rangle + \langle \Phi_3 | T_2 | \Phi \rangle + \langle \Phi_3 | T_1 G_0 T_2 + T_2 G_0 T_3 | \Phi_1 \rangle + \ldots$$

The first term corresponds to the spectator-stripping model, which arises naturally in the formalism. The second term is a displacement contribution in which atom B moves undisturbed, and so on. The first term may be conveniently calculated (Micha and McGuire, 1972) in the momentum representation, which gives

$$M_{31}(\text{S.S.}) = -\tfrac{1}{2}[W_1 - D_1 - p_1^2/(2m_1)]\tilde{\varphi}_3(\mathbf{p}_3)\tilde{\varphi}_1(\mathbf{p}_1)$$

where W_1, D_1, m_1 and $\mathbf{p}_1$ are internal energy, dissociation energy, reduced mass of BC and internal momentum of BC, respectively, and where $\tilde{\varphi}_k(\mathbf{p}_k)$ is the Fourier transform of the internal state in channel k. Fig. 12 shows integral reaction cross sections σ_r for selected $(n_1 j_1) = (0, 0) \to (n_3 j_3)$ transitions. The arrows indicate results from the kinematic spectator-stripping model, in

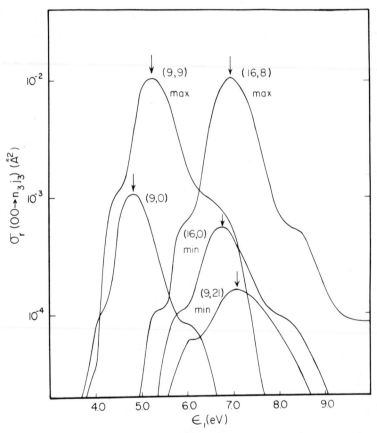

Fig. 12. Spectator-stripping cross sections for $T + H_2(0, 0) \to TH(n_3, j_3) + H$, calculated from the first term of the Faddeev–Watson multiple collision expansion (from D. A. Micha and P. McGuire, 1972).

striking agreement with the calculations. Product angular and energy distributions (rather than single values of $\theta_{c.m.}$ and ϵ') are seen to arise from the distribution of momenta in the reactant and product internal states.

The Faddeev–Watson equations are suitable to the study of permutational symmetry for identical nuclei. This has been done (Micha, 1974) for the three cases in which: (1) C = B'; (2) C = A''; and (3) B = A' and C = A'', to obtain transition amplitudes for direct, atom-exchange and dissociative processes. Nuclear spin variables were included, and amplitudes were found by successively reexpressing symmetrized amplitudes in terms of unsymmetrized ones, reducing nuclear-spin dependences and uncoupling the equations required for calculations. For example in case (2), the terms in the total wavefunction

$$\Psi_1 = \sum_{j=1}^{3} Y^{(j)}\Phi_1$$

may be obtained from only two coupled equations, although the operators $Y^{(j)}$ satisfy to begin with a set of three coupled equations similar to those for $T^{(j)}$. Decoupling may be achieved by using the permutation operator $Q = (AA'')$, and gives

$$Y^{(1)} = Y_1 + G_0 T_1 (Y^{(2)} + Q Y^{(1)} Q^{-1})$$
$$Y^{(2)} = Y_2 + G_0 T_2 (Y^{(1)} + Q Y^{(1)} Q^{-1})$$

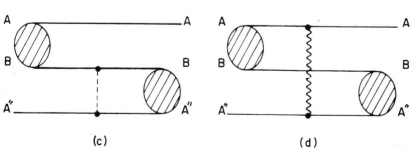

Fig. 13. The four terms contributing to the direct process in A + BA'' in the single-collision approximation (from D. A. Micha, 1974).

and $Y^{(3)} = QY^{(1)}Q^{-1}$. The pair of equations must be solved twice, once for each of the initial states Φ_1 and $Q\Phi_1$. Results may be analysed, e.g. in the single-collision approximation, which gives the four terms shown in Fig. 13. The dashed line represents $^1E_{BA''}$ and the wavy ones $^{1,3}T_j$. It is seen that, because of permutational symmetry, the spectator-stripping mechanism, Fig. 13c, contributes also to the direct process. This study adds new results to previous ones obtained by Miller (1969 op. cit) within a wave-mechanical formulation. Other aspects of permutational symmetry have been recently discussed by Scoles (1973).

V. DISCUSSION

A. Summary of Results

The material covered in the preceding sections is proof of the wealth of results obtained during the last few years within the quantum theory of reactive collisions. A great deal of the work has been quantitative, and has produced cross sections to compare with experiments on molecular beams or chemiluminescence. But, rather than listing numerical results, it is perhaps more important to summarize how our conceptual understanding of the field has been helped by theoretical developments.

Computational modelling has provided us with a glimpse to the variety we may expect in reactive collisions. This is in large part due to development of efficient computational approaches for collinear motion, such as the exponential method, the boundary-value method, and the method of constant piecewise potentials in two-dimensional (s, ρ) sectors. It has been found that we may expect formation of long-lived states, both of the shape-resonance type at low energies, and of the compound-state resonance type. In connection with local vibrational transitions, we may find both adiabatic behaviour at low collision energy and, more likely, a change into statistical behaviour as energy increases. The shape of cross sections at threshold is affected not only by tunnelling, but also by the dynamical action of path-curvature. There are indications that agreement of semiclassical results with quantum mechanical ones depends on the nature of the species involved. Numerical results are available on collision energy dependence, product vibrational distribution and isotope effects in the collinear case for systems involving hydrogen and halogens. Collision-energy and impact-parameter results are available for coplanar motion, but angular distributions are not yet reliable.

Among statistical models, the phase-space theory has been significantly extended to include closed channels and to calculate angular distributions. Studies have made clear limitations of the theory in predicting product energy and angular distributions, and have shown that fission and transition-state models are derivable from each other. Many neutral- and ion-reactions have been treated by means of the phase-space theory, including, more recently, collision-induced dissociation. Microscopic reversibility, which has in the past provided a useful check of numerical calculations, has now been put to work in the description of endoergic reactions that may be related to simple (reverse) exoergic ones. A suitable language, that of information theory, has been found to describe results of chemiluminescence measurements where reactions are direct and yet apparent temperatures may be defined for product distributions. Some understanding of this problem has been attained by using model Hamiltonians that include non-adiabatic effects, and by introducing dynamical constraints. Transition-state theories have been used to analyse a large number of experiments, and to obtain information on conformation of transition states and on energy-randomization during decomposition. They have proved applicable to atom–polyatom and diatom–diatom systems. Optical-potential models have also been very useful for analysing measured elastic cross sections in reactive systems, and have incorporated anisotropy terms, or the electron-jump mechanism, where necessary.

With regard to coupled-channels methods, there is now an extensive study on $H + H_2$ which has provided reliable three-dimensional integral cross sections (but not angular distributions) to compare with approximate approaches. Among these, distorted-wave and two-coupled-channel approximations appear useful for predicting angular distributions, at least for some simple, direct reactions. Some encouraging results have been obtained by applying a set of two coupled Lippmann–Schwinger equations to a number of model potentials. The procedure suggests that it might be possible to obtain reliable results, below the break-up threshold in atom–diatom reactions, by including the dissociation continuum in only an approximate manner.

Impulsive models have shown that many features of reaction cross sections are due to kinematic effects, and only in small degrees to interactions. The spectator-stripping model has been very useful in ion- and hot-atom-reactions. It has been extended by including long-range forces, by representing short-range (chemical) forces by hard core interactions, and by intuitively introducing effects due to internal molecular motions. Analysis of trajectories for alkali–halogen molecule reactions proceeding by electron-jump led to the DIPR model, which has also been subsequently extended to include anisotropy effects. These models provide distribution densities once integration over collision parameters and average over molecular orientations are carried out. A many-body approach to reactions has been developed in terms of the Faddeev–Watson equations and potentials written as sums of pairs of interactions. It shows the role of both attractive and repulsive interactions in successions of pair-collisions. It has provided an *a priori* foundation for the spectator-stripping model, and a natural way of handling permutational exchange symmetry.

B. Expectations and Future Needs

Considering the directions taken by the reviewed research in past years and the present interests of workers in the field, it is possible to mention some areas where new results may soon be expected. One of these areas is computational modelling of coplanar motion, where most of the background work has been done and in which different computational procedures are available. Within analytical models, studies of vibrationally diabatic transitions should continue to clarify our understanding of product energy distributions. In the fields of optical potentials and coupled-channels, efforts are being made to develop approximations suitable for many coupled channels (at present only for non-reactive systems), and to understand product angular distributions by analysing angular momentum coupling schemes for three bodies. Work is being done on ways of handling the continuum (dissociation) contribution to reactive collisions. The Faddeev–Watson equations are being used for high-energy collisions, to clarify the implications of different multiple-collision models.

The rapid pace of experimental developments has created a need for more theoretical work. This is also required to provide more satisfactory predictive models. We are at present lacking approximations capable of providing reliable product angular distributions, even for simple reactions. Models are needed that would apply to reactions with polyatomics, or for pairs of diatomics. The available statistical theories would have to be extended, to include more dynamical constrains and more details of potential interactions. Much remains to be done on energy randomization in decomposition following isolated bimolecular encounter and on collision-induced dissociation. We are lacking satisfactory definitions of optical potentials for reactive systems, which would untangle inelastic and reactive contributions to cross sections. More computational work is needed to find out the effect of permutational exchange of identical nuclei on reaction probabilities. Reactions involving electronic transitions have only recently been under investigation. Much analytical and computational work will have to be done in this area to identify relevant mechanisms, to begin with. Diverse semiempirical impulse models have been proposed in the past to explain a certain observed cross section. By providing an *a priori* foundation for impulsive models, we might be able to learn about their range of validity and of applicability to different species.

Quantum chemical information is needed in several areas. For example repulsive potentials are missing for many pairs of atoms that have otherwise been well studied. These repulsive potentials enter in the diatomics-in-molecules approach and in dynamical models. Whenever potential surfaces cross (or pseudo-cross) we need the intersurface couplings in order to understand whether electronic motions are adiabatic or diabatic. *A priori* information on conformations of long-lived complexes would be helpful, because they determine to a large extent product distributions. Finally, studies of changes in molecular orbitals during reactions are essential to a more basic understanding of the whole field, which will eventually require a simultaneous description of both electronic and nuclear motions.

Acknowledgements

This work has been supported by a grant from the National Science Foundation and by a fellowship from the Alfred P. Sloan Foundation. The author thanks the many researchers that made preprints and reprints available to him, and the authors of figures for allowing their reproduction. He appreciates comments on the manuscript made by Professors R. D. Levine, J. Light, R. Marcus and R. Wyatt.

References, Section I

B. Alder, S. Fernbach and M. Rotenberg (1971). (Ed.) *Methods in Computational Physics*, Vol. 10, Academic Press, New York.

R. Bersohn (1972). *Accounts Chem. Res.*, **5**, 200.

R. B. Bernstein (1971). *Israel J. Chem.*, **9**, 615.

D. L. Bunker (1971). In B. Alder *et al.* (eds.), *Methods in Computational Physics*, Vol. 10, Academic Press, New York, p. 287.

P. R. Certain and L. W. Bruch (1972). In W. Byers Brown (ed.), *MTP International Review of Science*, Physical Chemistry, Vol. 1, University Park Press, Baltimore, p. 113.

S. G. Christov (1972). *Ber. Bunsenges. Phys. Chem.* **76**, 507.

S. G. Christov (1974). *Ber. Bunsenges. Phys. Chem.* **78**, 537.

R. Daudel (1973), *Quantum Theory of Chemical Reactivity* (D. Reidel, Boston, 1973).

J. Dubrin (1973). *Ann. Rev. Phys. Chem.*, **24**, 97.

M. S. Dzhidzhoev, V. T. Platonenko and R. V. Khokhlov (1970). *Soviet Physics Uspekhi.* **13**, 247.

F. O. Ellison (1963). *J. Amer. Chem. Soc.*, **85**, 3540.

M. A. D. Fluendy (1970). *Ann. Report Chem. Soc. (London)*, **67**, A151.

M. A. D. Fluendy and K. P. Lawley (1973). *Chemical Applications of Molecular Beam Scattering*, Chapman and Hall, London.

W. Forst (1973). *Theory of Unimolecular Reactions*, Wiley, New York.

J. L. Franklin (1972) (ed.) *Ion-Molecule Reactions* (2 vols.), Plenum, New York.

L. Friedman and B. A. Reuben (1971). *Advan. Chem. Phys.*, **19**, 33.

T. F. George and J. Ross (1973). *Ann. Rev. Phys. Chem.*, **24**, 263.

P. E. Hodgson (1971). *Nuclear Reactions and Nuclear Structure*, Clarendon Press, Oxford.

M. Karplus (1970). In Ch. Schlier (ed.), *Molecular Beams and Reaction Kinetics*, Academic Press, New York, p. 320.

K. L. Kompa (1973). *Fortschr. Chem. Forsch.*, **37**.

D. J. Kouri (1973). In D. W. Smith and W. B. McRae (eds.), *Energy, Structure and Reactivity*, Wiley, New York, p. 26.

M. Krauss (1970). *Ann. Rev. Phys. Chem.*, **21**, 39.

P. J. Kuntz and A. C. Roach (1972). *Faraday Soc. Trans. II*, **68**, 259.

Y. T. Lee (1971). In T. R. Govers and F. J. de Heer (eds.), *Invited Papers VII ICPEAC*, 357. North-Holland.

R. D. Levine (1969). *Quantum Mechanics of Molecular Rate Processes*, Oxford University Press, London.

R. D. Levine (1972). In W. Byers Brown (ed.), MTP International Review of Science, Physical Chemistry, vol. 1, University Park Press, Baltimore, p. 229.

R. D. Levine and R. B. Bernstein (1974). *Molecular Reaction Dynamics*, Oxford University Press, New York.

W. H. Lester Jr. (1971). (ed.), *Potential Energy Surfaces in Chemistry*, IBM Research Lab, San Jose.

J. C. Light (1971). *Advan. Chem. Phys.*, **19**, 1.

R. A. Marcus (1971). In W. A. Lester (ed.), *Potential Energy Surfaces in Chemistry*, IBM Research Lab, San Jose. p. 58.

E. W. McDaniel, V. Cermák, A. Dalgarno, E. E. Ferguson and L. Friedman (1970). *Ion-Molecule Reactions*, Wiley-Interscience, New York.

D. A. Micha, (1973). *Accounts Chem. Res.*, **6**, 138.

D. A. Micha (1974). *Advan. Quantum Chem.*, **8**.

W. H. Miller (1971). *Accounts Chem. Res.*, **4**, 161.

C. B. Moore (1971). *Ann. Rev. Phys. Chem.*, **22**, 387.

C. Moore and P. F. Zittel (1973). *Science*, **182**, 541.

E. E. Nikitin (1968). *Ber. Bunsenges. Phys. Chem.*, **72**, 949.

J. C. Polanyi (1971). *Appl. Optics*, **10**, 1717.

J. C. Polanyi (1972). (ed.), *MTP International Review of Science*, Physical Chemistry, Vol. 9. University Park Press, Baltimore.

J. C. Polanyi and J. L. Schreiber (1974). In H. Eyring *et al.* (eds.), *Advanced Treatise of Physical Chemistry*, Vol. 6. Academic Press, New York.

P. J. Robinson and K. A. Holbrook (1972). *Unimolecular Reactions*, Wiley.

H. F. Schaefer (1973). In D. W. Smith and W. B. McRae (eds.), *Energy, Structure and Reactivity*, Wiley, New York, p. 148.

Ch. Schlier (1970). (Ed.) *Molecular Beams and Reaction Kinetics*, Academic Press, New York.

D. Secrest (1973). *Ann. Rev. Phys. Chem.*, **24**, 379.

L. D. Spicer and B. S. Rabinovich (1970). *Ann. Rev. Phys. Chem.*, **21**, 349.

J. I. Steinfeld and J. L. Kinsey (1970). *Progr. Reaction Kinetics*, **5**, 1.

J. P. Toennies (1974). In H. Eyring, D. Henderson and W. Jost (eds.), *Physical Chemistry An Advanced Treatise*, **VI**, Academic Press, New York, p. 228.

J. Troe and H. G. Wagner (1972). *Ann. Rev. Phys. Chem.*, **23**, 311.

J. C. Tully (1973a). *J. Chem. Phys.*, **58**, 1396.

J. C. Tully (1973b). *J. Chem. Phys.*, **59**, 5122.

A. A. Westenberg (1973). *Ann. Rev. Phys. Chem.*, **24**, 77.

R. E. Weston and H. A. Schwarz (1972). *Chemical Kinetics*, Prentice-Hall, Englewood Cliffs, N.J.

References, Section II

M. Baer (1974). *J. Chem. Phys.*, **60**, 1057.

M. V. Basilevsky (1972). *Mol. Phys.*, **23**, 161.

M. V. Basilevsky (1973). *Mol. Phys.*, **26**, 765.

J. M. Bowman and A. Kuppermann (1971). *Chem. Phys. Lett.*, **12**, 1.

J. M. Bowman and A. Kuppermann (1973). *J. Chem. Phys.*, **59**, 6524.

J. M. Bowman, A. Kuppermann, J. T. Adams and D. G. Truhlar (1973). *Chem. Phys. Lett.*, **20**, 229.

S. Chan, J. Light and J. Lin (1968). *J. Chem. Phys.*, **49**, 86.

J. N. L. Connor and M. S. Child (1970). *Mol. Phys.*, **18**, 653.
O. H. Crawford (1971a). *J. Chem. Phys.*, **55**, 2563.
O. H. Crawford (1971b). *J. Chem. Phys.*, **55**, 2571.
C. F. Curtiss and F. T. Adler (1952). *J. Chem. Phys.*, **20**, 249.
D. H. Diestler (1971). *J. Chem. Phys.*, **54**, 4547.
D. J. Diestler (1972). *J. Chem. Phys.*, **56**, 2092.
D. J. Diestler and M. Karplus (1971). *J. Chem. Phys.*, **55**, 5832.
D. J. Diestler and V. McKoy (1968). *J. Chem. Phys.*, **48**, 2951.
D. J. Diestler, D. G. Truhlar and A. Kuppermann (1972). *Chem. Phys. Lett.*, **13**, 1.
D. R. Dion, M. B. Milleur and J. O. Hirschfelder (1970). *J. Chem. Phys.*, **52**, 3179.
J. W. Duff and D. G. Truhlar (1973a). *Chem. Phys. Lett.*, **23**, 327.
J. W. Duff and D. G. Truhlar (1973b). *Chem. Phys.*, **4**, 1.
S. F. Fischer and M. A. Ratner (1972). *J. Chem. Phys.*, **57**, 2769.
F. K. Fong and S. Diestler (1972). *J. Chem. Phys.*, **57**, 4953.
R. G. Gilbert and T. F. George (1973). *Chem. Phys. Lett.*, **20**, 187.
S. H. Harms and R. E. Wyatt (1972). *J. Chem. Phys.*, **57**, 2722.
G. L. Hofacker and R. D. Levine (1971). *Chem. Phys. Lett.*, **9**, 617.
G. L. Hofacker and R. D. Levine (1972). *Chem. Phys. Lett.*, **15**, 165.
G. L. Hofacker and N. Rösch (1973). *Ber. Bunsenges. phys. Chem.*, **77**, 661.
P. B. James and G. R. North (1972). *J. Chem. Phys.*, **57**, 4415.
B. R. Johnson (1972). *Chem. Phys. Lett.*, **13**, 172.
G. W. Koeppl (1973). *J. Chem. Phys.*, **59**, 2168.
A. Kuppermann (1971). In W. A. Lester, Jr. (ed.), *Proc. Conf. Potential Energy Surfaces in Chem.*, IBM, S. Jose, p. 121.
R. D. Levine and B. R. Johnson (1970). *Chem. Phys. Lett.*, **7**, 404.
R. D. Levine and S-F. Wu (1971). *Chem. Phys. Lett.*, **11**, 557.
J. C. Light (1971a). *Methods Comput. Phys.*, **10**, 111.
J. C. Light (1971b). *Advan. Chem. Phys.*, **19**, 1.
J. C. Light (1971c). In W. A. Lester, Jr. (ed.), *Proc. Conf. Potential Energy Surfaces in Chem.*, IBM, S. Jose, p. 74.
J. Lin and J. C. Light (1966). *J. Chem. Phys.*, **45**, 2545.
R. A. Marcus (1966). *J. Chem. Phys.*, **45**, 4493.
R. A. Marcus (1967). *Discussion Faraday Soc.*, **44**, 7.
R. A. Marcus (1968). *J. Chem. Phys.*, **49**, 2610.
E. A. McCullough and R. E. Wyatt (1971a). *J. Chem. Phys.*, **54**, 3578.
E. A. McCullough and R. E. Wyatt (1971b). *J. Chem. Phys.*, **54**, 3592.
P. B. Middleton and R. E. Wyatt (1972). *J. Chem. Phys.*, **56**, 2702.
P. B. Middleton and R. E. Wyatt (1973). *Chem. Phys. Lett.*, **21**, 57.
C. G. Miller (1973). *J. Chem. Phys.*, **59**, 267.
G. Miller and J. Light (1971a). *J. Chem. Phys.*, **54**, 1635.
G. Miller and J. Light (1971b). *J. Chem. Phys.*, **54**, 1643.
W. H. Miller (1970). *J. Chem. Phys.*, **53**, 1949.
E. M. Mortensen and L. D. Gucwa (1969). *J. Chem. Phys.*, **51**, 5695.
E. M. Mortensen and K. S. Pitzer (1962). *Chem. Soc. (London)*, **16**, 57.
J. T. Muckerman (1971). *J. Chem. Phys.*, **54**, 1155.
J. T. Muckerman (1972). *J. Chem. Phys.*, **56**, 2997.
A. Persky and M. Baer (1974). *J. Chem. Phys.*, **60**, 133.
R. N. Porter and M. Karplus (1964). *J. Chem. Phys.*, **40**, 1105.
C. Rankin and J. Light (1969). *J. Chem. Phys.*, **51**, 1701.
P. D. Robinson (1970). *J. Chem. Phys.*, **52**, 3175.
J. D. Russell and J. C. Light (1971). *J. Chem. Phys.*, **54**, 4881.

R. P. Saxon and J. C. Light (1971). *J. Chem. Phys.*, **55**, 455.
R. P. Saxon and J. C. Light (1972a). *J. Chem. Phys.*, **56**, 3874.
R. P. Saxon and J. C. Light (1972b). *J. Chem. Phys.*, **56**, 3885.
R. P. Saxon and J. C. Light (1972c). *J. Chem. Phys.*, **57**, 2758.
G. C. Schatz and A. Kuppermann (1973). *J. Chem. Phys.*, **59**, 964.
G. C. Schatz, J. M. Bowman and A. Kuppermann (1973). *J. Chem. Phys.*, **58**, 4023.
I. Shavitt, R. M. Stevens, F. L. Minn and M. Karplus (1968). *J. Chem. Phys.*, **48**, 2700.
E. J. Shipsey (1969). *J. Chem. Phys.*, **50**, 2685.
E. J. Shipsey (1972). *J. Chem. Phys.*, **56**, 3843.
E. J. Shipsey (1973). *J. Chem. Phys.*, **58**, 232.
K. Tang, B. Kleinman and M. Karplus (1969). *J. Chem. Phys.*, **50**, 1119.
D. G. Truhlar and A. Kuppermann (1970). *J. Chem. Phys.*, **52**, 3841.
D. G. Truhlar and A. Kuppermann (1971). *Chem. Phys. Lett.*, **9**, 269.
D. G. Truhlar and A. Kuppermann (1972). *J. Chem. Phys.*, **56**, 2232.
D. G. Truhlar, A. Kuppermann and J. T. Adams (1973). *J. Chem. Phys.*, **59**, 395.
J. J. Tyson, R. P. Saxon and J. C. Light (1973). *J. Chem. Phys.*, **59**, 363.
R. B. Walker and R. E. Wyatt (1972a). *Chem. Phys. Lett.*, **16**, 52.
R. B. Walker and R. E. Wyatt (1972b). *J. Chem. Phys.*, **57**, 2728.
R. B. Walker and R. E. Wyatt (1974a). *J. Chem. Phys.*, **61**, 4839.
R. B. Walker and R. E. Wyatt (1974b). *Mol. Phys.*, **28**, 101.
S. F. Wu, B. R. Johnson and R. D. Levine (1973a). *Molec. Phys.*, **25**, 609.
S. F. Wu, B. R. Johnson and R. D. Levine (1973b). *Molec. Phys.*, **25**, 839.
S. F. Wu and R. D. Levine (1971). *Molec. Phys.*, **22**, 881.
S. F. Wu and R. D. Levine (1973). *Molec. Phys.*, **25**, 937.
S. F. Wu and R. A. Marcus (1970). *J. Chem. Phys.*, **53**, 4026.
R. E. Wyatt (1969). *J. Chem. Phys.*, **51**, 3489.
R. E. Wyatt (1972). *J. Chem. Phys.*, **56**, 390.

References, Section III

A. C. Allison (1972). *Computer Phys.*, **3**, 173.
K. G. Anlauf, D. H. Maylotte, J. C. Polanyi and R. B. Bernstein (1969). *J. Chem. Phys.*, **51**, 5716.
M. Baer (1973). *Molec. Phys.*, **26**, 369.
M. Baer and D. J. Kouri (1971a). *Phys. Rev. A*, **4**, 1924.
M. Baer and D. J. Kouri (1971b). *Mol. Phys.*, **22**, 289.
M. Baer and D. J. Kouri (1972a). *J. Chem. Phys.*, **56**, 4840.
M. Baer and D. J. Kouri (1972b). *J. Chem. Phys.*, **57**, 3441.
M. Baer and D. J. Kouri (1972c). *J. Chem. Phys.*, **56**, 1758.
A. Ben-Shaul (1973). *Chem. Phys.*, **1**, 244.
A. Ben-Shaul, R. D. Levine and R. B. Bernstein (1972a). *Chem. Phys. Lett.*, **15**, 160.
A. Ben-Shaul, R. D. Levine and R. B. Bernstein (1972b). *J. Chem. Phys.*, **57**, 5427.
R. B. Bernstein and R. D. Levine (1972b). *J. Chem. Phys.*, **57**, 434.
A. M. Brodsky and V. G. Levich (1973). *J. Chem. Phys.*, **58**, 3065.
D. L. Bunker and M. Pattengill (1968). *J. Chem. Phys.*, **48**, 772.
J. C. Y. Chen (1966). *Phys. Rev.*, **152**, 1454.
R. J. Cross, Jr. (1973). *J. Chem. Phys.*, **58**, 5178.
J. B. Delos and W. R. Thorson (1972). *Phys. Rev. Lett.*, **28**, 647.
L. M. Delves and A. C. Phillips (1969). *Rev. Mod. Phys.*, **41**, 497.
A. S. Dickinson, R. E. Roberts and R. B. Bernstein (1972). *J. Phys. B: Atom. Molec. Phys.*, **5**, 355.
G. V. Dubrovskii (1970). *Soviet Phys. JETP*, **31**, 577.

S. A. Evans and N. F. Lane (1973). *Phys. Rev. A*, **8**, 1385.

N. S. Evers and D. J. Kouri (1973). *J. Chem. Phys.*, **58**, 1955.

B. C. Eu (1970). *J. Chem. Phys.*, **52**, 3021.

B. C. Eu (1972). *J. Chem. Phys.*, **57**, 2531.

B. C. Eu (1973). *J. Chem. Phys.*, **58**, 472.

H. Feshbach (1962). *Ann. Phys. (New York)*, **19**, 28.

D. C. Fullerton and T. F. Moran (1971). *Chem. Phys. Lett.*, **10**, 626.

D. C. Fullerton and T. F. Moran (1972). *Int. J. Mass Spectrom Ion Phys.*, **9**, 15.

J. Geddes, H. F. Krause and W. L. File (1972). *J. Chem. Phys.*, **56**, 3298.

A. Gelb and R. J. Suplinskas (1970). *J. Chem. Phys.*, **53**, 2249.

T. F. George and J. Ross (1971). *J. Chem. Phys.*, **55**, 3851.

T. F. George and W. H. Miller (1972). *Phys. Rev. A*, **6**, 1885.

T. F. George and J. Ross (1972). *J. Chem. Phys.*, **56**, 5786.

E. A. Gislason (1972). *J. Chem. Phys.*, **57**, 3396.

M. L. Goldberger and K. M. Watson (1964. *Collision Theory*. Wiley, New York.

R. Grice and D. Herschbach (1974). *Molec. Phys.*, **27**, 159.

Y. Haas, R. D. Levine and G. Stein (1972). *Chem. Phys. Lett.*, **15**, 7.

Y. Hahn (1966). *Phys. Rev.* **142**, 603.

R. M. Harris and J. F. Wilson (1971). *J. Chem. Phys.*, **54**, 2088.

E. F. Hayes and D. J. Kouri (1971a). *J. Chem. Phys.*, **54**, 878.

E. F. Hayes and D. J. Kouri (1971b). *Chem. Phys. Lett.*, **11**, 233.

D. R. Herschbach (1966). *Advan. Chem. Phys.*, **10**, 319.

E. T. Jaynes (1963). In K. W. Ford (ed.), *1962 Brandeis Lectures*, Vol. 3, p. 181.

A. Katz (1967). *Principles of Statistical Mechanics*, Freeman, San Francisco.

G. M. Kendall and R. Grice (1972). *Molec. Phys.*, **24**, 1373.

J. L. Kinsey (1971a). *J. Chem. Phys.*, **54**, 1206.

J. L. Kinsey (1971b). In J. C. Polanyi (ed.), *MTP International Review of Science: Phys. Chem.*, Vol. 9, University Park Press, Baltimore, p. 173.

D. J. Kouri (1973). *J. Chem. Phys.*, **58**, 1914.

J. Krenos, R. Preston, R. Wolfgang and J. Tully (1971). *Chem. Phys. Lett.*, **10**, 17.

G. H. Kwei, A. B. Lees and J. A. Silver (1971). *J. Chem. Phys.*, **55**, 456.

A. Lee, R. L. LeRoy, Z. Herman and R. Wolfgang (1972). *Chem. Phys. Lett.*, **12**, 569.

A. B. Lees and C. H. Kwei (1973). *J. Chem. Phys.*, **58**, 1710.

W. A. Lester and R. B. Bernstein (1970). *J. Chem. Phys.*, **53**, 11.

R. D. Levine (1970). *Accounts Chem. Res.*, **3**, 273.

R. D. Levine (1972). *Israel J. Chem.*, **8**, 289.

R. D. Levine and R. B. Bernstein (1969). *Israel J. of Chem.*, **7**, 315.

R. D. Levine and R. B. Bernstein (1970). *J. Chem. Phys.*, **53**, 686.

R. D. Levine and R. B. Bernstein (1971). *Chem. Phys. Lett.*, **11**, 552.

R. D. Levine and R. B. Bernstein (1972). *J. Chem. Phys.*, **56**, 2281.

R. D. Levine and R. B. Bernstein (1973). *Discussions Faraday Soc.*, **55**, 100.

R. D. Levine, B. R. Johnson and R. B. Bernstein (1973). *Chem. Phys. Lett.*, **19**, 1.

R. D. Levine, F. A. Wolf and J. A. Maus (1971). *Chem. Phys. Lett.*, **10**, 2.

J. C. Light (1967). *Discussions Faraday Soc.*, **44**, 14.

Y. W. Lin, T. F. George and K. Morokuma (1973). *Chem. Phys. Lett.*, **22**, 547.

R. A. Marcus (1968). In H. Hartmann (ed.), *Chemische Elementarprozesse*. Springer-Verlag, Berlin, p. 109.

R. A. Marcus (1970). *J. Chem. Phys.*, **53**, 604.

R. A. Marcus (1971). *J. Chem. Phys.*, **54**, 3965.

R. Marriott and D. A. Micha (1969). *Phys. Rev.*, **180**, 120.

E. W. McDaniel, V. Cermak, A. Dalgarno, E. E. Ferguson and L. Friedman (1970). *Ion-Molecule Reactions*, Wiley-Interscience, New York. Chap. 3.

P. McGuire and D. A. Micha (1973). *Molec. Phys.*, **25**, 1335.
P. McGuire and C. R. Mueller (1971). *Phys. Rev. A*, **3**, 1358.
D. A. Micha (1965). *Arkiv Fysik*, **30**, 411, 425, 437.
D. A. Micha (1967). *Phys. Rev.*, **162**, 88.
D. A. Micha (1973). *Accounts Chem. Res.*, **6**, 138.
D. A. Micha (1974). *Advan. Quantum Chem.*, **8**.
D. A. Micha (1969). *J. Chem. Phys.*, **50**, 722.
W. B. Miller, S. A. Safron and D. R. Herschbach (1972). *J. Chem. Phys.*, **56**, 3581.
W. H. Miller (1970). *J. Chem. Phys.*, **52**, 543.
W. H. Miller (1969). *J. Chem. Phys.*, **50**, 407.
W. H. Miller and T. F. George (1972). *J. Chem. Phys.*, **56**, 5637.
T. F. Moran and D. C. Fullerton (1972). *J. Chem. Phys.*, **56**, 21.
T. F. Moran and D. C. Fullerton (1971). *J. Chem. Phys.*, **54**, 5231.
H. Nakamura (1973). *Molec. Phys.*, **26**, 673.
R. G. Newton (1966). *Scattering Theory of Waves and Particles*, McGraw-Hill, New York.
E. E. Nikitin and M. Ya. Ovchinnikova (1972). *Uspekhi*, **14**, 394.
R. E. Olson (1972). *J. Chem. Phys.*, **56**, 2979.
J. M. Parson and Y. T. Lee (1972). *J. Chem. Phys.*, **56**, 4658.
J. M. Parson, K. Shobatake, Y. T. Lee and S. A. Rice (1973). *J. Chem. Phys.*, **59**, 1402.
L. L. Paulsen, J. Ross and J. I. Steinfeld (1972). *J. Chem. Phys.*, **57**, 1592.
C. Rebick and R. D. Levine (1973). *J. Chem. Phys.*, **58**, 3942.
M. Redmon and D. A. Micha (1974). *Intern. J. Quantum Chem.*, Symposium **8**, 253.
O. K. Rice (1967). *Statistical Mechanics, Thermodynamics and Kinetics*, W. H. Freeman & Co., San Francisco.
S. J. Riley and D. R. Herschbach (1973). *J. Chem. Phys.*, **58**, 27.
B. Ritchie (1974). *J. Chem. Phys.*, **60**, 1386.
R. E. Roberts, R. B. Bernstein and C. F. Curtiss (1969). *J. Chem. Phys.*, **50**, 5163.
R. E. Roberts and R. B. Bernstein (1970). *Chem. Phys. Lett.*, **6**, 282.
R. E. Roberts and J. Ross (1970). *J. Chem. Phys.*, **52**, 1464.
I. Rusinek and R. E. Roberts (1973). *Chem. Phys.*, **1**, 392.
I. Rusinek and R. E. Roberts (1974). *Chem. Phys.*, **3**, 268.
S. A. Safron, N. D. Weinstein and D. R. Herschbach (1972). *Chem. Phys. Lett.*, **12**, 564.
Ch. Schlier (1970). (ed.) *Molecular Beams and Reaction Kinetics*, Academic Press, New York.
M. Shapiro (1972). *J. Chem. Phys.*, **56**, 2582.
E. J. Shipsey (1973). *J. Chem. Phys.*, **58**, 5368.
K. Shobatake, Y. T. Lee and S. A. Rice (1973). *J. Chem. Phys.*, **59**, 1435.
F. T. Smith (1969). In A. R. Hochstim (ed.), *Kinetic Processes in Gases and Plasmas*, Academic Press, New York, p. 321.
K. T. Tang (1972). *J. Chem. Phys.*, **57**, 1808.
K. T. Tang and M. Karplus (1971). *Phys. Rev. A*, **4**, 1844.
D. G. Truhlar (1971). *J. Chem. Phys.*, **54**, 2635.
D. G. Truhlar (1972). *J. Chem. Phys.*, **56**, 1481.
D. G. Truhlar and A. F. Wagner (1972). *J. Chem. Phys.*, **57**, 4063.
J. C. Tully and R. K. Preston (1971). *J. Chem. Phys.*, **55**, 562.
J. C. Tully (1973). *Ber. Bunsenges. phys. Chem.*, **77**, 557.
R. A. Van Santen (1972). *J. Chem. Phys.*, **57**, 5418.
E. V. Waage and B. S. Rabinovich (1970). *Chem. Rev.*, **70**, 377.
R. A. White, A. Altenberger-Siczek and J. Light (1973). *J. Chem. Phys.*, **59**, 200.
R. A. White and J. C. Light (1971). *J. Chem. Phys.*, **55**, 379.

F. A. Wolf and J. L. Haller (1970). *J. Chem. Phys.*, **52**, 5910.
G. Wolken, Jr. (1972). *J. Chem. Phys.*, **56**, 2591.
G. Wolken and M. Karplus (1974). *J. Chem. Phys.*, **60**, 351.

References, Section IV

D. T. Chang and J. C. Light (1970). *J. Chem. Phys.*, **52**, 5687.
J. C. Y. Chen (1974). In E. W. McDaniel and M. R. C. McDowell (eds.), *Case Studies in Atomic Collision Physics*, Vol. III, North-Holland Publishing Co., Amsterdam, p. 307.
C. M. Connolly and E. A. Gislason (1972). *Chem. Phys. Lett.*, **14**, 103.
L. M. Delves (1972). *Advan. Nucl. Phys.*, **5**, 1. Barauger ed. Plenum, New York.
L. D. Faddeev (1961). *Sov. Phys. JETP*, **12**, 1014.
L. D. Faddeev (1963). *Sov. Phys. Dokl.*, **7**, 600.
T. F. George and R. J. Suplinskas (1969). *J. Chem. Phys.*, **51**, 3666.
T. F. George and R. J. Suplinskas (1971a). *J. Chem. Phys.*, **54**, 1037.
T. F. George and R. J. Suplinskas (1971b). *J. Chem. Phys.*, **54**, 1046.
J. Gillespie and J. Nutall (1968). (eds.) *Three-Particle Scattering in Quantum Mechanics.* Benjamin, New York.
R. Grice (1970). *Molec. Phys.*, **19**, 501.
R. Grice and D. R. Hardin (1971). *Molec. Phys.*, **21**, 805.
K. T. Gillen, B. H. Mahan and J. S. Winn (1973). *J. Chem. Phys.*, **59**, 6380.
A. Henglein (1974). In H. Eyring *et al.* (eds.), *Physical Chemistry, an Advanced Treatise*, Vol. VI. Academic Press, New York.
A. Henglein, K. Lacmann and G. Jacobs (1965). *Ber Bunsenges. Physik. Chem.*, **69**, 279.
Z. Herman, J. Kerstetter, T. Rose and R. Wolfgang (1967). *Discussions Faraday Soc.*, **44**, 123.
G. M. Kendall (1973). *J. Chem. Phys.*, **58**, 3523.
P. J. Kuntz (1970). *Trans. Faraday Soc.*, **66**, 2980.
P. J. Kuntz (1972). *Molec. Phys.*, **23**, 1035.
P. J. Kuntz (1973). *Chem. Phys. Lett.*, **19**, 319.
P. J. Kuntz, M. H. Mok and J. C. Polanyi (1969). *J. Chem. Phys.*, **50**, 4623.
J. C. Light and S. Chan (1969). *J. Chem. Phys.*, **51**, 1008.
B. H. Mahan (1970). *Accounts Chem. Res.*, **3**, 393.
D. J. Malcolm-Lawes (1972a). *Faraday Soc. Trans.*, II **68**, 1613.
D. J. Malcolm-Lawes (1972b). *Faraday Soc. Trans.*, II **68**, 2051.
D. J. Malcolm-Lawes (1972c). *J. Chem. Phys.*, **57**, 5522.
M. T. Marron (1973). *J. Chem. Phys.*, **58**, 153.
D. A. Micha (1971). *Proc. VII ICPEAC*, North-Holland, Amsterdam, p. 217.
D. A. Micha (1972). *J. Chem. Phys.*, **57**, 2184.
D. A. Micha (1974). *J. Chem. Phys.*, **60**, 2480.
D. A. Micha and P. McGuire (1972). *Chem. Phys. Letts.*, **17**, 207.
R. E. Minturn, S. Datz and R. L. Becker (1966). *J. Chem. Phys.*, **44**, 1149.
G. Scoles (1973). *Proc. VIII Int. Conf. Phys. Elec. At. Collisions* (Invited Papers), Belgrade, Yugoslavia.
A. G. Sitenko (1971). *Lectures in Scattering Theory*, Pergamon Press, New York.
R. J. Suplinskas (1968). *J. Chem. Phys.*, **49**, 5046.
K. M. Watson and J. Nutall (1967). *Topics in Several Particle Dynamics*, Holden-Day, S. Francisco.

THE CLASSICAL *S*-MATRIX IN
MOLECULAR COLLISIONS*

WILLIAM H. MILLER†

*Department of Chemistry and Inorganic Materials Research Division,
Lawrence Berkeley Laboratory, University of California, Berkeley,
California 94720*

CONTENTS

I. Introduction. 77
II. Fundamental Correspondence Relations 79
  A. Summary of General Formulae 80
  B. Example: Franck–Condon Factors 82
  C. Matrix Elements 85
III. Classical *S*-matrix: Classically Allowed Processes 86
  A. Basic Formulae 86
  B. Applications 90
  C. Complex Formation and Scattering Resonances 99
  D. Atom-Surface Scattering 105
  E. Photodissociation 109
IV. Classical *S*-matrix: Classically Forbidden Processes 114
  A. Introductory Discussion 114
  B. Applications 118
  C. Partial Averaging 122
  D. Numerical Integration of Complex-Valued Classical Trajectories . . 129
V. Concluding Remarks 131
Notes and References 132

I. INTRODUCTION

The last three years have seen considerable interest in the development of semiclassical methods for treating complex molecular collisions, i.e. those which involve inelastic or reactive processes. One of the reasons for this activity is that the recent work, primarily that of Miller[1] and that of Marcus,[2] has shown how numerically computed classical trajectories can be used as input to the semiclassical theory, so that it is not necessary to make any dynamical approximations when applying these semiclassical approaches

* Supported in part by the National Science Foundation under grant GP-34199X and by the U.S. Atomic Energy Commission.
† Camille and Henry Dreyfus Teacher-Scholar.

to complex collision processes. There is thus the possibility of being able to augment purely classical (i.e. Monte Carlo) trajectory calculations,[3,4] which have proved extremely powerful and useful in their own right, with many of the quantum effects that may be important in molecular collision phenomena.

Another motivation for pursuing these semiclassical approaches to inelastic and reactive scattering is the well-known success that semiclassical theory has had in describing quantum effects in simpler elastic (potential) scattering.[5,6] Here one now knows that essentially all quantum effects can be adequately described in a semiclassical framework.

This paper reviews this 'classical S-matrix' theory, i.e. the semiclassical theory of inelastic and reactive scattering which combines exact classical mechanics (i.e. numerically computed trajectories) with the quantum principle of superposition. It is always possible, and in some applications may even be desirable, to apply the basic semiclassical model with approximate dynamics; Cross[7] has discussed the simplifications that result in classical S-matrix theory if one treats the dynamics within the sudden approximation, for example, and shown how this relates to some of his earlier work[8] on inelastic scattering. For the most part, however, this review will emphasize the use of exact classical dynamics and avoid discussion of various dynamical models and approximations, the reason being to focus on the nature and validity of the basic semiclassical idea itself, i.e., classical dynamics plus quantum superposition. Actually, all quantum effects—being a direct result of the superposition of probability amplitudes—are contained (at least qualitatively) within the semiclassical model, and the primary question to be answered regards the quantitative accuracy of the description.

Since I have reviewed certain aspects of semiclassical, or classical-limit quantum mechanics only a year ago,[9] this presentation will summarize the general theory only briefly and concentrate more on specific applications. The results of various calculations utilizing classical S-matrix theory are reviewed and the semiclassical description of several different physical processes—scattering resonances, scattering of atoms from surfaces and photodissociation of polyatomic molecules—is developed to illustrate more fully how one can translate between classical, semiclassical and quantum mechanical versions of a theory. The semiclassical theory of elastic scattering itself will not be discussed explicitly since this has been the subject of a recent review by Berry and Mount.[10] An understanding of the semiclassical techniques used in elastic scattering has, of course, been essential in extending semiclassical ideas to more general collision processes.

It should be noted that there are a number of other treatments of inelastic scattering, which will not be reviewed here, to which the term 'semiclassical' is also applied. The oldest and most common of these is the 'classical path model':[11] here a trajectory is assumed for the translational motion, this causing a time-dependent perturbation on the internal degrees of freedom which are treated quantum mechanically, i.e. via the time-dependent Schrödinger equation. The simplest version of this approach[11] assumes a straight line, constant velocity trajectory and applies first-order perturbation theory to solve the time-dependent Schrödinger equation for the internal degrees of freedom; neither of these assumptions is necessary, however, and there have been applications that invoke neither of them.[12] The fundamental distinction between this class of approximations and classical S-matrix theory is that in the former some degrees of freedom, namely translation, are treated classically and the others quantum mechanically, while classical S-matrix theory treats *all* degrees of freedom classically, superposition being the only element of quantum mechanics contained in the model. In classical S-matrix theory, therefore, it is completely straightforward to include the full dynamics exactly—by calculation of classical trajectories—while classical path models have inherent dynamical approximations embedded in them.

The remarks in the previous paragraph apply, of course, only to the case of electronically adiabatic molecular collisions for which all degrees of freedom refer to the motion of nuclei (i.e. translation, rotation and vibration); if transitions between different electronic states are also involved, then there is no way to avoid dealing with an explicit mixture of a quantum description of some degrees of freedom (electronic) and a classical description of the others.[9] The description of such non-adiabatic electronic transitions within the framework of classical S-matrix theory has been discussed at length in the earlier review[9] and is not included here.

II. FUNDAMENTAL CORRESPONDENCE RELATIONS

The basic semiclassical idea is that one uses a quantum mechanical *description* of the process of interest but then invokes classical mechanics to determine all dynamical relationships. A transition from initial state i to final state f, for example, is thus described by a transition *amplitude*, or S-matrix element S_{fi}, the square modulus of which is the transition probability: $P_{fi} = |S_{fi}|^2$. The semiclassical approach uses classical mechanics to construct the classical-limit approximation for the transition amplitude, i.e. the 'classical S-matrix'; the fact that classical mechanics is used to construct an amplitude means that the quantum principle of superposition is incorporated in the description, and this is the only element of quantum mechanics in the model. The *completely* classical approach would be to use classical mechanics to construct the transition probability directly, never alluding to an amplitude.

One thus needs a prescription for constructing the classical-limit approximation to quantum mechanical amplitudes or transformation elements. This is given most generally by establishing the correspondence of canonical transformations between various coordinates and momenta in classical mechanics to unitary transformations between various sets of states in quantum mechanics. These correspondence relations have been derived earlier[1,9] and are summarized below.

A. Summary of General Formulae

Let (p, q) be one set of canonically conjugate coordinates and momenta (the 'old' variables) and (P, Q) be another such set (the 'new' variables).[13] $(P, Q, p$ and q are N-dimensional vectors for a system with N degrees of freedom, but for the sake of clarity multidimensional notation will not be used; the explicitly multidimensional expressions are in most cases obvious.) In classical mechanics P and Q may be considered as functions of p and q, or inversely, P and Q may be chosen as the independent variables with p and q being functions of them. To carry out the canonical transformation between these two sets of variables, however, one must rather choose one 'old' variable and one 'new' variable as the independent variables, the remaining two variables then being considered as functions of them. The canonical transformation is then carried out with the aid of a generating function, or generator, which is some function of the two independent variables, and two equations which express the dependent variables in terms of the independent variables.[13]

If, for example, the 'old' coordinate q and the 'new' coordinate Q are chosen as the independent variables, and if $F_1(q, Q)$ is the generator, then the two equations which define p and P are

$$p(q, Q) = \frac{\partial F_1(q, Q)}{\partial q} \tag{1a}$$

$$P(q, Q) = -\frac{\partial F_1(q, Q)}{\partial Q} \tag{1b}$$

To express P and Q explicitly in terms of p and q it would be necessary to solve (1a), i.e.

$$p = \frac{\partial F_1(q, Q)}{\partial q} \tag{1a'}$$

for $Q(p, q)$ and then substitute this into (1b) to obtain $P(p, q)$. There are clearly three other combinations of 'one old variable and one new variable': (q, P), (p, Q) and (p, P). Equivalently, the generators $F_2(q, P)$, $F_3(p, Q)$ or $F_4(p, P)$ may be used in a similar manner, along with the appropriate pair of differential equations analogous to (1), to effect the transformation.

Quantum mechanically, the objects of interest are the elements of the unitary transformation from the 'old' states $|q\rangle$ and $|p\rangle$ to the 'new' states $|Q\rangle$ and $|P\rangle$. [The unitary transformation elements relating any canonically conjugate pair is always given by

$$\langle q|p \rangle = (2\pi i\hbar)^{-1/2} \exp(ipq/\hbar) \tag{2a}$$

$$\langle Q|P \rangle = (2\pi i\hbar)^{-1/2} \exp(iPQ/\hbar)] \tag{2b}$$

There are, just as in classical mechanics, four ways of choosing 'one old variable and one new variable', so there are four equivalent sets of unitary transformation elements connecting the 'old' and 'new' representations: $\langle q|Q \rangle$, $\langle q|P \rangle$, $\langle p|Q \rangle$ and $\langle p|P \rangle$. The fundamental correspondence relations express the classical limit of these unitary transformation elements in terms of the classical generating functions for the related classical canonical transformation:

$$\langle q|Q \rangle = \left[-\frac{\partial^2 F_1(q, Q)}{\partial q \partial Q} \Big/ 2\pi i\hbar \right]^{1/2} \exp\left[iF_1(q, Q)/\hbar \right] \tag{3a}$$

$$\langle q|P \rangle = \left[\frac{\partial^2 F_2(q, P)}{\partial q \partial P} \Big/ 2\pi i\hbar \right]^{1/2} \exp\left[iF_2(q, P)/\hbar \right] \tag{3b}$$

$$\langle p|Q \rangle = \left[\frac{\partial^2 F_3(p, Q)}{\partial p \partial Q} \Big/ 2\pi i\hbar \right]^{1/2} \exp\left[iF_3(p, Q)/\hbar \right] \tag{3c}$$

$$\langle p|P \rangle = \left[-\frac{\partial^2 F_4(p, P)}{\partial p \partial P} \Big/ 2\pi i\hbar \right]^{1/2} \exp\left[iF_4(p, P)/\hbar \right] \tag{3d}$$

In applications it is usually convenient to make use of the derivative relations of the generator[13] to express the preexponented factors above in a less symmetrical, but more useful form. If use is made of (1b), for example, it is easy to show that (3a) can be written as

$$\langle q|Q \rangle = \left[2\pi i\hbar \left(\frac{\partial q}{\partial P} \right)_Q \right]^{-1/2} \exp\left[iF_1(q, Q)/\hbar \right] \tag{3a'}$$

The derivation of these fundamental correspondence relations, (3), has been given previously,[9] and one should see Ref. 9 for a more detailed discussion. To obtain the results it is necessary to assume only (2) (which is essentially a statement of the uncertainty principle), make use of classical mechanics itself, and invoke the stationary phase approximation[14] to evaluate all integrals for which the phase of the integrand is proportional to $\hbar^{-1}$. Since the stationary phase approximation[14] is an asymptotic approximation which becomes exact as $\hbar \to 0$, this is the nature of the classical-limit approximation in (3). In a very precise sense, therefore, classical-limit quantum mechanics is the stationary phase approximation to quantum mechanics.

B. Example: Franck–Condon Factors

Several examples of the basic correspondence relations (3) have been worked out in Ref. 9, and here another one is considered. For brevity the following discussion assumes some familiarity of Section II of Ref. 9.

Let $V_a(x)$ and $V_b(x)$ be two one-dimensional potential wells with eigenstates (labeled by their vibrational quantum numbers) $|n_a\rangle$ and $|n_b\rangle$, respectively. A Franck–Condon factor is the square modulus of the amplitude $\langle n_b|n_a\rangle$, i.e. $|\langle n_b|n_a\rangle|^2$. Thinking semiclassically, one notes that n_a is the generalized momentum of the action-angle variables[15] (n_a, q_a) for the potential V_a; n_b is similarly the generalized momentum of the action-angle variables (n_b, q_b) that are defined with respect to V_b. The amplitude $\langle n_b|n_a\rangle$ is thus a matrix element between momentum states of different representations, so that its classical limit is given by (3d):

$$\langle n_b|n_a\rangle = \left[-\frac{\partial^2 F_4(n_b, n_a)}{\partial n_b \partial n_a} \bigg/ 2\pi i\hbar \right]^{1/2} \exp\left[iF_4(n_b, n_a)/\hbar \right] \tag{4}$$

where $F_4(n_b, n_a)$ is the F_4-type generator of the $(n_a, q_a) \leftrightarrow (n_b, q_b)$ classical canonical transformation.

To discover the appropriate F_4 generator for (4) it is useful first to consider the canonical transformation from (n_a, q_a) to ordinary cartesian variables (p, x), and then from (p, x) to the canonical set (n_b, q_b). As shown before,[9] the F_2-type generator for the $(n_a, q_a) \leftrightarrow (p, x)$ transformation is

$$F_2^a(x, n_a) = \int dx \{ 2m[\varepsilon_a(n_a) - V_a(x)] \}^{1/2} \tag{5}$$

and of a similar form for the $(n_b, q_b) \leftrightarrow (p, x)$ transformation:

$$F_2^b(x, n_b) = \int dx \{ 2m[\varepsilon_b(n_b) - V_b(x)] \}^{1/2} \tag{6}$$

where $\varepsilon_a(n_a)$ and $\varepsilon_b(n_b)$ are the WKB eigenvalue functions. Since

$$\langle n_b|n_a\rangle = \int dx \langle n_b|x\rangle \langle x|n_a\rangle \tag{7}$$

use of (3b) for the matrix elements in the integrand of (7) and stationary phase evaluation of the integral over x shows that

$$F_4(n_b, n_a) = F_2^a(x, n_a) - F_2^b(x, n_b) \tag{8}$$

where $x \equiv x(n_b, n_a)$ is defined implicitly by the stationary phase condition

$$0 = \frac{\partial F_2(x, n_a)}{\partial x} - \frac{\partial F_2^b(x, n_b)}{\partial x} \tag{9}$$

Because of the following derivation relation for an F_2-type

$$\frac{\partial F_2(x, n)}{\partial x} = p(x, n)$$

(9) may also be written as

$$0 = p(x, n_a) - p(x, n_b) \tag{10}$$

Thus the position x at which the $n_a \leftrightarrow n_b$ Franck–Condon transition occurs is the one (or ones) for which the Cartesian momentum is conserved; this is, of course, a statement of the Franck–Condon principle.[16] Since

$$p(x, n) = \{2m[\varepsilon(n) - V(x)]\}^{1/2}$$

$x(n_b, n_a)$ is equivalently defined as the solution of

$$\varepsilon_a(n_a) - V_a(x) = \varepsilon_b(n_b) - V_b(x) \tag{11}$$

The preexponential factor in (4) can be evaluated explicitly by making use of (8) and (9). Differentiation of (8) gives

$$\frac{\partial F_4(n_b, n_a)}{\partial n_a} = \left[\frac{\partial F_2^a(x, n_a)}{\partial x} - \frac{\partial F_2^b(x, n_b)}{\partial x}\right]\frac{\partial x(n_b, n_a)}{\partial n_a} + \frac{\partial F_2^a(x, n_a)}{\partial n_a} \tag{12}$$

and by virtue of (9) the first term vanishes. Differentiation of (12) with respect to n_b thus gives

$$\frac{\partial^2 F_4(n_b, n_a)}{\partial n_b \partial n_a} = \frac{\partial^2 F_2^a(x, n_a)}{\partial x \partial n_a}\frac{\partial x(n_b, n_a)}{\partial n_b} \tag{13}$$

and from (5) one can show that

$$\frac{\partial^2 F_2^a(x, n_a)}{\partial x \partial n_a} = \frac{m\varepsilon_a'(n_a)}{p} \tag{14}$$

where $p \equiv p(x, n_a) = p(x, n_b)$. Furthermore, differentiation of (11) with respect to n_b gives

$$[V_b'(x) - V_a'(x)]\frac{\partial x(n_b, n_a)}{\partial n_b} = \varepsilon_b'(n_b)$$

or

$$\frac{\partial x(n_b, n_a)}{\partial n_b} = \varepsilon_b'(n_b)/\Delta V'(x) \tag{15}$$

where $\Delta V(x) = V_b(x) - V_a(x)$. Equations (13)–(15) thus give the preexponential factor as

$$\frac{\partial^2 F_4(n_b, n_a)}{\partial n_b \partial n_a} \bigg/ 2\pi\hbar = \frac{m\varepsilon_a'(n_a)\varepsilon_b'(n_b)}{2\pi\hbar p \, \Delta V'(x)} \tag{16}$$

If there is one value of x which satisfies (11), then there will be two terms contributing to (4), one corresponding to $p > 0$ and one to $p < 0$. With (8) and (16), (4) finally gives the Franck–Condon amplitude as

$$\langle n_b | n_a \rangle = \left[\frac{m\varepsilon_a'(n_a)\varepsilon_b'(n_b)}{2\pi\hbar|p|\,|\Delta V'(x)|} \right]^{1/2} 2\cos\left[\frac{\pi}{4} + F_2^a(x, n_a)/\hbar - F_2^b(x, n_b) \times 2 \right] \quad (17)$$

where the F_2-generators are given by (5) and (6) and x is evaluated at the 'crossing point' [the root of (11)]. This result has been obtained before[17–19] from other more traditional approaches, but it is interesting to see that it results directly from the general correspondence relations. If there is more than one value of x which satisfies (11), then (17) is a sum of similar terms, one for each such value of x.

If one discards the interference term between the two terms that contribute to (17), then the *classical Franck–Condon factor* is obtained:

$$|\langle n_b | n_a \rangle|^2 = 2\frac{m\varepsilon_a'(n_a)\varepsilon_b'(n_b)}{2\pi\hbar|p|\,|\Delta V'(x)|} \quad (18)$$

the factor of 2 appearing because of the two terms that contribute equally. It is interesting (and useful) to show that this purely classical expression can also be obtained from simpler phase space considerations. The Franck–Condon factor in (18) is the joint probability that n_a and n_b have certain specific values; it can thus be written as the following phase space integral:

$$P(n_b, n_a) = (2\pi\hbar)^{-1} \int dP \int dQ\, \delta[n_b - n_b(P, Q)]\, \delta[n_a - n_a(P, Q)] \quad (19)$$

where P and Q are any set of canonically conjugate variables (since phase space integrals are invariant to a canonical transformation), and $n_b(P, Q)$ and $n_a(P, Q)$ are these variables expressed in terms of P and Q. Choosing (P, Q) to be the variables (n_a, q_a), for example, leads to

$$P(n_b, n_a) = (2\pi\hbar)^{-1} \int dq_a\, \delta[n_b - n_b(n_a, q_a)]$$

$$= (2\pi\hbar)^{-1} \left| \left(\frac{\partial n_b}{\partial q_a} \right)_{n_a}^{-1} \right|$$

$$= (2\pi\hbar)^{-1} \left| \frac{\partial q_a(n_b, n_a)}{\partial n_b} \right|. \quad (20)$$

Since one of the derivative relations for an F_4-type generator is[13]

$$q_a(n_b, n_a) = \frac{\partial F_4(n_b, n_a)}{\partial n_a}$$

one sees that

$$\frac{\partial q_a(n_b, n_a)}{\partial n_b} = \frac{\partial^2 F_4(n_b, n_a)}{\partial n_b \partial n_a} \tag{21}$$

so that (20) thus becomes

$$P(n_b, n_a) = \left| \frac{\partial^2 F_4(n_b, n_a)}{\partial n_b \partial n_a} \middle/ 2\pi\hbar \right|, \tag{22}$$

which is the same preexponential factor as in (4) and thus the same classical Franck–Condon factor as (18).

C. Matrix Elements

The correspondence relations summarized in Section IIA show how any unitary transformation element can be evaluated within the classical limit. Sometimes, however, one is interested in matrix elements of operators which are not unitary. Consider, for example, the one-dimensional system discussed in the previous section; if A is some operator, the question is how does one obtain the classical limit of the matrix element $\langle n_b|A|n_a \rangle$.

Let $|\xi\rangle$ be some yet unspecified set of states; then

$$\langle n_b|A|n_a \rangle = \int d\xi \int d\xi' \langle n_b|\xi' \rangle \langle \xi'|A|\xi \rangle \langle \xi|n_a \rangle \tag{23}$$

To evaluate this semiclassically one should choose the basis $|\xi\rangle$ so that the representation of A is local and multiplicative,

$$\langle \xi'|A|\xi \rangle = \delta(\xi' - \xi)A(\xi) \tag{24}$$

and where $A(\xi)$ is non-singular as $\hbar \to 0$. If A is an operator with a simple classical analog, as is usually the case, then the choice of the basis $|\xi\rangle$ is obvious. If A is the kinetic energy, $A = p^2/2m$, for example, the basis should be chosen to be the Cartesian momentum states $|p\rangle$. If A is a simple function of the Cartesian coordinates, $A = A(x)$, then one should use a Cartesian coordinate representation.

In this latter case, for example, (23) becomes

$$\langle n_b|A|n_a \rangle = \int dx \langle n_b|x \rangle A(x) \langle x|n_a \rangle \tag{25}$$

stationary phase evaluation of which clearly gives

$$\langle n_b|A|n_a \rangle = A(x) \langle n_b|n_a \rangle \tag{26}$$

where $x = x(n_b, n_a)$ and $\langle n_b|n_a \rangle$ is the Franck–Condon factor discussed in the previous section. Probably the most common example of this result is when A is the dipole operator, $A(x) = ex$, e being the electron charge; this is the application made in discussing photodissociation in Section III.E below.

III. CLASSICAL S-MATRIX: CLASSICALLY ALLOWED PROCESSES

In a collision system such as an atom A colliding with a diatomic molecule BC, one is interested in the transition amplitudes, or S-matrix elements, which describe transitions between specific quantum states of the molecule BC. From the S-matrix elements one can construct scattering amplitudes for any collision process resulting from A + BC, the square modulus of the amplitudes being the cross sections.

Several derivations of the classical limit of the S-matrix, the 'classical S-matrix', for complex collisions (i.e. those for which the collision partners have internal degrees of freedom) have been given;[1,2,9] the results follow almost directly from the fundamental correspondence relations, (3), the only modifications being those required to factor out an energy-conserving delta function. This section first summarizes the general expressions and then discusses their application.

A. Basic Formulae

For ease of presentation a non-reactive collision is considered first; the modifications required to include reactive processes are straightforward and simple.[1,9] The general system consists of N degrees of freedom, one being relative translation of the collision partners and the other $(N - 1)$ being internal degrees of freedom which are quantized in the asymptotic regions. The translational degree of freedom is described by the center of mass coordinate R and momentum P, while the internal degrees of freedom are described by their action angle variables $(\mathbf{n}, \mathbf{q}) \equiv (n_i, q_i)$, $i = 1, 2, \ldots, N - 1$. The action variables $\{n_i\}$ are the classical counterpart of the quantum numbers for these degrees of freedom and will thus be referred to simply as the 'quantum numbers' although classically, of course, they are continuous functions of time like any other (generalized) momenta; in the asymptotic regions $(R \to \infty)$ before and after collision they are required to be integers.[20]

The classical Hamiltonian function for the system is given in terms of these variables by

$$H(P, R, \mathbf{n}, \mathbf{q}) = P^2/2\mu + \varepsilon(\mathbf{n}) + V(R, \mathbf{n}, \mathbf{q}) \tag{27}$$

where μ is the reduced mass of relative motion, $\varepsilon(\mathbf{n})$ is the WKB eigenvalue function for the internal degrees of freedom and V is an interaction which vanishes as $R \to \infty$. It is the dependence of V on R and $\mathbf{q}$ which prevents P and $\mathbf{n}$ from being constants of the motion; since $V \to 0$ as $R \to \infty$, P and $\mathbf{n}$ are conserved asymptotically, and as noted above the asymptotically constant values of $\mathbf{n}$ must be integers.

The quantities of interest are the on-shell S-matrix elements,

$$S_{\mathbf{n}_2, \mathbf{n}_1}(E)$$

which are the probability amplitudes for the $\mathbf{n}_1 \to \mathbf{n}_2$ transition. Their classical-limit approximation, the classical S-matrix, is constructed from the classical trajectory (or trajectories) with initial conditions[21] at time t_1 $(t_1 \to -\infty)$

$$\mathbf{n}(t_1) = \mathbf{n}_1 \text{ (a specific set of integers)} \tag{28a}$$

$$R(t_1) = \text{large} \tag{28b}$$

$$P(t_1) = -\{2\mu[E - \varepsilon(\mathbf{n}_1)]\}^{1/2} \tag{28c}$$

$$\mathbf{q}(t_1) = \bar{\mathbf{q}}_1 + \frac{\partial\varepsilon(\mathbf{n}_1)}{\partial\mathbf{n}_1}\mu R(t_1)/P(t_1) \tag{28d}$$

and with final conditions at time t_2 $(t_2 \to +\infty)$

$$\mathbf{n}(t_2) = \mathbf{n}_2 \text{ (another set of integers)} \tag{29a}$$

$$R(t_2) = \text{large} \tag{29b}$$

$$P(t_2) = +\{2\mu[E - \varepsilon(\mathbf{n}_2)]\}^{1/2} \tag{29c}$$

$$\mathbf{q}(t_2) = \text{anything} \tag{29d}$$

For a system with several degrees of freedom the trajectories must, of course, be determined by numerical integration of Hamilton's equations step by step in time. To find the trajectory (or trajectories) which obey the boundary conditions in (28) and (29) it is convenient to introduce the function $\mathbf{n}_2(\bar{\mathbf{q}}_1, \mathbf{n}_1; E)$, the final value of the quantum numbers that result from a trajectory with the initial conditions of (28); in general, of course, $\mathbf{n}_2(\bar{\mathbf{q}}_1, \mathbf{n}_1; E)$ is non-integral. For a given total energy E and a given set of initial integral quantum numbers $\mathbf{n}_1$, the task is to find the particular values of the angle variables $\bar{\mathbf{q}}_1$ for which $\mathbf{n}_2(\bar{\mathbf{q}}_1, \mathbf{n}_1; E)$ turn out to be the specific set of integers $\mathbf{n}_2$; i.e., suppressing the arguments $\mathbf{n}_1$ and E, one must solve the equations

$$\mathbf{n}_2(\bar{\mathbf{q}}_1) = \mathbf{n}_2 \tag{30}$$

where $\mathbf{n}_2$ on the RHS is a given set of integers. This is a set of $(N-1)$ equations in $N-1$ unknowns.

The classical S-matrix element for the $\mathbf{n}_1 \to \mathbf{n}_2$ transition is then given by[22]

$$S_{\mathbf{n}_2,\mathbf{n}_1}(E) = \left[(-2\pi i)^{N-1}\frac{\partial\mathbf{n}_2(\bar{\mathbf{q}}_1,\mathbf{n}_1;E)}{\partial\bar{\mathbf{q}}_1}\right]^{-1/2} \exp\left[i\Phi(\mathbf{n}_2, \mathbf{n}_1; E)/\hbar\right] \tag{31}$$

where $\partial\mathbf{n}_2/\partial\bar{\mathbf{q}}_1$ is the determinant of the $N-1$ Jacobian, $\bar{\mathbf{q}}_1$ is evaluated at the root of (30) and Φ is the classical action integral

$$\Phi(\mathbf{n}_2, \mathbf{n}_1; E) = -\int dt[R(t)\dot{P}(t) + \mathbf{q}(t)\cdot\dot{\mathbf{n}}(t)] \tag{32}$$

evaluated along the trajectory which satisfies the above double-ended boundary conditions. If there is more than one trajectory at this energy corresponding to the same initial and final quantum numbers $\mathbf{n}_1$ and $\mathbf{n}_2$, (31) is a sum of similar terms, one for each such trajectory.

Before proceeding to discuss more substantive examples it is interesting to see that the above expressions do reproduce the standard WKB results for one-dimensional dynamical systems. For a system with only a translational degree of freedom, i.e. no internal degrees of freedom, the pre-exponential factor in (31) is unity and the phase Φ is

$$\Phi(E) = -\int_{t_1}^{t_2} dt \, R(t)\dot{P}(t)$$

$$= -(R_2 P_2 - R_1 P_1) + \int_{t_1}^{t_2} dt \, P(t)\dot{R}(t) \tag{33}$$

If the interaction potential $V(R)$ has a repulsive core and $R_<$ is the classical turning point, then (33) becomes

$$\Phi(E) = -2RP + 2\int_{R_<}^{R} dR' \, P(R') \tag{34}$$

where $R \equiv R_1 \equiv R_2 \to \infty, P \equiv -P_1 = P_2 = (2mE)^{1/2}$ and $P(R) = \{2m[E - V(R)]\}^{1/2}$. The S-matrix is thus

$$S(E) = \exp[i\Phi(E)/\hbar]$$

$$= \exp[2i\eta(E)] \tag{35}$$

where the phase shift $\eta(E)$ is given by the usual WKB expression:

$$\eta(E) = \lim_{R \to \infty} \left[-kR + \int_{R_<}^{R} dR' \, k(R') \right] \tag{36}$$

where

$$k = (2mE/\hbar^2)^{1/2}$$

and

$$k(R) = \{2m[E - V(R)]/\hbar^2\}^{1/2}$$

For a single particle moving in three dimensions under the influence of a spherically symmetric potential $V(R)$, the classical Hamiltonian is

$$H(P, R, l, q_l, m, q_m) = (2\mu)^{-1}(P^2 + l^2/R^2) + V(R) \tag{37}$$

and one sees that the quantum numbers l (orbital angular momentum) and m (its z-component) for the two 'internal' degrees of freedom are conserved since their conjugate variables q_l and q_m do not appear in the Hamiltonian.

The dynamical system thus reduces to a one-dimensional one with a Hamiltonian

$$H_l(P, R) = (2\mu)^{-1}(P^2 + l^2/R^2) + V(R) \tag{38}$$

that depends parametrically on l. Because of the centrifugal term $l^2/2\mu R^2$, however, the Hamiltonian in (38) is not precisely of the form in (27). To remedy this one transforms from variables P and R to the new variables $\bar{P}$ and $\bar{R}$, where

$$\bar{P}^2 = P^2 + l^2/R^2 \tag{39}$$

The classical generator which effects this canonical transformation is

$$\begin{aligned}
F_2(R, \bar{P}) &= \int dR\, (\bar{P}^2 - l^2/R^2)^{1/2} \\
&= R(\bar{P}^2 - l^2/R^2)^{1/2} - l\cos^{-1}(l/\bar{P}R)
\end{aligned} \tag{40}$$

By invoking the derivation relations for an F_2-type generator,

$$P = \frac{\partial F_2(R, \bar{P})}{\partial R} \tag{41a}$$

$$\bar{R} = \frac{\partial F_2(R, \bar{P})}{\partial \bar{P}} \tag{41b}$$

one can easily verify that (39) is fulfilled and that R is given in terms of the new variables by

$$R = (\bar{R}^2 + l^2/\bar{P}^2)^{1/2} \tag{42}$$

The Hamiltonian thus takes the desired form in terms of the new variables

$$H_l(\bar{P}, \bar{R}) = \bar{P}^2/2\mu + V([\bar{R}^2 + l^2/\bar{P}^2]^{1/2}) \tag{43}$$

and the S-matrix in this angular momentum representation has the one-dimensional form

$$S_l(E) = \exp[i\Phi_l(E)/\hbar] \tag{44}$$

where

$$\Phi_l(E) = -\int_{t_1}^{t_2} dt\, \bar{R}(t)\frac{d\bar{P}(t)}{dt} \tag{45}$$

Use of (41) shows that

$$\begin{aligned}
\frac{d}{dt}F_2(R, \bar{P}) &= \frac{\partial F_2(R, \bar{P})}{\partial R}\dot{R} + \frac{\partial F_2(R, \bar{P})}{\partial \bar{P}}\dot{\bar{P}} \\
&= P\dot{R} + \bar{R}\dot{\bar{P}}
\end{aligned} \tag{46}$$

so that the phase in (45) is equivalently given by

$$\Phi_l(E) = -F_2(R, \bar{P})|_{t_1}^{t_2} + \int_{t_1}^{t_2} dt \, P(t)\dot{R}(t) \tag{47}$$

By eliminating t in favour of R in the usual way it is then easy to show that

$$\Phi_l(E) = 2\hbar\eta_l(E)$$

i.e.,

$$S_l(E) = \exp[2i\eta_l(E)] \tag{48}$$

where the phase shift is given by the standard WKB expression

$$\eta_l(E) = \frac{\pi}{2}l - kR + \int_{R_<}^{R} dR' \, k_l(R') \tag{49}$$

where $R \to \infty$ and $k_l(R) = \{2\mu[E - V(R)] - l^2/R^2\}^{1/2}/\hbar$; in practice[6] the replacement $l \to l + 1/2$ is usually made.

B. Applications

The first calculations using the theory described in the preceding section were carried out by Miller[23] for the non-reactive collinear A + BC collision system for which Secrest and Johnson[24] had earlier obtained accurate quantum mechanical transition probabilities. Since this is a system of two degrees of freedom, and thus only one internal degree of freedom, (27)–(32) applying with the vector designation removed from the pair of action-angle variables (n, q).

Thus the $n_1 \to n_2$ vibrational transition is constructed from those trajectories which satisfy

$$n_2(\bar{q}_1) = n_2 \tag{50}$$

where $n_2(\bar{q}_1)$ (with the arguments n_1 and E suppressed) is the final vibrational quantum number, not necessarily integral, which result from the trajectory with the initial conditions in (28). Fig. 1 shows this function for the case $n_1 = 1$, $E = 10\hbar\omega$ and for the potential parameters chosen to correspond to a He + H$_2$ collision. It is clear from the figure that there are two roots to (50), when $n_2 = 2$ for example, i.e. there are two classical trajectories that contribute to the $1 \to 2$ vibrational transition. The classical S-matrix element (31) for it is thus the sum of two terms

$$S_{n_2,n_1}(E) = [2\pi|n_2'(\bar{q}_1)|]^{-1/2} \exp\left(i\frac{\pi}{4} + i\Phi_I/\hbar\right)$$

$$+ [2\pi|n_2'(\bar{q}_{II})|]^{-1/2} \exp\left(-i\frac{\pi}{4} + i\Phi_{II}/\hbar\right) \tag{51}$$

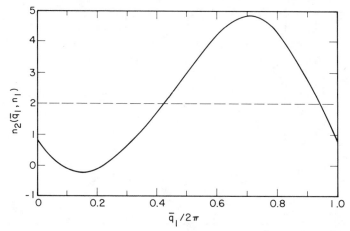

Fig. 1. An example of the quantum number function $n_2(\bar{q}_1, n_1)$, here for a collision of He and H_2 at a total energy $E = 10\hbar\omega$ and with $n_1 = 1$. The ordinate is the final value of the vibrational quantum number as a function of the initial phase $\bar{q}_1$ of the oscillator, along a classical trajectory with the initial conditions in (28). The dashed line at $n_2 = 2$ indicates the graphical solution for the two roots of the equation $n_2(\bar{q}_1, 1) = 2$.

where $\bar{q}_I$ and $\bar{q}_{II}$ are the two roots of (50), and Φ_I and Φ_{II} are the action integrals for these two trajectories. The transition probability,

$$P_{n_2,n_1}(E) \equiv |S_{n_2,n_1}(E)|^2 \tag{52}$$

is therefore the sum of the probabilities associated with the two trajectories plus an interference between the two:

$$P_{n_2,n_1}(E) = p_I + p_{II} + 2(p_I p_{II})^{1/2} \sin[(\Phi_{II} - \Phi_I)/\hbar], \tag{53}$$

where

$$p_K = [2\pi|n_2'(\bar{q}_K)|]^{-1}$$

$K = I, II$. The situation is quite analogous to the usual discussion[25] of the 'two slit experiment', and just as there, it is not proper to say that the $n_1 \to n_2$ transition takes place via trajectory I or trajectory II, for logic would then demand that the probabilities add; rather there is a probability *amplitude* for the trajectory being I or II, and these amplitudes add.[25]

The calculations[23,26] show the semiclassical results to be in excellent agreement with the accurate quantum mechanical values. Furthermore, the interference term in (53) is quite significant in that the completely classical transition probability,

$$P_{n_2,n_1}^{CL}(E) \equiv p_I + p_{II}, \tag{54}$$

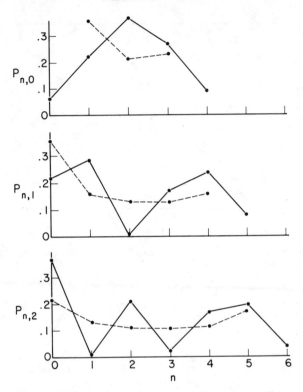

Fig. 2. Vibrational transition probabilities for collinear He + H_2 at total energy $E = 10\hbar\omega$ for an initial vibrational state $n_1 = 0$ (top), 1, 2 (bottom). The dashed lines connect results of the completely classical approximation, (54), and the solid lines connect the uniform semiclassical values (which on the scale of the drawing are essentially the same as the exact quantum mechanical values of Ref. 24).

gives poor results for individual transition probabilities. On the average, however, the classical transition probability (54) is correct, as seen in Fig. 2, so that the purely classical treatment is adequate if one is interested in collision properties that involve an average over some initial and/or final quantum states.

Cohen and Alexander[26A] have applied the semiclassical theory to collinear collisions of D_2 with H_2/D_2 and calculated transition probabilities for several inelastic processes of the type

$$H_2(n_1^a) + D_2(n_1^b) \rightarrow H_2(n_2^a) + D_2(n_2^b)$$

They find reasonably good agreement between the semiclassical transition probabilities and the corresponding quantum mechanical values.

The atom-rigid rotor collision in three-dimensions has also been treated by Miller,[27] and the comparison with accurate quantum mechanical values is similar to the collinear results discussed above: individual S-matrix elements are described accurately only by proper inclusion[28] of interference between the various trajectories which contribute to the transition. If the transition probabilities are summed over several quantum states, however, the interference effects are quenched and the purely classical probability gives accurate results.

The conclusion which seems to be emerging from these examples, therefore, is that individual S-matrix elements, and thus the transition probability between a complete set of initial and final quantum numbers, cannot be described accurately without proper inclusion[28] of the interference terms provided by the classical S-matrix approach. If the transition probabilities or cross sections of interest are summed and/or averaged over some of the final or initial quantum numbers, however, the interference terms tend to average to zero so that the completely classical treatment becomes adequate.

The fact which makes interference effects so 'fragile' is that there can be interference only between those processes which are *in principle indistinguishable*.[25] This means that only the transition probability, or cross section, for a *completely state-selected* quantity is given by the square modulus of an amplitude; an averaged cross section is not given by the square modulus of an average amplitude but rather by the average of the square modulus of the completely state-selected amplitude—i.e.

$$\langle \sigma \rangle = \langle |f|^2 \rangle$$
$$\neq |\langle f \rangle|^2$$

With regard to simple potential (elastic) scattering, for example, since there are no internal degrees of freedom the 'completely state-selected' quantity is the differential cross section; thus

$$\sigma_E(\theta) = |f_E(\theta)|^2 \qquad (55)$$

and $f_E(\theta)$ is constructed semiclassically[5,6,29] from the trajectories which satisfy the appropriate double-ended boundary conditions (i.e. that the energy be E and that the scattering angle be θ). Interference structure is thus readily observed in $\sigma_E(\theta)$, whereas it is usually quenched in the total cross section:

$$\sigma(E) \equiv 2\pi \int_0^\pi d\theta \sin \theta \sigma_E(\theta)$$
$$= \frac{\pi}{k^2} \sum_{l=0}^{\infty} (2l + 1)|S_l(E) - 1|^2 \qquad (56)$$

i.e., even though the individual 'transition probabilities'

$$|S_l(E) - 1|^2 \equiv 4 \sin^2 [\eta_l(E)] \qquad (57)$$

show prominent interference between the scattered particle and its 'shadow', the sum over l quenches it. (Under certain conditions a residual oscillatory term in $\sigma(E)$ does survive the average over l; this is the 'glory' effect.[6])

For three-dimensional A + BC collisions[1] the completely state-selected quantity is the differential cross section from a given initial state $|n_1 j_1 m_1\rangle$ to a specific final state $|n_2 j_2 m_2\rangle$; it is thus given by the square modulus of an amplitude

$$\sigma_{n_2 j_2 m_2 \leftarrow n_1 j_1 m_1}(\theta) = |f_{n_2 j_2 m_2 \leftarrow n_1 j_1 m_1}(\theta)|^2 \qquad (58)$$

where the 'classical amplitude' is constructed from the trajectories with the appropriate initial and final boundary conditions.[1,9] There should undoubtedly be significant interference effects in these completely state-selected differential cross sections, but any less detailed quantity, being an average of (58), will have the interference more or less quenched.

Although the practical difficulty of observing quantum interference effects in complex collisions is discouraging from one point of view (because a completely state-selected experiment is clearly quite difficult), it is encouraging to one who desires a relatively easy way of calculating observed scattering properties, for if interference is neglected the semiclassical expressions degenerate to purely classical ones. For classically allowed processes which involve an average over some of the initial or final quantum numbers and/or the scattering angle, one therefore expects a purely classical (e.g. Monte Carlo) treatment to be adequate. This conclusion has been one of the reasons that most of the recent semiclassical work has dealt with *classically forbidden* processes (see Section IV) for which purely classical treatments are inapplicable.

With regard to reactive processes, there have been several applications of classical S-matrix theory to the

$$H + H_2(n_1 = 0) \rightarrow H_2(n_2 = 0) + H$$

reaction in the energy region above the classical threshold.[30] Bowman and Kupperman,[31] Wu and Levine,[32] Duff and Truhlar,[32A] and Stine and Marcus[32B] have treated it within the collinear model, and Tyson, Saxon, and Light[33] have considered the co-planar model. (George and Miller[34] and Doll, George, and Miller[35] have treated this same process with a collinear and fully three-dimensional models, respectively, in the energy region *below* the classical threshold for reaction, but this 'classically forbidden' case will be discussed in Section IV.) The first three groups of workers[31–32A] find only rough agreement between their semiclassical and quantum mechanical calculations for the reaction probability; in particular, their semiclassical results show no evidence of the non-adiabatic resonance[32] which appears near the threshold for vibrationally excited product. In a more careful treatment, however, Stine and Marcus[32B] find additional roots to (50)—i.e., additional reactive trajectories which satisfy the correct boundary conditions—and with these terms included the semiclassical results reproduce the non-adiabatic resonance quite well. These additional trajectories found by Stine and Marcus[32B] correspond to extra oscillations in the vicinity of the saddle point—i.e., to a short-lived collision complex—and Section III. C shows in more detail how this leads to resonances in the semiclassical scattering amplitude. The coplanar calculations of Tyson et al.[33] are at one collision energy, for $j_1 = j_2 = 0$, and the differential, as well as total cross section has been considered; they also find rough agreement between the 'primitive' semiclassical and quantum mechanical cross sections.

In neither the collinear nor the coplanar cases, however, is the agreement as quantitative as for the non-reactive applications discussed above. The difficulty lies in applying the appropriate *uniform* semiclassical expression, for the small differences between action integrals of the various trajectories which contribute to the $0 \rightarrow 0$ transition makes the "primitive" semiclassical expressions inaccurate. For the collinear case, for example, Fig. 3 shows the

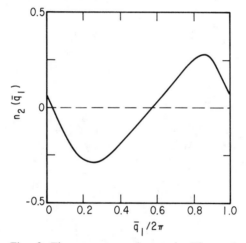

Fig. 3. The same quantity as in Figure 1, except for the *reactive* process $H + H_2(n_1 = 0) \rightarrow H_2(n_2) + H$, for a total energy $E = 14.7$ kcal/mole, as computed by Wu and Levine (reference 32). There are two roots of the equation $n_2(\bar{q}_1, 0) = 0$.

reactive quantum number function $n_2(\bar{q}_1)$ of Wu and Levine[32] for the case $n_1 = 0$ and a total energy of 14.7 kcal/mole. Because the function is so 'flat', i.e.

$$|n_2(\bar{q}_1) - n_1| \ll 1$$

for all $\bar{q}_1$, the uniform semiclassical expression based on Airy functions[23] (which was used in Refs. 31–32A) is not applicable. (This is also true for an elastic non-reactive transition when all inelastic transitions are classically forbidden; this is of little concern, however, since one is usually not interested in the elastic transition probability in such cases.) A uniform semiclassical expression based on Bessel functions[36,37] is the appropriate one for this case, and one would expect it to give more satisfactory results.

[Briefly, the Bessel function uniform expression is generated from the primitive semiclassical expression

$$P_{n_2,n_1} = (p_{\mathrm{I}}^{1/2} + p_{\mathrm{II}}^{1/2})^2 \sin^2\left(\frac{\pi}{4} + \frac{\Delta\Phi}{2}\right)$$

$$+ (p_{\mathrm{I}}^{1/2} - p_{\mathrm{II}}^{1/2})^2 \cos^2\left(\frac{\pi}{4} + \frac{\Delta\Phi}{2}\right) \tag{59}$$

by the replacement

$$\sin^2\left(\frac{\pi}{4} + \frac{\Delta\Phi}{2}\right) \rightarrow \frac{\pi}{2}(z^2 - n^2)^{1/2} J_n(z)^2 \tag{60a}$$

$$\cos^2\left(\frac{\pi}{4} + \frac{\Delta\Phi}{2}\right) \rightarrow \frac{\pi}{2}(z^2 - n^2)^{1/2} J_n'(z)^2 \bigg/ \left(1 - \frac{n^2}{z^2}\right) \tag{60b}$$

where $n = |n_2 - n_1|$ and z is defined by the equation

$$(z^2 - n^2)^{1/2} - n\cos^{-1}(n/z) = \frac{\Delta\Phi}{2} \tag{61}$$

See Ref. 37 for a derivation. Where the Airy function expression is valid, the two uniform semiclassical expressions are essentially equivalent. For the case of present interest, $n_1 = n_2 = 0$, (59) and (60) become

$$P_{n_2,n_1}^{\mathrm{UN}} = \frac{\pi}{4}\Delta\Phi\left[(p_{\mathrm{I}}^{1/2} + p_{\mathrm{II}}^{1/2})^2 J_0\left(\frac{\Delta\Phi}{2}\right)^2 \right.$$

$$\left. + (p_{\mathrm{I}}^{1/2} - p_{\mathrm{II}}^{1/2})^2 J_1\left(\frac{\Delta\Phi}{2}\right)^2\right] \tag{62}$$

where the fact has been used that $J_0'(z) = -J_1(z)$.]

Fig. 4 shows another example of the collinear reactive quantum number function $n_2(\bar{q}_1)$ from the paper of Wu and Levine,[32] this one also for $n_1 = 0$ and for a total energy of 13·7 kcal. Here the function is considerably more complicated than those of Figs. 1 or 3, there being four trajectories that contribute to the $0 \rightarrow 0$ reaction. At first glance one might expect no semiclassical treatment to be possible; Connor,[38] however, has developed more general uniform semiclassical formulae which take account of four terms, and it would clearly be desirable to see if these expressions give more accurate results for this application.

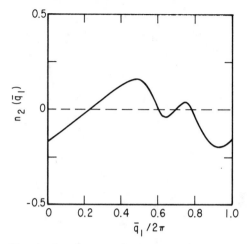

Fig. 4. The same as Fig. 3, except for a total energy $E = 13.7$ kcal/mole. Here there are four roots to the equation $n_2(\bar{q}_1, 0) = 0$.

In highly quantum-like situations such as these, therefore, it is necessary to use the appropriate uniform semiclassical expressions to obtain quantitative results for transitions between individual quantum states. These will, too, undoubtedly be cases for which the quantum number function is too highly structured for any semiclassical treatment to be quantitatively useful.

Finally, I would like to discuss briefly the bimodal structure that has been observed recently in product vibrational state distributions in three-dimensional classical trajectory calculations.[39] It is illustrative to see how this arises even in the simplest situation, the non-reactive collinear model. Within a completely classical framework, neglecting semiclassical interference terms, the $n_1 \rightarrow n_2$ vibrational transition probability of (54) is

$$P^{CL}_{n_2,n_1} = [2\pi|n'_2(\bar{q}_1)|]^{-1} + [2\pi|n'_2(\bar{q}_{II})|]^{-1} \tag{63}$$

where $\bar{q}_1$ and $\bar{q}_{II}$ are the roots of (50). Considering n_2 for the moment to be a continuous variable, as n_2 approaches the maximum or the minimum of $n_2(\bar{q}_1)$ (see Fig. 1) it is clear that the transition probability in (63) becomes infinite [because $n'_2(\bar{q}_1) \rightarrow 0$]. As a continuous function of n_2, the $n_1 \rightarrow n_2$ transition probability will thus have the qualitative shape sketched in Fig. 5 (which also shows the semiclassical interference).

In a Monte Carlo calculation the classical probability of (63) is actually not the quantity calculated, but rather this transition probability averaged over the final quantum number:

$$P^{AV}_{n_2 \leftarrow n_1} \equiv \int_{n_2 - 1/2}^{n_2 + 1/2} dn_2 \, P^{CL}_{n_2,n_1} \tag{64}$$

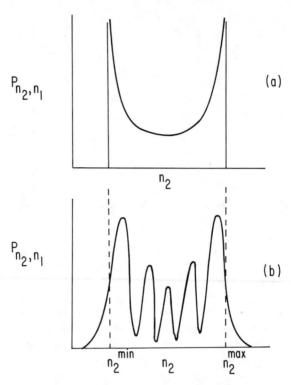

Fig. 5. A qualitative sketch of the classical (upper) and uniform semiclassical (lower) transition probability P_{n_2,n_1} as a continuous function of n_2 (with n_1 fixed). $n_2^{\max}$ and $n_2^{\min}$ indicate the extrema of the function $n_2(\bar{q}_1, n_1)$ as a function of $\bar{q}_1$ (as seen in Fig. 1, for example).

and one can easily see that this gives

$$P_{n_2 \leftarrow n_1}^{AV} = \Delta q_1/2\pi \tag{65}$$

where Δq_1 is the increment of the $\bar{q}_1$ interval for which

$$n_2 - \tfrac{1}{2} \leq n_2(\bar{q}_1) \leq n_2 + \tfrac{1}{2} \tag{66}$$

This averaging procedure rounds off the two classical infinities, leaving finite peaks; for the case $n_1 = 0$ the two peaks usually overlap so that there is just one peak, but this need not always be the case.

The bimodal structure of the vibrational state distribution is thus simply a result of the fact that $n_2(\bar{q}_1)$ has one maximum and one minimum. The effect is entirely analogous to the classical rainbow[5,6] in elastic scattering which results because the deflection function $\Theta(b)$ has a minimum. It is possible too, that $n_2(\bar{q}_1)$ could have more than one relative maximum and minimum, hence the classical vibrational state distribution would have more than two maxima. If semiclassical interference is taken account of, then there can of course be any number of peaks in the vibrational state distribution even though $n_2(\bar{q}_1)$ has just one maximum and one minimum (cf. Fig. 2).[86]

C. Complex Formation and Scattering Resonances

As noted in the Introduction, all quantum effects are a direct result of the superposition principle and must therefore be contained, at least qualitatively within the semiclassical model. This section discusses more explicitly how scattering resonances arise semiclassically.

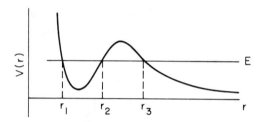

Fig. 6. A potential curve $V(r)$ and collision energy E for which potential (i.e. single-particle) resonances exist.

Consider first the case of a potential, or 'single particle' resonance which results from one-dimensional tunnelling through a potential barrier (see Fig. 6). T, the probability of tunnelling through the barrier, is given by[40]

$$T = e^{-2\theta}/(1 + e^{-2\theta}) \tag{67}$$

where θ is the barrier penetration integral

$$\theta = \int_{r_2}^{r_3} dr \, \{2m[V(r) - E]/\hbar^2\}^{1/2} \tag{68}$$

$R \equiv 1 - T$, the probability of reflection from the barrier, is

$$R = (1 + e^{-2\theta})^{-1}$$

The one-dimensional S-matrix is given by a sum of amplitudes, one for each possible trajectory. (There is more than one trajectory because tunnelling is being allowed for.) The simplest possible trajectory is one that comes in from large r and is reflected at $r = r_3$; the amplitude associated with this is

$$R^{1/2} \, e^{-i(\pi/2)} \, e^{2i\eta_0} \tag{69}$$

where

$$\eta_0 = \lim_{r \to \infty} \left[-kr + \int_{r_3}^{r} dr'\, k(r') \right] \tag{70}$$

$$k(r) = \{2m[E - V(r)]/\hbar^2\}^{1/2}$$

$$k = k(\infty)$$

The preexponential factor is the square root of the probability for this trajectory (a reflection at the barrier), the phase $-\pi/2$ results any time a reflection occurs, and the phase η_0 is the usual semiclassical phase shift for motion from $r = \infty$ to $r = r_3$ and back to $r = \infty$.

Another possible trajectory is one that comes in from infinity, travels through the barrier, moves across the potential well and back, and then tunnels back out through the barrier. The amplitude associated with this trajectory is

$$(T^{1/2})^2\, e^{-i(\pi/2)}\, e^{2i\phi}\, e^{2i\eta_0} \tag{71}$$

where ϕ is the phase integral across the potential well

$$\phi = \int_{r_1}^{r_2} dr\, k(r)$$

the probability factor corresponds to the fact that two tunnellings are required. A third possible trajectory is similar to the previous one except that instead of tunnelling back out of the potential well, it is reflected at the barrier and makes an additional passage back and forth across the well before tunnelling out. This trajectory involves two tunnellings, one barrier reflection and two reflections at $r = r_1$, so that the amplitude associated with it is

$$(T^{1/2})^2 R^{1/2}(e^{-i(\pi/2)})^3\, e^{4i\phi}\, e^{2i\eta_0} \tag{72}$$

These three trajectories are depicted schematically in Fig. 7. Clearly there are an infinite number of possible trajectories arising from the various number of oscillations the particle may make back and forth across the well before it tunnels back out through the barrier. The amplitude for the trajectory that makes N round trips across the well is

$$(T^{1/2})^2 (R^{1/2})^{N-1} (e^{-i(\pi/2)})^{2N-1}\, e^{2iN\phi}\, e^{2i\eta_0} \tag{73}$$

The S-matrix, being the sum of the amplitudes for all possible trajectories is thus given by

$$S = R^{1/2}\, e^{-i\pi/2}\, e^{2i\eta_0} + (T^{1/2})^2\, e^{2i\eta_0} \sum_{N=1}^{\infty} (R^{1/2})^{N-1} (e^{-i(\pi/2)})^{2N-1}\, e^{2iN\phi} \tag{74}$$

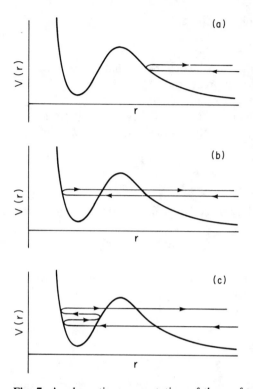

Fig. 7. A schematic representation of three of the trajectories which contribute to the elastic scattering from the potential shown in Fig. 6. There are an infinite sequence of other trajectories which differ from (c) only in the number of oscillations made between r_1 and r_2.

and this geometric series can be summed to give

$$S = -i\, e^{2i\eta_0}\left[R^{1/2} + \frac{T\, e^{2i\phi}}{1 + R^{1/2}\, e^{2i\phi}} \right] \tag{75}$$

$$= -i\, e^{2i\eta_0}\left[\frac{R^{1/2} + e^{2i\phi}}{1 + R^{1/2}\, e^{2i\phi}} \right] \tag{76}$$

Furthermore, it is not difficult to show that

$$\frac{R^{1/2} + e^{2i\phi}}{1 + R^{1/2}\, e^{2i\phi}} = e^{2i\eta_r} \tag{77}$$

where

$$\eta_r = \tan^{-1}\left\{ \tan\phi \left[\frac{(1 + e^{-2\theta})^{1/2} - 1}{(1 + e^{-2\theta})^{1/2} + 1} \right] \right\} \tag{78}$$

so that the net phase shift is

$$\eta = \eta_0 + \eta_r \tag{79}$$

(78) and (79) is the semiclassical result that has been obtained previously by a number of other approaches.[5,41,42]

A 'resonance' thus occurs from the constructive interference of the many trajectories which contribute to the process (elastic scattering in this case). To see this more explicitly, suppose the energy E is far below the top of the barrier so that

$$R \simeq 1$$

$$T \ll 1$$

the sum of multiply reflected terms in (74) is then given approximately by

$$iT\,e^{2i\eta_0} \sum_{N=1}^{\infty} e^{2iN(\phi - \pi/2)} \tag{80}$$

Each term contributes very little (since $T \ll 1$), but there are many of them. If $\Phi \equiv \Phi(E)$ is such that

$$\Phi(E) - \frac{\pi}{2} = n\pi \tag{81}$$

n being an integer, then

$$e^{2iN(\phi - \pi/2)} = 1 \tag{82}$$

for all N, and the sum in (80) is infinite. (81) is thus the condition that all the terms in (80) add up in phase and cause a resonance; it is also recognized as the WKB (Bohr–Sommerfeld) quantum condition[43] for the potential well.

Proceeding more formally,[44] the definition of a resonance is that the S-matrix, considered as a function of the (complex-valued) energy, has a pole; the real part of this complex pole is the energy at which the resonance occurs, and its imaginary part is the width of the resonance, or reciprocal lifetime of the collision complex. Referring to (76), the semiclassical S-matrix has a pole if

$$1 + R^{1/2}\,e^{2i\phi} = 0$$

or

$$(1 + e^{-2\theta})^{1/2} + e^{2i\phi} = 0$$

or

$$\phi = \frac{1}{2i} \ln\left[-(1 + e^{-2\theta})^{1/2}\right]$$

or

$$\phi(E) = (n + \tfrac{1}{2})\pi - \frac{i}{4} \ln\left[1 + e^{-2\theta(E)}\right] \tag{83}$$

where it has been emphasized that ϕ and θ are both functions of E. For the case that the energy is far below the top of the barrier,

$$e^{-2\theta} \ll 1$$

and one can write

$$E = E_r - i\Gamma/2$$

with

$$\Gamma \ll E_r$$

thus

$$\phi(E) = \phi\left(E_r - i\frac{\Gamma}{2}\right)$$

$$\simeq \phi(E_r) - i\frac{\Gamma}{2}\phi'(E_r)$$

and

$$\ln(1 + e^{-2\theta}) \simeq e^{-2\theta}$$

so that from these approximations and (83) one identifies E_r as determined by the quantum condition of (81) with Γ given by

$$\Gamma = \tfrac{1}{2}e^{-2\theta}/\phi'(E_r) \tag{84}$$

The potential resonances discussed above are 'classically forbidden' processes (see Section IV) in that they involve tunnelling. 'Classically allowed' complex formation is possible only for systems that have internal degrees of freedom in addition to translation; i.e. classically allowed resonances must be 'multiparticle resonances'. These internal excitation or Feshbach resonances result from an energy transfer mechanism: if the interaction between A and BC is attractive, when they collide more energy may be transferred into excitation of the internal degrees of freedom than is energetically possible when A and BC are infinitely separated. As A and BC attempt to separate, therefore, a translational turning point is encountered so that A and BC suffer another collision, and so forth, until the internal degrees of freedom loose sufficient energy for A and BC (or AB and C, for example, if reactive processes are possible) to separate.

Classical complex formation such as outlined above has been observed in a number of classical Monte Carlo trajectory studies,[45] and Brumer and Karplus[46] have recently reported an extensive study of alkali halide–alkali halide reactions which involve long-lived collision complexes. These purely classical studies cannot, of course, describe the resonance structure in the energy dependence of scattering properties, but rather give an average energy dependence; the resonance structure, a quantum effect, is described only by a theory which contains the quantum principle of superposition.

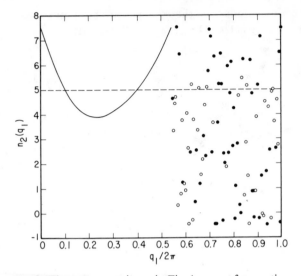

Fig. 8. The same quantity as in Fig. 1, except for reactive (solid line and solid points) and non-reactive (open points) collisions of $H + Cl_2(n_1 = 0) \rightarrow H + Cl_2(n_2)$, $HCl(n_2) + Cl$ as calculated by Rankin and Miller (Ref. 47). The total energy (referred to the saddle point) is $0\cdot3$ eV. The dashed line at $n_2 = 5$ indicates that two 'direct' trajectories and many 'snarled' ones contribute to the $0 \rightarrow 5$ reactive transition.

To see how Feshbach resonances appear in classical S-matrix theory, consider the collinear $H + Cl_2$ collision as studied by Rankin and Miller.[47] Fig. 8 shows the quantum number function $n_2(\bar{q}_1)$; for one region of $\bar{q}_1$ the function is smooth, these trajectories being 'direct'. The remaining interval of $\bar{q}_1$ leads to complex trajectories, those which spend a number of additional vibrational periods in the interaction region; for this region of $\bar{q}_1$ values the final vibrational quantum number changes dramatically with small changes in $\bar{q}_1$. The S-matrix for the particular transition indicated in Fig. 8 thus has the form

$$S = S_D + S_R \tag{85}$$

where S_D is the 'direct' contribution which is constructed in the usual semi-classical fashion from the two direct trajectories; the 'complex' contribution is the sum of many terms

$$S_R = \sum_K [2\pi i n_2'(\bar{q}_K)]^{-1/2} \exp(i\Phi_K/\hbar) \tag{86}$$

Analogous to (80) in the discussion of potential resonances, each term in (86) makes a small contribution [because $n_2'(\bar{q}_K)$ is very large], but there are many such terms. Since the various complex trajectories differ from one another essentially by the number of oscillations of the collision complex before it decomposes, the action integrals Φ_K differ roughly by integer multiples of the action integral for an oscillation of the complex. Thus if the energy is such that

$$\Phi_K - \Phi_{K'} = 2\pi \times \text{integer}$$

the terms in (86) will add up in phase to cause a resonance in the scattering; at other energies the interference is destructive and $S_R \simeq 0$.

The reader will recognize that the semiclassical origin of the resonance, namely the constructive interference of the amplitudes associated with the many trajectories that arise from a collision complex, is the same for Feshbach (multi-particle) resonances as for potential (single-particle) resonances. The physical mechanism causing the collision complex is quite different, however, being barrier tunnelling for the case of a single degree of freedom and an energy transfer between degrees of freedom for the Feshbach case.

D. Atom–Surface Scattering

As another application of classical S-matrix theory it is interesting to see how the scattering of atoms from a solid surface is described. (The extension to scattering of molecules should also be clear.) This has been worked out by Doll[49] and closely parallels Wolken's[50] quantum mechanical formulation of the problem.

By solid surface one means that the surface is being represented by a potential, $V(\mathbf{r})$, so that the Hamiltonian for the particle is

$$H(\mathbf{p}, \mathbf{r}) = \mathbf{p}^2/2\mu + V(\mathbf{r}) \tag{87}$$

where $\mathbf{r} = (x, y, z)$ and $\mathbf{p} = (p_x, p_y, p_z)$. z is the direction perpendicular to the surface, i.e.,

$$\lim_{z \to \infty} V(x, y\, z) = 0$$

$$\lim_{z \to -\infty} V(x, y, z) = +\infty$$

and the surface periodicity is

$$V(x, y, z) = V(x + ma_x, y + na_y, z) \tag{88}$$

where m and n are integers and (a_x, a_y) are the unit cell dimensions.

If $\mathbf{k}_1$ is the initial wave vector of the particle, i.e. $\hbar\mathbf{k}_1$ is the initial momentum, then the quantity of interest is

$$P_{\mathbf{k}_2 \leftarrow \mathbf{k}_1}$$

the probability that the final wave vector is $\mathbf{k}_2$; quantum mechanically, this is given by the square modulus of a probability amplitude, or S-matrix element

$$P_{\mathbf{k}_2 \leftarrow \mathbf{k}_1} = |S_{\mathbf{k}_2 \leftarrow \mathbf{k}_1}|^2$$

Since internal degrees of freedom of the surface are being neglected, the scattering of the atom must be elastic, i.e.

$$\mathbf{k}_1 \cdot \mathbf{k}_1 = \mathbf{k}_2 \cdot \mathbf{k}_2 \tag{89}$$

so that only two components of the wave vector can change independently, k_x and k_y, say. Thus one actually seeks the S-matrix on the 'energy shell'

$$S_{k_{x_2} k_{y_2} \leftarrow k_{x_1} k_{y_1}}(E) \tag{90}$$

Since the desired S-matrix element is in a cartesian momentum representation, it is clear from the general description outlined in Section III.A that the *classical* S-matrix is given by

$$S_{k_{x_2} k_{y_2} \leftarrow k_{x_1} k_{y_1}}(E) = C \left[\frac{\partial(k_{x_2}, k_{y_2})}{\partial(\bar{x}_1, \bar{y}_1)} \right]^{1/2} \exp\left[i\Phi(\mathbf{k}_2, \mathbf{k}_1)/\hbar\right] \tag{91}$$

where the action integral Φ is

$$\Phi(\mathbf{k}_2, \mathbf{k}_1) = -\int_{t_1}^{t_2} dt \, (\mathbf{r} \cdot \dot{\mathbf{p}})$$

$$= -\int_{t_1}^{t_2} dt \, (x\dot{p}_x + y\dot{p}_y + z\dot{p}_z) \tag{92}$$

The meaning of (91) is analogous to (31): the initial conditions for a trajectory are specified as[51]

$$p_{x_1} = \hbar k_{x_1}$$

$$p_{y_1} = \hbar k_{y_1}$$

$$p_{z_1} = -(2mE - p_{x_1}^2 - p_{y_1}^2)^{1/2}$$

$$z_1 = \text{large and positive}$$

$$y_1 = \bar{y}_1 + p_{y_1} z_1 / p_{z_1}$$

$$x_1 = \bar{x}_1 + p_{x_1} z_1 / p_{z_1} \tag{93}$$

$K_{x_2}(\bar{x}_1, \bar{y}_1)$ and $K_{y_2}(\bar{x}_1, \bar{y}_1)$ are the final values of k_x and k_y that result from this trajectory, and the S-matrix element in (94) is constructed from the trajectory (or trajectories) which satisfies

$$K_{x_2}(\bar{x}_1, \bar{y}_1) = k_{x_2}$$

$$K_{y_2}(\bar{x}_1, \bar{y}_1) = k_{y_2} \tag{94}$$

The constant C in (91) is a normalization factor that will be specified below. As usual, if there is more than one root to (94), (91) is a sum of terms, one for each such trajectory.

Ler $S^0_{k_{x_2}k_{y_2} \leftarrow k_{x_1}k_{y_1}}(E)$ be the S-matrix of (91) that is constructed from all roots of (94) for which $\bar{x}_1$ and $\bar{y}_1$ lie in one unit cell:

$$0 \leq \bar{x}_1 \leq a_x$$
$$0 \leq \bar{y}_1 \leq a_y \tag{95}$$

If $\bar{x}_1$ and $\bar{y}_1$ satisfy (94), then it is clear from the symmetry of the potential energy function (88) that a root to (94) will also result if one makes the replacement

$$\bar{x}_1 \rightarrow \bar{x}_1 + ma_x$$
$$\bar{y}_1 \rightarrow \bar{y}_1 + na_y \tag{96}$$

where m and n are integers. The resulting trajectory is identical to the original one, simply shifted an integral number of unit cell dimensions parallel to the surface:

$$x(t) \rightarrow x(t) + ma_x$$
$$y(t) \rightarrow y(t) + na_y$$
$$z(t) \rightarrow z(t)$$
$$\mathbf{p}(t) \rightarrow \mathbf{p}(t)$$

for all t. The preexponential factor in (91) is the same for the two symmetrically related trajectories, but the phase, defined by (92), is changed according to

$$\Phi(\mathbf{k}_2, \mathbf{k}_1) \rightarrow -\int_{t_1}^{t_2} dt\{[x(t) + ma_x]\dot{p}_x + [y(t) + na_y]\dot{p}_y + z(t)\dot{p}_z\}$$
$$= \Phi(\mathbf{k}_2, \mathbf{k}_1) - ma_x(p_{x_2} - p_{x_1}) - na_y(p_{y_2} - p_{y_1}) \tag{97}$$

The contribution to the S-matrix from this new trajectory is therefore given by

$$S_{k_{x_2}k_{y_2} \leftarrow k_{x_1}k_{y_1}}(E) = S^0_{k_{x_2}k_{y_2} \leftarrow k_{x_1}k_{y_1}}(E)\exp\left(-ima_x \Delta k_x - ina_y \Delta k_y\right) \tag{98}$$

where $\Delta k_x = k_{x_2} - k_{x_1}$, $\Delta k_y = k_{y_2} - k_{y_1}$.

Since the above arguments are valid for any integers m and n, the S-matrix element that results from all roots,

$$-\infty < \bar{x}_1 < \infty$$
$$-\infty < \bar{y}_1 < \infty$$

is

$$S_{k_{x_2}k_{y_2}\leftarrow k_{x_1}k_{y_1}}(E) = S^0_{k_{x_2}k_{y_2}\leftarrow k_{x_1}k_{y_1}} \sum_{m,n=-\infty}^{\infty} \exp[-ima_x\,\Delta k_x - ina_y\,\Delta k_y]$$

$$= S^0_{k_{x_2}k_{y_2}\leftarrow k_{x_1}k_{y_1}}(E)\left(\sum_{m=-\infty}^{\infty}\exp(-ima_x\,\Delta k_x)\right)$$

$$\times\left(\sum_{n=-\infty}^{\infty}\exp(-ina_y\,\Delta k_y)\right) \tag{99}$$

If

$$a_x\,\Delta k_x = 2\pi \times \text{(integer)}$$

then the sum over m will be infinite (since all the terms are unity); it vanishes otherwise. Stated more precisely, the Poisson sum[52] formula implies that

$$\sum_{m=-\infty}^{\infty}\exp(-ima_x\,\Delta k_x) = \sum_{m=-\infty}^{\infty}\delta\left(m - \frac{a_x\,\Delta k_x}{2\pi}\right) \tag{100}$$

and similarly for the sum over n:

$$\sum_{n=-\infty}^{\infty}\exp(-ina_y\,\Delta k_y) = \sum_{n=-\infty}^{\infty}\delta\left(n - \frac{a_y\,\Delta k_y}{2\pi}\right) \tag{101}$$

The interference of all symmetrically related trajectories, i.e. the quantum principle of superposition, thus leads to the Bragg diffraction law which allows only certain discrete changes in the x and y components of momentum. The S-matrix element S^0, which is constructed from those trajectories with initial values $\bar{x}_1$ and $\bar{y}_1$ restricted to one cell, is the S-matrix on the 'diffraction spot shell'.

To summarize in more convenient notation, let E be the initial translational energy and (θ_i, ϕ_i) the polar and azimuthal angles of incidence. The initial conditions for a trajectory are specified by

$$p_{x_1} = \hbar k \sin\theta_i\cos\phi_i$$

$$p_{y_1} = \hbar k \sin\theta_i\sin\phi_i$$

$$p_{z_1} = -\hbar k \cos\theta_i \tag{102}$$

where $\hbar k = (2mE)^{1/2}$, and with (x_1, y_1, z_1) given as in (93). From the final values of the x and y components of momentum one defines the 'diffraction order' functions

$$M(\bar{x}_1, \bar{y}_1) = (k_{x_2} - k_{x_1})a_x/2\pi$$

$$N(\bar{x}_1, \bar{y}_1) = (k_{y_2} - k_{y_1})a_y/2\pi \tag{103}$$

The relative intensity of the (m, n) diffraction spot is then given by

$$P_{mn}(E) = |S_{mn}(E)|^2 \tag{104}$$

where

$$S_{mn}(E) = \sum \left[a_x a_y \frac{\partial(M, N)}{\partial(\bar{x}_1, \bar{y}_1)} \right]^{-1/2} \exp(i\Phi_{mn}/\hbar) \qquad (105)$$

with $\bar{x}_1$ and $\bar{y}_1$ determined by the 'quantum conditions'

$$M(\bar{x}_1, \bar{y}_1) = m$$

$$N(\bar{x}_1, \bar{y}_1) = n \qquad (106)$$

and where Φ_{mn} is given by (92); the sum in (105) is over all the roots $\bar{x}_1$ and $\bar{y}_1$ in the intervals $(0, a_x)$ and $(0, a_y)$, respectively. The proper constants have been supplied in (105) so that the relative intensities are normalized to unity within the usual classical limit:

$$\sum_{m,n} P_{mn} \simeq \int dm \int dn |S_{mn}|^2$$

$$\simeq \int dm \int dn (a_x a_y)^{-1} \left| \frac{\partial(M, N)}{\partial(\bar{x}_1, \bar{y}_1)} \right|^{-1}$$

$$= (a_x a_y)^{-1} \int_0^{a_x} d\bar{x}_1 \int_0^{a_y} d\bar{y}_1 \quad (1)$$

$$= 1$$

With a semiclassical description such as this it is possible to discuss rainbow phenomena[53] in a manner parallel to the treatment in elastic scattering.[5,6] Doll[49] has also discussed the 'quenching' of the diffraction spots which results when imperfect periodicities of the lattice are taken into account.

E. Photodissociation

The final application of classical S-matrix theory to be discussed is the description of photodissociation of a complex (e.g. triatomic) molecule. The completely classical description, essentially the 'half-collision' model of Holdy, Klutz and Wilson,[54] is discussed first, and then the semiclassical version of the theory is presented. A completely quantum mechanical description of the process has been developed in detail recently by Shapiro.[55]

The quantity of interest is the transition dipole,

$$\langle E_2 \mathbf{n}_2 | \boldsymbol{\mu} | \mathbf{N}_1 \rangle \qquad (107)$$

which describes the process

$$ABC(\mathbf{N}_1) + h\nu \rightarrow A + BC(\mathbf{n}_2) \qquad (108)$$

where the total energy of the final (dissociated) state, E_2, is related to the photon's energy by

$$E_2 - E_1(\mathbf{N}_1) = hv \qquad (109)$$

from this dipole matrix element the absorption coefficient is given by standard formulae.[56] The classical Hamiltonian for the ground electronic state is of the form

$$H_1(\mathbf{p}, \mathbf{r}) = \mathbf{p}^2/2m + V_1(\mathbf{r}) \qquad (110)$$

where $(\mathbf{p}, \mathbf{r})$ are the Cartesian coordinates and momenta of the system. To define the initial state $|\mathbf{N}_1\rangle$ semiclassically, however, it is necessary to introduce the action-angle variables $(\mathbf{N}, \mathbf{Q})$ for the ground state potential energy surface V_1: this requires that $V_1(\mathbf{r})$ be separable, and in most applications one would probably go farther and assume it to be harmonic. A particular set of integer values of the action variables $\mathbf{N}, \mathbf{N}_1$, is the classical, or semiclassical equivalent of the quantum state $|\mathbf{N}_1\rangle$, and the energy $E_1(\mathbf{N}_1)$ is simply the Hamiltonian H_1 expressed in terms of the action-angle variables:

$$E_1(\mathbf{N}_1) = H_1(\mathbf{p}(\mathbf{N}_1, \mathbf{Q}_1), \mathbf{r}(\mathbf{N}_1, \mathbf{Q}_1))$$
$$= H_1(\mathbf{N}) \qquad (111)$$

(Note that rotational effects are being neglected in this simplified discussion; if one assumes that rotation and vibration are separable, then it is a trivial matter to incorporate rotation explicitly.) The state $|E_2\mathbf{n}_2\rangle$ in (107) is a scattering state on the excited electronic potential surface $V_2(\mathbf{r})$, the state corresponding to total energy E_2 and to BC being asymptotically in state $|\mathbf{n}_2\rangle$; since the quantum numbers $\mathbf{n}_2$ refer to the diatomic fragment BC with A infinitely separated, no assumptions about the potential surface $V_2(\mathbf{r})$ are required.

It is simplest first to use the phase space approach discussed in Section II.B to construct the purely classical expression for the square modulus of the dipole matrix element in (107). Thus if M denotes the total number of degrees of freedom, one has

$$|\langle E_2\mathbf{n}_2|\boldsymbol{\mu}|\mathbf{N}_1\rangle|^2 = (2\pi\hbar)^{-M} \int d\bar{\mathbf{q}}_1 \int d\bar{\mathbf{p}}_1 \, \mu[\mathbf{r}(\bar{\mathbf{p}}_1, \bar{\mathbf{q}}_1)]^2 \, \delta_M[\mathbf{N}_1 - \mathbf{N}(\bar{\mathbf{p}}_1, \bar{\mathbf{q}}_1)]$$
$$\times \, \delta_{M-1}[\mathbf{n}(\bar{\mathbf{p}}_1, \bar{\mathbf{q}}_1) - \mathbf{n}_2] \, \delta[E_2 - H_2(\bar{\mathbf{p}}_1, \bar{\mathbf{q}}_1)] \qquad (112)$$

where $(\bar{\mathbf{p}}_1, \bar{\mathbf{q}}_1)$ is an arbitrary set of canonical variables (since phase space integrals are invariant to a canonical transformation), $\mathbf{n}(\bar{\mathbf{p}}_1, \bar{\mathbf{q}}_1)$ is the set of $M - 1$ quantum numbers of BC that result from a (dissociative) trajectory that begins on the excited potential surface $V_2(\mathbf{r})$ with initial conditions $(\bar{\mathbf{p}}_1, \bar{\mathbf{q}}_1)$, and $\mathbf{N}(\bar{\mathbf{p}}_1, \bar{\mathbf{q}}_1)$ is the set of M vibrational quantum numbers of the ground state potential surface expressed in terms of the variables $(\bar{\mathbf{p}}_1, \bar{\mathbf{q}}_1)$; the subscripts on the delta functions denote their dimensionality. $H_2(\bar{\mathbf{p}}_1, \bar{\mathbf{q}}_1)$

is the classical Hamiltonian for the excited potential surface expressed in terms of $(\bar{\mathbf{p}}_1, \bar{\mathbf{q}}_1)$, and $\mu(\mathbf{r})$ is the transition dipole function.[57] In terms of Cartesian coordinates and momenta the Hamiltonian H_2 is

$$H_2(\mathbf{p}, \mathbf{r}) = \mathbf{p}^2/2m + V_2(\mathbf{r}) \tag{113}$$

but in order to evaluate (86) it is most convenient to choose $(\bar{\mathbf{p}}, \bar{\mathbf{q}})$ to be the action-angle variables $(\mathbf{N}, \mathbf{Q})$; (112) then becomes

$$|\langle E_2 \mathbf{n}_2 | \mu | \mathbf{N}_1 \rangle|^2 = (2\pi\hbar)^{-M} \int d\mathbf{Q}_1 \, \mu[\mathbf{r}(\mathbf{N}_1, \mathbf{Q}_1)]^2$$

$$\times \delta_{M-1}[\mathbf{n}(\mathbf{N}_1, \mathbf{Q}_1) - \mathbf{n}_2] \, \delta[E_2 - H_2(\mathbf{N}_1, \mathbf{Q}_1)] \tag{114}$$

Since

$$\begin{aligned} H_2 &= \mathbf{p}^2/2m + V_2(\mathbf{r}) \\ &= \mathbf{p}^2/2m + V_1(\mathbf{r}) + V_2(\mathbf{r}) - V_1(\mathbf{r}) \\ &= H_1 + \Delta V(\mathbf{r}) \\ &= E_1(\mathbf{N}_1) + \Delta V[\mathbf{r}(\mathbf{N}_1 \mathbf{Q}_1)] \end{aligned}$$

where $\Delta V = V_2 - V_1$, (114) becomes

$$|\langle E_2 \mathbf{n}_2 | \mu | \mathbf{N}_1 \rangle|^2 = (2\pi\hbar)^{-M} \int d\mathbf{Q}_1 \, \mu(\mathbf{r}_1)^2$$

$$\times \delta_{M-1}[\mathbf{n}(\mathbf{N}_1, \mathbf{Q}_1) - \mathbf{n}_2] \, \delta[E_2 - E_1 - \Delta V(\mathbf{r}_1)] \tag{115}$$

where $\mathbf{r}_1 = \mathbf{r}(\mathbf{N}_1, \mathbf{Q}_1)$. Because of the M delta function factors in (115), the M-fold integral over Q_1 can be carried out, giving the final result

$$|\langle E_2 \mathbf{n}_2 | \mu | \mathbf{N}_1 \rangle|^2 = \mu(\mathbf{r}_1)^2 \left[(2\pi\hbar)^M \left| \frac{\partial(\mathbf{n}, \Delta V)}{\partial \mathbf{Q}_1} \right| \right]^{-1} \tag{116}$$

where the M-dimensional Jacobian determinant is evaluated at the values of $\mathbf{Q}_1 \equiv \{Q_i(t_1)\}$ determined by the M equations

$$\mathbf{n}(\mathbf{N}_1, \mathbf{Q}_1) \tag{117a}$$

$$\Delta V[\mathbf{r}(\mathbf{N}_1, \mathbf{Q}_1)] = E_2 - E_1(\mathbf{N}_1) \tag{117b}$$

$\mathbf{n}(\mathbf{N}_1, \mathbf{Q}_1)$ is the set of $M - 1$ asymptotic quantum numbers of BC that result from a dissociative trajectory beginning on potential surface $V_2(\mathbf{r})$ with initial conditions

$$\mathbf{r}_1 = \mathbf{r}(\mathbf{N}_1, \mathbf{Q}_1) \tag{118a}$$

$$\mathbf{p}_1 = \mathbf{p}(\mathbf{N}_1, \mathbf{Q}_1) \tag{118b}$$

If there is more than one root to (117), then (116) is a sum of such terms over all roots.

The physical picture which may be attached to (116) and (91) is as follows:[54] up until time t_1, say, the system is in state $|N_1\rangle$, i.e. the action variables have these particular integer values and the conjugate angle variables at time t_1, Q_1, have random values in the internal $(0, 2\pi)$; the corresponding values of the Cartesian coordinates and momenta at t_1 are $r_1 \equiv r(N_1, Q_1)$, $p_1 = p(N_1, Q_1)$. A photon is absorbed at time t_1, which changes the potential function from $V_1(r)$ to $V_2(r)$, but which conserves the instantaneous values of the Cartesian coordinates and momenta. Since the absorption process conserves the Cartesian variables, the kinetic energy, $p^2/2m$, cannot change, so that the photon's energy $h\nu \equiv E_2 - E_1$ must be matched exactly by the instantaneous change in the potential energy (117b). A classical trajectory begins at time t_1 on the excited potential surface with initial conditions (r_1, p_1), leading to dissociation of ABC into A + BC(n). In order for the $M - 1$ quantum numbers of BC to turn out to be the specific integer values n_2, and for the potential energy difference ΔV to be exactly $h\nu$, the M variables $Q_1 \equiv \{Q_i(t_1)\}$, $i = 1,\ldots, M$, must be chosen to be certain specific values (117). The intensity of the transition is the square modulus of the transition dipole function at r_1, weighted by the Jacobian which maps the initial random variables Q_1 onto specific final values of n and ΔV.

It would not be practical or even desirable, however, to carry out a classical calculated in the above framework. The practical difficulty would be related to finding the roots of (117), the usual multi-dimensional root-search problem, and the result would be undesirable because zeros in the Jacobian determinant cause singularities, classical 'rainbows', in the classical probability distribution in (116). To remedy both of these features one averages the classical expression over a quantum number increment about n_2 and over some increment about E_2:

$$\overline{|\langle E_2 n_2|\mu|N_1\rangle|^2} \equiv \frac{1}{\varepsilon} \int_{E_2 - \frac{1}{2}\varepsilon}^{E_2 + \frac{1}{2}\varepsilon} dE_2 \int_{n_2 - \frac{1}{2}}^{n_2 + \frac{1}{2}} dn_2 |\langle E_2 n_2|\mu|N_1\rangle|^2 \qquad (119)$$

and from (114) this is seen to give

$$|\langle E_2 n_2|\mu|N_1\rangle|^2 = (2\pi\hbar)^{-M} \int dQ_1 \frac{\mu(r_1)^2}{\varepsilon} \chi(Q_1; N_1, n_2, E_2) \qquad (120)$$

where

$$\chi = 1$$

if

$$E_2 - E_1 - \frac{\varepsilon}{2} \leq \Delta V(r_1) \leq E_2 - E_1 - \frac{\varepsilon}{2}$$

and

$$n_2 - \tfrac{1}{2} \leq n(N_1, Q_1) \leq n_2 + \tfrac{1}{2}$$

and $\chi = 0$ otherwise. A Monte Carlo procedure would probably be the most efficient way to carry out such a calculation;[54] thus with N_1 fixed, Q_1 would be chosen at random

$$Q_i(t_1) = 2\pi\xi_i$$

where $\{\xi_i\}$ are random numbers in (0, 1). $\Delta V[\mathbf{r}(N_1, Q_1)]$ and $\mathbf{n}(N_1, Q_1)$ are then determined and the value of $\mu(\mathbf{r}_1)^2/\varepsilon$ added to the appropriate quantum number and energy 'box', with the procedure repeated many times.

The semiclassical version of the above classical description is fairly obvious; the matrix element itself (i.e. the amplitude) is constructed first and then the square modulus formed. The classical limit of the amplitude is the square root of the classical expression in (116) times a phase factor, and analogous to the discussion in Section II.B one can see that the appropriate phase is

$$\Phi(E_2\mathbf{n}_2, N_1) = F_2^{(1)}(\mathbf{r}_1, N_1) - F_2^{(2)}(\mathbf{r}_1, \mathbf{n}_2 E_2) \tag{121}$$

where $F_2^{(i)}$ is the F_2-type generator for potential surface i. Since $\mathbf{r}_1$ and N_1 refer to the same time, t_1, and since $V_1(\mathbf{r})$ is separable, the first term in (121) is simply the sum of one dimensional generators of the form in (5) for each vibrational mode. In the second term of (121), however, $\mathbf{r}_1$ and $\mathbf{n}_2 E_2$ refer to different times, so it is convenient to think first of a canonical transformation from $\mathbf{r}_1$ to $\mathbf{r}_2$, the Cartesian coordinates at time t_2, and then a transformation from $\mathbf{r}_2$ to $(\mathbf{n}_2, E_2)$; by the general prescription for combining successive canonical transformations,[1,9] one has

$$F_2^{(2)}(\mathbf{r}_1, \mathbf{n}_2 E_2) = F_1^{(2)}(\mathbf{r}_1, \mathbf{r}_2) + F_2^{(2)}(\mathbf{r}_2, \mathbf{n}_2 E_2) \tag{122}$$

and the first term here, the generator of the dynamical transformation from $\mathbf{r}_1$ to $\mathbf{r}_2$, is known[58] to be the action integral along the trajectory

$$F_1^{(2)}(\mathbf{r}_1, \mathbf{r}_2) = -F_1^{(2)}(\mathbf{r}_2, \mathbf{r}_1)$$

$$= -\int_{t_1}^{t_2} dt \, \mathbf{p} \cdot \dot{\mathbf{r}} \tag{123}$$

In the second term of (122) $\mathbf{r}_2$ and $(\mathbf{n}_2, E_2)$ both refer to time t_2, so that it is simply the F_2-generator for potential surface $V_2(\mathbf{r})$ at time t_2; i.e.

$$F_2^{(2)}(\mathbf{r}_2, \mathbf{n}_2 E_2) = P_2 R_2 + f_2^{BC}(\mathbf{r}_2^{M-1}, \mathbf{n}_2) \tag{124}$$

where f_2^{BC} is the F_2-type generator for diatomic molecule BC.

The semiclassical expression for the dipole matrix element is therefore given by

$$\langle E_2\mathbf{n}_2|\boldsymbol{\mu}|N_1\rangle = \boldsymbol{\mu}(\mathbf{r}_1)\left[(2\pi i)^M \frac{\partial(\mathbf{n}, \Delta V)}{\partial Q_1}\right]^{-1/2} \exp\left[i\Phi(\mathbf{n}_2 E_2, N_1)/\hbar\right] \tag{125}$$

where

$$\Phi(\mathbf{n}_2 E_2, N_1) = \int_{t_1}^{t_2} dt \, \mathbf{p} \cdot \dot{\mathbf{r}} + F_2^{(1)}(\mathbf{r}_1, N_1) - F_2^{(2)}(\mathbf{r}_2, \mathbf{n}_2 E_2) \tag{126}$$

There will typically be more than one trajectory that obeys the appropriate double-ended boundary conditions, i.e. more than one root to (117), so that (125) will be the sum of several such terms.

If the excited state potential surface is repulsive, so that the dissociative trajectories are 'direct', the dipole matrix element will be a smooth function of E_2, i.e. the absorption spectrum is 'continuous'. If, on the other hand $V_2(\mathbf{r})$ is attractive so that the complex ABC lives a long time before dissociating into A and BC, the quantum number function $\mathbf{n}(\mathbf{N}_1, \mathbf{Q}_1)$ will be highly structured (cf. Fig. 8) and thus a large number of terms will contribute to (125). Analogous to the semiclassical discussion of resonances in Section III.C these many terms will interfere destructively at all but certain specific values of E_2 at which the interference is constructive and the matrix element extremely large. In such cases, therefore, there will be a 'line spectrum' with the width of the absorption lines related to the time the excited state lives before dissociating.

IV. CLASSICAL S-MATRIX: CLASSICALLY FORBIDDEN PROCESSES

Classically forbidden processes are those that do not take place via ordinary classical dynamics. The simplest example of such a process is one dimensional tunneling through a potential barrier, and the 'classically forbidden' concept is essentially a generalization of tunneling to dynamical systems of more than one degree of freedom. In addition to being one of the most intrinsically interesting aspects of classical S-matrix theory, the ability to describe classically forbidden processes—for which a completely classical theory is obviously inadequate—provides an extension of classical trajectory methods that may have practical utility, particularly the 'partial averaging' mode of calculation discussed in Section IV.C.

A. Introductory Discussion

The essential idea is that classically forbidden transitions are treated by *analytic continuation*. To motivate the approach, consider evaluation of the definite integral

$$I = \int_{-\infty}^{\infty} dt\, g(t) \exp\left[if(t)/\hbar\right] \tag{127}$$

by the method of stationary phase;[14] this is an asymptotic approximation which becomes exact as $\hbar \to 0$ and which is the basic semiclassical approximation. If t_0 is the point of stationary phase, i.e. the root of the equation

$$f'(t) = 0 \tag{128}$$

then this approximation gives[14]

$$I \simeq g(t_0)\left[\frac{2\pi i\hbar}{f''(t_0)}\right]^{1/2} \exp\left[if(t_0)/\hbar\right] \tag{129}$$

If there is more than one point of stationary phase, i.e. more than one root to (128), then (129) is a sum of similar terms, one for each such root.

If there are no roots to (128), then the 'primitive' stationary phase approximation implies $I \simeq 0$; although it is true that in such cases the value of the integral is small, one often wishes to know *how* small—10^{-2}, say, or 10^{-4}. To determine the asymptotic approximation to the integral in such cases one analytically continues (129), the mathematical apparatus for which is the 'method of steepest descent'.[59] This approach notes that although there are no *real* values of t which satisfy (128), there will in general be complex values which do so—provided, of course, that it is possible to analytically continue the function $f(t)$ into the complex t-plane. The method of steepest descent then deforms the path of integration in (127) from the real t-axis,

$$I = \int_C dt \, g(t) \exp \left[i f(t)/\hbar \right] \qquad (130)$$

where C is a contour in the complex t-plane which passes through the 'complex point of stationary phase' t_0. The resulting approximation[59] for the integral is *exactly the same* as (129), the only difference being that t_0, the root of (128), is now complex, and $g(t_0)$ and $f(t_0)$ are the (unique) analytic continuations of $g(t)$ and $f(t)$. Since t_0 is complex, $g(t_0)$ and $f(t_0)$ are in general also complex, so that the square modulus of the integral, which is usually the quantity of interest, is given by

$$|I|^2 \simeq |g(t_0)|^2 \frac{2\pi\hbar}{|f''(t_0)|} \exp \left[-2 \, \mathscr{Im} \, f(t_0)/\hbar \right] \qquad (131)$$

The original integral in (127) requires the functions $f(t)$ and $g(t)$ only at real values of t, and it is the asymptotic approximation to the integral which introduces their analytic continuation to complex t.

To illustrate how classically forbidden processes are described with this type of approximation, consider tunnelling through a one-dimensional potential barrier as sketched in Fig. 9. Apart from some irrelevant constants, the amplitude for the particle going from the left to the right of the barrier at fixed energy E is a matrix element of the Green's function, which in term is a Fourier transform of the propagator:

$$\langle x_2 | G(E) | x_1 \rangle = (i\hbar)^{-1} \int_0^\infty dt \, e^{iEt/\hbar} \langle x_2 | e^{-iHt/\hbar} | x_1 \rangle \qquad (132)$$

$$\simeq (i\hbar)^{-1} \int_0^\infty dt \left[\frac{\partial^2 \phi(x_2, x_1; t)}{\partial x_2 \, \partial x_1} \middle/ (-2\pi i\hbar) \right]^{-1/2}$$

$$\times \exp \left\{ \frac{i}{\hbar} [Et + \phi(x_2, x_1; t)] \right\} \qquad (133)$$

where the classical-limit propagator has been invoked in (133). Consistent

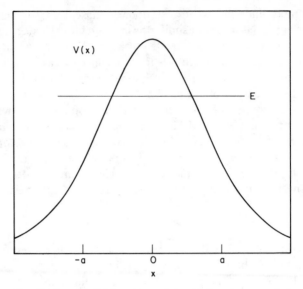

Fig. 9. A potential barrier and translational energy for
which tunnelling occurs.

with classical-limit quantum mechanics,[9] the integral over t is evaluated by
stationary phase, the time of stationary phase being determined by

$$0 = E + \frac{\partial \phi(x_2, x_1; t)}{\partial t} \tag{134}$$

But classical mechanics implies that the time derivative of the action integral
along a trajectory is the negative of the energy of the trajectory:

$$\frac{\partial \phi(x_2, x_1; t)}{\partial t} \equiv -E(x_2, x_1; t) \tag{135}$$

$E(x_2, x_1; t)$ being the energy of the trajectory that goes from x_1 to x_2 in time t.
It is clear intuitively (and can be shown rigorously) that the trajectory which
connects x_1 and x_2 in a very short time must correspond to a very high
energy, and the one that takes a long time to go from x_1 to x_2 corresponds
to an energy just slightly above V_{max}; i.e.,

$$\lim_{t \to 0} E(x_2, x_1; t) = +\infty \tag{136a}$$

$$\lim_{t \to \infty} E(x_2, x_1; t) = V_{max} \tag{136b}$$

Thus for any value of E on the range

$$V_{max} < E < \infty$$

there is a real value of t, $0 < t < \infty$, which satisfies the stationary phase condition, (134); this corresponds to the ordinary classical trajectory which goes from x_1 to x_2 if $E > V_{max}$. For $E < V_{max}$, however, it is clear from (136) that no real value of t satisfies (134) and the transition is thus classically forbidden, i.e. there is no real-valued trajectory at this energy which obeys the appropriate double-ended boundary conditions. It is possible to analytically continue $\phi(x_2, x_1; t)$, however, and find a complex value of t which satisfies (134); the stationary phase approximation then proceeds in the manner discussed above, and the resulting expression from the transition amplitude is the same as if the transition were classically allowed:

$$\langle x_2|G(E)|x_1 \rangle \sim \exp\left\{ \frac{i}{\hbar}[Et_0 + \phi(x_2, x_1; t_0)] \right\} \tag{137}$$

except that the time t_0, the root of (134), is complex.

The action integral $\phi(x_2, x_1; t)$ is the time integral of the Lagrangian

$$\phi(x_2, x_1; t) = \int_0^t dt' \left[\tfrac{1}{2}\mu \dot{x}(t')^2 - V[x(t')] \right] \tag{138}$$

where $x(t')$ is the trajectory determined by solving the equations of motion with the boundary conditions $x(0) = x_1$, $x(t) = x_2$. In order to analytically continue ϕ to complex time, therefore, it is necessary to analytically continue the trajectory $x(t')$ itself to complex time. For systems with more than one degree of freedom trajectories must of course be determined by numerical integration of the classical equations of motion step by step in time, so that the analytic continuation of a trajectory (and therefore the action integral) to complex values of time in general proceeds as follows: rather than incrementing the time variable along the real t-axis, one increments it along the desired contour in the complex t plane. Since numerical integration of the equations of motion[3] amounts to approximating the coordinates and momenta at each integration step by polynomials in t—a manifestly analytic representation—it is clear that numerical integration step by step along a complex time contour does indeed generate the analytic continuation of the coordinates and momenta, and thus the action integral, from which the classical S-matrix is constructed. The expression for the classical S-matrix element for a classically forbidden process is the same as for a classically allowed one (31), the only difference being that the appropriate trajectory is complex-valued.

Analogous to the present use of complex time to construct amplitudes which refer to a definite energy is the more familiar use of complex energy to describe time dependence. The most common example of this is when considering the decay of a prepared state.[60] The probability amplitude that the system has not decayed from its initial state $|\phi\rangle$ is a diagonal matrix element of the propagator, which is conveniently written as a transform of the Green's function [essentially the inverse of (132)]:

$$\langle\phi|e^{-iHt/\hbar}|\phi\rangle = (-2\pi i\hbar)^{-1}\int_{-\infty}^{\infty} dE\, e^{-iEt/\hbar}\langle\phi|G^{+}(E)|\phi\rangle \tag{139}$$

Clearly only real values of E appear in this expression, but since the analytic continuation of the Green's function $G^{+}(E)$ into the lower half E-plane often has a pole in that region,

$$\langle\phi|G^{+}(E)|\phi\rangle \simeq A/(E - E_r + i\Gamma/2) \tag{140}$$

it is convenient to convert the above real integral into a contour integral enclosing the lower half E-plane.[60] If there is only the one such pole, then evaluation of this contour integral is trivial, giving

$$\langle\phi|e^{-iHt/\hbar}|\phi\rangle \simeq A\, e^{-iE_r t/\hbar}\, e^{-\Gamma t/2\hbar} \tag{141}$$

the square modulus of which gives the probability that the system has not yet left its initial state:

$$|\langle\phi|e^{-iHt/\hbar}|\phi\rangle|^2 \simeq |A|^2\, e^{-\Gamma t/\hbar} \tag{142}$$

Thus although energy and time are obviously real physical quantities, it is a useful mathematical devise to invoke the idea of complex energy (a 'complex eigenvalue' or complex pole of the Green's function) when considering time evolution, and conversely, the notion of complex time when constructing amplitudes that refer to processes at a definite energy.

B. Applications

Classical S-matrix theory has been applied to classically forbidden processes in A + BC collision systems, both collinear and three-dimensional models, reactive as well as non-reactive processes having been studied. This section discusses some of these results.

Consider first the simplest case, the non-reactive collinear system, for which there is just one internal degree of freedom. To construct the classical S-matrix for the $n_1 \rightarrow n_2$ transition one must find the roots of the equation

$$n_2(\bar{q}_1) = n_2 \tag{143}$$

where the meaning of the quantities is the same as discussed in Section III.B. Fig. 10 shows the function $n_2(\bar{q}_1)$ for $n_1 = 1$ and for an energy such that *all inelastic transitions are classically forbidden*, meaning simply that they have small probabilities; this is actually the typical situation for vibrationally inelastic processes in thermal energy kinetics. For $n_2 \neq n_1$ there are thus no

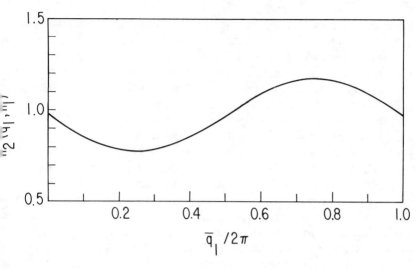

Fig. 10. The same quantity as in Fig. 1, except for a total energy $E = 3\hbar\omega$. Here all inelastic transitions are classically forbidden.

eal values of $\bar{q}_1$ which satisfy (143); there are, however, complex values of $\bar{q}_1$ which do so—provided of course that one can analytically continue the unction $n_2(\bar{q}_1)$ to find them.

To evaluate $n_2(\bar{q}_1)$ for complex values of $\bar{q}_1$ one must integrate the classical equations of motion with complex-valued initial conditions. During the course of such a trajectory, all coordinates and momenta become complex-valued,[61] but this causes no difficulties since the objects of physical meaning, the quantum numbers in the asymptotic regions, are real-valued for the trajectories which satisfy the appropriate double-ended boundary conditions; Section IV.B of Ref. 9 discusses these points in some detail. The classical S-matrix is still given by (51), but there is now typically just one complex root of (143) for which $\mathscr{I}m\ \Phi > 0$ so that the vibrational transition probability is

$$P_{n_2,n_1}(E) = [2\pi|n_2'(\bar{q}_1)|]^{-1} \exp[-2\,\mathscr{I}m\ \Phi(n_2, n_1; E)/\hbar] \qquad (144)$$

where Φ is the action integral of (32). The exponential damping factor which multiplies the 'classical' probability factor (the reciprocal Jacobian) is the multidimensional generalization of the tunnelling probability for one-dimensional barrier penetration, i.e. classically forbidden processes are essentially a generalized kind of tunnelling.

The first calculations of the type outlined above were carried out by Miller and George;[62] similar calculations were carried out independently by

Stine and Marcus.[63] These results are in excellent agreement (as few %)
with the accurate quantum mechanical values obtained by Secrest and
Johnson[24] even for extremely weak transitions with a probability as small as
10^{-11}. It is thus encouraging that the semiclassical model is able to describe
such quantum-like phenomena for which ordinary (i.e. real-valued) classical
trajectory methods would clearly be inapplicable.

It should be noted in passing that for this non-reactive collinear system a
completely quantum mechanical (i.e. coupled channel) calculation may
actually be no more difficult—i.e. require no more computer time—than
these semiclassical calculations. Whether this is true or not is beside the
point, of course, for the obvious interest in the semiclassical model is that it
can be applied to physically realistic three-dimensional systems (see Section
IV.C) for which coupled channel calculations are usually unreasonable
unless simplifying approximations are introduced. The purpose for carrying
out semiclassical calculations for collinear systems is to obtain definitive
comparisons with reliable quantum mechanical values (which exist only for
collinear systems).

Doll[64] has applied classical S-matrix theory to the collinear A + BC
collision where atoms A and B interact via a hard sphere collision; this is the
model studied quantum mechanically by Shuler and Zwanzig.[65] Doll treats
classically allowed and forbidden processes and finds good agreement
between semiclassical and quantum mechanical transition probabilities.
This is a remarkable achievement for the semiclassical theory, for the hard
sphere interaction is far from the 'smooth' potential that one normally
assumes to be necessary for the dynamics to be classical-like.

George and Miller[34] have also treated classically forbidden transitions
in the collinear reactive system $H + H_2(n_1 = 0) \rightarrow H_2(n_2 = 0) + H$ at
collision energies below the classical threshold for reaction. The reaction
probability is given by (144) for this case also, and the only new feature of the
calculation is that the complex time path must be chosen to insure that the
reaction does occur; choice of a purely real time path would of course lead
to a non-reactive trajectory at these energies. Section IV.D discusses some of
these aspects of the calculation in more detail.

Since this calculation by George and Miller[34] there have been several
extensive quantum mechanical calculations of the reaction probability in
this energy region on the Porter–Karplus[66] potential surface; the results of
Bowman and Kuppermann[67] and Duff and Truhlar[68] are in excellent
agreement and can be considered to be the numerically exact quantum
mechanical values. Stimulated by this work, Hornstein and Miller[68a]
re-analyzed the semiclassical calculation[34] and concluded that since only
one final vibrational state of H_2 is energetically open in the threshold region
a better value for the reaction probability would be obtained if the pre
exponential factor in (144) were set equal to unity. Fig. 11 shows the com

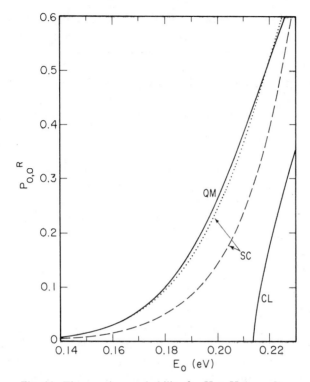

Fig. 11. The reaction probability for $H + H_2(n_1 = 0) \rightarrow$ $H_2(n_2 = 0) + H$ as a function of initial translational energy E_0. QM and CL designate the quantum mechanical (Refs. 67, 68) and quasi-classical [D. J. Diestler and M. Karplus, *J. Chem. Phys.*, **55**, 5832 (1971)] results, while the dashed line gives the results of the original semiclassical calculation (Ref. 34) and the dotted line the values obtained with the modification introduced in Ref. 68A.

parison of the original semiclassical,[34] quantum mechanical,[67,68] and modified semiclassical[68A] reaction probabilities in the threshold region; one sees that the modified semiclassical results are in quite good agreement with the quantum mechanical values.

With regard to three dimensional A + BC collision systems, Doll and Miller[69] have calculated a few specific S-matrix elements for classically forbidden vibrational excitation in He + H_2 collisions. Agreement with quantum mechanical values is good, within the uncertainty of the correct quantum mechanical values. A number of S-matrix elements have also been calculated by Doll, George, and Miller[35] for reactive collisions of

$$H + H_2(n_1, j_1 = 0, 0) \rightarrow H_2(n_2, j_2 = 0, 1) + H$$

in three dimensions on the Porter–Karplus[66] potential surface, again in the tunnelling region. Agreement with the quantum mechanical calculations of Wolken and Karplus[70] is quite reasonable (within a factor of 2), but here again the quantum mechanical values, although the most serious treatment thus far, are probably not the numerically exact quantum results for this potential surface.

In the three-dimensional $A + BC$ examples discussed above the S-matrix elements

$$S_{n_2 j_2 l_2, n_1 j_1 l_1}(J, E) \tag{145}$$

were calculated semiclassically. Thus with the initial quantum numbers n_1, j_1, l_1—and the total angular momentum J and total energy E—held fixed, the conjugate angle variables $\bar{q}_{n_1}, \bar{q}_{j_1}, \bar{q}_{l_1}$ must be chosen iteratively so that the final quantum numbers take on their desired integer values; i.e. one must solve the three equation

$$n_2(\bar{q}_{n_1}, \bar{q}_{j_1}, \bar{q}_{l_1}) = n_2$$

$$j_2(\bar{q}_{n_1}, \bar{q}_{j_1}, \bar{q}_{l_1}) = j_2 \tag{146}$$

$$l_2(\bar{q}_{n_1}, \bar{q}_{j_1}, \bar{q}_{l_1}) = l_2$$

simultaneously. (See Section III.E of Ref. 9 for more discussion of the semiclassical description of three-dimensional $A + BC$ collision systems.) Because of the difficulty of finding the roots of (146), a multi-dimensional root-search problem, it is not practical to calculate the large number of individual S-matrix elements that would be necessary in order to construct actual cross sections. (Although Tyson, Saxon and Light[33] have, with considerable effort, carried out such calculations for coplanar $H + H_2$ collisions.) The value of these three-dimensional calculations has been to show that the semiclassical model, to the extent that it can be applied, provides a reasonably accurate description of the quantum effects in molecular collision dynamics. As a practical means for carrying out calculations for three-dimensional $A + BC$ collision processes the 'partial averaging' approach described in the following section is fortunately much more useful.

C. Partial Averaging

When considering three-dimensional collision systems one is of course not interested in individual S-matrix elements, but rather cross sections that involve sums over many of them. Under realistic conditions, too, the cross sections of interest are a sum and average over some of the final and initial quantum states. For the still quite idealized process

$$A + BC(n_1 j_1) \rightarrow A + BC(n_2 j_2)$$

$$\rightarrow AB(n_2 j_2) + C$$

for example, where the m-components of the rotational states are not observed, the integral cross section is a sum over three quantum numbers:

$$\sigma_{n_2 j_2 \leftarrow n_1 j_1}(E_1) = \frac{\pi}{k_1^2(2j_1 + 1)} \sum_{J, l_1, l_2} (2J + 1)|S_{n_2 j_2 l_2, n_1 j_1 l_1}(J, E)|^2 \quad (147)$$

As discussed in Section III.B, quantum interference structure tends to be quenched by these sums, and if the transition is classically allowed, the semiclassical theory then effectively degenerates to a completely classical result.

To see explicitly how the Monte Carlo classical procedure emerges, note that if interference terms are discarded the square modulus of the classical S-matrix (31) for a classically allowed transition is given by (setting $\hbar = 1$)

$$S_{n_2 j_2 l_2, n_1 j_1 l_1}(J, E) = \left[(2\pi)^3 \frac{\partial(n_2 j_2 l_2)}{\partial(\bar{q}_{n_1} \bar{q}_{j_1} \bar{q}_{l_1})} \right]^{-1} \quad (148)$$

so that (147) becomes

$$\sigma_{n_2 j_2 \leftarrow n_1 j_1}(E_1) = \frac{\pi}{k_1^2(2j_1 + 1)} \int dJ \int dl_1 \int dl_2 \frac{2J + 1}{(2\pi)^3} \left| \frac{\partial(n_2 j_2 l_2)}{\partial(\bar{q}_{n_1} \bar{q}_{j_1} \bar{q}_{l_1})} \right|^{-1} \quad (149)$$

where it has been assumed that enough quantum numbers contribute to the sums in (147) to justify replacing them by integrals. If there are at least a few values of n_2 and j_2 that are classically allowed transitions from the initial values (n_1, j_1), then it is permissible to average (149) over a quantum number width[71] about the integer values of n_2 and j_2:

$$\sigma_{n_2 j_2 \leftarrow n_1 j_1}(E_1) \simeq \int_{n_2 - 1/2}^{n_2 + 1/2} dn_2 \int_{j_2 - 1/2}^{j_2 + 1/2} dj_2 \, \sigma_{n_2 j_2 \leftarrow n_1 j_1}(E_1) \quad (150)$$

and with (149) this becomes

$$\sigma_{n_2 j_2 \leftarrow n_1 j_1}(E_1) = \frac{\pi}{k_1^2(2j_1 + 1)} \int dJ \int dl_1 \int dn_2 \int dj_2 \int dl_2$$

$$\times \frac{2J + 1}{(2\pi)^3} \left| \frac{\partial(n_2 j_2 l_2)}{\partial(\bar{q}_{n_1} \bar{q}_{j_1} \bar{q}_{l_1})} \right|^{-1} \quad (151)$$

The advantageous feature of this averaging process is that (151) now involves an integral over all the final quantum numbers n_2, j_2 and l_2, so that a change of variables of integration from (n_2, j_2, l_2) to their conjugate initial values $(\bar{q}_{n_1}, \bar{q}_{j_1}, \bar{q}_{l_1})$—i.e.

$$\int dn_2 \int dj_2 \int dl_2 = \int d\bar{q}_{n_1} \int d\bar{q}_{j_1} \int d\bar{q}_{l_1} \left| \frac{\partial(n_2 j_2 l_2)}{\partial(\bar{q}_{n_1} \bar{q}_{j_1} \bar{q}_{l_1})} \right|$$

—introduces a Jacobian factor which *exactly cancels* the one in (151). The resulting expression for the cross section is considerably simplified:

$$\sigma_{n_2 j_2 \leftarrow n_1 j_1}(E_1) = \frac{\pi}{k_1^2(2j_1 + 1)} \int dl_1 \int dJ \int d\bar{q}_{n_1} \int d\bar{q}_{j_1} \int d\bar{q}_{l_1} \frac{2J + 1}{(2\pi)^3}$$

$$\times \ h[\tfrac{1}{2} - |n_2(\bar{q}_{n_1}, \bar{q}_{j_1}, \bar{q}_{l_1}) - n_2|]$$

$$\times \ h[\tfrac{1}{2} - |j_2(\bar{q}_{n_1}, \bar{q}_{j_1}, \bar{q}_{l_1}) - j_2|] \qquad (152)$$

where $h(x)$ is the unit step function,

$$h(x) = \begin{array}{ll} 1, & x \geq 0 \\ 0, & x < 0 \end{array}$$

i.e. the product of the two step functions is 1 if

$$n_2 - \tfrac{1}{2} \leq n_2(\bar{q}_{n_1}, \bar{q}_{j_1}, \bar{q}_{l_1}) \leq n_2 + \tfrac{1}{2}$$

$$j_2 - \tfrac{1}{2} \leq j_2(\bar{q}_{n_1}, \bar{q}_{j_1}, \bar{q}_{l_1}) \leq j_2 + \tfrac{1}{2}$$

and zero otherwise. To carry out such an integral in practice one simply sweeps the integration variables—now all initial conditions—through their complete ranges and assigns the final values $n_2(\bar{q}_{n_1}, \bar{q}_{j_1}, \bar{q}_{l_1})$ and $j_2(\bar{q}_{n_1}, \bar{q}_{j_1}, \bar{q}_{l_1})$ to the appropriate quantum number 'boxes', thereby generating in one calculation the cross sections from (n_1, j_1) to all classically allowed final states.

The limits of integration are $l = 0 \to \infty$ and $J = |j_1 - l_1| \to (j_1 + l_1)$ but (152) is cast in a more obvious Monte Carlo form by replacing l_1 by the impact parameter b,

$$b = (l_1 + \tfrac{1}{2})/k_1 \qquad (153)$$

and replacing J by the variable z,

$$J + \tfrac{1}{2} = [(l_1 - j_1)^2 + (2l_1 + 1)(2j_1 + 1)z]^{1/2} \qquad (154)$$

for which the integration limits are $z = 0 \to 1$. It is also customary to cut the impact parameter integration off at some value B beyond which there are no trajectories which lead to the transition of interest; since

$$\pi \int_0^B db \, 2b = \pi B^2 \int_0^1 d\xi$$

where

$$\xi = (b/B)^2 \qquad (155)$$

the changes of variables implied by (153), (154) and (155) lead to the desired result:

$$\sigma_{n_2 j_2 \leftarrow n_1 j_1}(E_1) = \pi B^2 \int_0^1 d\xi \int_0^1 dz \int_0^1 d(\bar{q}_{n_1}/2\pi) \int_0^1 d(\bar{q}_{j_1}/2\pi) \int_0^1 d(\bar{q}_{l_1}/2\pi)$$

$$\times h[\tfrac{1}{2} - |n_2(\bar{q}_{n_1}, \bar{q}_{j_1}, \bar{q}_{l_1}) - n_2|]h[\tfrac{1}{2} - |j_2(\bar{q}_{n_1}, \bar{q}_{j_1}, \bar{q}_{l_1}) - j_2|]$$

$$(156)$$

where

$$l_1 + \tfrac{1}{2} = k_1 B \sqrt{\xi}$$

$$J + \tfrac{1}{2} = [(l_1 - j_1)^2 + (2l_1 + 1)(2j_1 + 1)z]^{1/2}$$

The cross section which is summed over final rotational states,

$$\sigma_{n_2 \leftarrow n_1 j_1}(E_1) \equiv \sum_{j_2} \sigma_{n_2 j_2 \leftarrow n_1 j_1}(E_1)$$

$$\simeq \int dj_2 \, \sigma_{n_2 j_2 \leftarrow n_1 j_1}(E_1)$$

is given by a similar expression:

$$\sigma_{n_2 \leftarrow n_1 j_1}(E_1) = \pi B^2 \int_0^1 d\xi \int_0^1 dz \int_0^1 d(\bar{q}_{n_1}/2\pi) \int_0^1 d(\bar{q}_{j_1}/2\pi) \int_0^1 d(\bar{q}_{l_1}/2\pi)$$

$$\times h[\tfrac{1}{2} - |n_2(\bar{q}_{n_1}, \bar{q}_{j_1}, \bar{q}_{l_1}) - n_2|]$$

$$(157)$$

Since all five integrals in (156) and (157) have limits $0 \to 1$, implementation of Monte Carlo integration procedures is straightforward.

If the $n_1 j_1 \to n_2 j_2$ transition is classically forbidden, then although the above development is obviously inapplicable as it stands, it is still possible to follow it to some extent. Thus it is still a good approximation to neglect interference between various trajectories which contribute to the same S-matrix element—since they will be quenched by the sums—so that (148) is modified only by the addition of the exponential damping factor:

$$|S_{n_2 j_2 l_2, n_1 j_1 l_1}(J, E)|^2 = \left[(2\pi)^3 \left| \frac{\partial(n_2 j_2 l_2)}{\partial(\bar{q}_{n_1} \bar{q}_{j_1} \bar{q}_{l_1})} \right| \right]^{-1} \exp(-2 \, \mathscr{I}m \, \Phi) \quad (158)$$

Since only a very few vibrational states are involved—because the transition is weak—it is not possible to average over the vibrational quantum number as was done above. However there will still typically be a reasonable number of final rotational states that have comparable transition probabilities. For the H + H$_2$ ($n_1 = 0, j_1 = 0$) reaction in three dimensions, for example, the classical trajectory results of Karplus, Porter and Sharma[72] show that final rotational states $j_2 = 0 \to 5$ all have comparable probability for energies just above the classical threshold even though $n_2 = 0$ is the only energetically open vibrational state.

The 'partial averaging' procedure is thus to average over the final rotational state but not the final vibrational state. (The reason one wishes to average over as many final quantum numbers as possible is that boundary conditions for those degrees of freedom can be replaced by *initial conditions*, thereby eliminating the root-search problem.) Since

$$\int dl_2 \int dj_2 = \int d\bar{q}_{j_1} \int d\bar{q}_{l_1} \left| \frac{\partial(j_2 l_2)}{\partial(\bar{q}_{j_1}, \bar{q}_{l_1})} \right|_{n_2, n_1}$$

$$= \int d\bar{q}_{j_1} \int d\bar{q}_{l_1} \left| \frac{\partial(n_2 j_2 l_2)}{\partial(\bar{q}_{n_1} \bar{q}_{j_1} \bar{q}_{l_1})} \right|_{\bar{q}_{n_1}, n_1} \left| \frac{\partial n_2}{\partial \bar{q}_{n_1}} \right|_{n_1}^{-1}$$

the partially averaged expression is

$$\sigma_{n_2 j_2 \leftarrow n_1 j_1}(E_1) = \frac{\pi}{k_1^2(2j_1 + 1)} \int dJ \int dl_1 \int d(\bar{q}_{j_1}/2\pi) \int d(\bar{q}_{l_1}/2\pi)$$

$$\times h[\tfrac{1}{2} - |j_2(\bar{q}_{n_1}, \bar{q}_{j_1}, \bar{q}_{l_1}) - j_2|] \left[2\pi \left| \frac{\partial n_2}{\partial \bar{q}_{n_1}} \right| \right]^{-1} \exp(-2 \mathcal{I}m \, \Phi) \tag{159}$$

where $\bar{q}_{n_1}$ is not integrated over as in (156) but rather must be chosen to be that specific (complex) value for which

$$n_2(\bar{q}_{n_1}; \bar{q}_{j_1}, \bar{q}_{l_1}, l_1, j_1, J, E_1) = n_2 \tag{160}$$

The root-search problem has not been eliminated, but has been reduced to a one-dimensional one which must be carried out many times.[73]

The same changes of variables as introduced above in (156) and (157) can also be made in (159), so that the more useful expression for the cross section, summed over j_2, is

$$\sigma_{n_2 \leftarrow n_1 j_1}(E_1) = \pi B^2 \int_0^1 d\xi \int_0^1 dz \int_0^1 d(\bar{q}_{j_1}/2\pi) \int_0^1 d(\bar{q}_{l_1}/2\pi) P_{n_2, n_1}(j_1 l_1 \bar{q}_{j_1} \bar{q}_{l_1}; JE_1) \tag{161}$$

where

$$P_{n_2, n_1}(j_1 l_1 \bar{q}_{j_1} \bar{q}_{l_1}; JE_1) = \left[2\pi \left| \frac{\partial n_2}{\partial \bar{q}_{n_1}} \right| \right]^{-1} \exp(-2 \mathcal{I}m \, \Phi) \tag{162}$$

Dimensionally, P_{n_2, n_1} defined by (162) is a collinear-like vibrational transition probability [cf. (144)] which depends parametrically on the initial conditions of the other degrees of freedom. The analogy to a collinear collision is purely formal, however, for there are no dynamical approximations which have been introduced; the only approximations involved, beyond that of classical S-matrix theory itself, are the neglect of interference terms between different trajectories that contribute to the same S-matrix element and the assumption that enough j_2-values have comparable probability for a sum over them to be replaced by an integral.

The classically allowed version of (161), namely (157), can also be written in the form of (161) by defining the classically allowed vibrational transition probability as

$$P_{n_2,n_1}(j_1 l_1 \bar{q}_{j_1} \bar{q}_{l_1}; J E_1) = (2\pi)^{-1} \int d\bar{q}_{n_1} h[\tfrac{1}{2} - |n_2(\bar{q}_{n_1}; \bar{q}_{j_1} \bar{q}_{l_1} j_1 l_1 J E_1) - n_2|]$$

$$= \Delta q_{n_1}/2\pi \tag{163}$$

which is also recognized as the form of the averaged vibrational transition probability for a collinear model [cf. (65)]. (For truly collinear systems it is a poor approximation, of course, to ignore the interference terms, however, since there are no averages over other variables to quench them.) In both the allowed and forbidden cases the cross section can be written in the phenomenological form often used in energy transfer theory,[74]

$$\sigma_{n_2 \leftarrow n_1 j_1}(E_1) = \pi B^2 \langle P_{n_2 \leftarrow n_1 j_1}(E_1) \rangle \tag{164}$$

where the 'average transition probability' is the average over the four initial conditions as in (161), with the integrand given by (163) and (162) for the classically allowed and forbidden cases, respectively.

Although only integral cross sections have been discussed, it should be clear that differential cross sections, i.e. angular distributions, can also be generated within the Monte Carlo framework; these are defined by

$$\sigma_{n_2 \leftarrow n_1 j_1}(E_1) = \int d(\cos\theta) \sigma_{n_2 \leftarrow n_1 j_1}(\cos\theta, E_1) \tag{165a}$$

$$= \int dj_2 \int d(\cos\theta) \sigma_{n_2 j_2 \leftarrow n_1 j_1}(\cos\theta, E_1) \tag{165b}$$

to obtain the $n_1 j_1 \to n_2$ cross section differential in final rotational state and in scattering angle—i.e. $\sigma_{n_2 j_2 \leftarrow n_1 j_1}(\cos\theta)$—therefore, one simply defines a set of 'j_2-boxes' and 'cos θ-boxes', and with the integration variables in (161) chosen by Monte Carlo the numerical value of the integrand, i.e. the vibrational transition probability, is assigned to the j_2- and cos θ-box which corresponds to the final values of j_2 and cos θ for the trajectory which satisfies (160), i.e. the one from which the transition probability in (72) is constructed. The distributions in j_2 and cos θ are thus obtained simultaneously with the computation of the integral cross section $\sigma_{n_2 \leftarrow n_1 j_1}$, the only limitations being the usual Monte Carlo ones—that is, the more differential the quantities desired, the more Monte Carlo points required. Thus it might require only 50 Monte Carlo points, for example, to evaluate the integral cross section $\sigma_{n_2 \leftarrow n_1 j_1}$ to within 10% statistical error, but a larger number of points would be required to obtain the distribution of final rotational states, $\sigma_{n_2 j_2 \leftarrow n_1 j_1}$, and a still larger number of points to obtain the 'doubly differential' cross section, $\sigma_{n_2 j_2 \leftarrow n_1 j_1}(\cos\theta)$, differential in j_2 and $\cos\theta$, to within 10% statistical error. It sounds very much like the experimental situation: The more detailed the information desired, more is the effort which is required.

Preliminary results of calculations such as these have been reported by Miller and Raczkowski[75] for the $0 \to 1$ vibrational excitation of H_2 and He. Calculations have also been made for the $1 \to 0$ vibrational de-activation of H_2 by Li^+, and comparison with the quantum mechanical coupled channel calculations of Schaefer and Lester[77] are quite encouraging. Fig. 12 shows

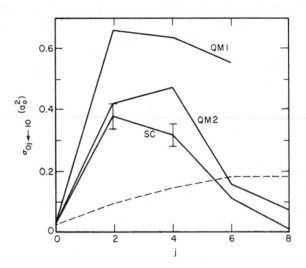

Fig. 12. Cross sections for vibrational de-activation, $Li^+ + H_2(n = 1, j = 0) \to Li^+ + H_2(n = 0, j)$, as a function of final rotational state, for an initial translational energy of 0·684 eV. The semiclassical (SC) and quantum mechanical (QM1, QM2) values of those of Ref. 76 and 77, respectively, and the dashed line the statistical distribution of (166).

the cross sections $\sigma_{n_2 j_2 \leftarrow n_1 j_1}$, for $(n_1, j_1) = (1, 0)$ and $n_2 = 0$, as a function of the final rotational state, for an initial translational energy of 0·684 eV. QM1 in Fig. 12 refers to the results obtained with the basis set

$$v = 0, \qquad j = 0, 2, \ldots, 10$$

$$v = 1, \qquad j = 0, 2, 4$$

$$v = 2, \qquad j = 0, 2$$

for the coupled channel calculation,[77] and QM2 to the results obtained by adding one more j-state to each vibrational manifold. The difference between

QM1 and QM2 indicates that the coupled channel expansion may not have converged for these vibrationally inelastic processes;[78] within this level of uncertainty, however, there is good agreement between the semiclassical and quantum calculations. Also shown in Fig. 12 is a phase space distribution

$$(2j_2 + 1)[E_{tot} - \varepsilon(0, j_2)] \tag{166}$$

$\varepsilon(n_2, j_2)$ being the vibrational–rotational energy levels of H_2, which has been normalized to the total cross section of the quantum and semiclassical calculations. Thus although there is considerable rotational excitation in the $1 \rightarrow 0$ vibrational deactivation, it is not nearly so much as a completely random redistribution of the rotational energy.

D. Numerical Integration of Complex-Valued Trajectories

Although the formalism of classical S-matrix theory deals with initial and final values of action-angle variables, it is actually most convenient to carry out the numerical integration of Hamilton's equations in Cartesian coordinates and momenta. The procedure is that one specifies initial conditions in terms of action-angle variables (e.g., $n_1, q_{n_1}, j_1, q_{j_1}, l_1, q_{l_1} \ldots$), transforms these into initial conditions for the Cartesian variables, carries out the numerical integration of the trajectory in Cartesian variables, and at the end of the trajectory transforms the final values of the Cartesian variables into final values of the action-angle variables (e.g., $n_2, j_2, l_2 \ldots$). Appendix C of Ref. 27 gives the expressions for the initial values of the Cartesian variables in terms of the action-angle variables (see also Section II.B of Ref. 69). With regard to the transformation at the end of the trajectory, the final angular momentum variables j_2 and l_2 are easily determined from the Cartesian variables by using the classical relations

$$j_2 = [(\mathbf{r}_2 \times \mathbf{p}_2) \cdot (\mathbf{r}_2 \times \mathbf{p}_2)]^{1/2}$$
$$l_2 = [(\mathbf{R}_2 \times \mathbf{P}_2) \cdot (\mathbf{R}_2 \times \mathbf{P}_2]^{1/2}$$

where $(\mathbf{r}_2, \mathbf{p}_2)$ are the Cartesian variables of the diatom and $(\mathbf{R}, \mathbf{P})$ the Cartesian variables for the atom–diatom separation. The final vibrational quantum number is determined from the Cartesian variables by first computing the total energy of the diatom

$$\varepsilon_2 = (\mathbf{p}_2 \cdot \mathbf{p}_2)/2m + v(r_2) \tag{167}$$

$v(r)$ being the vibrational potential of the diatom, and then solving the equation

$$\varepsilon(n_2, j_2) = \varepsilon_2 \tag{168}$$

for n_2 (since j_2 is known), where $\varepsilon(n, j)$ is the WKB energy level formula for the diatom; it is usually known as a Dunham expansion. Alternatively, with ε_2

known from (167) n_2 can be computed directly from the WKB quantum condition:

$$(n_2 + \tfrac{1}{2})\pi = \int_{r_<}^{r_>} dr \, \{2m[\varepsilon_2 - v(r)] - j_2^2/r^2\}^{1/2} \tag{169}$$

To a large extent the actual numerical integration of complex-valued trajectories is the same as for ordinary real-valued ones; this is possible by taking advantage of the complex arithmetic capabilities of FORTRAN IV. Thus it is only necessary to declare all the coordinates and momenta, and the time increment, to be COMPLEX variables and use essentially the same numerical integration algorithm[78]—e.g. Runge–Kutta, Adams–Moulton, etc.—as used for real-valued trajectories. Since it is often convenient, however, to vary the direction in the complex time plane of the complex time increment, Miller and George[62] developed a variable step-size predictor–corrector algorithm; it has the variable step-size and self-starting advantages of Runge–Kutta routines with the efficiency of a predictor–corrector (e.g. Adams–Moulton) method. Appendix C of Ref. 62 gives the predictor and corrector formulae for the fifth order [error $\sim O(h^6)$] version of the algorithm; used in the PECE mode,[79] the integrator has excellent stability characteristics.

The principal feature which distinguishes the numerical integration of complex-valued trajectories from real-valued ones lies in the flexibility one has in choosing the complex time path along which time is incremented. Although the quantities from which the classical S-matrix is constructed are *analytic* functions and thus independent of the particular time path,[9] there are practical considerations that restrict the choice. Thus although translational coordinates behave as low order polynomials in time, so that nothing drastic happens to them when t becomes complex, the vibrational coordinate is oscillatory—

$$r(t) - r_{eq} \sim \cos(\omega t + \eta)$$

—so that it can become exponentially large along a complex time path. The complex time path must be chosen, therefore, in order to stabilize the vibrational motion.

There are a variety of ways of stabilizing the vibrational motion, but the most satisfactory procedure we have found to date is to head the oscillation always toward its next equilibrium position. Thus at time t_n the values $r_n, \dot{r}_n, \ddot{r}_n [r_n = r(t_n),$ etc.] are known, so that for t near t_n one has the approximation

$$r(t) = r_n + \dot{r}_n(t - t_n) + \tfrac{1}{2}\ddot{r}_n(t - t_n)^2 \tag{170}$$

and one wishes to choose the next time, t_{n+1}, so that

$$r(t_{n+1}) = r_{eq} \tag{171}$$

solving (170) and (171) gives

$$\Delta t \equiv t_{n+1} - t_n$$
$$= (\ddot{r}_n)^{-1}\{-\dot{r}_n \pm [\dot{r}_n^2 + 2\ddot{r}_n(r_{eq} - r_n)]^{1/2}\} \tag{172}$$

with the $\pm$ sign chosen to insure $\mathscr{R}e\,(\Delta t) > 0$.[80] Actually one wishes only to cause $r(t)$ to head in the *direction* of r_{eq}; thus the new time increment is chosen to have the *phase* of that in (172) but the magnitude determined by the truncation error estimate of the integrator.[62] If Δt is given by (172), then the new time increment is chosen as

$$h(\Delta t)/|\Delta t|$$

where $|\Delta t|$ is the complex absolute value of Δt and h is the magnitude of time increment allowed by the integrator.

The above algorithm for choosing the complex time path applies to non-reactive A + BC collisions, collinear or three-dimensional, throughout the entire trajectory. It is currently being used, for example, in the 'partial averaging' calculations described in the previous section.

To describe tunnelling in reactive systems,[34,35] A + BC → AB + C, the above procedure must be modified somewhat. If r_a is the vibrational co-ordinate of diatom BC, then the above procedure for choosing the time path is followed with regard to the variable r_a until A and BC reach their distance of closest approach. At this point the complex time path is chosen to cause the reaction to occur; i.e. one wants $r_c(t)$, the vibrational coordinate of AB, to head toward its equilibrium value. Thus the same procedure is used to choose the time path but with regard to r_c, the vibrational coordinate of the new diatom.

The procedures described above for choosing the complex time path are the most generally satisfactory ones we have found thus far, but they should not be considered the final answer to the problem. Various different approaches are still being actively pursued, and it appears that there is still much to be learned about analytically continued classical mechanics.

V. CONCLUDING REMARKS

One has at hand, therefore, a completely general semiclassical mechanics which allows one to construct the classical-limit approximation to any quantum mechanical quantity, incorporating the complete classical dynamics with the quantum principle of superposition. As has been emphasized, and illustrated by a number of examples in this review, all quantum effects— interference, tunnelling, resonances, selection rules, diffraction laws, even quantization itself—arise from the superposition of probability amplitudes and are thus contained at least qualitatively within the semiclassical description. The semiclassical picture thus affords a broad understanding and clear insight into the nature of quantum effects in molecular dynamics.

In many cases, too, the semiclassical model provides a quantitative description of the quantum effects in molecular systems, although there will surely be situations for which it fails quantitatively or is at best awkward to apply. From the numerical examples which have been carried out thus far— and more are needed before a definitive conclusion can be reached—it appears that the most practically useful contribution of classical S-matrix theory is the ability to describe classically forbidden processes; i.e. although completely classical (e.g. Monte Carlo) methods seem to be adequate for treating classically allowed processes, they are not meaningful for classically forbidden ones. (Purely classical treatments will not of course describe quantum interference effects which are present in classically allowed processes, but under most practical conditions these are quenched.) The semiclassical approach thus widens the class of phenomena to which classical trajectory methods can be applied.

Two common examples of classically forbidden processes have been discussed in Section IV: vibrationally inelastic transitions in atom–diatom collisions (V–T energy transfer) and tunnelling near the threshold of chemical reactions. Another important example, which has been discussed in detail previously,[9] is electronically non-adiabatic transitions, i.e. transitions from one potential energy surface to another.[81] Miller and George[82] have formulated this problem semiclassically in such a way that incorporates the exact classical dynamics of the heavy particle motion (i.e. classical trajectories) and the quantum principle of superposition; the electronic transition is accounted for within a Stueckelberg-like model, i.e. by considering complex-valued classical trajectories which change from one adiabatic surface to another at a complex point of intersection.[83] (Such processes are 'classically forbidden', therefore, since only complex-valued classical trajectories can reach these complex intersection points in order to change adiabatic potential energy surfaces.) The principle physical requirement of the model is that the electronic transition be localized in space and time, but it is important to recognize that this does *not* require that the adiabatic potential curves or surfaces have an 'avoided intersection' for real coordinates; see Section V of Ref. 9. Recent calculations based on this theory have been carried out by Lin, George and Morokuma[84] (for $H^+ + D_2 \rightarrow HD^+ + D$) and by Preston, Sloane and Miller[85] (for $F(^2P_{3/2}) + Xe \rightarrow F(^2P_{1/2}) + Xe$), and it seems clear that there will be much more activity in the following years regarding the general topic of non-adiabatic transitions in low energy molecular collisions.

Notes and References

1. W. H. Miller, *J. Chem. Phys.*, **53**, 1949 (1970), and other references below.
2. R. A. Marcus, *Chem. Phys. Lett.*, **7**, 525 (1970); *J. Chem. Phys.*, **54**, 3965 (1971) and other references below.
3. D. L. Bunker, *Methods Compnt. Phys.*, **10**, 287 (1971).

4. J. C. Polanyi, *Accounts Chem. Res.*, **5**, 161 (1972).
5. K. W. Ford and J. A. Wheeler, *Ann. Phys. (N.Y.)*, **7**, 259, 287 (1959).
6. R. B. Bernstein, *Advan. Chem. Phys.*, **10**, 75 (1966).
7. R. J. Cross, Jr., *J. Chem. Phys.*, **58**, 5178 (1973).
8. R. J. Cross, Jr., *J. Chem. Phys.*, **47**, 3724 (1967); **48**, 4338 (1968); **51**, 5163 (1969).
9. W. H. Miller, *Advan. Chem. Phys.*, **25**, 69 (1974).
10. M. V. Berry and K. E. Mount, *Rep. Prog. Phys.*, **35**, 315 (1972).
11. See, for example, (a) N. F. Mott and H. S. W. Massey, *The Theory of Atomic Collisions*, Oxford U.P., N.Y., 1965, pp. 802–808; (b) R. J. Cross, Jr., Proceedings of the International School of Physics, Enrico Fermi, Course XLIV *Molecular Beams and Reaction Kinetics*, Academic Press, N.Y., 1970, p. 50; (c) R. W. Fenstermacker, C. F. Curtiss and R. B. Bernstein, *J. Chem. Phys.*, **51**, 2439 (1969); (d) M. D. Pattengill, C. F. Curtiss and R. B. Bernstein, *J. Chem. Phys.*, **54**, 2197 (1971); (e) R. D. Levine and G. G. Balint-Kurti, *Chem. Phys. Lett.*, **6**, 101 (1970).
12. For a recent example, see W. B. Nielsen and R. G. Gordon, *J. Chem. Phys.*, **58**, 4131, 4149 (1973).
13. For a clear and concise discussion of canonical transformations see H. Goldstein, *Classical Mechanics*, Addison-Wesley, Reading, Mass., 1950, pp. 237–247.
14. A. Erdélyi, *Asymptotic Expansions*, Dover, N.Y., 1956, pp. 51–56.
15. See, for example, reference 13, pp. 288–307.
16. See, for example, G. Herzberg, *Molecular Spectra and Molecular Structure. I. Spectra of Diatomic Molecules*, Van Nostrand, Princeton, 1950, p. 194 *et seq.*
17. L. D. Landau and E. M. Lifshitz, *Quantum Mechanics*, Pergamon, N.Y., 1965, pp. 177–181.
18. M. S. Child, *Mol. Phys.*, **8**, 517 (1964).
19. W. H. Miller, *J. Chem. Phys.*, **48**, 464 (1968).
20. This restricts the treatment to collisions of diatomic molecules or to polyatomic molecules which are described initially by non-interacting normal modes.
21. The second term in the expression for $\mathbf{q}(t)$ subtracts out the asymptotic 'free particle' time dependence of the angle variables; i.e. since asymptotically $\mathbf{n}$ and $\mathbf{P}$ are constant and

$$R(t) \sim (P/\mu)t + \text{constant}$$

$$\mathbf{q}(t) \sim \frac{\partial \varepsilon(\mathbf{n})}{\partial \mathbf{n}} t + \text{constant}$$

it is clear that the phase shift variables,

$$\bar{\mathbf{q}}(t) \equiv \mathbf{q}(t) - \frac{\partial \varepsilon(\mathbf{n})}{\partial \mathbf{n}} \mu R(t)/P$$

are asymptotically constant.
22. Equation (31) takes a manifestly symmetrical form by noting that

$$\left[\frac{\partial \mathbf{n}_2(\bar{\mathbf{q}}_1, \mathbf{n}_1; E)}{\partial \bar{\mathbf{q}}_1} \right]^{-1} = \frac{\partial^2 \Phi(\mathbf{n}_2, \mathbf{n}_1; E)}{\partial \mathbf{n}_2 \, \partial \mathbf{n}_1}$$

since $\Phi(\mathbf{n}_2, \mathbf{n}_1; E) \equiv \Phi(\mathbf{n}_1, \mathbf{n}_2; E)$, the classical S-matrix is thus identically symmetric, i.e. microscopically reversible. The form of the pre-exponential factor in (31), however, is usually easier to evaluate numerically.
23. W. H. Miller, *J. Chem. Phys.*, **53**, 3578 (1970); *Chem. Phys. Lett.*, **7**, 431 (1970).
24. D. Secrest and B. R. Johnson, *J. Chem. Phys.*, **45**, 4556 (1966).
25. R. P. Feynman and A. R. Hibbs, *Quantum Mechanics and Path Integrals*, McGraw-Hill, N.Y., 1965, pp. 2–9.

26. W. H. Wong and R. A. Marcus, *J. Chem. Phys.*, **55**, 5663 (1971).
26a. A. Cohen and B. C. Alexander, *J. Chem. Phys.*, **61**, 3967 (1979).
27. W. H. Miller, *J. Chem. Phys.*, **54**, 5386 (1971).
28. Sometimes the 'proper inclusion' of interference terms means that an appropriate *uniform* semiclassical expression must be utilized rather than the 'primitive' semiclassical expressions of (31) and (51).
29. P. Pechukas, *Phys. Rev.*, **181**, 166 (1969).
30. The 'classical threshold' for any process is the energy below which there are no ordinary (i.e. real-valued) classical trajectories which lead to the process. For the collinear $H + H_2$ reaction on the Porter–Karplus potential surface, for example, the classical threshold is ~ 0.20–0.21 eV collision energy; quantum mechanically, of course, there is no absolute energetic threshold for the $0 \rightarrow 0$ reaction.
31. J. M. Bowman and A. Kuppermann, *Chem. Phys. Lett.*, **19**, 166 (1973).
32. S.-f. Wu and R. D. Levine, *Mol. Phys.*, **25**, 937 (1973).
32a. J. W. Duff and D. G. Truhlar, *Chem. Phys. Lett.*, **23**, 327 (1973).
32b. J. R. Stine and R. A. Marcus, *Chem. Phys. Lett.*, **29**, 575 (1974).
33. J. J. Tyson, R. P. Saxon and J. C. Light, *J. Chem. Phys.*, **59**, 363 (1973).
34. T. F. George and W. H. Miller, *J. Chem. Phys.*, **56**, 5722 (1972); **57**, 2458 (1972).
35. J. D. Doll, T. F. George and W. H. Miller, *J. Chem. Phys.*, **58**, 1343 (1973).
36. W. H. Miller, unpublished work.
37. J. R. Stine and R. A. Marcus, *J. Chem. Phys.*, **59**, 5145 (1973).
38. J. N. L. Connor, *Faraday Disc. Chem. Soc.*, **55**, 51 (1973); *Mol. Phys.*, **25**, 181 (1973).
39. A. M. G. Ding, L. J. Kirsch, D. S. Perry, J. C. Polanyi and J. L. Schreiber, *Faraday Disc. Chem. Soc.*, **55**, 252 (1973); C. A. Parr, J. C. Polanyi, W. H. Wong and D. C. Tardy, *Faraday Disc. Chem. Soc.*, 308 (1973); J. C. Polanyi, J. L. Schreiber, and J. J. Sloan, *Faraday Disc. Chem. Soc.*, 124 (1973).
40. N. Fröman and P. O. Fröman, *JWKB Approximation*, North-Holland, Amsterdam, 1965.
41. W. H. Miller, *J. Chem. Phys.*, **48**, 1651 (1968).
42. J. N. L. Connor, *Mol. Phys.* **15**, 621 (1968).
43. See, for example, L. I. Schiff, *Quantum Mechanics*, McGraw-Hill, N.Y., 1968, pp. 275–277.
44. See, for example, R. G. Newton, *Scattering Theory of Waves and Particles*, McGraw-Hill, N.Y., 1966, pp. 316–317; also, J. R. Taylor, *Scattering Theory*, Wiley, N.Y., 1972, pp. 240–244, 407–413.
45. See, for example, G. H. Kwei, B. P. Boffardi and S. F. Sun, *J. Chem. Phys.*, **58**, 1722 (1973).
46. P. Brumer and M. Karplus, *Faraday Disc. Chem. Soc.*, **55**, 80 (1973).
47. C. C. Rankin and W. H. Miller, *J. Chem. Phys.*, **55**, 3150 (1971).
48. R. A. Marcus, *Faraday Disc. Chem. Soc.*, **55**, 34 (1973).
49. J. D. Doll, *J. Chem. Phys.*, **61**, 954 (1974).
50. G. Wolken, *J. Chem. Phys.*, **58**, 3047 (1973).
51. The asymptotic free-particle time dependence of $x(t)$ and $y(t)$ is subtracted out in the usual way; see Ref. 21.
52. See, for example, P. M. Morse and H. Feshbach, *Methods of Theoretical Physics*, McGraw-Hill, N.Y., 1953, pp. 466–467.
53. J. D. McClure, *J. Chem. Phys.*, **57**, 2823 (1972).
54. K. E. Holdy, L. C. Klotz and K. R. Wilson, *J. Chem. Phys.*, **52**, 4588 (1970).
55. M. Shapiro, *J. Chem. Phys.*, **56**, 2582 (1972).
56. See, for example, Ref. 43, pp. 404–405.

57. $\mu(\mathbf{r})$ is the electronic matrix element of the dipole operator between two Born–Oppenheimer electronic wavefunctions and depends on nuclear coordinates through the parametric dependence of the wavefunctions on nuclear coordinates.

58. See, for example, Ref. 13, pp. 273–284.

59. See Ref. 14, pp. 39–41; or Ref. 52, pp. 437–441.

60. M. L. Goldberger and K. M. Watson, *Collision Theory*, Wiley, N.Y., 1964, p. 431 *et seq.*

61. When dealing with one-dimensional systems it is often convenient to eliminate t in favour of x as the independent variable and then require x to be real [cf. D. W. McLaughlin, *J. Math. Phys.*, **13**, 1099 (1972)]. This is no restriction for one-dimensional systems because it is always possible to choose the time path to keep $x(t)$ real; i.e. for any value of $\mathcal{R}e\,t$ one can choose $\mathcal{I}m\,t$ to satisfy the equation $\mathcal{I}m\,x(t) = 0$. One sees, however, that this is possible only for one-dimensional systems, for as soon as there are two coordinates, say, one cannot in general choose $\mathcal{I}m\,t$ to make the imaginary parts of both coordinates vanish; i.e. one cannot satisfy more than one equation with only one variable.

62. W. H. Miller and T. F. George, *J. Chem. Phys.*, **56**, 5668 (1972).

63. J. Stine and R. A. Marcus, *Chem. Phys. Lett.*, **15**, 536 (1972).

64. W. Eastes and J. D. Doll, *J. Chem. Phys.*, **60**, 297 (1974).

65. K. E. Shuler and R. Zwanzig, *J. Chem. Phys.*, **33**, 1778 (1960).

66. R. N. Porter and M. Karplus, *J. Chem. Phys.*, **40**, 1105 (1964).

67. J. M. Bowman and A. Kuppermann, *J. Chem. Phys.*, **59**, 6524 (1973).

68. J. W. Duff and D. G. Truhlar, *Chem. Phys. Lett.*, **23**, 327 (1973).

68a. S. M. Hornstein and W. H. Miller, *J. Chem. Phys.*, **61**, 745 (1974).

69. J. D. Doll and W. H. Miller, *J. Chem. Phys.*, **57**, 5019 (1972).

70. G. Wolken and M. Karplus, *J. Chem. Phys.*, **60**, 351 (1974).

71. If the diatom is homonuclear so that Δj is required to be even, then the width of the quantum number box is of course taken to be 2, rather than 1.

72. M. Karplus, R. N. Porter and R. D. Sharma, *J. Chem. Phys.*, **43**, 3259 (1965).

73. The analysis is slightly more complicated because the integrals over $\bar{q}_{l_1}$ and $\bar{q}_{j_1}$ are actually *contour* integrals chosen to insure that $\mathcal{I}m\,l_2 = \mathcal{I}m\,j_2 = 0$. This does not cause any essential difficulty, however, and will be discussed in more detail in Ref. 76.

74. See, for example, K. F. Herzfeld and T. A. Litovitz, *Absorption and Dispersion of Ultrasonic Waves*, Academic Press, N.Y., 1959.

75. W. H. Miller and A. W. Raczkowski, *Faraday Disc. Chem. Soc.*, **55**, 45 (1973).

76. A. W. Raczkowski and W. H. Miller, *J. Chem. Phys.*, **61**, 5413 (1974).

77. J. Schaefer and W. A. Lester, Jr., *Chem. Phys. Lett.*, **20**, 575 (1973).

78. Schaefer and Lester (ref. 77) were primarily concerned with pure rotational excitation and for this process the expansion does seem to have converged.

79. See, for example, L. Lapidus and J. H. Seinfeld, *Numerical Solution of Ordinary Differential Equations*, Academic Press, N.Y., 1971.

80. This prevents the possibility of the trajectory backing up to a prior time at which the equilibrium position was passed.

81. E. E. Nikitin, in *Chemische Elementarprozesse* (Eds. J. Heidberg, H. Heydtmann and G. H. Kohhmaier), Springer, N.Y., 1968, p. 43; E. E. Nikitin, *Advan. Quantum Chem.*, **5**, 135 (1970).

82. W. H. Miller and T. F. George, *J. Chem. Phys.*, **56**, 5637 (1972).

83. E. C. G. Stueckelberg, *Helv. Phys. Acta*, **5**, 369 (1932). Also see Ref. 17, pp. 185–187.

84. Y.-W. Lin, T. F. George and K. Morokuma, *Chem. Phys. Lett.*, **22**, 547 (1973).

85. R. K. Preston, C. Sloane and W. H. Miller, to be published.

86. Very recently there have been a number of other applications of classical S-matrix theory to molecular collision phenomena. P. A. Whitlock and J. T. Muckerman, *J. Chem. Phys.* **61**, 4618 (1974), have carried out such calculations for the collinear $F + D_2 \rightarrow FD + D$ reaction, finding reasonable agreement with the corresponding quantum mechanical scattering calculations of J. M. Bowman, G. C. Schatz, and A. Kuppermann, *Chem. Phys. Lett.* **24**, 378 (1974). H. Kreek, R. L. Ellis, and R. A. Marcus, *J. Chem. Phys.* **62**, 913 (1975), have made a study of the atom-rigid rotor collision system, emphasizing that care must be exercised in constructing the appropriate "uniform" semiclassical approximation to individual S-matrix elements. J. W. Duff and D. G. Truhlar, in 'Tests of Semiclassical Treatments of Vibrational-Translational Energy Transfer in Collinear Collisions of Helium with Hydrogen Molecules', (preprint) have made a very thorough study of this collision process, and they find the "uniform" version of classical S-matrix theory to be quantitatively accurate in comparison to essentially exact quantum results, with the H_2 molecule described either as a harmonic or as a Morse oscillator.

POTENTIAL ENERGY SURFACES FOR CHEMICAL REACTIONS

G. G. BALINT-KURTI

School of Chemistry, Bristol University

CONTENTS

I. Introduction. 137
II. Methods for *Ab Initio* Calculations 140
  A. General Theoretical Considerations Common to all *Ab Initio* Methods . . 141
  B. The Hartree–Fock and Self-Consistent Field (SCF) Methods 143
  C. Orbital Approximations other than Hartree–Fock: The Spin Optimized Self-Consistent Field Method 146
  D. Correlated Wavefunctions 148
    1. Configuration Interaction 148
    2. Perturbation Methods 152
    3. Valence-Bond Methods. 154
    4. Pair-Functions or Germinals 156
III. Semi-Theoretical Methods 157
  A. Atoms-in-Molecules Methods 158
  B. Pseudo-Potentials 160
  C. Closed-Shell Interaction Model 162
IV. Semi-Empirical Methods 163
  A. The Diatomics-in-Molecules (DIM) Method 163
  B. Semi-Empirical Valence-Bond Treatments (not based on DIM formalism) . 166
  C. Other Semi-Empirical Methods 168
V. Entirely Empirical Methods 170
  A. The London Equation and its Generalizations 170
  B. Ionic Models 172
  C. Convenient Analytic Formulae 172
VI. Summary 174
Acknowledgements 176
References 176

I. INTRODUCTION

Experiments using the crossed molecular beams technique are yielding increasingly detailed information concerning the outcome of reactive and non-reactive molecular collisions.[1,2] The theoretical description of such experiments may be broken up into three steps as follows:

Step 1: Obtain the potential energy function or surface, which describes the interaction of the colliding atoms and molecules with each other.

Step 2: Solve the scattering or dynamic problem, so as to obtain the probabilities of the different reactive and inelastic processes which can take
place in the system.

Step 3: Average the probabilities calculated in the previous step (step 2)
over the various velocity and internal state distributions which
characterize the experimental situation.

The results of Step 3 should yield a quantity which can be directly compared with an experimental measurement. The problems with carrying out
this program begin with Step 1, in that for the vast majority of chemically
interesting systems we are unable, as yet, to define or calculate reliable
potential surfaces. There is, furthermore, no direct way of inverting the
process and finding the potential from the experimental measurements in
the case of a non-spherically symmetric potential.

The indirect inversion process which is often attempted is to assume an
analytic form for the potential energy surface, solve the scattering problem
numerically using either classical mechanics (i.e., Hamilton's equations of
motion)[3-7] or quantum mechanics,[8,9] and after performing the averaging
procedures to compare the calculated and experimental results. If the two
do not agree, then the assumed form of the potential is altered and the
process repeated until a potential is found which will satisfactorily reproduce
the observed results. This procedure does not yield a unique potential, in
that it will often be possible to find several significantly different potentials
which all give rise to calculated observables compatible with presently available experimental results.[10] Ab initio and semi-empirical calculations of
potential energy surfaces therefore have a very important role to play, not
primarily in providing absolutely accurate potentials, but rather in limiting
as far as possible, the detailed form of the potentials essayed in the rather ad
hoc inversion procedure outlined above.

There have recently been several books and review articles on interatomic
forces and potential energy surfaces[11-18] and the proceedings of a conference
on the topic has been published.[19] The purpose of the present chapter will
be to outline and discuss the various methods available for calculating
potential energy surfaces for simple chemical reactions. These methods can
be subdivided into four categories:

1. *ab initio* or rigorous theoretical
2. semi-theoretical
3. semi-empirical
4. entirely empirical.

The *ab initio* methods attempt to calculate the potential energy surface by
solving the relevant Schrödinger equation as accurately as possible. The
semi-theoretical methods also involve a reasonably rigorous solution of the
basic Schrödinger equation. They are, however, distinguished from the

completely *ab initio* methods by the fact that they either include some experimental parameters in an attempt to correct recognized shortcomings of the *ab initio* calculation, or by the fact that they make some specific and reasonable approximations so as to simplify the computations involved in the *ab initio* method. The semi-empirical methods, while generally being based on some *ab initio* approach, make rather gross approximations so as to reduce the complexity of the problem, and then use experimental diatomic potential energy curves as a means of avoiding the computation of most of the remaining integrals. Despite the many approximations made in the semi-empirical methods, they have a correct description of both the reactants and the products built into them and normally provide a good first idea of what a given surface might look like. The entirely empirical methods are either based on an electrostatic-type physical model or are just analytic functional forms for potential energy surfaces which include adjustable parameters, whose choice determines the shape of the surface.

The various methods to be discussed in this chapter are listed in Table I. Clearly, the list of methods is not exhaustive. Among the possible *ab initio*

TABLE I
Methods for calculating potential energy surfaces

Ab Initio

Hartree–Fock or Self-Consistent Field (SCF) Method
Spin Optimized Self-Consistent Field Method
Configuration Interaction
Iterative Natural Orbital Method
Multi-Configuration SCF
Many Body Perturbation Theory
Valence-Bond Method
Pair-Function or Geminal Method

Semi-Theoretical

Atoms-in-Molecules Methods
Rigorous Pseudo-Potential Theories
Closed Shell Interaction Model

Semi-Empirical

Diatomics-in-Molecules (DIM) Method
Semi-Empirical Valence-Bond Methods
Approximate Pseudo-Potential Theories

Entirely Empirical

London Equation and its Generalizations (LEPS)
Ionic (Rittner type) Models
Convenient Analytic Formulae

---

methods only those which have been applied to at least small triatomic systems, or which show promise of being useful in the description of such systems, are included. Also excluded are the various methods based on perturbation theory which are applicable only at large separations of the interacting molecules.[20,21]

The methods listed in Table I are considered one by one in the next four sections. In discussing the *ab initio* and semi-theoretical methods (Sections II and III), the relationships of the difference methods to each other are examined and the physical models on which they are based, outlined. The reliability of a given potential energy surface depends on two aspects: (a) The intrinsic limitations of the method used and (b) The details of how a specific calculation is executed (e.g. the basis set used). Both these aspects of the reliability question are examined. The application and theoretical background of the semi-empirical methods is considered in Section IV. While some of the empirical potential energy functions which have been proposed or used are discussed in Section V. References to illustrative examples of all the different methods are given in the relevant sections. Only references available to the author before the end of 1973 are included in the review. In Section VI a brief summary is given and some important aspects of potential energy surfaces which are not covered in the chapter are mentioned.

II. METHODS FOR *AB INITIO* CALCULATIONS

In order to calculate a potential energy surface from first principles we first write down a Hamiltonian operator $\mathbf{H}_e$ for the electrons of the system with the nuclei fixed at predetermined positions, and then attempt to find as good an approximation as possible to the solution of the Schrödinger equation

$$\mathbf{H}_e(\vec{r}; \vec{R})\Psi_e(\vec{r}; \vec{R}) = E(\vec{R})\Psi_e(\vec{r}; \vec{R}) \tag{1}$$

where $\vec{r}$ represents collectively all the coordinates of the electrons, $\vec{R}$ those of the nuclei and $\mathbf{H}_e(\vec{r}; \vec{R})$ does not contain any terms arising from the nuclear kinetic energy. Approximate solutions to the Schrödinger equation are found using either the variational theorem or a combination of it and perturbation theory.[22] Behind each of the methods of finding approximate solutions to the Schrödinger equation is a model for the wavefunction of the system. Having found approximate eigenvalues for the Schrödinger equation with the nuclei fixed, we then repeat the process at different values of the nuclear coordinates, thus mapping out the eigenvalues of the electronic Hamiltonian $(E_i(\vec{R}))$ as a function of the nuclear positions. These eigenvalues correspond to the ground and excited state potential energy surfaces in the standard Born–Oppenheimer approximation,[23,24] [often denoted by $V(R)$] and they can be used in the context of either a quantum or classical mechanical calculation as the potential functions governing the motion of the nuclei.

A. General Theoretical Considerations Common to All *Ab Initio* Methods

The variational theorem, which is used as a basis for all the *ab initio* and semi-theoretical methods, states that for any approximate electronic wavefunction $\Psi_e(\vec{r}_1, \vec{r}_2 \ldots \vec{r}_i \ldots \vec{r}_N; \vec{R})$ (where $\vec{r}_i$ represents the space and spin coordinates of the ith electron and $\vec{R}$ represents collectively all of the nuclear coordinates) the calculated expectation value of the energy

$$E_{\text{approx}} = \frac{\int \Psi_e^*(\vec{r}_1, \ldots \vec{r}_N; \vec{R}) \mathbf{H}_e(\vec{r}_1, \ldots \vec{r}_N; \vec{R}) \Psi_e(\vec{r}_1, \ldots \vec{r}_N; \vec{R}) \, d\vec{r}_1 \ldots d\vec{r}_N}{\int \Psi_e^*(\vec{r}_1, \ldots \vec{r}_N; \vec{R}) \Psi_e(\vec{r}_1, \ldots \vec{r}_N; \vec{R}) \, d\vec{r}_1 \ldots d\vec{r}_N} \tag{2}$$

is always greater than the exact value for the ground electronic state of the system. The basic idea is, therefore, to choose a flexible functional form for Ψ_e and vary the adjustable parameters in it so as to minimize E_{approx} in (2), thus obtaining the best possible approximation, within the limitations of the chosen functional form, to the lowest electronic eigenfunction. This provides us with an upper bound to the lowest electronic eigenvalue $E_0(\vec{R})$. As will be discussed below, the variation theorem can also provide approximate eigenvalues for the excited electronic states $E_i(\vec{R})$. These can either be rigorous upper bounds or not, depending on particular method used.

The exact electronic wavefunction must possess certain properties. These are:

1. It must obey the Pauli principle (i.e. the wavefunction must change sign on interchange of any two electronic coordinates $\vec{r}_i$ and $\vec{r}_j$).
2. The wavefunction must be an eigenfunction of the operators corresponding to the square of the total spin angular momentum (S^2) and to its z component $\mathbf{S}_z$.†
3. If the geometry of the nuclear framework possesses any symmetry then the electronic wavefunction must possess the symmetry of an irreducible representation of the symmetry group in question.

As the exact electronic eigenfunction is known to possess these properties it is clearly best to ensure that our approximate eigenfunction also possesses them. Most of the methods do this; the exception being a variant of the Hartree–Fock method called the unrestricted Hartree–Fock method (UHF) which yields wavefunctions which are not eigenfunctions of S^2 (see below). The Pauli principle is of central importance, and the problem of constructing approximate wavefunctions which obey it is considered below.

The Pauli Principle

The simplest possible analytic form for a many electron wavefunction is just a product of one electron wavefunctions or orbitals

$$\Psi(\vec{r}_1, \vec{r}_2 \ldots \vec{r}_N; \vec{R}) = \phi_1(\vec{x}_1)\alpha_1 \phi_2(\vec{x}_1)\beta_2 \ldots \phi_N(\vec{x}_N)\beta_N \tag{3}$$

† This applies in the approximation that we ignore relativistic terms such as the spin-orbit coupling.[22]

where $\vec{r}_i$ represent both the space and spin coordinates of electron i, $\vec{\chi}_i$ are just its spacial coordinates and α_i and β_i are the spin functions associated with electron i. The electrons have been assigned spin functions in an arbitrary manner just to indicate that they must not be forgotten. The many-electron function in (3) does not obey the Pauli principle. Let us now define spin-orbitals ψ_j, which incorporate both the space and spin functions of the electron i.e.

$$\psi_1(\vec{r}_i) = \phi_1(\vec{x}_i)\alpha_i$$
$$\psi_2(\vec{r}_i) = \phi_2(\vec{x}_i)\beta_i$$

(4)

The simplest way to form functions which obey the Pauli principle is to build up a determinant for the spin-orbitals ψ_j:

$$\Psi(\vec{r}_1,\ldots\vec{r}_N;\vec{R}) = (N!)^{-1/2} \begin{vmatrix} \psi_1(\vec{r}_1) & \psi_1(\vec{r}_2) & \ldots & \psi_1(\vec{r}_N) \\ \psi_2(\vec{r}_1) & \psi_2(\vec{r}_2) & \ldots & \psi_2(\vec{r}_N) \\ \cdot & \cdot & & \cdot \\ \cdot & \cdot & & \cdot \\ \psi_N(\vec{r}_1) & \psi_N(\vec{r}_2) & \ldots & \psi_N(\vec{r}_N) \end{vmatrix}$$

(5)

Such a determinant is called a Slater determinant.[25] The properties of determinants ensure that a wavefunction which is expressed either as a single Slater determinant or as a linear combination of Slater determinants will obey the Pauli principle. This is true for both orthogonal and non-orthogonal spacial orbitals. These Slater determinants are eigenfunctions of S_z but not in general of S^2. To form eigenfunctions of S^2, we normally have to form suitable linear combinations of Slater determinants.

There is a slightly different viewpoint from which we can approach the problem of satisfying the Pauli Principle. This is to separate the spacial and spin parts of the wavefunction, i.e.:

$$\Psi(\vec{r}_1,\ldots\vec{r}_N;\vec{R}) = \{\phi_1(\vec{x}_1)\phi_2(\vec{x}_2)\ldots\phi_N(\vec{x}_N)\}\{\alpha_1\beta_2\ldots\beta_N\}$$

(6)

This function is, of course, identical to that in (3). As the operators S_z and S^2 act only on the spin part of a function, the spin function in the second curly brackets may be replaced by an eigenfunction of S_z and S^2. These spin eigenfunctions, which we will denote by $\Theta_{S,M,k}$ can be generated using angular momentum coupling theory. The index k distinguishes spin eigenfunctions corresponding to the same eigenvalues of S^2 and S_z but which arise via different couplings of the electron spins. We now have a function which can be written as a spacial part multiplied by a spin function:

$$\Psi_{S,M,k}(\vec{r}_1,\ldots\vec{r}_N;\vec{R}) = G(\vec{x}_1,\ldots\vec{x}_N;\vec{R})\Theta_{S,M,k}(1\ldots N)$$

(7)

The spacial part of the function G is just a product of one electron orbitals $\phi_i(\vec{\chi})$. This still does not obey the Pauli principle. The simplest way to proceed is to antisymmetrize the above function by operating with the antisymmetrization operator:

$$\mathscr{A} = (N!)^{-1/2} \sum_p \epsilon_p \mathbf{P} \tag{8}$$

where $\mathbf{P}$ is a permutation operator for the space and spin coordinates of the N electrons, ϵ_p is $+1$ for an even permutation and -1 for an odd one and the summation $\sum_p$ is over all the $N!$ permutations. Thus our molecular functions which obey the Pauli principle can be written as

$$\Psi_{S,M,k}(\vec{r}_1, \vec{r}_2 \ldots \vec{r}_N; \vec{R}) = \mathscr{A}\{G(\vec{\chi}_1 \ldots \vec{\chi}_N; \vec{R})\Theta_{S,M,k}(1 \ldots N)\} \tag{9}$$

Just as with the Slater determinants of (5), a wavefunction which is expressed either as a single function of this type or as a linear combination of such functions is guaranteed to obey the Pauli principle. It is, furthermore, explicitly an eigenfunction of both S^2 and S_z. The use of this type of function has been reviewed by Gerratt[26] for the general case where the spacial functions G are constructed from non-orthogonal orbitals, and by Ruedenberg and Poshusta[27] for the case where orthogonal orbitals are used.

B. The Hartree–Fock and Self-Consistent Field (SCF) Methods

The idea behind the Hartree–Fock model is that each electron should be assigned for a single spin-orbital. These spin orbitals are then multiplied together and antisymmetrized to form a single Slater determinant. Most stable molecules have singlet ground states and may be described by a Hartree–Fock wavefunction in which each spacial orbital occurs twice; once with α spin and once with β. Such systems are called closed shell systems. The Hartree–Fock wavefunction for a closed shell system may be written in the form:

$$\Psi_{HF}(\vec{r}_1, \ldots \vec{r}_N; \vec{R}) = (N!)^{-1/2} \begin{vmatrix} \psi_1(\vec{r}_1) & \psi_1(\vec{r}_2) & \ldots & \psi_1(\vec{r}_N) \\ \overline{\psi}_1(\vec{r}_1) & \overline{\psi}_1(\vec{r}_2) & \ldots & \overline{\psi}_1(\vec{r}_N) \\ \psi_2(\vec{r}_1) & \psi_2(\vec{r}_2) & \ldots & \ldots \\ \vdots & \vdots & & \vdots \\ \overline{\psi}_{N/2}(\vec{r}_1) & \overline{\psi}_{N/2}(\vec{r}_2) & \ldots & \overline{\psi}_{N/2}(\vec{r}_N) \end{vmatrix} \tag{10}$$

where a bar above a spin orbital indicates that it is associated with β spin i.e. $\overline{\psi}_i(\vec{r}) = \phi_i(\vec{x})\beta$] and no bar indicates α spin [i.e. $\Psi_i(\vec{r}) = \phi(\vec{x})\alpha$]. The spin-orbitals in the Hartree–Fock method are taken to be orthogonal and normalized. This single Slater determinant is then used in the context of the variational method (2) to calculate an approximate energy for the ground state of the system, and the orbitals are varied so as to minimize this energy. The condition that the energy be a minimum is found to be that the spacial parts of the spin-orbitals be solutions to a pseudo-eigenvalue equation of the form

$$\mathbf{F}(\vec{x}; \vec{R})\phi_i(\vec{x}) = \epsilon_i\phi_i(\vec{x}) \tag{11}$$

This is termed a pseudo-eigenvalue equation rather than a proper eigenvalue equation because the operator $\mathbf{F}(\vec{x}; \vec{R})$ depends on the lowest $N/2$ eigenfunctions $\phi_i(\vec{x})$. Equation (11) is therefore solved using an iterative method by first guessing some orbitals $\phi_i(\vec{x})$, using these to evaluate a Fock operator $\mathbf{F}$, and then using this operator in (11) to calculate a new set of orbitals $\phi_i(\vec{x})$, which are used to calculate an improved Fock operator and so on. This procedure normally converges for closed shell molecules.

The best possible wavefunction of the form of (10) is called the Hartree–Fock wavefunction. For molecules it is difficult to solve (11) numerically. The most widely used procedure was proposed by Roothaan.[28] This involves expressing the molecular orbitals $\phi_i(\vec{x})$ as a linear combination of basis functions (normally atomic orbitals) and varying the coefficients in this expansion so as to find the best possible solutions to (11) within the limits of a given basis set. This procedure is called the self-consistent field (SCF) method. As the size and flexibility of the basis set is increased the SCF orbitals and energy approach the true Hartree–Fock ones.

The advantages of the Hartree–Fock method are that it provides a conceptually appealing orbital picture of the molecular wavefunction. This is especially useful in considering electronic spectra. Its main disadvantage is that it cannot properly describe the process of a closed-shell molecule dissociating into two open-shell fragments. Examples of this failing are found in SCF calculations of the potential energy functions for H_2,[29] F_2,[30] LiF,[31] H_2O,[32] FKrF[33] and many other systems.

The Hartree–Fock method is designed to approximate the ground electronic states of molecular systems. It has been used to examine excited states in some cases where there is a lower state of the same symmetry.[34,35] Despite the apparent success of these calculations there is, as yet, no theoretical justification for using it in this manner, and the results of such calculations must be treated with great caution (i.e. it must be checked that the excited states are orthogonal to all lower states). If the excited state in question is the lowest state of its symmetry, then the Hartree–Fock method can still be used.†

Many systems of current interest, especially those which occur in reactive crossed molecular beam experiments, have unpaired electrons and cannot be described by closed-shell type wavefunctions. The orbital equations which arise for such open-shell systems are considerably more complicated than those for the closed-shell case. Several different ways of extending the

† Some CI calculations have also been performed for excited states in which lower states of the same symmetry are ignored. While these calculations seem to yield good results they are not strictly justified on theoretical grounds.[36]

Hartree–Fock method to handle the open-shell problem have been proposed.[37–43] It is generally more difficult to make the iterative SCF method converge for open-shell problems than for closed-shell ones, and this is one of the main reasons why relatively few SCF calculations are performed on open-shell systems. One way of trying to handle this situation is to allow the spatial orbitals associated with α and β spins to be different and to solve different orbital equations for each of them.[44] This method is called the unrestricted Hartree–Fock (UHF) method. Its main disadvantage is that it leads to a wavefunction which is not an eigenfunction of S^2. The orbitals arising from a UHF calculation can, however, be used to form a wavefunction which is an eigenfunction of S^2. This wavefunction is in general a linear combination of Slater determinants. Some recent applications of the UHF method, both with and without projecting out the correct eigenfunction of S^2, have been made to the systems HNO,[45] H_2NO[46] and NH_2, BH_2, CH_2^+.[47]

The Hartree–Fock method is an orbital approximation in that we attempt to calculate the motion of each individual electron in the averaged field of the other electrons. No allowance is made for the correlation of the motion of different electrons with each other. As we will see below, the Hartree–Fock method is not the most general orbital approximation. This is because it imposes the unnecessary but highly convenient condition that the orbitals be orthogonal to each other, and also imposes a specific spin-coupling scheme on the wavefunction.

The energy difference between the Hartree–Fock energy and the exact solution to the non-relativistic Schrödinger equation is called the correlation energy. The correlation energy is normally at least as large as energies of chemical interest such as dissociation energies or potential energy barriers. A potential energy surface calculated by the Hartree–Fock method can, therefore, only be useful if the correlation energy does not vary greatly over the region of the surface which is of interest. This generally implies that the number and nature of the chemical bonds in the system remain the same. Thus, the Hartree–Fock method may be used to examine: (a) the region close to the equilibrium configuration in polyatomic systems (e.g. H_2O;[32] CH_2;[48] BeF_2, MgF_2 and CaF_2,[49] CO_2[50] and LiO_2[51]); (b) interactions between non-bonded atoms such as energy barriers to internal rotation;[52,53] (c) the energy barrier to changes in molecular geometry such as the inversion of NH_3;[54] and (d) hydrogen bonding.[55,56] It may also be used with reasonable confidence to examine interaction potentials between closed shell atoms and ions and closed shell molecules in regions of the potential where no chemical bonds are broken. Two recent examples of such applications are to the He + H_2 and Li^+ + H_2 potential energy surfaces. For He and H_2 system Gordon and Secrest[57] found little difference between the SCF and the more exact CI method, while for Li^+ + H_2 the SCF calculations of Lester[58,59] compare reasonably well with the calculation of Kutzelnigg,

Staemmler and Hoheisel[60] who first performed an SCF calculation and then estimated the correlation error by using perturbation theory. These calculations were limited to H–H separations close to the equilibrium distance. At larger H–H separations the SCF results for the ground state of H_2 become progressively worse,[61,62] and the SCF method becomes unreliable.

The SCF method is the most convenient of all the *ab initio* methods to use, in that computer programs to perform SCF calculations are quite widely available. It is often applied to situations which are not as favourable as those described above. In such cases great care must be taken in assessing the value of the results. Some recent applications which fall into this category are to the systems; $H^- + HF \rightarrow H_2 + F^-$,[63] $Li + HF$,[64] $HeBeH_2$,[65] $He + H_2^+ \rightarrow HeH^+ + H$,[66] CH_5^+,[67] $CH_2 + H_2$,[68] F^- and $CN^- + CH_3F$,[69] Na^+ and $F^- + H_2O$ etc.,[70] $Cl + H_2$,[71] CaF_2,[72] Li_3,[73] CH_3NC[74] and H_2O.[256,257] It is especially inappropriate to use the SCF method for discussing energy differences between different electronic states, curve crossing processes or the breaking of covalent bonds.

C. Orbital Approximations other than Hartree–Fock: The Spin Optimized Self-Consistent Field Method

An orbital or independent particle approximation is defined as one in which we can, in a loose way, associate each electron with a given space orbital. The most general wavefunction of this form is:[26]

$$\Psi_{S,M}(\vec{r}_1, \ldots \vec{r}_N; \vec{R}) = \sum_k d_{S,k} \Psi_{S,M,k}(\vec{r}_1, \ldots \vec{r}_N; \vec{R}) \tag{12}$$

where $\Psi_{S,M,k}$ is defined in (6) to (9), S and M are the quantum numbers associated with S^2 and S_z, and the subscript k labels the different ways of coupling the electron spins. The wavefunction of (12) has a single spacial orbital for every electron.

The best possible variationally determined wavefunction of this form is that in which both the spacial orbitals ϕ_i and the coefficients $d_{S,k}$ are allowed to vary freely so as to minimize the energy, and the only constraints are that the orbitals and the total wavefunction be normalized. For such a wavefunction, the variational method leads to orbital equations similar to the Hartree–Fock equations (10). These are:[26]

$$F_i^s(\vec{x}; \vec{R})\phi_i(\vec{x}) = \epsilon_i \phi_i(\vec{x}) \tag{13}$$

These equations differ from the Hartree–Fock equations because the expressions for F_i^S are much more complicated than those for the Fock operator F of (11), and a different F_i^S operator is required for every orbital ϕ_i. The simple orbital picture is, however, still retained in that (13) describes the motion of an electron assigned to the orbital ϕ_i moving in the averaged field of all the other electrons. This general type of theory† has as yet been applied to only a few small systems (i.e. H_3,[75] $H_2 + D_2$,[76] He_2,[77] BH[78]). The form of the wavefunction in (12) is basically very versatile and is capable of correctly describing the dissociation of a covalent bond. It does, however, lead to complicated orbital equations and we may, if we wish, choose to limit the generality of the wavefunction in different ways. Each of these restrictions leads to a simplification of the orbital equations. Below are outlined some of the possible simplifications which involve the choice of specific spin coupling schemes. There are, in fact, two reasonably obvious choices for the spin coupling:

1. The orbitals may be taken two at a time and coupled to form singlets of the form:

$$[\phi_1(\vec{x}_1)\phi_2(\vec{x}_2) + \phi_2(\vec{x}_1)\phi_1(\vec{x}_2)][\alpha_1\beta_2 - \beta_1\alpha_2]2^{-1/2} \tag{14}$$

Such a singlet might be thought of as representing a covalent bond. If no orthogonality constraints are introduced between the orbitals, this corresponds to the method which Goddard calls G1[78] and is called maximally paired Hartree–Fock by Kaldor.

2. If the same spin coupling is used as in (14) above, but if the orbitals forming any given pair are made orthogonal to those of all other pairs, then we obtain what Goddard calls a Generalized Valence–Bond (GVB) wavefunction. This type of wavefunction was first discussed by Hurley, Lennard-Jones and Pople[79,80] in 1953. Hunt, Hays and Goddard have recently applied this method to H_3, H_2O and some other molecules,[81] and Hay has applied it to NO_2[82] (see also Ref. 253).

3. Another possible choice of spin coupling scheme is a generalization of that used in the closed shell Hartree–Fock method. In this scheme we first couple the spins of half of the electrons to give a spin state with the maximum possible total spin angular momentum, and then couple the spins of the other electrons one by one to this spin function, reducing the total spin quantum number by half at each step. In the case where each spacial orbital appears twice in the spacial part of the wavefunction (6) and when the different spacial orbitals are restricted to be mutually orthogonal, this wavefunction reduces to the Hartree–Fock function. As neither of these restrictions is imposed, the method is, in effect, a generalization of the Hartree–Fock method. It is called the spin extended Hartree–Fock method by Kaldor[83] and the GF method by Goddard.[84] The spin coupling scheme used in this method is not in keeping with our intuitive idea of electron-pair bonds, as are those of approximations 1 and 2 above. This lack of intuitive basis for the method is reflected in poorer energy values.

† It is called the Spin Optimized Self-Consistent field method by Kaldor and SOGI by Goddard.

The wavefunctions corresponding to all the methods described in this Section (II.C) are more flexible than the Hartree–Fock wavefunction and without exception lead to lower energies than the equivalent Hartree–Fock calculation.

D. Correlated Wavefunctions

Correlated wavefunctions are ones which go beyond the independent particle model. In these we can no longer think of the electrons as being assigned to orbitals which describe the motion of one electron in the averaged field of the others. For most problems involving the energetics of the break-ing or rearrangement of chemical bonds, it is essential to use correlated wavefunctions if quantitative results of chemical accuracy are to be attained. Many authors have come to this conclusion.[13] As an example, consider the spin optimized self consistent field (SOSCF) calculations of Ladner and Goddard[75] on the barrier to reaction in the H + H$_2$ system. These calcula-tions correspond to the best possible independent particle results (within the constraints of their chosen basis set) and gave a barrier height o ~ 17 kcals/mole. This differs significantly from 9·8 kcals/mole which is the result obtained with the best available correlated wavefunction.[85]

Correlated wavefunctions can be split into two main categories; those based on molecular orbitals and those based on valence-bond (VB) theory. In general one can regard the molecular orbital methods as improvements of the Hartree–Fock method. The valence-bond method is based on regard-ing a molecule as being built up from its constituent atoms. The orbitals or the different atoms are not orthogonal to each other and VB wavefunctions therefore, generally involve the use of non-orthogonal orbitals. Besides these two main types of methods there are methods based on pair functions which describe the correlated motion of pairs of electrons at a time.

1. Configuration Interaction

The Hartree–Fock method approximates the true wavefunction by a single Slater determinant (10). A better approximation to the true wave function is to take a linear combination of many Slater determinants so that the total wavefunction is expressed as:

$$\Psi(\vec{r}_1, \ldots \vec{r}_N; \vec{R}) = \sum_{K=1}^{n} C_K \Psi_K(\vec{r}_1, \ldots \vec{r}_N; \vec{R}) \tag{15}$$

where the Ψ_K's are Slater determinants

$$\Psi_K(\vec{r}_1 \ldots \vec{r}_N; \vec{R}) = (N!)^{-1/2} \begin{vmatrix} \psi_{k_1}(\vec{r}_1) & \psi_{k_1}(\vec{r}_2) & \ldots & \psi_{k_1}(\vec{r}_N) \\ \psi_{k_2}(\vec{r}_1) & & & \cdot \\ \cdot & & & \vdots \\ \cdot & & & \\ \psi_{k_N}(\vec{r}_1) & \ldots & \ldots & \psi_{k_N}(\vec{r}_N) \end{vmatrix} \tag{16}$$

The molecular spin-orbitals $\psi_{k_i}(\vec{r})$ are chosen from a set of orthogonal molecular spin-orbitals $\psi_j(\vec{r})$. Every Slater determinant is built up using different members of the complete set of molecular orbitals. The Slater determinants Ψ_K are referred to as configurations, and a wavefunction such as that in (15) is called a configuration interaction wavefunction.

Having expanded the wavefunction in the form of (15), the coefficients C_K are determined according to the variational principle. This gives rise to a set of so-called secular equations, and a secular determinant whose zeros determine the approximate energy levels E_i.[29,86] The equation which determines the energy levels is:

$$
\begin{vmatrix}
(H_{11} - E) & H_{12} & H_{13} & \cdots & H_{1n} \\
H_{21} & (H_{22} - E) & H_{23} & \cdots & H_{2n} \\
\vdots & & & & \vdots \\
H_{n1} & & \cdot & \cdot & (H_{nn} - E)
\end{vmatrix} = 0 \qquad (17)
$$

or

$$
\det(\mathbf{H} - E\mathbf{1}) = 0
$$

where

$$
H_{ij} = \int \Psi_i(\vec{r}; \vec{R}) \mathbf{H}(\vec{r}; \vec{R}) \Psi_j(\vec{r}; \vec{R}) \, d\vec{r}
$$

and the molecular spin orbitals are normalized and orthogonal. This equation has n solutions E_i. It can be shown that the lowest solution is an upper bound to the exact ground state solution of the Schrödinger equation and that the higher solutions are also upper bounds to the respective exact solutions.[87] Thus, the configuration interaction (CI) method provides approximations not only to the ground state energy but also to the energies of the excited states. It is, in principle, possible to express the exact wavefunction in the form of (15). As more terms are added to the expansion on the right-hand side the calculated energy levels converge on the exact solutions of the Schrödinger equation.

Clearly, it is desirable to make the length of the expansion in (15) as short as possible. The number of Slater determinants needed to obtain a given accuracy depends on the molecular orbital basis set used to construct the Slater determinants. There have been several proposals in the recent literature for suitable molecular orbital basis sets.[81,88-90] It is generally agreed that the virtual Hartree–Fock orbitals (i.e. those Hartree–Fock orbitals which result from solving the standard Hartree–Fock orbital equations, but which are not used in the Hartree–Fock determinant) are a poor basis for expanding the wavefunction. A very important concept in connection with both the analysis of complex wavefunctions and with the implementation of con figuration interaction calculations was introduced by Löwdin.[91] This is the concept of natural orbitals. The natural orbitals can be found if the exact wavefunction for the system is known. Löwdin showed that the configura tion interaction expansion (15) converges fastest if these orbitals are used to construct the Slater determinants. Bender and Davidson have used this property of natural orbitals to formulate the iterative natural orbital method (INO).[92,93] The method involves performing a configuration interaction (CI) calculation using a basis set of Slater determinants of a given size. The natural orbitals corresponding to the CI wavefunction are then determined, and the calculation is repeated using the newly determined natural orbitals in place of the original molecular orbitals. This process is repeated iteratively until convergence is attained.

Several systems have recently been studied using both the CI method (He + H_2,[57,94] H_3^+,[62,95] H + H_2,[96,97] H_2 + H_2,[98,99,258] O^+ + N_2,[100] O_3 and O_3^-,[101] C_3,[102] CO_2,[103] KrF_2,[33] NH_3,[104] HeH^+ + H_2[105] and CH[106]) and the INO method (H + H_2,[85] F + H_2,[107,108] H + F_2,[109] HO_2,[110] CH_2,[111] O_2[112] and NH[113]). These calculations, sometimes involv ing in the order of 5000 configurations for the CI calculations and 500 for the INO calculations, indicate the capabilities of currently available computer programs. Without exception, they correctly predict all the important qualitative features of the potential energy surfaces examined. Several of the calculations in which both the SCF and correlated wavefunctions were calculated have emphasized the need to include correlation explicitly in a cal culation if even qualitatively correct results are to be realized.[33,96,100,107,108] The error estimates associated with the above calculations vary considerably. In Table II rough indications of the maximum deviations of the calculated surfaces from the exact ones are given for a few selected systems. These calculations demonstrate that for triatomic systems composed of first row atoms, CI or INO methods can now yield reliable (within, say 5 kcal/mol of exact) potential surfaces. The quality of the final results clearly depends on the atomic orbital basis set used. In general, to obtain results of the accuracy indicated above, it is necessary to use a carefully selected basis set of a quality referred to as double-zeta plus polarization.[86]

The CI and INO calculations discussed above yield very unwieldy wave functions which do not provide any simple qualitative picture of the bond ing. An alternative approach is to pick a reasonably small number of con figurations and to try to find, not only the best coefficients C_K in the CI wavefunction of (15), but also to vary the molecular orbitals so as to obtain the optimum orbitals for the chosen form of the wavefunction. This method is known as the multi-configuration SCF method (MC–SCF). The con figurations chosen will normally include the Hartree–Fock configuration plus those additional configurations which add the most important types of

TABLE II

Rough estimates of the maximum deviations of some calculated (*ab initio*) potential energy surfaces from the exact surfaces[a]

System	Method of calculation	Estimate of maximum error[b] (kcal/mole)	Reference
H_3^+	CI (120 configs; $5s2p$ Gaussians on each atom)	2	62 (95)
H_3	CI (680 configs; double zeta + polarization basis, exponents optimized)	3	97
H_3 (linear)	INO (672 configs; very extended orbital basis)	0·8	85
$He + H_2$ H_2 distance close to equilibrium)	CI (80 configs; double zeta + polarization functions on He)	4	57
$Li^+ + H_2$ H_2 distance close to equilibrium)	SCF + estimate of correlation energy (extended sp basis of Gaussian lobe functions)	2	60
$F + H_2$	INO (>214 configs; double zeta + polarization basis)	4	108
$H + F_2$	INO (555 configs; double zeta basis)	15	109

[a] It is the shape of the surfaces which is being compared, not their absolute energies. The zero of energy is taken to be the experimental energies of the separated atoms for the exact surface and the calculated energies for the calculated surface.
[b] These are in most cases just crude estimates.

lexibility to the Hartree–Fock wavefunction. If this small configuration nteraction wavefunction is taken and the orbitals are varied so as to ninimize the energy, subject to the condition that they remain orthonormal, hen a set of equations which the orbitals must satisfy is obtained.[114] These quations are:

$$F_i(\vec{x}, \vec{R})\phi_i(\vec{x}) = \epsilon_i\phi_i(\vec{x}) + \sum_{j \neq i} \epsilon_{ij}\phi_j(\vec{x}) \tag{18}$$

They differ from the standard Hartree–Fock equations because there is a lifferent Fock operator F_i for each orbital and because of the summation •n the right-hand side. The ϵ_{ij}'s in this summation are non-diagonal Lagrange nultipliers which arise from the orthogonality constraint on the orbitals. Their presence has in the past made the orbital equations difficult to solve ind created convergence problems. It appears now that these problems may •e overcome by using alternative means of imposing the orthogonality •onstraints. Applications of the MC–SCF method have, till now, been mainly to diatomic systems, the exceptions being some calculations on $CH_2 + CH_2$[115] and some recent work on $Li + H_2$.[116] The results obtained using the MC–SCF method with relatively few configurations have been very encouraging.[114,117] A disadvantage of the MC–SCF method is that it does not give approximate energies for the excited states in the same straightforward manner that the CI or INO methods do. An MC–SCF method specifically designed to calculate energies of excited states has recently been proposed,[118] but the method may be open to some criticism in that it does not seem to guarantee that the excited state wavefunction be orthogonal to the ground state one.

The simplest type of MC–SCF wavefunction is one composed of just two configurations. It is often possible to choose the two configurations in such a way as to ensure that the dissociation products are correctly described. Das and Wahl first introduced this method[119] arguing that it corrected the most important deficiency of the Hartree–Fock method. They called it the optimized double configuration (ODC) method. It has recently been used to study the system $F + Li_2$[120] and was also used as a starting point in the CI calculations on KrF_2.[33]

2. Perturbation Methods

We will consider the application of the perturbation approach only to a closed-shell system. We take the Hartree–Fock wavefunction as our zeroth order wavefunction, see (10). The individual Hartree–Fock spin-orbitals are eigenfunctions of the Fock operator:

$$\mathbf{F}(\vec{r}_i; \vec{R})\psi_j(\vec{r}_i) = \epsilon_j \psi_j(\vec{r}_i) \tag{19}$$

The Hartree–Fock wavefunction is itself an eigenfunction of a zeroth order Hamiltonian matrix $\mathbf{H}_0$:

$$\mathbf{H}_0\Psi_0 = E_0\Psi_0 \tag{20}$$

where

$$\mathbf{H}_0 = \sum_{i=1}^{N} \mathbf{F}(\vec{r}_i; \vec{R})$$

and

$$E_0 = \sum_{i=1}^{N} \epsilon_i$$

The Fock operator $\mathbf{F}(\vec{r}_i; \vec{R})$ contains two parts. The first, $\mathbf{h}(\vec{r}_i)$, is just the same as the sum of the one-electron operators in the full Hamiltonian, while the second, which we denote as $\mathbf{G}(\vec{r}_i) = \sum_{j=1}^{N} \mathbf{g}_j(\vec{r}_i)$, is the average effect of the electron–electron repulsion between the ith electron and all the others.†

†
$$\mathbf{g}_j(\vec{r}_i) = \int \phi_j(\vec{r}_\mu)\frac{(1 - \mathbf{P}_{\mu i})}{r_{\mu i}}\phi_j(\vec{r}_\mu)\, d\vec{r}_\mu$$

where $\mathbf{P}_{\mu i}$ is a permutation operator.

i.e.

$$F(\vec{r}_i; \vec{R}) = h(\vec{r}_i) + G(\vec{r}_i) \tag{21}$$

The total Hamiltonian is:

$$H = \sum_i h(\vec{r}_i) + \sum_{i>j} \frac{1}{r_{ij}} \tag{22}$$

We can rearrange the total Hamiltonian in such a way as to split it into a zeroth order part (20) plus a perturbation.

$$H = \sum_i \{h(\vec{r}_i) + G(\vec{r}_i)\} + \left\{ \sum_{i>j} \frac{1}{r_{ij}} - \sum_i G(\vec{r}_i) \right\} \tag{23}$$

$$= H_0 + V$$

where

$$H_0 = \sum_i \{h(\vec{r}_i) + G(\vec{r}_i)\} = \sum_i F(\vec{r}_i; \vec{R})$$

and

$$V = \sum_{i>j} \frac{1}{r_{ij}} - \sum_i G(\vec{r}_i)$$

The standard Hartree–Fock energy corresponds to the zeroth order energy (20) plus the first order correction

$$E_{HF} = \langle \Psi_0 | H | \Psi_0 \rangle$$

$$= \langle \Psi_0 | H_0 + V | \Psi_0 \rangle$$

$$= E_0 + \langle \Psi_0 | V | \Psi_0 \rangle \tag{24}$$

Higher order corrections to the energy can be obtained by perturbation theory. Two distinct paths are available for obtaining them, namely:

(a) To identify the higher order terms in the wave function, and to minimize the energy arising from them using a variation-perturbation approach. This method has been advocated by Sinanoğlu[121,122] and Nesbet[123] and has so far been applied mainly to atoms. The knowledge gained from these atomic calculations has, however, greatly increased our understanding of electron correlation and has permitted the development of methods for estimating the correlation energy. One such method is based on estimating the energy arising from the correlation of two electrons at a time, and then adding together all the pair correlation energies. This method ignores higher order effects but has proved highly successful in the cases of $Li^+ + H_2$[60] and $He + H_2$,[124] and has been applied to some boron and berylium hydrides.[125–127] A more extended form of the theory, which takes the coupling of different excitations into account, has been applied to BH_3.[254]

(b) Alternatively straightforward Rayleigh–Schrödinger perturbation theory can be used to evaluate the higher order perturbation corrections. In order to do this it has been found convenient to employ the formalism developed by Brueckner and Goldstone in connection with the study of nuclear matter.

An introductory text has been written on this many-body perturbation theory approach.[128] Its application to atoms and molecules has been reviewed[129] and the relationship of the two different types of perturbation expansions to each other has been analysed.[130] Methods of estimating the correlation energies of molecules based on this approach are now starting to be used.[131,132]

It is also possible to use perturbation theory in conjunction with CI and MC–SCF calculations. The more important configurations are handled according to the CI or MC–SCF method and the contribution of the less important configurations is estimated using perturbation theory.[133–135]

3. Valence-Bond Methods

There are two slightly different valence-bond approaches. The first approach, which is more the conventional one and is sometimes referred to as the Heitler–London–Pauling–Slater method, involves preparing the atoms in valence states, which are not eigenfunctions of the atoms, before bringing them together to form electron pair bonds. This type of approach has been discussed extensively by Van Vleck.[136,137] It forms the basis for the various types of hybridization schemes chemists constantly invoke, and for the discussion of resonances between different valence-bond structures such as the Kekulé structures for benzene. The alternative valence-bond approach builds up the molecular wavefunction using atomic eigenfunctions. This approach has been called the method of spin-valence[138] and is equivalent to a method introduced by Moffitt.[139,140]

We can illustrate Moffitt's approach by considering a diatomic molecule A—B. The first step is to form functions which have the correct spacial and spin symmetry to be approximate eigenfunctions for the states of the isolated atoms A and B. These approximate atomic eigenfunctions can be denoted as $\Phi_i^A(\vec{r}_1, \ldots \vec{r}_{N_A})$. They are antisymmetric with respect to the exchange of electrons, and are eigenfunctions of the various angular momentum operators (i.e. $\mathbf{L}^2$, $\mathbf{S}^2$ and $\mathbf{S}_z$). In order to form functions which may be used as a basis for expanding the molecular wavefunction, we multiply together approximate atomic eigenfunctions for the two atoms A and B and antisymmetrize the product with respect to interchange of electrons originally assigned to the different atoms. These basis functions are called Composite Functions (CF's) and for a diatomic AB they take the form

$$\Phi_{ij}^{AB} = \mathscr{A}'[\Phi_i^A \Phi_j^B] \tag{25}$$

where $\mathscr{A}'$ is a partial antisymmetrizer and completes the antisymmetrization process. The CF's are not eigenfunctions of S^2 for the whole molecule. Eigenfunctions of S^2 can be formed by taking simple linear combinations of CF's, and it is these spin eigenfunctions (Φ_k^{SM}) which correspond to the basis set used in the method of spin-valence. The total molecular wavefunction can be expressed as a linear combination of spin eigenfunctions

$$\Psi^{SM} = \sum_k C_k \Phi_k^{SM} \tag{26}$$

The coefficients C_k are determined using the variational principle. This gives rise to a set of secular equations and a secular determinant very similar to those encountered in the CI method (17). The solutions to this secular problem provide approximate eigenvalues and eigenfunctions to both the ground and excited states of the system.

Composite functions, or linear combinations of them taken to form spin eigenfunctions, have the highly desirable property that at large internuclear separations they provide good descriptions of specific states of the constituent atoms of the molecule. Thus, wavefunctions which are expressed as linear combinations of such basis functions are guaranteed to dissociate properly, and the composite function basis set is, therefore, ideally suited to discussing scattering problems and situations involving the breaking of chemical bonds.

The valence-bond approach plays a very important role in the qualitative discussion of chemical bonding. It provides the basis for the two most important semi-empirical methods of calculating potential energy surfaces (LEPS and DIM methods, see below), and is also the starting point for the semi-theoretical atoms-in-molecules method. This latter method attempts to use experimental atomic energies to correct for the known atomic errors in a molecular calculation. Despite its success as a qualitative theory the valence-bond method has been used only rarely in quantitative applications. The reason for this lies in the so-called non-orthogonality problem, which refers to the difficulty of calculating the Hamiltonian matrix elements between valence-bond structures.

The valence-bond structures or the CF's can be represented as linear combinations of non-orthogonal Slater determinants, i.e. Slater determinants built up from non-orthogonal orbitals. Thus, the problem of calculating Hamiltonian matrix elements between them reduces to that of calculating matrix elements between non-orthogonal Slater determinants. This problem has been discussed by Löwdin,[91] who presented formulae for the Hamiltonian matrix elements in terms of long summations of products of minors of the overlap matrix with one- and two-electron integrals over the non-orthogonal orbitals. As there are so many of them it is essential that these minors be calculated efficiently, and two different schemes for their efficient calculation have been proposed.[141,142]

Alternatively, the non-orthogonality problem may be completely side-stepped by transforming the non-orthogonal orbitals to an orthogonal orbital basis set as an intermediate step. This implies that the Slater determinants built up from non-orthogonal orbitals are expanded as linear combinations of Slater determinants built up from orthogonal orbitals, and after the Hamiltonian matrix elements have been evaluated in the basis of the orthogonal Slater determinants they are then transformed back into the basis of the non-orthogonal Slater determinants. The formalism needed for this approach was worked out, in connection with a different problem, by Moffitt[143] and its use in performing valence-bond calculations was pioneered by Hurley,[144] and has been used by the present author.[145–147] The disadvantage of this latter approach is that the two-electron integrals must be transformed from the original basis of non-orthogonal orbitals in which they are calculated, to the orthogonalized orbital basis. This step must also, however, be performed in all CI calculations, as in that case the two-electron integrals are needed in the basis set of the molecular orbitals.

It is now possible to perform valence-bond calculations using either of the above techniques to handle the non-orthogonality problem. Indeed, the number of valence-bond calculations being performed is increasing rapidly, and it is clear that multi-structure valence-bond calculations no longer present any insuperable problems. Recent work in the field has been reviewed by Gerratt[148] who discusses the different valence-bond approaches and lists most of the recent calculations. Much less effort has so far been put into the computation of *ab initio* valence-bond wavefunctions than into molecular orbital ones. The presently available valence-bond calculations are, therefore, somewhat less sophisticated than their molecular-orbital counterparts. One of the big advantages of the multi-structure valence-bond method is that it is possible to utilize chemical intuition to pick out many of the important structures and to reject the unimportant ones.[149] Some systems of interest which have been recently studied using valence-bond techniques are BeH_2,[150–153] BeH_2^+,[154] He_2H^+,[155] CH_2,[156] HeF_2,[157] NeF_2,[158] H_4^+[159] and benzene.[160]

4. Pair-Functions or Geminals

The electron-pair bond plays a central role in the qualitative understanding of molecular structure. The pair-function or geminal approach attempts to put this electron-pair concept into a more quantitative form. In this approach the total wavefunction is assumed to be of the form of an antisymmetrized product of pair-functions:

$$\Psi(\vec{r};\vec{R}) = \mathscr{A}'[\Lambda_1(\vec{r}_1,\vec{r}_2)\Lambda_2(\vec{r}_3,\vec{r}_4)\ldots\Lambda_{N/2}(\vec{r}_{N-1},\vec{r}_N)] \tag{27}$$

where $\Lambda_n(\vec{r}_i,\vec{r}_{i+1})$ are the pair functions or geminals, and are antisymmetric functions of the space and spin coordinates of electrons i and $i+1$. $\mathscr{A}'$ is a partial antisymmetrizer which makes the product of the pair-functions Λ_n antisymmetric with respect to exchange of electrons originally assigned to different pair-functions. The pair-functions are normally restricted to have singlet or triplet spin symmetry.

In its simplest form, where either one or two atomic or hybrid orbitals are used to form the pair-functions, the wavefunction corresponds to a single valence-bond structure. The singlet pair-functions can then be identified as closed shells, electron-pair bonds or lone pairs. The original approximate valence-bond theories were based on functions of this type.[136] The first step in generalizing this wavefunction is clearly to allow each pair-function to be built up from two orbitals which are no longer restricted to be atomic or hybrid orbitals. When the form of the orbitals is generalized in this manner, but the orbitals used to build up different pair-functions are restricted to be orthogonal to each other, we obtain what Goddard calls the Generalized Valence-Bond (GVB) method.[79,81] The GVB method falls within the independent particle methods and has been discussed above.

If the pair-functions are further generalized so that they are built up from more than two orbitals, i.e.

$$\Lambda_n(\vec{r}_1, \vec{r}_2) = \sum_{kl} a_{kl}^n \psi_k(\vec{r}_1) \psi_l(\vec{r}_2) \tag{28}$$

but the pair-functions are still restricted to be strongly orthogonal,

$$\int \Lambda_n(\vec{r}_1, \vec{r}_2) \Lambda_m(\vec{r}_1, \vec{r}_3) \, d\vec{r}_1 = 0 \quad \text{if } n \neq m \tag{29}$$

Then the so-called separated pair approximation is obtained.[79] This approximation has been applied to several molecules.[161-164] Further generalization of the method is possible by removing the strong orthogonality condition on the pair-functions. Such a completely generalized pair-function method has recently been applied to some very small diatomic systems.[165]

To some extent the pair function methods seem to fall between two stools. They attempt to utilize chemical intuition in solving the complex quantum mechanical problem, and they do indeed possess an appealing simplicity of interpretation in terms of electron-pair bonds, especially at the GVB or separated pair level. The price that must be paid for a wavefunction of such an easily interpretable form is that such methods are unlikely, in general, to yield chemically accurate potential energy surfaces.

III. SEMI-THEORETICAL METHODS

The methods designated here as semi-theoretical all have in common the fact that they are based on an *ab initio* approach and deviate from it only in that they introduce a limited number of corrections or approximations. The atoms-in-molecules method involves building up the molecular wavefunction from approximate atomic wavefunctions and then applying corrections to the calculated *ab initio* Hamiltonian matrix elements so as to correct for known atomic errors. The various properties of the atoms-in-molecules method are discussed below. Its two most important properties are that as the basis set is extended the correction terms vanish and the method reduces to the *ab initio* approach (in the limit of a highly extended basis set) and that the process of breaking chemical bonds is always correctly described. The pseudo-potential methods attempt to allow for the effect of inner 'core' electrons without including them in the full *ab initio* manner. Methods of this type will clearly be of the greatest importance in calculations on systems containing heavier atoms. The closed-shell interaction model is a recently introduced method for the calculation of forces between closed-shell atoms and molecules. It is somewhat out of place in this review in that it cannot describe the breaking of chemical bonds. On the other hand, it appears to be computationally very simple and to yield reliable results. It should prove useful for the calculation of potential surfaces to describe vibrationally and rotationally inelastic molecular collisions.

A. Atoms-in-Molecules Methods

Moffit noted in 1951 that the cause of many of the errors arising in molecular calculations could be traced back to errors which were already present in the atomic calculations.[139] By choosing to use composite functions [i.e. antisymmetrized products of approximate atomic eigenfunctions, see (25) above] as basis functions for the molecular calculation, he was able to identify the atomic contributions to the molecular energy. He then proposed two ways in which experimental atomic energy levels might be used to correct for the known errors arising from the approximate atomic energies. One of these methods involved calculating an *ab initio* Hamiltonian matrix and then applying corrections to its individual matrix elements. Thus in the case of a diatomic the corrected Hamiltonian matrix elements take the form:

$$H_{ij,kl} = \tilde{H}_{ij,kl} + \tfrac{1}{2}\tilde{S}_{ij,kl}[\Delta E_i^A + \Delta E_j^B + \Delta E_k^A + \Delta E_l^B] \qquad (30)$$

where $\tilde{H}_{ij,kl}$ is an *ab initio* Hamiltonian matrix element between a composite function built up from approximate atomic eigenfunctions for the ith state of atom A and the jth state of atom B and one built up from atomic functions for the kth state of atom A and the lth state of atom B; $\tilde{S}_{ij,kl}$ is the corresponding overlap matrix element and

$$\Delta E_i^A = E_i^A - \tilde{E}_i^A \qquad (31)$$

where E_i^A is the experimental energy of the ith state of atom A and $\tilde{E}_i^A$ is the energy of that state calculated using the same approximate atomic eigenfunction as was used to construct the composite function. The energy levels of the system are then computed by finding the roots of the secular determinant in the standard manner;

$$\det (\mathbf{H} - E\tilde{\mathbf{S}}) = 0 \qquad (32)$$

It is immediately clear that as we improve the description of the atoms and use progressively more sophisticated approximate atomic wavefunctions to construct the composite functions, the correction terms ΔE_i^A will decrease in magnitude and in the limit, when exact atomic eigenfunctions are used, they will vanish. It can however be shown[146] that the *ab initio* method and Moffitt's AIM method become identical long before this limit is reached. They become identical when all the corrections for the states of atom A become equal to each other and similarly for those of atom B (i.e. $\Delta E_i^A = \Delta E_k^A$ and $\Delta E_j^B = \Delta E_l^B$). This situation arises because in the standard *ab initio* procedure, energies such as dissociation energies are calculated relative to the calculated atomic energies and not relative to the experimental ones, as in the AIM method. Another important property of Moffit's AIM method is that as the atoms are pulled apart and the composite functions become mutually orthogonal, the calculated energies tend to the experimental energies of the separated atoms, both for the ground and excited states. This correct asymptotic behaviour is of great importance for scattering calculations, especially when curve-crossings are involved.

When Moffitt's AIM method was tried out using H_2 as a test case, it was found to be unsatisfactory and two modifications of it were proposed. Hurley's ICC method[166,167] uses calculated atomic energies ($\tilde{E}_i^A$) in (31) which are obtained using the best possible atomic wavefunctions of the same functional form as those used in constructing the composite functions. This normally means that different orbital exponents are used in calculating $\tilde{E}_i^A$ and $\tilde{H}_{ij,kl}$. While this produces the desired effect of reducing the magnitude of the correction terms in (30), it destroys the asymptotic behaviour of the potential energy curves for all but a few states, and most of the curves no longer tend to the experimental atomic energies at large internuclear separations. The ICC method is also found to exhibit some of the unsatisfactory properties of Moffitt's original method at small separations.[146] Another modification of the AIM method was proposed by Arai.[168] In his deformed atoms-in-molecules method Arai attempts to allow for the deformation of the atomic eigenfunction in the molecular environment. The resulting formulae are, however, very complex and no attempt has been made to apply them to anything larger than Li_2.

Moffitt's AIM method has recently been reexamined[145–147] and an alternative modification proposed. The new modification, which is called the Orthogonalized Moffitt (OM) method, preserves all the desirable attributes of Moffitt's original method while at the same time preventing the unreasonable behaviour of Moffitt's method at small internuclear separations.

The OM method consists of orthogonalizing the composite functions to each other using a Schmidt orthogonalization procedure, and then applying the corrections, as in (30), to the Hamiltonian matrix elements calculated in the orthogonalized basis. The OM method has been applied to some diatomics[145-147] and to the triatomic systems LiF_2,[169] Li_2F[146,147] and $LiHF$.[146,147] In all of these cases $1s$, $2s$ and $2p$ orbitals were used on all of the atoms. From the diatomic calculations it is clear that the application of the correction improves the calculated dissociation energies, that is the OM method consistently leads to at least as good or better results than the *ab initio* method using the same basis set. The question remains whether the atomic orbital basis sets were sufficiently large to provide an adequate description of the potential surfaces examined. Calculations on some of the above systems using extended basis sets are now in progress.[170]

B. Pseudo-Potentials

The pseudo-potential idea was introduced by Phillips and Kleinman[171,172] in 1959. They tried to formulate a method for calculating the valence orbitals in molecules and solids without taking explicit account of the core electrons. The basic assumption is that the core orbitals are unaffected by the chemical environment and are the same in the molecule or solid as in the isolated atom. In Hartree–Fock theory the valence and core orbitals are required to be orthogonal. If the presence of the core orbitals is simply ignored, then there is the danger that when we attempt to calculate a valence orbital by solving the Hartree–Fock equations, the solution of the equations will 'collapse' to give a core orbital. Philips and Kleinman showed that the orthogonality constraint between core and valence orbitals could be replaced by an effective, non-local potential which was repulsive in nature and served to exclude the valence electrons from the inner core region.

The Hartree–Fock equation for a valence orbital can be written as

$$\mathbf{F}\phi_v = \varepsilon_v \phi_v \tag{33}$$

where $\mathbf{F}$ is the Fock operator and ϕ_v is orthogonal to the lower lying core orbitals. Alternatively ϕ_v can be expressed as a pseudo-orbital χ, which has a smooth radial dependence and is not orthogonal to the core orbitals, plus some terms which orthogonalize χ to the core orbitals:

$$\phi_v = \chi - \sum_c \left[\int \phi_c \chi \, d\tau \right] \phi_c \tag{34}$$

where ϕ_c are the core orbitals. Equation (34) can be reexpressed using projection operators $\mathbf{P}_c$ such that,

$$\mathbf{P}_c \chi = \left[\int \phi_c \chi \, d\tau \right] \phi_c \tag{35}$$

Using (35) we can rewrite (34) as

$$\phi_v = \left(1 - \sum_c \mathbf{P}_c\right)\chi \tag{36}$$

By substituting for ϕ_y in (33), using either of the two equivalent expressions 34) or (36), we obtain an equation for the pseudo-orbital χ.

$$\mathbf{F}\chi - \sum_c \left[\int \phi_c\chi \, d\tau\right]\varepsilon_c\phi_c = \varepsilon_v\left\{\chi - \sum_c \left[\int \phi_c\chi \, d\tau\right]\phi_c\right\} \tag{37}$$

where we have used the fact that $\mathbf{F}\phi_c = \varepsilon_c\phi_c$. The above can be written in the orm:

$$\mathbf{F}\chi - \sum_c \varepsilon_c\mathbf{P}_c\chi = \varepsilon_v\left\{\chi - \sum_c \mathbf{P}_c\chi\right\}$$

which can be rearranged to give:

$$\left\{\mathbf{F} + \sum_c \mathbf{P}_c(\varepsilon_v - \varepsilon_c)\right\}\chi = \varepsilon_v\chi \tag{38}$$

Equation (38) is now an equation for the pseudo-orbital χ which is no longer constrained to be orthogonal to the core orbitals. The term $\sum_c \mathbf{P}_c(\varepsilon_v - \varepsilon_c)$ is referred to as the pseudo-potential and arises directly from the orthogonality constraint between the core and valence Hartree–Fock orbitals. In deriving (38) all we have done is to recast the Hartree–Fock equations in a slightly different form. To solve it rigorously we would still need to calculate all the core orbitals and all the core–core and core–valence integrals. The importance of (38) is that it provides a basis for further approximations.

The most obvious approximation is to replace the terms in the effective Fock operator $\{\mathbf{F} + \sum_c \mathbf{P}_c(\varepsilon_v - \varepsilon_c)\}$ which depend on the core orbitals, by an analytic, local repulsive model potential. One commonly used model potential of this type is the Hellmann potential

$$V_H = -\frac{z}{r} + \frac{A\,e^{-2Kr}}{r} \tag{39}$$

This potential is used to replace the effect of all the terms involving core orbitals. The normal way of determining the parameters A and K in the Hellmann potential has been to choose them such that the ground state and one excited state of the atoms involved have their correct experimental energies.[173,174] This is not in keeping with an *ab initio* approach, but it could be considered that this procedure incorporates an empirical correction. The Hellmann potential, or similar local potentials, have been used recently to replace the effect of the core orbitals in several studies of molecules with one or two valence electrons.[175–180] In these, as in the other calculations mentioned below, model potentials, which approximate the effect of the core electrons, are first calculated for each of the atoms in the molecule and the model potential for the molecule is then taken to be the sum of the atomic potentials.

The Hellmann potential is spherically symmetric. On the other hand the pseudo-potential and the core-valence exchange potential in the effective Fock operator of (38) involve angular functions which are not spherically symmetric. A better approximation to the effective Fock operator would be to include a different local potential for the different angular components of the valence orbitals. This means that an angular projection operator is included in the model potential which replaces the effect of the core-electrons, and that this potential is now non-local in the angular variables. Model potentials of this type have recently been employed in several calculations on reasonably large systems (H_2S_2 and Cl_2S_2,[181] HCHO and NaCl,[182] LiH_2, Li_3, BH_2 etc.[183] TiF^{3+} [184]). In some of these calculations[181,182,184] the model potentials used specifically included radial projection operators. This made them non-local also in the radial variable and provided a much better method for allowing for the effect of the core orbitals. While the results of these calculations are encouraging, especially from the point of view that the valence electron only and the all electron calculations yielded very similar results,[181,183] pseudo-potential techniques are still in the development stage. One of the developments to be expected in the future is the use of the pseudo-potential method within a configuration interaction or valence-bond[184] framework. This will permit the correct description of correlation effects among the valence electrons and of the formation of chemical bonds.

C. Closed-Shell Interaction Model

A model (designated here as the Closed-Shell Interaction model) for describing the interaction of closed-shell atoms and molecules, which do not form strong chemical bonds, has recently been proposed by Gordon and Kim.[185] The model assumes that the total electron density of the system may be well approximated as the sums of the electron densities of its two separate parts. In the case of an atom–atom interaction, for instance, we would take the total electron density to be just the sums of the two atomic densities. The atomic densities themselves are normally taken from SCF calculations. All the necessary contributions to the interaction energy are then calculated directly in terms of the electron density. The Coulomb energy, including the electron–electron repulsion is calculated using electrostatics. The other energy terms are calculated by assuming that the charge cloud may be approximated locally as a uniform electron gas. The analytic formulae which has been developed to represent the kinetic and exchange energy of a uniform electron gas as a function of its density, are used to estimate the local contributions to these energies from infinitesimal volume elements and these contributions are then integrated. An approximate expression is also available for the correlation energy of a uniform electron gas in terms of its density, and this permits the correlation energy to be estimated in the same manner. It is the inclusion of the correlation energy which in fact permits this approximate method to often yield better interaction energies than the Hartree–Fock method itself. Another novel feature of the method is that the interaction energies are calculated directly rather than as differences between large numbers. For most of the diatomic systems to which the method has so far been applied, it describes the Van der Waals well quantitatively, giving the well depth within 10 % or so. The Hartree–Fock method does not predict any well for such systems. As the method is conceptually very simple and computationally very fast, there would seem to be great potential for its future application.

IV. SEMI-EMPIRICAL METHODS

Semi-empirical methods are characterized by the fact that, while they are formally based on a rigorous quantum mechanical approach, they incorporate approximations which use empirical data to replace most of the difficult computation. Most of the generally applicable semi-empirical theories are based on the valence-bond formalism. One of the strong points of these theories is that they have the correct asymptotic limits built into them; i.e. in a reaction of the type $A + BC \rightarrow AB + C$, when A is far from BC the potential energy surface reduces to the experimental BC potential energy curve, and a similar situation holds for the other two diatomics. Depending on the particular method in question, it may be possible only to calculate the ground state surface or several excited state surfaces as well. The correct asymptotic behaviour is, however, assured for all of the surfaces. Because of this built in behaviour, approximate methods based on valence-bond theory form an ideal basis for entirely empirical schemes, where some of the parameters in the theory are chosen to give the potential energy surface a desired form in the region where the reactants are close to each other.

Besides these theories, which are applicable to general systems, some other theories of less general applicability have been proposed. These are outlined separately below. A group of empirical methods which has been omitted from the present review are the semi-empirical molecular orbital methods known by acronyms such as MINDO, INDO and CNDO. The reader is referred to a book by Murrell and Harget[259] for a description of these methods and to articles by Chutjian and Segal[260] and by MacGregor and Berry[261] for examples of their use.

A. The Diatomics-in-Molecules (DIM) Method

The most aesthetically satisfying and generally applicable semi-empirical valence-bond approximation is the so-called diatomics-in-molecules (DIM) method developed by Ellison.[186] This method is not, as its name might imply, a generalization of Moffitt's atoms-in-molecules method, but is rather an approximate method by which information on the ground and excited states of diatomic molecules may be used to replace all the difficult computation in polyatomic calculations. The basis functions for the method are composite functions (i.e. antisymmetrized products of approximate atomic eigenfunctions). Basis sets are first constructed for all the diatomics involved. These may or may not include ionic functions or functions built up from excited atomic states, depending on the atomic functions used. The diatomic problems are then solved. While this will, in general, involve the solution of a set of secular equations, the necessity of solving these can be avoided by choosing a sufficiently limited basis set of atomic eigenfunctions. The approximate eigenfunctions resulting from the diatomic calculations are used to construct triatomic or polyatomic basis sets by combining them with approximate atomic eigenfunctions for the other atoms. For a triatomic it is thus possible to build up three equivalent basis sets (if ionic states are not included) which may be designated as [AB]C, [BC]A and [AC]B. These basis sets are related to each other, and to the composite function basis, by simple transformation matrices. The total Hamiltonian is now written in terms of atomic and diatomic Hamiltonians. For a triatomic calculation involving no ionic functions the Hamiltonian can be written in the form;

$$\mathbf{H} = \mathbf{H}_{AB} + \mathbf{H}_{AC} + \mathbf{H}_{BC} - \mathbf{H}_A - \mathbf{H}_B - \mathbf{H}_C \qquad (40)$$

The matrix elements of the Hamiltonian between two composite functions can thus be expressed as a sum of the matrix elements of the six diatomic or atomic Hamiltonians on the right hand side of (40). Each of these matrix elements may in turn be expressed as a product of experimental atomic or diatomic energies and an overlap term. In order to accomplish this it is necessary in the case of an atomic Hamiltonian to assume that the approximate atomic eigenfunctions are exact. For a diatomic Hamiltonian (i.e. $\mathbf{H}_{AB}$) it is necessary to relate the composite function on the right hand side of the matrix element, to the basis set built up from the relevant approximate diatomic eigenfunctions (i.e. [AB]C). If we then assume that the approximate diatomic eigenfunctions are in fact exact, we can replace the matrix elements of the diatomic Hamiltonian by a sum of terms which each contain a product of an experimental diatomic energy, times an overlap matrix element. It is this last approximation which is the crucial one, in that it leads to the disappearance of all three-centre integrals.

In order to apply the DIM method in the form discussed above, we have to know experimental potential energy curves for the states of the diatomics involved in the calculation, and we must also calculate overlap matrix elements between the composite functions. If experimental diatomic curves are not available, then the best available calculated curves may be used. Ellison noted in his first applications of the method to H_3 and H_3^+ [187] that the results of the calculation were hardly changed if, in calculating the overlap matrix elements, the orbitals were assumed to be orthogonal to each other. It was later pointed out by Companion[188] that, for a triatomic system which possessed three valence s electrons, the DIM method with neglect of overlap gave rise to the same energy expression as that proposed by London,[189] which may be derived[190] using a different approximate valence-bond approach (see below). Indeed for this case the DIM method is identical to the variation of the London–Eyring–Polyanyi–Sato (LEPS) method proposed by Cashion and Herschbach[191] (see below). The DIM method, normally with neglect of overlap, has been applied to the systems H_2X and X_2H (linear, X = H, F, Cl, Br, I),[192] Li_3^+,[193] $H_n^+ (n = 3 \text{ to } 6)$,[194] Li_2H^+ and LiH_2^+,[195] BeH_2,[196] H_3^+ (ground and excited states),[197–199] H_4,[200] LiH_2 and Li_2H,[188] Li_3,[201] Li_4,[202] the first row triatomic hydrides (i.e. AH_2)[203] and small lithium atom clusters.[255] A more thorough calculation, including both overlap and p orbitals has been performed on ArH_2^+ [204] and the results have been used to rationalize the available experimental data on the rare gas plus H_2^+ and rare gas ions plus H_2 systems.

Recently Steiner, Certain and Kuntz[205] have reconsidered the test case of H_3 and have investigated the effect of including ionic terms in the composite function (or valence-bond structure) basis set. They conclude that the DIM method is liable to give unreasonably shaped potential energy surfaces if overlap is included, and that the basis functions should therefore be assumed to be orthogonal. The difficulties arising from the inclusion of overlap are most clearly seen in the region of small internuclear separations. Steiner, Certain and Kuntz also systematize Ellison's original formulations of the DIM method and cast it in a matrix form which should be useful when applying it to more complex systems.

The great attraction of the DIM method is that it provides a very general method for smoothly connecting up the potential energy curves of reactants and products in chemical reactions. Despite the fact that it agrees remarkably well with good *ab initio* calculations for several systems (H_3,[205] H_3^+,[95] HeH_2^+ [206]), it should not be regarded as possessing any reliable predictive value for otherwise unexplored systems. It may prove extremely useful to use the method in conjunction with limited *ab initio* calculations. The philosophy being to fit the available *ab initio* results by varying some of the excited diatomic potential energy curves, which are of no direct interest, and to use the resulting potential as a means of extrapolating the *ab initio* results so as to yield a complete surface. This use of the DIM method has been proposed by Kuntz and applied by him to the HeH_2^+ system.[206] The only major disagreement to emerge till now between DIM and *ab initio* calculations has arisen for the $H_2 + H_2$ exchange reaction where the DIM method[200] predicts an activation energy of 68 kcal/mole in contrast to the *ab initio*[98] prediction of 115 kcal/mole.

B. Semi-Empirical Valence-Bond Treatments (not based on DIM formalism)

For very simple systems such as H_3 or H_4, when only $1a$ orbitals are used and only covalent valence-bond structures are considered, it is possible to write down explicitly all the terms needed in an *ab initio* valence-bond treatment. For H_3 there are only two independent valence-bond structures in this approximation. Porter and Karplus[207] have considered all the terms arising from these structures. Most of the two-centre integrals were related to theoretical energy expressions for the lowest $^1\Sigma_g^+$ and $^3\Sigma_u^+$ states of H_2, evaluated in the same approximation, and then these energy expressions were replaced by very accurate theoretical potential energy curves for the two states. There only remained overlap integrals, which were evaluated, and some two and three centre integrals, which were approximated in a systematic manner. The resulting analytic potential energy surface for H_3 agrees well with the best available *ab initio* results.[85,97]

The various semi-empirical valence-bond calculations on H_3 have been critically examined by Pedersen and Porter,[208] who also propose their own modification of the Porter–Karplus surface. They show that the calculated barrier height is sensitive to the assumptions made about the variation of the $1s$ orbital exponents in calculating the overlap integrals, and that for H_3 semi-empirical methods of this type cannot be relied upon to predict the surface more accurately than within 5 kcal/mole or so. The Porter–Karplus method has been applied to the $H_2 + H_2$ reaction for which it predicts an activation energy of 63 kcal/mole.[209] This agrees with the DIM calculation[200] but is much lower than the best current *ab initio* value.[98] A semi-theoretical VB calculation for H_4,[210] in which only $1s$ orbitals (with variable exponents) were used and three and four centre integrals were evaluated using the Mulliken approximation, agrees well with the *ab initio* calculation.

As an alternative to attempting to approximate all the terms arising in the simple valence-bond treatment of H_3, we can make the rather drastic, but traditional, approximation of assuming orthogonal orbitals. This leads to an energy expression for the lowest electronic state of H_3 of the form:

$$E = Q_{ab} + Q_{bc} + Q_{ac} - [\tfrac{1}{2}(J_{ab} - J_{bc})^2 + \tfrac{1}{2}(J_{ab} - J_{ac})^2 + \tfrac{1}{2}(J_{bc} - J_{ac})^2]^{1/2}$$
(41)

where Q_{ab} is a diatomic Coulomb integral for hydrogen atoms A and B:

$$Q_{ab} = \int \phi_a(\vec{x}_1)\phi_b(\vec{x}_2)\mathbf{H}_{ab}\phi_a(\vec{x}_1)\phi_b(\vec{x}_2)\, d\vec{x}_1\, d\vec{x}_2$$

and J_{ab} is a diatomic exchange integral:

$$J_{ab} = \int \phi_a(\vec{x}_1)\phi_b(\vec{x}_2)\mathbf{H}_{ab}\phi_b(\vec{x}_1)\phi_a(\vec{x}_2)\, d\vec{x}_1\, d\vec{x}_2 \tag{43}$$

Such an expression was first proposed by London[189] and is easily derived from those of Porter and Karplus by setting all the overlap integrals to zero. In the same approximation, the energies of the singlet and triplet states of H_2 are given by:

$$^1E_{ab} = Q_{ab} + J_{ab}$$
$$^3E_{ab} = Q_{ab} - J_{ab} \tag{44}$$

Thus the Coulomb and exchange integrals in (41) may be estimated in terms of the best available singlet and triplet curves for H_2, i.e.

$$Q_{ab} = (^1E_{ab} + {}^3E_{ab})/2$$
$$J_{ab} = (^1E_{ab} - {}^3E_{ab})/2 \tag{45}$$

This is essentially the method proposed by Cashion and Herschbach[191] and yields results of comparable accuracy to those of Porter and Karplus.[207] Several more empirical methods based on the London equation (41) have been proposed and these are discussed in the next section. It has been shown that the Cashion–Herschbach energy expression may be represented as a sum of pair potentials which are spin operators, the correct energy expression being obtained after the doublet spin eigenfunction has been projected out. This spin dependent type of potential has been used to discuss reactive scattering within the framework of the Faddeev formalism.[211]

If the diatomic overlap is retained in evaluating the diatomic energies then (44) becomes

$$^1E_{ab} = (Q_{ab} + J_{ab})/(1 + S_{ab}^2)$$
$$^3E_{ab} = (Q_{ab} - J_{ab})/(1 - S_{ab}^2) \tag{46}$$

These equations may be used to obtain expressions for the diatomic Coulomb and exchange integrals in terms of the singlet and triplet energies, and the integrals determined in this manner may be substituted in London's equation. Whitehead and Grice[212] have recently applied this method and the Cashion–Herschbach method (no overlap) to alkali atom-dimer exchange reactions involving Li and Na atoms.

The Cashion–Herschbach treatment of H_3 can be extended to more complicated systems. This has been done recently by Blais and Truhlar[213] for the $F + H_2$ system. All the valence electrons were explicitly considered (i.e. $2p^5$ on F and $1s$ on each of the hydrogen atoms). There are four covalent valence-bond structures of the correct symmetry to contribute to the lowest electronic state, and the integrals involved in the calculation can be expressed in terms of the energies of the $^1\Sigma^+$ (ground state), $^3\Sigma^+$, $^1\Pi$ and $^3\Pi$ states of HF, and the $^1\Sigma_g^+$ and $^3\Sigma_u^+$ states of H_2. Reasonably good potential energy curves are available for all these states, but Blais and Truhlar found that they had to slightly modify (by 5 kcal/mole) the calculated excited states of HF in order to obtain a potential energy barrier of the size anticipated on the basis of the experimental results. The method used in this treatment of H_2F should be identical to a DIM treatment without overlap.

C. Other Semi-Empirical Methods

Roach and Child[214] have developed a method, based on the pseudo-potential formalism, for the calculation of potential energy surfaces appropriate to exchange reactions of the type $K + NaCl \rightarrow KCl + Na$. The system is treated as a single valence electron moving in the field of K^+, Na^+ and Cl^- ions. The electrostatic interaction of the ions is calculated using an extension of Rittner's method,[215] which has proved highly successful in describing the potential energy curves of the alkali halides. The method includes terms corresponding to the charge–charge, charge–induced dipole, induced dipole–induced dipole, Van der Waals and core–core repulsion energies and also to the quasi-elastic energy of dipole formation. The Schrödinger equation must be solved for a single valence electron moving in the field of these ions. However, as none of the other electrons in the problem are explicitly taken into account, the molecular orbital occupied by the valence electron must be kept orthogonal to those of the core electrons. This is accomplished by using a crude model pseudo-potential for the alkali metal ions. The form of this pseudo-potential is taken to be:

$$
\left.
\begin{aligned}
V(r) &= 0 && r < \sigma \\
&= -\frac{1}{r} && r > \sigma
\end{aligned}
\right\}
\tag{47}
$$

where σ is the ionic radius of the ion. The molecular orbital for the valence electron is now expanded in terms of a basis set of s and p orbitals on the alkali metal atoms, thus giving rise to a set of secular equations. The kinetic energy and one centre integrals are estimated using the alkali metal ionization potentials and the other integrals are explicitly evaluated. Finally, after the secular equations have been solved, the effect of the polarization of the ionic cores by the valence electron is estimated as a perturbation.

While the method is complicated it seems to lead to very reasonable results. It yields not only the ground state but also five excited state surfaces. Struve[216] has applied the method to several other alkali metal–alkali halide systems (Li_2F, Na_2Cl, $LiNaCl$, $NaKCl$, K_2Cl, Li_2Cl and $LiKCl$). His results for Li_2F are in qualitative agreement with theoretically more rigorous calculations,[120,146,147] and his conclusions as to the mechanism of the vibrational to electronic energy exchange process which occurs in these systems is essentially the same as that arrived at on the basis of a simple atoms-in-molecules calculation.[146,147] The method has also been applied to some alkali diatomic molecules and their positive ions (Na_2, Na_2^+, K_2, K_2^+, NaK and NaK$^+$) and gives remarkably good results.[217]†

Another semi-empirical method which utilizes the pseudo-potential concept and seems to yield reasonable results is that developed by Woolley and Child[218] to treat the reaction $Cl^- + CH_3Br \rightarrow CH_3Cl + Br^-$. The system is treated as a four electron problem, all the other electrons being regarded as core electrons. The wavefunction of the four 'participating' electrons is treated using a three structure valence-bond type function. Careful consideration is given to the many contributions to the energy, and extensive use is made of ionization potentials and spectroscopic data. The method gives remarkably good agreement with the experimentally observed dissociation energies of CH_3Cl and CH_3Br, and the calculated bond lengths are also very good.

Nyeland and Ross[219] have estimated the potential energy surface of K–Cl–Cl. Their method is based on an accurate SCF wavefunction for Cl_2^-. The energy of the system $K^+Cl_2^-$ is calculated using perturbation theory and wavefunctions for $K + Cl_2$ and $KCl + Cl$ are then approximated, again using the Cl_2^- SCF calculation. The diagonal and off diagonal Hamiltonian matrix elements between these 'configurations' are then estimated and the 3×3 secular determinant is solved. A large number of approximations are made in this calculation, the result, however, seems very reasonable and the qualitative features of the potential are similar to those found in the atoms-in-molecules treatment of LiF_2.[169]

The last semi-empirical calculation we consider is an approximate valence-bond treatment of the six lowest potential energy surfaces describing the decomposition of CH_4 into $CH_3 + H$ and $CH_2 + H_2$.[220] The calculation is based on the traditional semi-empirical valence-bond techniques (see for example Refs 136 and 137). In these the matrix elements between valence-bond structures are expanded in terms of molecular Coulomb and exchange integrals. These integrals, in contrast to those of the London method (42) and (43), involve all the orbitals and the total molecular Hamiltonian. The Hamiltonian matrix elements are then estimated using empirically determined values for the integrals. Seven valence-bond structures were used in the calculation and the resulting potential energy surfaces permitted a discussion of the products of the photo-dissociation of methane.

---

† The results for these diatomics are perhaps too good when the restricted basis set used is taken into account. An *ab initio* calculation using the same basis set, but including the core electrons would not give as good agreement with experiment.

V. ENTIRELY EMPIRICAL METHODS

Empirical methods are those which are not based on quantum mechanical valence theory, or if their basic form is derived from valence theory, the parameters in the potential are adjusted so as to give the potential a desired shape. By far the most widely used potentials in trajectory calculations are those based on the London equation and its empirical generalizations. For systems containing alkali halides electrostatic models based on the inter-action of polarizable ions have been used and often yield very good surfaces. The other potentials discussed below are merely convenient analytic func-tions. All of the empirical surfaces are designed to yield reasonable diatomic limits. No attempt has been made to describe all the potential energy surfaces which have been used. It is hoped, however, that most of the surfaces which have a general applicability have been included, along with one or two illustrative references.

A. The London Equation and its Generalizations

The London equation (41) and the method proposed by Cashion and Herschbach for evaluating the Coulomb and exchange integrals which occur in it have been discussed in the preceding section. These methods were developed to describe the interaction of three hydrogen atoms and they can also be used to describe other triatomic systems composed of hydrogen and alkali metal atoms. When the London equation, or its generalizations, are used to describe systems containing atoms with more valence electrons, the resulting potential energy surface must be regarded as merely a convenient empirical functional form, because the equation no longer resembles the correct valence-bond description of the system.[212] Its use, however, normally assures a reasonably correct description of the surface in the asymptotic limits when the reactants or products are well separated from each other

The London equation may be used directly as an empirical equation by evaluating the Coulomb and exchange integrals in the manner used by Cashion and Herschbach, but choosing some of the parameters in the triplet diatomic potentials (i.e. the dissociation energy) so as to give the potential energy surface some desired property (e.g. barrier height).[221] A method based on similar approximations to those used in the Cashion–Herschbach treatment has been applied to the surface for the reaction $H_2 + I_2 \rightarrow 2HI$.[222] The system was treated as composed of four s orbitals coupled to form two covalent bonds. Due to a complete lack of knowledge about the lowest triplet states of HI and I_2 these had to be crudely estimated. The estimates involved a single adjustable parameter, which arises in the formalism as the charge on the halogen atom core, and was determined so as to yield a potential energy surface with an activation energy in agreement with experiment. White[223] has applied the same method to the H_2Br_2

system. He estimates the adjustable parameter so as to give the correct activation energy for the $Br + H_2 \rightarrow H + HBr$ reaction and then uses this parameter to calculate surfaces for the $H + Br_2 \rightarrow HBr + Br$ and $H + HBr \rightarrow H_2 + Br$ reactions. The trajectory calculations using these surfaces are in reasonable agreement with experiment.

In the London–Eyring–Polanyi–Sato (LEPS) method[224] the original London equation is multiplied by an empirical factor which is supposed to account for the effect of overlap.

$$E_{LEPS} = \frac{1}{1+k}\{Q_{ab} + Q_{bc} + Q_{ac} - [\tfrac{1}{2}(J_{ab} - J_{bc})^2 + \tfrac{1}{2}(J_{ab} - J_{ac})^2$$

$$+ \tfrac{1}{2}(J_{bc} - J_{ac})^2]^{1/2}\} \tag{48}$$

The Coulomb and exchange integrals are now estimated from the equations

$$^1E_{ab} = \frac{Q_{ab} + J_{ab}}{1+k}$$

$$^3E_{ab} = \frac{Q_{ab} - J_{ab}}{1-k} \tag{49}$$

The singlet curves are taken to be Morse curves and the triplets to be anti-Morse[224] curves

$$^1E_{ab} = D_e[e^{-2\beta(r-r_e)} - 2\,e^{-\beta(r-r_e)}]$$

$$^3E_{ab} = \tfrac{1}{2}D_e[e^{-2\beta(r-r_e)} + 2\,e^{-\beta(r-r_e)}] \tag{50}$$

The parameter k in (48) can then be altered to give the potential energy surface some desired property (such as a potential energy barrier of a specific height).[225–227] Kuntz[228] has shown that varying the parameter k in the LEPS equation has the same effect as varying the multiplier D_e in the expression for the triplet energy $^3E_{ab}$. There is, therefore, no point in varying both k and the D_e in the triplet energy expression.

More recently an empirical extension of the LEPS surface has been introduced.[229] This extension introduces three overlap type parameters;

$$E_{ELEPS} = \frac{Q_{ab}}{(1+k_{ab})} + \frac{Q_{bc}}{(1+k_{bc})} + \frac{Q_{ac}}{(1+k_{ac})} - \left[\frac{J_{ab}^2}{(1+k_{ab})^2} + \frac{J_{bc}^2}{(1+k_{bc})^2} \right.$$

$$+ \frac{J_{ac}^2}{(1+k_{ac})^2} - \frac{J_{ab}J_{bc}}{(1+k_{ab})(1+k_{bc})} - \frac{J_{ab}J_{ac}}{(1+k_{ab})(1+k_{ac})}$$

$$\left. - \frac{J_{bc}J_{ac}}{(1+k_{bc})(1+k_{ac})} \right]^{1/2} \tag{51}$$

The integrals being evaluated using (49) and (50). The three empirical parameters k_{ab}, k_{bc} and k_{ac} give the potential energy function a greater flexibility while still assuring the correct dissociation limits. This type of potential has recently been extensively used in trajectory calculations.[7,230,231]

The original LEPS surface, the extended LEPS surface and several variations of both of them have recently been used by Jonathan, Okuda and Timlin[232] in a trajectory study of the $H + F_2$ system. One of the proposed variations of the extended LEPS surface involved using the squares of actual overlap integrals to replace the k_{ab} parameters in (49) and (51). The introduction of distance dependent parameters in this manner preserves the correct asymptotic behaviour of the surface, while some of the other variations discussed in the paper do not. A thorough review of LEPS and related potential energy surfaces for triatomic reactions involving hydrogen and halogen atoms has recently been published.[16]

B. Ionic Models

The Rittner model[215] describes the binding of alkali halides in terms of the electrostatic interaction of two mutually polarizable ions. It has proved to be highly successful for the diatomic alkali halides, and has been used as a basis for the semi-empirical pseudo-potential treatment of M_2X molecules, where M is an alkali metal and X a halogen atom (see Section IV.C above). Recently Lin, Wharton and Grice[233] have used a slightly simplified version of the Rittner model to discuss the potential energy surfaces of several M_2X^+ systems. Their model is the same as Rittner's except that they omit the induced dipole–induced dipole and Van der Waal's interactions. A Rittner type model has also been applied to the potential surface of the four atom alkali halide exchange reaction[234] $NaBr + KCl \rightarrow NaCl + KBr$. In this model the repulsions between the ions were dependent on their polarization. The model led to good dissociation energies for the alkali halide dimers [$(NaCl)_2$ and $(KCl)_2$], and a satisfactory value for the exothermicity of the exchange reaction.

C. Convenient Analytic Formulae

In order to investigate the dynamics of chemical reactions by means of classical mechanical trajectory calculations it is desirable to have an analytic form for the potential energy surface so as to permit efficient calculation of the potential and its derivatives. All the empirical and most of the semi-empirical surfaces mentioned so far have been of this form, but all of them have been based on theoretical or physical models. In attempting either to find potential surfaces to describe specific reactions, or to investigate the effect of different features of a surface on the energy and angular distribution of the reaction products, several convenient and flexible functional forms for potential energy surfaces have been proposed.

Perhaps the most flexible of these surfaces are the hyperbolic map function surfaces proposed by Bunker and Blais[235] and later modified by

Bunker and Parr.[236] For a triatomic reaction of the form $A + BC \rightarrow AB + C$ let $\chi = r_{AB} - r_{AB}^{eq}$ (i.e. x is the deviation of the AB distance from its equilibrium value in the AB diatomic) and $y = r_{BC} - r_{BC}^{eq}$. We then define the reaction path, which is the minimum energy path from reactants to products, as a rectangular hyperbola

$$xy = u_0 \qquad (52)$$

The parameter u_0 controls the sharpness of the bend in the reaction path. We can now define the family of hyperbolas conjugate to the reaction path (52) by

$$v = \tfrac{1}{2}(y^2 - x^2) \qquad (53)$$

These hyperbolas interesect the reaction path at right angles. The point of intersection, which depends on v, is denoted by the coordinates x_0, y_0. For v very negative the conjugate hyperbola crosses the reaction path in the entrance channel (i.e. the A–B distance is very large), while for v very large the crossing point is in the exit channel. The value of v can therefore be used to measure the progress along the reaction path. S is now defined as the distance from the point (x, y) to the point (x_0, y_0), where (x_0, y_0) is the point at which the conjugate hyperbola which passes through (x, y) intersects the reaction path. The potential is finally defined as a generalized Morse potential:

$$V = F(\alpha, v)D(v)\{\exp[-2\beta(v)S] - 2\exp[-\beta(v)S]\} \qquad (54)$$

where α is the angle between the AB and BC bonds. The angular function $F(\alpha, v)$ goes to unity in the limits of very negative or positive v. The Morse parameters $D(v)$ and $\beta(v)$ go smoothly, though not necessarily monotonically, from the values appropriate to the reactants to those appropriate to the products as v increases. Clearly the variation of the functions $F(\alpha, v)$, $D(v)$ and $\beta(v)$ as well as that of the reaction path, within the framework outlined above, provides a very flexible functional form.[6] An analytic surface with a similar functional form,[237] but defined in terms of Marcus's reaction coordinates, has been fitted to the ab initio H_3 surface of Shavitt et al.[97]

The other main type of analytic potential energy surface, which does not have any theoretical or physical basis is the switching function type of surface first proposed by Blais and Bunker.[238] This surface is of the form

$$V = D_{AB}\{1 - \exp[-\beta_{AB}(r_{AB} - r_{AB}^{eq})]\}^2 + D_{BC}\{1 - \exp[-\beta_{BC}(r_{BC} - r_{BC}^{eq})]\}^2$$
$$+ D_{BC}\{1 - \tanh(ar_{AB} + c)\}\exp[-\beta_{BC}(r_{BC} - r_{BC}^{eq})]$$
$$+ D_{AC}\exp[-\beta_{AC}(r_{AC} - r_{AC}^0)] \qquad (55)$$

The first two terms are Morse potentials for the product and reactant diatomics, respectively. The third term is a switching or attenuation factor which reduces the BC attraction as the A atom approaches. The fourth term is a repulsive interaction between A and C. It is clearly the last two terms which govern the form of the potential energy surface in the region of strong interaction. Raff and Karplus[4] have proposed a modification to (55) which is to include an additional term of the same form as the third term but with the roles of r_{AB} and r_{BC} interchanged. This term reduces the AB attraction at short BC bond lengths.

Besides the above potentials, several potentials have been proposed to describe reactions such as $K + Br_2$ which proceed by means of an electron-jump mechanism.[239–242] Of these surfaces, the one with the greatest physical content is that of Karplus and Godfrey[240] which incorporates an ionic surface based on the Rittner model and a reasonable covalent surface (for large $K–Br_2$ separations). The jump from the covalent to the ionic surface is accomplished in a continuous manner by means of an energy dependent switching function. None of these surfaces are completely satisfactory for the description of $K + Br_2$ type reactions.[169]

VI. SUMMARY

Perhaps the main thing to realize about potential energy surfaces for chemical reactions, is that we know very little about them. The large amount of experimental data which is being accumulated can, as yet, be used to furnish only rather indirect and imprecise information on the surface involved. *Ab initio* calculations in this field are, therefore, in the rare position of playing a vital complementary role to experimental results in aiding the choice of potential energy surface used to interpret specific reactions.

The *ab initio* calculations which have so far been performed have, on the whole, been rather limited. The most extensive treatments have been for the systems H_3^+,[62,95] H_3[85,97] and H_2F.[108] The *ab initio* method which at present seems to offer the most practical means of calculating reliable surfaces is the iterative natural orbital (INO) method.[92,108] Relatively little effort has as yet been put into the application of valence-bond methods which, from a theoretical point of view, provide the most natural description of processes involving the breaking and making of chemical bonds. The Hartree–Fock or SCF method, which is the most convenient and widely used of all the *ab initio* methods, gives an inadequate description of these processes and often yields qualitatively incorrect results.[13,107]

Because of the difficulty of performing reliable *ab initio* calculations, it is often necessary to resort to more approximate methods especially for larger systems. These methods could be just the standard *ab initio* methods using basis sets of limited size or they could be one of the semi-theoretical methods (atoms-in-molecules or pseudo-potential) outlined in Section III.

If such methods are applied with care they should yield a reliable picture of the general form of the potential energy surface. The quantitative results obtained from them cannot be expected to be as reliable as those obtained from a really good *ab initio* calculation.

The semi-empirical methods, especially the diatomics-in-molecules (DIM) method, provide a simple way of obtaining an idea of what a surface might look like. They possess the great advantage that the correct limits are built into them. That is, they are guaranteed to give a correct description of the separated reactants and the separated products. They may be regarded as providing a means of smoothly interpolating the potential surface from the reactant to the product region. For small systems (i.e. H_3^+ [95]) they often yield results which agree remarkably well with vastly more difficult *ab initio* calculations. For larger systems, however, (i.e. H_4 [209]) they may yield unreliable results and should therefore be treated with caution.

In order to use a calculated surface it must be possible to evaluate the potential, and often its derivatives, at any arbitrary nuclear configuration. In practice this means that the calculated points must be fitted to an analytic functional form. The process of making this analytic fit is far from trivial and relatively little work has so far been done to find a good fitting procedure. [237,243–246]. The best way of calculating a complete potential energy surface might well involve the use of *ab initio*, semi-theoretical and semi-empirical methods to complement each other. Thus we could perform very thorough *ab initio* calculations at a few nuclear geometries. These calculations would then be extended by performing some semi-theoretical calculations at a much larger number of geometries and using the results to interpolate between the *ab initio* points. This yields a large number of points on a potential energy surface, which is a good estimate of what a thorough *ab initio* calculation would have given. DIM theory could be used to provide an analytic form which would approximately fit these results. [206] The excited states of the diatomic molecules needed in the DIM method, may be chosen so as to ensure that the DIM surface for the ground state fits the calculated points as well as possible.

There are two important topics which have not been discussed in this chapter. The first is the crossing or avoided crossing of potential energy surfaces, and the second, spin–orbit coupling. Potential energy surfaces can cross if there exists a nuclear configuration of high symmetry, whose symmetry group possesses a degenerate irreducible representation. When the system is distorted out of the highly symmetric situation, those surfaces which are degenerate in the symmetric nuclear configuration will split. Thus, as the nuclei traverse the symmetric nuclear configuration two potential energy surfaces touch or cross each other. The Born–Oppenheimer approximation breaks down in the region of the crossing and there is the possibility that the system will hop from one potential to the other. Such a symmetry related crossing is called the Jahn–Teller Renner effect.[247] There have been recent theoretical investigations of the Jahn–Teller effects in H_3[248] and CH_4^+.[249]

A more general situation in which the Born–Oppenheimer approximation breaks down is that associated with avoided crossings of potential energy surfaces, which are not normally symmetry related. Such avoided crossings often, but not always, arise from charge exchange processes (i.e. $Li + F_2$,[169] $Li + LiF$,[146,147,216] $H^+ + H_2$[198,199]). Only for $H^+ + H_2$ has the dynamics of a system involving an avoided crossing of surfaces and the possibility of surface hopping been considered.[198,199,250] More work on this problem is clearly needed. One of the difficulties is the calculation of the matrix elements of the nuclear kinetic energy operator. It is these matrix elements which couple the two surfaces in the quantum mechanical treatment of the scattering problem. Preston and Tully used DIM theory in conjunction with the Hellman–Feynman theorem to estimate the matrix elements.[198,199] A method for calculating them using *ab initio* CI type wavefunctions has also been proposed[251] but has not yet been applied.

Spin–orbit coupling terms have not been considered in any potential energy surface calculation. For heavier atoms they are often so large that they cannot validly be ignored. However, even for the relatively light system $F + H_2$, they are comparable in magnitude to energies of chemical interest such as the potential energy barrier.[252] Clearly these terms, which have till now been totally neglected, demand proper consideration.

Acknowledgements

I am happy to acknowledge the benefit of many useful discussions with Dr. N. C. Pyper, Mr. S. Wilson, Dr. J. Gerratt, Dr. G. Duxbury and Professor R. N. Dixon. I am also grateful to Dr. J. N. L. Connor for sending me a copy of his review article on 'The Theory of Molecular Collisions and Reactive Scattering' which is to appear in Annual Reports of the Chemical Society vol. 70A, 1973, before publication.

References

1. *Faraday Disc. Chem. Soc.*, **55** (1973); and other chapters of this volume.
2. J. L. Kinsey, *MTP Int. Rev. Sci.*, Series 1, **9**, 173 (1972).
3. M. Karplus, R. N. Porter and R. D. Sharma, *J. Chem. Phys.*, **43**, 3259 (1965).
4. L. M. Raff and M. Karplus, *J. Chem. Phys.*, **44**, 1212 (1966).
5. R. A. La Budde and R. B. Bernstein, *J. Chem. Phys.*, **55**, 5499 (1971).
6. T. B. Borne and D. L. Bunker, *J. Chem. Phys.*, **55**, 4861 (1971).
7. C. A. Parr, J. C. Polyani and W. H. Wong, *J. Chem. Phys.*, **58**, 5 (1973).
8. J. Schaefer and W. A. Lester, Jr., *Chem. Phys. Lett.*, **20**, 575 (1973).
9. W. Eastes and D. Secrest, *J. Chem. Phys.*, **56**, 640 (1972).
10. M. Karplus in *Structural Chemistry and Molecular Biology* (Eds. A. Rich and N. Davidson), W. H. Freeman and Co., San Francisco, California, 1968.

11. J. O. Hirschfelder, Ed., *Advances in Chemical Physics*, Vol. 12, John Wiley, New York, 1967.

12. C. Schlier, *Ann. Rev. Phys. Chem.*, **20**, 191 (1969).

13. M. Krauss, *Ann. Rev. Phys. Chem.*, **21**, 39 (1970).

14. M. Karplus in *Proceedings of the International School of Physics, Enrico Fermi, Course XLIV* (Ed. Ch. Schlier), Academic Press, New York, 1970.

15. A. D. Buckingham and B. D. Utting, *Ann. Rev. Phys. Chem.*, **21**, 287 (1970).

16. C. A. Parr and D. G. Truhlar, *J. Phys. Chem.*, **75**, 1844 (1971).

17. H. Margenau and N. R. Kestner, *Theory of Intermolecular Forces*, 2nd ed., Pergamon Press, New York, 1971.

18. P. R. Certain and L. W. Bruch, *MTP Int. Rev. Sci.*, Series 1, 113 (1972).

9. W. A. Lester, Ed., *Potential Energy Surfaces in Chemistry*, IBM Research Laboratory, San Jose, 1970.

20. D. M. Chipman, J. D. Bowman and J. O. Hirschfelder, *J. Chem. Phys.*, **59**, 2830 (1973).

21. J. N. Murrell in *Orbital Theories of Molecules and Solids* (Ed. N. H. March), Clarendon Press, Oxford, England, 1974.

22. See, for example, P. W. Atkins, *Molecular Quantum Mechanics*, Clarendon Press, Oxford, England, 1970.

23. M. Born and R. Oppenheimer, *Ann. Physik*, **84**, 457 (1927).

24. R. B. Gerber, *Proc. Roy. Soc. (London)*, **A309**, 221 (1969).

25. J. C. Slater, *Quantum Theory of Atomic Structure*, Vol. 1, McGraw-Hill Book Co. Inc., 1960.

26. J. Gerratt, *Advan. At. Mol. Phys.*, **7**, 141 (1971).

27. K. Ruedenberg and R. D. Poshusta, *Advan. Quantum Chem.*, **6**, 267 (1972).

28. C. C. J. Roothaan, *Rev. Mod. Phys.*, **23**, 69 (1951).

29. C. A. Coulson, *Valence*, Oxford University Press, England, 1961.

30. A. C. Wahl, *J. Chem. Phys.*, **41**, 2600 (1964).

31. A. D. McLean, *J. Chem. Phys.*, **39**, 2653 (1963).

32. T. H. Dunning, Jr., R. M. Pitzer and S. Aung, *J. Chem. Phys.*, **57**, 5044 (1972).

33. P. S. Bagus, B. Liu and H. F. Schaefer III, *J. Am. Chem. Soc.*, **94**, 6635 (1972).

34. M. P. Melrose and D. Russel, *J. Chem. Phys.*, **57**, 2586 (1972).

35. M. E. Schwartz, *Chem. Phys. Lett.*, **5**, 50 (1970).

36. I. H. Hillier, V. R. Saunders and M. H. Wood, *Chem. Phys. Lett.*, **7**, 323 (1970).

37. C. C. J. Roothaan, *Rev. Mod. Phys.*, **32**, 179 (1960).

38. S. Huzinaga, *Phys. Rev.*, **122**, 131 (1961).

39. R. K. Nesbet, *Rev. Mod. Phys.*, **33**, 28 (1961).

40. W. J. Hunt, T. H. Dunning, Jr. and W. A. Goddard III, *Chem. Phys. Lett.*, **3**, 606 (1969).

41. R. McWeeny and B. T. Sutcliffe, *Methods of Molecular Quantum Mechanics*, Academic Press, 1969.

42. E. R. Davidson, *Chem. Phys. Lett.*, **21**, 565 (1973).

43. G. A. Segal, *J. Chem. Phys.*, **52**, 3530 (1970).

44. J. A. Pople and R. K. Nesbet, *J. Chem. Phys.*, **22**, 571 (1954).

45. A. W. Salotto and L. Burnelle, *J. Chem. Phys.*, **52**, 2936 (1970).

46. A. W. Salotto and L. Burnelle, *J. Chem. Phys.*, **53**, 333 (1970).

47. R. D. Brown and G. R. Williams, *Mol. Phys.*, **25**, 673 (1973).

48. J. E. Del Bene, *Chem. Phys. Lett.*, **9**, 68 (1971).

49. J. L. Gole, A. K. Q. Siu and E. F. Hayes, *J. Chem. Phys.*, **58**, 857 (1973). See also J. L. Gole, *J. Chem. Phys.*, **58**, 869 (1973).

50. M. Vucelic, T. Ohrn and J. R. Sabin, *J. Chem. Phys.*, **59**, 3003 (1973).

51. S. V. O'Neil and H. F. Schaefer III and C. F. Bender, *J. Chem. Phys.*, **59**, 3608 (1973).

52. W. H. Fink and L. C. Allen, *J. Chem. Phys.*, **46**, 3266 (1967).

53. L. Pedersen and K. Morokuma, *J. Chem. Phys.*, **46**, 3941 (1967).

54. R. M. Stevens, *J. Chem. Phys.*, **55**, 1725 (1971).

55. J. E. Del Bene, *J. Chem. Phys.*, **57**, 1899 (1972).

56. K. Morokuma and L. Pedersen, *J. Chem. Phys.*, **48**, 3275 (1968).

57. M. D. Gordon and D. Secrest, *J. Chem. Phys.*, **52**, 120 (1970); erratum *J. Chem. Phys.*, **53**, 4408 (1970).

58. W. A. Lester, Jr., *J. Chem. Phys.*, **53**, 1511 (1970).

59. W. A. Lester, Jr., *J. Chem. Phys.*, **54**, 3171 (1971); erratum *J. Chem. Phys.*, **57**, 3028 (1972).

60. W. Kutzelnigg, V. Staemmler and C. Hoheisel, *Chem. Phys.*, **1**, 27 (1973).

61. W. Kolos and C. C. J. Roothaan, *Rev. Mod. Phys.*, **32**, 219 (1960).

62. I. G. Csizmadia, R. E. Kari, J. C. Polanyi, A. C. Roach and M. A. Robb, *J. Chem. Phys.*, **52**, 6205 (1970).

63. C. D. Ritchie and H. F. King, *J. Am. Chem. Soc.*, **88**, 1069 (1966).

64. W. A. Lester, Jr. and M. Krauss, *J. Chem. Phys.*, **52**, 4775 (1970).

65. J. J. Kaufman and L. M. Sachs, *J. Chem. Phys.*, **52**, 3534 (1970).

66. P. J. Brown and E. F. Hayes, *J. Chem. Phys.*, **55**, 922 (1971).

67. M. F. Guest, J. N. Murrell and J. B. Pedley, *Mol. Phys.*, **20**, 81 (1971).

68. J. N. Murrell, J. B. Pedley and S. Durmaz, *J.C.S. Faraday II*, **69**, 1370 (1973).

69. A. J. Duke and R. F. W. Bader, *Chem. Phys. Lett.*, **10**, 631 (1971).

70. H. Kistenmacher, H. Popkie and E. Clementi, *J. Chem. Phys.*, **58**, 1689 (1973); **58**, 5627 (1973).

71. S. Rothenberg and H. F. Schaefer III, *Chem. Phys. Lett.*, **10**, 565 (1971).

72. D. R. Yarkong, W. J. Hunt and H. F. Schaefer III, *Mol. Phys.*, **26**, 941 (1973).

73. D. W. Davies and G. del Conde, *Faraday Disc. Chem. Soc.*, **55**, 369 (1973).

74. D. H. Liskow, C. F. Bender and H. F. Schaefer III, *J. Chem. Phys.*, **57**, 4509 (1972).

75. R. C. Ladner and W. A. Goddard III, *J. Chem. Phys.*, **51**, 1073 (1969).

76. C. Woodrow Wilson, Jr. and W. A. Goddard III, *J. Chem. Phys.*, **56**, 5913 (1972).

77. D. Kunik and U. Kaldor, *J. Chem. Phys.*, **56**, 1741 (1972).

78. R. J. Blint and W. A. Goddard III, *J. Chem. Phys.*, **57**, 5296 (1972).

79. A. C. Hurley, J. E. Lennard-Jones and J. A. Pople, *Proc. Roy. Soc. (London)*, **A220**, 446 (1953).

80. A. C. Hurley, *Proc. Roy. Soc. (London)*, **A235**, 224 (1956).

81. W. J. Hunt, P. J. Hays and W. A. Goddard III, *J. Chem. Phys.*, **57**, 738 (1972).

82. P. J. Hays, *J. Chem. Phys.*, **58**, 4706 (1973).

83. U. Kaldor, *Phys. Rev.*, **176**, 19 (1968).

84. S. L. Guberman and W. A. Goddard III, *J. Chem. Phys.*, **53**, 1803 (1970).

85. B. Liu, *J. Chem. Phys.*, **58**, 1925 (1973). This paper uses a method which is not the INO method, but shares with it the feature that approximate, natural orbitals for the system are estimated and progressively improved.

86. H. F. Schaefer III, *The Electronic Structure of Atoms and Molecules*, Addison-Wesley Publishing Co., 1972.

87. J. K. L. McDonald, *Phys. Rev.*, **43**, 830 (1933).

88. S. Lunell, *Chem. Phys. Lett.*, **15**, 27 (1972).

89. R. L. Chase, H. P. Kelly and H. S. Köhler, *Phys. Rev. A.*, **3**, 1550 (1970).

90. D. McWilliams and S. Huzinaga, *J. Chem. Phys.*, **55**, 2604 (1971).

91. P. O. Löwdin, *Phys. Rev.*, **97**, 1474 (1955).

92. C. F. Bender and E. R. Davidson, *J. Phys., Chem.*, **70**, 2675 (1966).
93. E. R. Davidson, *Advan. Quantum Chem.*, **6**, 235 (1972).
94. H. F. Schaefer III, D. Wallach and C. F. Bender, *J. Chem. Phys.*, **56**, 1219 (1972).
95. C. W. Bauschlicher, Jr., S. V. O'Neil, R. K. Preston and H. F. Schaefer III, *J. Chem. Phys.*, **59**, 1286 (1973).
96. C. Edmiston and M. Krauss, *J. Chem. Phys.*, **49**, 192 (1968).
97. I. Shavitt, R. M. Stevens, F. L. Minn and M. Karplus, *J. Chem. Phys.*, **48**, 2700 (1968). Erratum *J. Chem. Phys.*, **49**, 4048 (1968).
98. D. M. Silver and R. M. Stevens, *J. Chem. Phys.*, **59**, 3378 (1973).
99. D. M. Silver, *Chem. Phys. Lett.*, **14**, 105 (1972).
00. A. Pipano and J. J. Kaufman, *J. Chem. Phys.*, **56**, 5258 (1972).
01. M. M. Heaton, A. Pipano and J. J. Kaufman, *Int. J. Quant. Chem.*, **S6**, 181 (1972).
02. D. H. Liskow, C. F. Bender and H. F. Schaefer III, *J. Chem. Phys.*, **56**, 5075 (1972).
03. N. W. Winter, C. F. Bender and W. A. Goddard III, *Chem. Phys. Lett.*, **20**, 489 (1973).
04. R. E. Kari and I. G. Csizmadia, *J. Chem. Phys.*, **56**, 4337 (1972).
05. M. J. Benson and D. R. McLaughlin, *J. Chem. Phys.*, **56**, 1322 (1972).
06. G. C. Lie, J. Hinze and B. Liu, *J. Chem. Phys.*, **59**, 1872 (1973).
07. C. F. Bender, P. K. Pearson, S. V. O'Neil and H. F. Schaefer III, *J. Chem. Phys.*, **56**, 4626 (1972).
08. C. F. Bender, P. K. Pearson, S. V. O'Neil and H. F. Schaefer III, *Science*, **176**, 1412 (1972).
09. S. V. O'Neil, P. K. Pearson, H. F. Schaefer III and C. F. Bender, *J. Chem. Phys.*, **58**, 1126 (1973).
10. D. H. Liskow, H. F. Schaefer III and C. F. Bender, *J. Am. Chem. Soc.*, **93**, 6734 (1971).
11. S. V. O'Neil, H. F. Schaefer III and C. F. Bender, *J. Chem. Phys.*, **55**, 162 (1971).
12. H. F. Schaefer III, *J. Chem. Phys.*, **54**, 2207 (1971).
13. S. V. O'Neil and H. F. Schaefer III, *J. Chem. Phys.*, **55**, 394 (1971).
14. A. C. Wahl and G. Das, *Advan. Quantum Chem.*, **5**, 261 (1970).
15. H. Basch, *J. Chem. Phys.*, **55**, 1700 (1971).
16. A. C. Wahl, A. Wagner and A. Karo, 'Electronic and Atomic Collisions', in *Abstracts of papers VIIIth ICPEAC Conf.* (Eds., B. C. Čobić and M. V. Kurepa), Institute of Physics, Beograd, 1973. Also private communications.
17. M. Krauss and D. Neumann, *Mol. Phys.*, **27**, 917 (1974).
18. G. Das, *J. Chem. Phys.*, **58**, 5104 (1973).
19. G. Das and A. C. Wahl, *J. Chem. Phys.*, **44**, 876 (1966).
20. P. K. Pearson, W. J. Hunt, C. F. Bender and H. F. Schaefer III, *J. Chem. Phys.*, **58**, 5358 (1973).
21. O. Sinanoğlu and K. A. Brueckner, *Three Approaches to Electron Correlation in Atoms*, Yale University Press, 1970.
22. R. G. Parr, *Quantum Theory of Molecular Electronic Structure*, W. A. Benjamin, 1964.
23. R. K. Nesbet, *Advan. Chem. Phys.*, **14**, 1 (1969). This volume is devoted to a discussion of correlation effects in atoms and molecules, and has several articles on perturbation methods of treating the problem.
24. B. Tsapline and W. Kutzelnigg, *Chem. Phys. Lett.*, **23**, 173 (1973).
25. M. Jungen, *Chem. Phys. Lett.*, **5**, 241 (1970).
26. M. Jungen and R. Ahlrichs, *Theoret. Chim. Acta*, **17**, 339 (1970).
27. R. Ahlrichs, *Theoret. Chim. Acta*, **17**, 348 (1970).

128. S. Raimes, *Many Electron Theory*, North-Holland, 1972.
129. K. F. Freed, *Ann. Rev. Phys. Chem.*, **22**, 313 (1971).
130. K. F. Freed, *Phys. Rev.*, **173**, 1 (1968).
131. J. H. Miller and H. P. Kelly, *Phys. Rev. A*, **4**, 480 (1971).
132. M. A. Robb, *Chem. Phys. Lett.*, **20**, 274 (1973). Also private communications.
133. Z. Gershgorn and I. Shavitt, *Int. J. Quant. Chem.*, **2**, 751 (1968).
134. G. Das and A. C. Wahl, *Phys. Rev. Lett.*, **24**, 440 (1970).
135. B. Huron and R. Fancurel, *Chem. Phys. Lett.*, **13**, 515 (1972).
136. J. H. Van Vleck and A. Sherman, *Rev. Mod. Phys.*, **7**, 167 (1935).
137. J. H. Van Vleck, *J. Chem. Phys.*, **1**, 236 (1933); **2**, 20 (1934).
138. G. Nordheim-Pöschl, *Ann. Physik*, **26**, 258 (1936).
139. W. Moffitt, *Proc. Soc. (London)*, **A210**, 245 (1951).
140. W. Moffitt, *Rept. Prog. Phys.*, **17**, 173 (1954).
141. F. Prosser and S. Hagstrom, *Int. J. Quant. Chem.*, **2**, 89 (1968); *J. Chem. Phys.*, **48**, 4807 (1968).
142. H. F. King, R. E. Stanton, H. Kim, R. E. Wyatt and R. G. Parr, *J. Chem. Phys.*, **47**, 1936 (1967).
143. W. Moffitt, *Proc. Roy. Soc. (London)*, **A218**, 486 (1953).
144. A. C. Hurley, *Rev. Mod. Phys.*, **32**, 400 (1960).
145. G. G. Balint-Kurti and M. Karplus, *J. Chem. Phys.*, **50**, 478 (1969).
146. G. G. Balint-Kurti and M. Karplus in *Orbital Theories of Molecules and Solids* (Ed., N. H. March), Clarendon Press, Oxford, 1974.
147. G. G. Balint-Kurti, *Ph.D. Thesis*, Columbia University, 1969.
148. J. Gerratt in *Specialist Reports of The Chemical Society, No. 33, Vol. 1, Quantum Chemistry*, (Ed., R. N. Dixon), 1974.
149. W. J. Campion and M. Karplus, *Mol. Phys.*, **25**, 921 (1973).
150. R. G. A. R. MacLagan and G. W. Schnuelle, *J. Chem. Phys.*, **55**, 5431 (1971).
151. K. A. R. Mitchell and T. Thirunâmachandran, *Mol. Phys.*, **23**, 947 (1972).
152. R. P. Hosteny and S. A. Hagstrom, *J. Chem. Phys.*, **58**, 4396 (1973).
153. G. A. Gallup and J. M. Norbeck, *Chem. Phys. Lett.*, **21**, 495 (1973).
154. R. D. Poshusta, D. W. Klint and A. Liberles, *J. Chem. Phys.*, **55**, 252 (1971).
155. R. D. Poshusta and W. F. Siems, *J. Chem. Phys.*, **55**, 1995 (1971).
156. J. F. Harrison and L. C. Allen, *J. Am. Chem. Soc.*, **91**, 807 (1969).
157. L. C. Allen, R. M. Ardahl and J. L. Whitten, *J. Am. Chem. Soc.*, **87**, 3769 (1965).
158. L. C. Allen, A. M. Lesk and R. M. Erdahl, *J. Am. Chem. Soc.*, **88**, 615 (1966).
159. R. D. Poshusta and D. F. Zelik, *J. Chem. Phys.*, **58**, 118 (1973).
160. G. A. Gallup and J. M. Norbeck, *J. Am. Chem. Soc.*, **95**, 4460 (1973).
161. M. Klessinger and R. McWeeny, *J. Chem. Phys.*, **42**, 3343 (1965).
162. M. Klessinger, *J. Chem. Phys.*, **53**, 225 (1970).
163. D. M. Silver, E. L. Mehler and K. Ruedenberg, *J. Chem. Phys.*, **52**, 1174, 1181 (1970); and references quoted therein.
164. P. F. Franchini and C. Vergani, *Theoret. Chim. Acta*, **13**, 46 (1969).
165. P. J. Carrington and G. Doggett, *Mol. Phys.*, **26**, 641 (1973).
166. A. C. Hurley, *Proc. Phys. Soc. (London)*, **A69**, 49 (1956).
167. A. C. Hurley, *J. Chem. Phys.*, **28**, 532 (1958).
168. T. Arai, *Rev. Mod. Phys.*, **32**, 370 (1960).
169. G. G. Balint-Kurti, *Mol. Phys.*, **25**, 393 (1973).
170. R. N. Yardley and G. G. Balint-Kurti, work in progress.
171. J. C. Phillips and L. Kleinman, *Phys. Rev.*, **116**, 287 (1959).
172. For a review of pseudo-potentials see: J. D. Weeks, A. Hazi and S. A. Rice, *Advan. Chem. Phys.*, **16**, 283 (1969).

73. L. Szasz and G. McGinn, *J. Chem. Phys.*, **45**, 2898 (1966).
74. G. J. Iafrate, *J. Chem. Phys.*, **46**, 728 (1967).
75. L. Szasz and G. McGinn, *J. Chem. Phys.*, **48**, 2997 (1968).
76. W. Kutzelnigg, *Chem. Phys. Lett.*, **4**, 435 (1969).
77. G. Simons and A. Mazziotti, *J. Chem. Phys.*, **52**, 2449 (1970).
78. A. Dalgarno, C. Bottcher and G. A. Victor, *Chem. Phys. Lett.*, **7**, 265 (1970). This calculation on Li_2^+ included terms which allowed for the polarization of the core, by both the core and valence electrons. See also C. Bottcher, *Chem. Phys. Lett.*, **18**, 457 (1973).
79. C. Bottcher and K. Docken, in 'Electronic and atomic collisions', in *Abstracts of papers VIIIth ICPEAC Conf.* (Eds. B. C. Čobić and M. V. Kurepa), Institute of Physics, Beograd, 1973.
80. C. Bottcher, *J. Phys. B.*, **6**, 2368 (1973). A highly novel spin dependent pseudopotential is used in this reference.
81. R. N. Dixon and J. M. V. Hugo, *Mol. Phys.*, To be published.
82. M. Kleiner and R. McWeeny, *Chem. Phys. Lett.*, **19**, 476 (1973).
83. L. R. Kahn and W. A. Goddard III, *J. Chem. Phys.*, **56**, 2685 (1972).
84. P. J. Carrington and P. G. Walton, *Mol. Phys.*, **26**, 705 (1973).
85. R. G. Gordon and S. Y. Kim, *J. Chem. Phys.*, **56**, 3122 (1972).
86. F. O. Ellison, *J. Am. Chem. Soc.*, **85**, 3540 (1963).
87. F. O. Ellison, N. T. Huff and J. C. Patel, *J. Am. Chem. Soc.*, **85**, 3544 (1963).
88. A. L. Companion, *J. Chem. Phys.*, **48**, 1186 (1968).
89. F. London, *Z. Electrochem.*, **35**, 552 (1929).
90. J. C. Slater, *Phys. Rev.*, **38**, 1109 (1931); A. S. Coolidge and H. M. James, *J. Chem. Phys.*, **2**, 811 (1934).
91. J. K. Cashion and D. R. Herschbach, *J. Chem. Phys.*, **40**, 2358 (1964).
92. F. O. Ellison and J. C. Patel, *J. Am. Chem. Soc.*, **86**, 2115 (1964).
93. G. V. Pfeiffer and F. O. Ellison, *J. Chem. Phys.*, **43**, 3405 (1965).
94. G. V. Pfeiffer, N. T. Huff, E. M. Greenawalt and F. O. Ellison, *J. Chem. Phys.*, **46**, 821 (1967).
95. A. A. Wu and F. O. Ellison, *J. Chem. Phys.*, **47**, 1458 (1967).
96. A. A. Wu and F. O. Ellison, *J. Chem. Phys.*, **48**, 727 (1968).
97. A. A. Wu and F. O. Ellison, *J. Chem. Phys.*, **48**, 1491, 5032 (1968).
98. R. K. Preston and J. C. Tully, *J. Chem. Phys.*, **54**, 4297 (1971).
99. J. Krenos, R. K. Preston, R. Wolfgang and J. C. Tully, *Chem. Phys. Lett.*, **10**, 17 (1971).
100. R. B. Abrams, J. C. Patel and F. O. Ellison, *J. Chem. Phys.*, **49**, 450 (1968).
101. A. L. Companion, D. J. Steible, Jr. and A. J. Starshak, *J. Chem. Phys.*, **49**, 3637 (1968).
102. A. L. Companion, *J. Chem. Phys.*, **50**, 1165 (1969).
103. J. C. Tully, *J. Chem. Phys.*, **58**, 1396 (1973).
104. P. J. Kuntz and A. C. Roach, *J.C.S. Faraday II*, **68**, 259 (1972); erratum *J.C.S. Faraday II*, **69**, 926 (1973).
105. E. Steiner, P. R. Certain and P. J. Kuntz, *J. Chem. Phys.*, **59**, 47 (1973).
106. P. J. Kuntz, *Chem. Phys. Lett.*, **16**, 581 (1972). See also: *Report No. WIS-TCI-418*, Theoretical Chemistry Institute, University of Wisconsin.
107. R. N. Porter and M. Karplus, *J. Chem. Phys.*, **40**, 1105 (1964).
108. L. Pedersen and R. N. Porter, *J. Chem. Phys.*, **47**, 4751 (1967).
109. K. Morokuma, L. Pedersen and M. Karplus, *J. Am. Chem. Soc.*, **89**, 5064 (1967).
110. B. Freihaut and L. M. Raff, *J. Chem. Phys.*, **58**, 1202 (1973).
111. D. A. Micha, *J. Chem. Phys.*, **57**, 2184 (1972).

212. J. C. Whitehead and R. Grice, *Mol. Phys.*, **26**, 267 (1973).
213. N. C. Blais and D. G. Truhlar, *J. Chem. Phys.*, **58**, 1090 (1973).
214. A. C. Roach and M. S. Child, *Mol. Phys.*, **14**, 1 (1968).
215. E. S. Rittner, *J. Chem. Phys.*, **19**, 1030 (1951).
216. W. S. Struve, *Mol. Phys.*, **25**, 777 (1973).
217. A. C. Roach and P. Baybutt, *Chem. Phys. Lett.*, **7**, 7 (1970).
218. A. W. Woolley and M. S. Child, *Mol. Phys.*, **19**, 625 (1970).
219. C. Nyeland and J. Ross, *J. Chem. Phys.*, **54**, 1165 (1971).
220. S. Karplus and R. Bersohn, *J. Chem. Phys.*, **51**, 2040 (1969).
221. R. L. Jaffe and J. B. Andersen, *J. Chem. Phys.*, **54**, 2224 (1971).
222. L. M. Raff, L. Stivers, R. N. Porter, D. L. Thompson and L. B. Sims, *J. Chem Phys.*, **52**, 3449 (1970); erratum *J. Chem. Phys.*, **58**, 1271 (1973).
223. J. M. White, *J. Chem. Phys.*, **58**, 4482 (1973).
224. S. Sato, *J. Chem. Phys.*, **23**, 592, 2465 (1955).
225. J. C. Polanyi and S. D. Rosner, *J. Chem. Phys.*, **38**, 1028 (1963).
226. J. T. Muckerman, *J. Chem. Phys.*, **54**, 1151 (1971).
227. R. L. Wilkins, *J. Chem. Phys.*, **57**, 912 (1972); ibid., **58**, 2326 (1973).
228. P. J. Kuntz, *Report No. WIS-TCI-420*, Theoretical Chemistry Institute, Th University of Wisconsin.
229. P. J. Kuntz, E. M. Nemeth, J. C. Polanyi, S. D. Rosner and C. E. Young, *J. Chem Phys.*, **44**, 1168 (1966).
230. A. M. G. Ding, L. J. Kirsch, D. S. Perry, J. C. Polanyi and J. L. Schreiber, *Farada Disc. Chem. Soc.*, **55**, 252 (1973).
231. J. T. Muckerman, *J. Chem. Phys.*, **56**, 2997 (1972).
232. N. Jonathan, S. Okuda and D. Timlin, *Mol. Phys.*, **24**, 1143 (1972); erratum ibid **25**, 496 (1973).
233. S. M. Lin, J. G. Wharton and R. Grice, *Mol. Phys.*, **26**, 317 (1973).
234. P. Brumer and M. Karplus, *Faraday Disc. Chem. Soc.*, **55**, 80 (1973).
235. D. L. Bunker and N. C. Blais, *J. Chem. Phys.*, **41**, 2377 (1964).
236. D. L. Bunker and C. A. Parr, *J. Chem. Phys.*, **52**, 5700 (1970).
237. R. P. Saxon and J. C. Light, *J. Chem. Phys.*, **56**, 3885 (1972).
238. N. C. Blais and D. L. Bunker, *J. Chem. Phys.*, **37**, 2713 (1962); **39**, 315 (1963).
239. N. C. Blais, *J. Chem. Phys.*, **49**, 9 (1968).
240. M. Karplus and M. Godfrey, *J. Chem. Phys.*, **49**, 3602 (1968).
241. P. J. Kuntz, E. M. Nemeth and J. C. Polanyi, *J. Chem. Phys.*, **50**, 4607 (1969).
242. P. J. Kuntz, M. H. Mok and J. C. Polanyi, *J. Chem. Phys.*, **50**, 4623 (1969).
243. I. G. Csizmadia, J. C. Polanyi, A. C. Roach and W. H. Wong, *Can. J. Chem.*, **4** 4097 (1969).
244. D. J. Truhlar and A. Kuppermann, *J. Chem. Phys.*, **56**, 2232 (1972).
245. W. A. Lester, Jr., *J. Chem. Phys.*, **53**, 1611, 1970.
246. D. R. McLaughlin and D. L. Thompson, *J. Chem. Phys.*, **59**, 4393 (1973).
247. G. Herzberg, *Molecular Spectra and Molecular Structure*, Vol. III, D. Va Nostrand Co. Inc., Princeton, New Jersey, 1966.
248. R. N. Porter, R. M. Stevens and M. Karplus, *J. Chem. Phys.*, **49**, 5163 (1968).
249. R. N. Dixon, *Mol. Phys.*, **20**, 113 (1971).
250. J. C. Tully and R. K. Preston, *J. Chem. Phys.*, **55**, 562 (1971).
251. B. R. Johnson and R. D. Levine, *Chem. Phys. Letts.*, **13**, 168 (1972).
252. J. T. Muckerman and M. D. Newton, *J. Chem. Phys.*, **56**, 3191 (1972).
253. S. Wilson and J. Gerratt, to be published.
254. J. Paldus, J. Cizek and I. Shavitt, *Phys. Rev.*, **A5**, 50 (1972).
255. B. T. Pickup, *Proc. Roy. Soc. (London)*, **A333**, 69 (1973).

56. R. F. W. Bader and R. A. Gangi, *J. Am. Chem. Soc.*, **93**, 1831 (1971).
57. R. A. Gangi and R. F. W. Bader, *J. Chem. Phys.*, **55**, 5369 (1971). This calculation includes a very limited configuration interaction.
58. E. Kochanski, B. Roos, P. Siegbahn and M. H. Wood, *Theoret. Chim. Acta*, **32**, 151 (1973).
59. J. N. Murrell and A. J. Harget, *Semi-Empirical Self-Consistent-Field Molecular Orbital Theory of Molecules*, Wiley-Interscience, London, 1972.
60. A. Chutjian and G. A. Segal, *J. Chem. Phys.*, **57**, 3069 (1972).
61. M. MacGregor and R. S. Berry, *J. Phys.*, **B6**, 181 (1973).

SCATTERING OF POSITIVE IONS BY MOLECULES

WALTER S. KOSKI

Department of Chemistry, The Johns Hopkins University, Baltimore, Maryland 21218

CONTENTS

I. Introduction 186
II. Experimental Considerations 187
  A. Single Beam Spectrometers 188
  B. Crossed-Beam Spectrometers 189
  C. Internal Energies of Projectile Ions 191
III. Reactive Scattering of Positive Ions 193
  A. Three Atom Ion–Molecule System 194
    1. Reaction of H^+ with H_2 194
    2. Reaction of Ar^+ with D_2 199
    3. Reaction of Kr^+ with D_2 202
    4. Reaction of N^+ with H_2 203
    5. Reaction of N^+ with O_2 203
    6. Reaction of O^+ with N_2 204
    7. Reaction of O^+ with D_2 208
    8. Reaction of C^+ with D_2 209
  B. Four Atom Ion–Molecule Systems 210
    1. Reaction of H_2^+ with H_2 210
    2. Reaction of O_2^+ with D_2 210
    3. Reaction of N_2^+ with H_2 211
  C. Polyatomic Ion–Molecule Systems 212
    1. Reaction of $C_2H_4^+$ with C_2H_4 212
    2. Reactions of $C_2H_2^+ + C_2H_4$ and $I_2^+ + C_2H_4$ . . . 215
  D. Conditions for Complex Formation 216
IV. Inelastic Scattering of Positive Ions 218
  A. Experimental 218
  B. Electronic Excitation 219
    1. Vibrational and Rotational States in Electronic Transitions . . 223
  C. Vibration–Rotation Excitation 223
    1. Vibration–Rotation Excitation by Protons 224
    2. Vibration–Rotation Excitation by H_2^+ 225
    3. Vibration–Rotation Excitation by Other Ions . . . 225
  D. Pure Rotational Excitation 227
V. Elastic Scattering of Positive Ions 231
  A. Classical Considerations 232
  B. Quantum Mechanical Considerations 233
    1. Evaluation of the Potential Parameters 236

   2. Proton–Rare Gas Systems 237
   3. Polyatomic Systems 237
References 242

I. INTRODUCTION

For decades the main method of attack on ion–molecule reactions was to use the single-source conventional mass spectrometer. In this method electrons emitted by a heated filament pass through a reaction chamber into which a gas is flowing. The gas was ionized and a repeller pushed out the ions into the analysing region of the instrument. In the conventional study of mass spectra, these exiting ions are mass analysed and detected. The distance from the electron beam to the exit slit is small (mm) but if the pressure of the target molecules in the ionization chamber is sufficiently high the ions produced directly by the electrons collide with neutral molecules and undergo reactions with them. These new ions which exit with the ions that are directly produced by electron bombardment are mass analysed and detected. By varying the repeller voltage, one can vary the energy of the primary ions and one can study the yield of the ion–molecule reaction as a function of projectile ion kinetic energy. Many ion–molecule reactions have been studied by this method and some excellent work has been realized.[1] In spite of these successes, there are a number of serious objections to this method of studying ion–molecule reactions. A few that might be mentioned are the spread in the reactant ion energy, the range of kinetic energy is limited, confusion between the ionic product and the primary ions occurs at times, in some instances it may be difficult to distinguish between the projectile and the target and so on.

A solution to most of the difficulties inherent in a conventional single-source mass spectrometer for the study of ion–molecule reactions was realized with the development of the tandem mass spectrometer.[2] These instruments consist of two mass spectrometers in series. The first mass spectrometer mass analyses the primary ion beam, the ion energy is then adjusted to a desired value by an appropriate lens system which also focuses the projectile ion beam into a reaction chamber where the reactant gas is present. Alternatively, a molecular beam of target molecules can be used. The ionic products are then extracted and focused into a secondary mass spectrometer which mass analyses the products. Most of the cross section measurements and variation of cross section with projectile ion kinetic energy in recent years have been obtained with tandem mass spectrometers or by the ion cyclotron resonance[3] method. On the other hand, attempts to apply tandem machines to the study of mechanism of ion–molecule reactions have not met with outstanding success. In order to correct this deficiency investigators extended the techniques used in the study of nuclear reactions to the study of scattering of ions by molecules.

In this technique a beam of mass analysed ions of known kinetic energy and in some instances a known state is permitted to collide with target molecules either in a reaction chamber or in a molecular beam. The energy and direction of the resulting ionic products is then determined. Elastic, inelastic and reactive collisions are being studied in this manner. This is a comparatively recent development but it has already given us a considerable insight not only in understanding ionic processes but also into chemical kinetics in general. The types of information that result from such studies are:

1. Differential cross sections as a function of projectile ion energy. Total cross sections can also be determined but these are probably best done with a tandem machine.
2. Angular distribution of the products.
3. Velocity and total internal energy of the products.
4. Cross sections as a function of electronic states of the projectile ions have been realized in some cases.
5. In general, states of the products have not been determined in this approach but recent developments indicate that some success might be expected in this area in the near future.
6. Inelastic scattering studies are giving us internal energy distributions in target molecules.
7. Elastic scattering is being used to give potential energy parameters of simple systems.

It is clear from what has been said that we are far from realizing the ideal experiment where angular distributions, velocities, orientations and the detailed states of reactants and products are all determined. However, inroads are being made in this direction.

II. EXPERIMENTAL CONSIDERATIONS

It is clear that apparatus design is a key factor in the quality of the results obtained. In a recent review of beam studies of ion–molecule reactions Herman and Wolfgang[4] gave an excellent discussion of the apparatus and techniques used in this field. Techniques have not changed significantly during the short intervening period so this material will not be treated here. On the other hand some new spectrometers have appeared and although no dramatic instrumental breakthrough has occurred some improvements have been realized so it was considered appropriate to outline two spectrometers illustrating the type of equipment in use. One is a single beam machine, that is, the projectile ions are prepared as a beam and the target molecules are in a reaction chamber. This machine described below has been mainly used in this configuration but provisions have been made for replacing the chamber by a molecular beam. The second spectrometer is a crossed beam device in which a reaction chamber can be substituted for the beam if desired but it has been primarily used in the former configuration.

A. Single Beam Spectrometers

A single beam spectrometer recently put into operation in the author' laboratory[5] is shown in Fig. 1. This spectrometer differs from others used fo

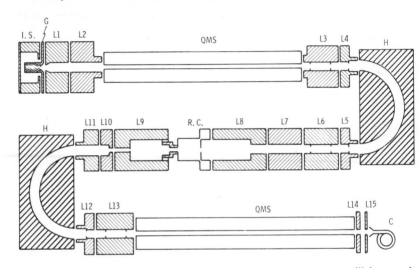

Fig. 1. Schematic diagram of single-beam scattering spectrometer utilizing quadrupole mass filters[5] I.S.—ion source, G—grid, L_1–L_{15}—lens elements, R.C.—reaction chamber, C—channeltron multiplier, QMS—quadrupole mass filters, H—hemispherical energy analysers.

the same purpose in that it consists of two quadrupole mass spectrometer and two hemispherical electrostatic energy analysers[6] in tandem. The whol assembly is mounted inside a vacuum chamber and the spectrometer tha contains the detector can be rotated using appropriate mechanical feed throughs about the reaction chamber in the plane perpendicular to the page The beam of ions is extracted from a high pressure ion source. The reactan ions are mass analysed, focused and energy adjusted to a desired value befor reaction with the neutral species. Careful attention is given to the design o the ion optics[7] in order to insure that the ion beam angular profile does no change appreciably over the energy range 0·5–50·0 eV. Ionic products ar collected as a function of scattering angle. The second hemispherical energ analyser with its associated ion optics measures the ion energy while th quadrupole mass spectrometer identifies the mass of the scattered ion. Th ions are detected with a channeltron. Pulse counting techniques are use throughout the data handling electronics. Data are stored in a 400 chann analyser and may be processed by computer to yield a product ion intensity contour map in barycentric velocity coordinates.[8] Initially the beam had an angular spread of 0·9° FWHM and an energy spread of 80 MV FWHM. In order to get a higher beam intensity the aperture sizes were increased and the present operating conditions are 1·4° angular spread and 160 MV FWHM energy spread. The reaction chamber can be substituted by a molecular beam which is especially important at low energies where the thermal spread of the target gas in a reaction chamber can be troublesome.

An example of a similar spectrometer is the one used by Champion et al.[9] Primary ions were produced by electron bombardment, accelerated and mass analysed by a 60°, 13·3 cm magnetic mass spectrometer. The ions were then retarded to the desired energy and energy selected by a 127° electrostatic cylindrical velocity selector. The energy resolution was 5%. The beam half angle is reported to be 1·8°. The ions then entered a reaction chamber at a pressure of about 10^{-4} torr. The chamber had an exit slit which could be rotated with the detecting system which consisted of another 127° velocity selector, a quadrupole mass spectrometer and an electron multiplier.

Still another example of a spectrometer that has produced a number of impressive contour maps of ion–molecule reactions is the one used by Mahan's group.[10,11,12] Ions are produced by electron impact and mass analysed magnetically, focused and adjusted to a desired energy and then focused into a slit of a cylindrical reaction chamber. The detecting system which samples the exiting ions as a function of angle consists of a 90° spherical electrostatic energy analyser (energy resolution 3%, angular resolution 2·5°), a quadrupole mass spectrometer, an ion detector and counting system.

These ion beam–reaction chamber instruments permit one to determine the energy and angular distribution of the ionic products of the reaction from which one can obtain detailed contour maps of the distribution of product ions. Although such instruments do have a disadvantage compared to crossed-beam devices in that they are used in the energy domain where the thermal energies of the target molecules are negligible they do have greater sensitivity and superior discrimination properties if used with electrostatic analysers. Mahan[13] attributes his success in observing the low intensity back-scattered product in the reaction $Ar^+(D_2, D)ArD^+$ even in the presence of a high intensity of a forward scattered product to the higher gas density in the reaction chamber and to the superior discrimination properties of the electrostatic analyser. Champion et al.[9] have also observed the back-scattered product but Wolfgang et al.[14] did not observe such products in the same reaction with their ion–molecular beam equipment.

B. Crossed-Beam Spectrometers

In Fig. 2 a schematic of a crossed-beam machine used by Herman's group in Prague is shown.[15] It will be noted that this spectromer uses magnetic

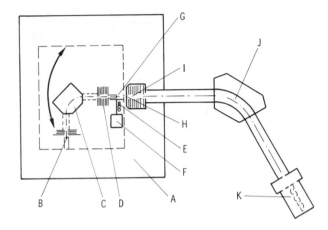

Fig. 2. Schematic diagram of a crossed-beam scattering spectrometer[15] A—vacuum chamber, B—ion source, C—primary mass selector, D—deceleration lens, E—molecular beam, F—chopper, G—collision zone, H—detection slit, I—energy analyser, J—analysing mass spectrometer, K—electron multiplier.

analysis in both primary and secondary mass spectrometers. A retarding grid assembly is used for product energy measurement and a Lindholm type lens system is used for adjusting the projectile ion energy. Lock in detection technique is used. The angular spread of the beam is reported to be about 1° and energy spread is a few tenths of an eV FWHM.

A very similar cross-beam apparatus is the one used by Wolfgang's group[16] at Yale University. The ions again are produced by electron impact. They are then accelerated and mass analysed by a 180° permanent magnet and then decelerated to a desired energy by a lens system designed by Gustafsson and Lindholm.[17] The ion beam then collides with a beam of target molecules. The ionic products are then energy analysed by a retarding grid analyser and then mass analysed with a 60° magnetic-sector mass spectrometer and registered by an electron multiplier. The neutral beam is chopped and phase sensitive detection is used to distinguish between reactions occurring in the crossed-beam zone and those occurring in the scattered background gas. The lowest useful laboratory projectile energy is 0·5 eV. The angular spread of the beam is less than 2° FWHM and the energy spread ranges from 0·2 to 0·5 eV.

The main advantage of the crossed-beam configuration is a better definition of the collision energy which is an important feature at low projectile energies since if a reaction chamber is used one introduces a complicating factor in the form of the random thermal motion of the target molecules.

C. Internal Energies of Projectile Ions

When projectile ions are prepared by electron bombardment internal excitation generally occurs. In monoatomic ions one has to deal only with electronic excitation, however, in polyatomic ions one is faced with a more difficult problem since rotational and vibrational excitation are also present and may be more difficult to eliminate or to take into account. In the past this complication has either been neglected or a rationale was established that the distribution of internal energy had a negligible influence on the final conclusion reached. In more recent times some attention has been given to this problem and some success realized in preparing monoatomic projectile ions in either the ground electronic state or in a known mixture of a few electronic states.

A technique that has been used successfully to determine beam composition, as far as electronic states are concerned, was first applied to the area of ion–molecule reactions by Turner, et al.[18] The method consists of measuring the attenuation of the beam in an appropriate gas as a function of the pressure of that gas. When an ion beam containing ions in a single electronic state (i) passes through a reaction chamber of length l containing a gas whose pressure can be varied, it will be attenuated in an exponential manner

$$I_i = I_0 \exp\left(-n\sigma_i l\right) \tag{1}$$

where I_i is the ion current at the detector, I_0 is the ion current reaching the detector when the gas number density (n) is zero and σ_i is the sum of the cross sections for all processes by which ions of the ith type are lost. A semilogarithmic plot of I/I_0 vs gas pressure gives a straight line. If another ion in the jth state is present, the attenuation of the ion beam is given by the sum of the attenuations of each state.

$$I = (1 - f)I_0 \exp\left(-n\sigma_i l\right) + fI_0 \exp\left(-n\sigma_j l\right) \tag{2}$$

A semilogarithmic plot of I/I_0 vs gas pressure now gives a curved line which can be resolved into its two components, and f, the fractional composition of the beam can be obtained by extrapolating the curves to zero pressure. Typical results[19] for H_2O^+ produced by electron bombardment of water vapour are shown in Fig. 3. In Fig. 3b, 70 eV electrons were used and the water vapour pressure was $0.40\,\mu$. The attenuating gas was nitrous oxide. The plot clearly indicates the presence of more than one electronic state of H_2O^+. In Fig. 3a, again for 70 eV electrons but for a higher pressure of H_2O vapour, a single electronic state is indicated for H_2O^+. Appearance potential measurements calibrated against known ionization potential of argon show that under such conditions the ground state ($^2B_1$) of H_2O^+ is produced. Likewise in Fig. 3c the $^2B_1$ state is also obtained when the electron energy is lowered to 14 eV. The results show that the composition of the H_2O^+ beam,

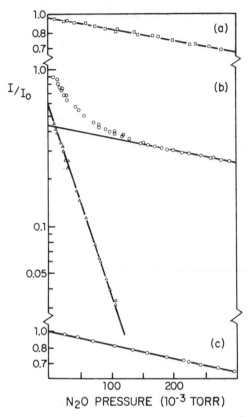

Fig. 3. Attenuation of H_2O^+ in N_2O (a) H_2O^+ kinetic energy is 50 eV, electron energy is 70 eV and H_2O pressure is $18.0\,\mu$; (b) same as (a) except H_2O pressure is $0.40\,\mu$, (c) same as (b) except electron energy is 14 eV.[19]

as far as the electronic states are concerned, is a function of electron bombarding energy and the water vapour pressure in the primary ion source. Consequently in this case one can vary the beam composition by varying the ion-source pressure or the energy of the bombarding electrons. A similar situation exists in the production of C^+ by electron bombardment of CO.[20] In some cases such as producing D_3O^+ by electron bombardment of D_2O it was not possible to eliminate the excited state of D_3O^+ merely by increasing the source pressure but it was possible to deactivate the higher state by adding a paramagnetic gas, NO, to the D_2O in the ion source.[21] Still in other cases such as the production B^+ in its ground state from a mixture of states, i.e. B^+ (^1S) and B^+ (^3P) it was not feasible to eliminate ^3P state by the addition of paramagnetic impurities or by collision deactivation. In this case it was found that electron bombardment of BF_3 produced a mixture of the ^1S and ^3P states of B^+ but in using BI_3 the B^+ was produced only in the ground state.[22] Using this technique it has been feasible to produce the projectile ion in a measurable mixture of electronic states or in the ground state so that the effects of each of the states could be determined. The technique in its present form does not help one with the vibrational or rotational states of diatomic and polyatomic ions. The same technique can in principle be applied to determining the electronic states of product ions but as far as is known by the author such applications have not yet been made.

Probably the most elegant way of preparing the projectile ions in well-defined electronic vibrational and rotational states is photoionization. Some impressive in–molecule work has been reported in this area.[23] However, widespread application of this technique to the scattering of ions by molecules will undoubtedly be delayed until the intensity problems are solved. Developments in the application of lasers to this area will probably play a decisive role in the problem.

The determination of the states of the product ions may be obtained from a study of the luminescence resulting from the collision of ions with molecules. Initial successes in this area have been realized by Brandt et al.[24] in the reactions $N^+(NO, N^+)NO^*$, $N^+(NO, O)N_2^+{}^*$, $Ne^+(CO, Ne)CO^+{}^*$, $Ne^+(CO, Ne, O)C^+{}^*$ and $C(NO, O)CN^*$. They were able to observe the N_2^+ first negative system corresponding to the transition $N_2^+{}^*(B\,^2\Sigma_u^+) \rightarrow N_2^+(X\,^2\Sigma_g^+) + h\nu$ in considerable detail. 10 Å resolution, photon counting and signal averaging were used. Again intensity problems will have to be solved before one can expect general application of this technique to the reactive scattering of ions.

III. REACTIVE SCATTERING OF POSITIVE IONS

Under the subject of reactive scattering of ions we will treat those binary collisions between ions and molecules in which chemical reaction takes place and for which some energy and angular distributions have been measured. No attempt will be made to make a complete coverage of the reported results but rather a representative coverage will be attempted. The overlap between this material and that covered by Herman and Wolfgang[4] will be kept to a minimum.

It has become customary to divide reactive collisions on the basis of mechanism of reaction, i.e. into persistent complex and direct mechanisms. The criteria for complex or direct mechanisms are purely operational since a convenient time scale is the rotational period of the system. Consequently if the projectile and reactant form a complex which lives for more than a rotational period the angular distribution of the products will be symmetrical about a line $\pm 90°$ to the beam direction and passing through the centre of mass of the system. On the other hand if the projectile undergoes reaction in a time span significantly smaller than the rotational period the product will be strongly peaked in the direction of the ion beam and its angular distribution will be asymmetric relative to the $\pm 90°$ line. There will, of course, be a grey area in which the collision complex stays together for a time less than the several rotational periods and consequently we will not be able to classify the reaction as direct or one going through a persistent complex.

A. Three Atom Ion–Molecule Systems

1. Reaction of H^+ with H_2

The simplest three atom ion–molecule reactions are those of H^+ with molecular hydrogen and its various isotopic variants. This is an interesting reaction not only because of its simplicity but the reaction involves a deep potential well represented by the stable H_3^+ ion and hence one has the interesting possibility that a long-lived collision complex may be involved. Some preliminary work on such reactions as $D^+(HD, H)D_2^+$, $D^+(HD, D)HD^+$ and $D^+(HD, D_2)H^+$ has been reported.[25] However, it is only recently that the complete details of the combined theoretical and experimental study of these reactions has become available.[26,27] The associated theoretical treatment of these reactions is the trajectory surface hopping TSH technique extended from the classical trajectory method.[28]

In the classical trajectory approach, if a potential energy surface is available, one prescribes initial conditions for a particular trajectory. The initial variables are selected at random from distributions that are representative of the collisions process. The initial conditions and the potential energy function define a classical trajectory which can be obtained by numerical integration of the classical equations of motion. Then another set of initial variables is chosen and the procedure is repeated until a large number of trajectories simulating real collision events have been obtained. The reaction parameters can be obtained from the final conditions of the trajectories. Details of this technique are given by Bunker.[29]

The trajectory surface hopping method is an additional extension of the classical trajectory method. Potential energy surfaces are constructed for each electronic state involved in the collision. In addition, a function has to be obtained that defines the locus of points at which 'hops' between surfaces can occur. Still another function is necessary which gives the probability of such jumps as a function of nuclear positions and velocities.[28] Diatomics-in-molecules surfaces approximated the two lowest singlet potential surfaces of H_3^+. The surfaces have been shown[30] to be in good agreement with accurate *ab initio* calculations by Conroy.[31]

Some typical experimental results from this study are given in Fig. 4 in which velocity contour diagrams for the reaction $D^+(HD, D_2)H^+$ were

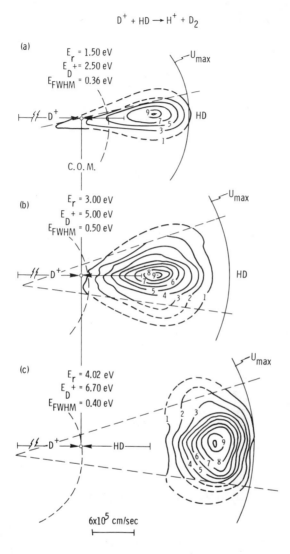

$$D^+ + HD \longrightarrow H^+ + D_2$$

(a)

$E_r = 1.50$ eV
$E_{D^+} = 2.50$ eV
$E_{FWHM} = 0.36$ eV

(b)

$E_r = 3.00$ eV
$E_{D^+} = 5.00$ eV
$E_{FWHM} = 0.50$ eV

(c)

$E_r = 4.02$ eV
$E_{D^+} = 6.70$ eV
$E_{FWHM} = 0.40$ eV

6×10^5 cm/sec

Fig. 4. Velocity contour plots for H^+ from the reaction $D^+(HD, D_2)H^+$ for the indicated collisions energies.[26]

obtained with a crossed-beam instrument. The product intensity is definitely forward of the centre of mass of the system at all energies studied indicating a direct reaction. In Fig. 5 the contour maps are shown for the reaction $D^+(HD, H)D_2^+$. Here the intensity distribution is symmetric to within experimental error about the centre of mass indicating a persistent complex intermediate with a life time longer than the rotational period of the system.

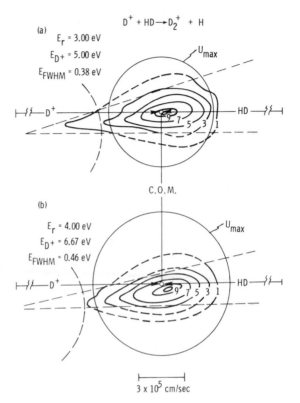

Fig. 5. Velocity contour plots for D_2^+ from the reaction $D^+(HD, H)D_2^+$ for the indicated collision energies.[26]

At higher energies the reaction was found to proceed by a predominantly direct, impulsive mechanism.

Total cross sections, translational energy distributions and velocity contour diagrams for many of the products for the $H^+ + D_2$ and $D^+ + HD$ reactions have been obtained from the crossed ion beam–molecular beam measurements and the TSH calculations making these reactions probably the best understood in the field. This combined experimental and theoretical investigation provided a unique opportunity for a detailed comparison of an *a priori* theory of chemical reactions with experiments. These comparisons are very encouraging since the TSH theory involves no empirical information or adjustable parameters.

A comparison of calculated cross sections with existing experimental results was good. The calculated (TSH) branching ratios were correctly predicted. There were some discrepancies. For example, the calculated cross section for the reaction $D^+(HD, D)H^+$ was larger than the experimental ones at $E_{rel} = 3\,eV$ and the calculated charge transfer cross sections were significantly larger than the observed ones at higher energies. Some of this disagreement was rationalized in terms of experimental difficulties. In general, however, the TSH cross sections were quantitatively correct to within probable experimental error and give strong support for the accuracy of the model used.

Product translational energy distributions are often more sensitive than cross sections to details of the collision dynamics. A typical result realized by the authors is illustrated in Fig. 6 in which the product relative energy distributions for the reaction $D^+(HD, D)HD^+$ are compared. The histograms represent the TSH results. It will be noted that at $E_{rel} = 3{\cdot}0\,eV$ and $4{\cdot}0\,eV$ the agreement is very good. The discrepancy in Fig. 6c is attributed to beam

PRODUCT RELATIVE ENERGY DISTRIBUTION

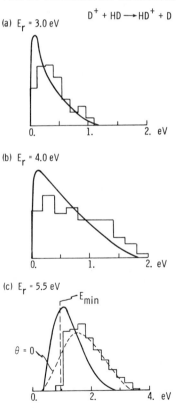

Fig. 6. Product relative energy distributions for the reaction $D^+(HD, D)HD^+$ for various relative energies.[26]

spread and the presence of a finite instrumental resolution. Product ion should not be observed with a translational energy lower than that indicated by the vertical dashed line in Fig. 6c since under such conditions the product ion, HD^+, has sufficient internal energy to lead to its dissociation. The authors concluded that within limitations imposed by the statistical uncertainty of the calculations and experimental resolution effects that there was complete agreement between theory and experiment in the product translational energy distributions.

Finally, a comparison of the experimental and theoretical velocity contour plots for the reaction $D^+(HD, H)D_2^+$ is given in Fig. 7. It will be noted that exact detailed agreement of such plots can not be expected because of instrumental beam spread and energy resolution. To within such limitations the agreement in Fig. 7 is good. It will be noted that the large barycentric

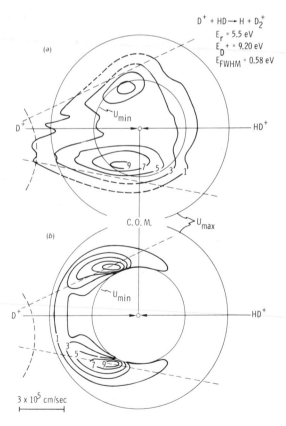

Fig. 7. Velocity contour maps for D_2^+ from the reaction $D^+(HD, H)D_2^+$ at $5\cdot5\,eV$ collision energy (a) experimental measurements (b) theoretical results.[26]

angle scatter of the product ion peaks are well accounted for by the TSH calculations.

The trajectory surface hopping model of ion–molecule reaction dynamics has realized an impressive agreement between theory and experiment in this reaction, i.e. $H^+ + H_2$, and it provides the experimentalist with a realistic and workable theory to use in the comparison with and interpretation of experimental results. As reliable potential energy surfaces become available for other ion–molecule systems, we can expect further tests of this theory and its applicability to more complicated reactions.

2. Reaction of Ar^+ with D_2

One of the most extensively studied ion–molecule reactions is the reaction of Ar^+ with molecular hydrogen and its stable isotopic variants.[4] Since the review of Herman and Wolfgang[4] other measurements on the reaction of Ar^+ with molecular hydrogen have been made and of particular interest are the results obtained by Mahan's[13] group in Berkeley. In this study both reactive and non-reactive scatterings were measured. Fig. 8 gives typical examples of the contour maps. It is clear from Fig. 8a that the product distribution is not symmetrical about the $\pm 90°$ line in agreement with earlier work. It shows a strong forward peak interpreted as arising from direct reactions and a practically isotropic contribution at large scattering angles. The latter has been interpreted as arising from head-on and near head-on collisions. Although these measurements were made at a relative energy of $2.72\,eV$ earlier measurements by Wolfgang[32] showed that the direct mechanism was present at energies as low as $0.05\,eV$. There is no reason to believe that going to lower energies would favour the persistent complex mechanism since it has been pointed out[32] that the long-range ion-induced dipole forces will always accelerate the reactants toward each other, so even at zero initial velocity the actual energy of impact would be a few tenths of an eV.

In Fig. 8b the non-reactive scattering exhibits a remarkable elasticity and it was shown to have a striking resemblance to Ar^+ scattered by He. Analyses of these results indicated that even the head-on collisions between Ar^+ and D_2 produce no reactions in the majority of cases at the relative energy of $2.72\,eV$. However, it is expected that the probability of reaction for head-on collisions is likely to tend to unity at sufficiently low energies.

Over the years as the reaction of Ar^+ with hydrogen molecules was studied there was a running controversy over whether the reaction proceeded by a direct mechanism or complex formation.[33] Part of this uncertainty stemmed from the fact that early investigations on cross sections which were carried out with a single-stage mass spectrometer showed an isotope effect which could be explained by assuming complex formation. This work was also confirmed by low energy measurements with tandem mass spectrometers. Later theoretical studies showed that such an isotope effect could also be

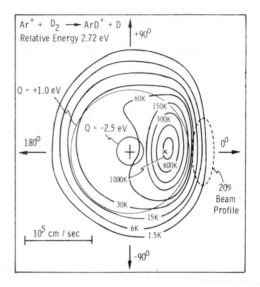

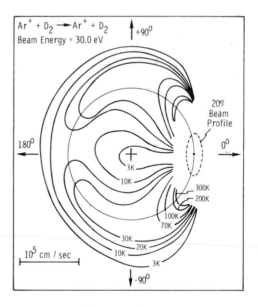

Fig. 8. The upper curve gives the contour map of the intensity of ArD^+ for the reaction $Ar^+(D_2, D)ArD^+$. The lower curve gives the contour map of Ar^+ scattered non-reactively from D_2. The circle passing through the beam profile is the locus of elastic scattering.[13]

expected from a direct mechanism. It is clear, however, that the experimental angular distributions of the product resolve the uncertainty in favour of the direct mechanism. Recently some theoretical studies having a bearing on the problem have been carried out and it is of interest to examine them.

Kuntz and Roach[34] carried out a series of classical trajectory calculations on reactions of the type $A^+ + BC \rightarrow AB^+ + C$. The key in this approach is the construction of a realistic potential function representing the reaction. Kuntz and Roach used a modification of a potential energy function previously reported by Polanyi and Kuntz.[35] A LEPS function[36] with spectroscopic parameters adapted to the isolectronic reaction $Cl(D_2, D)DCl$ is added to a potential which contains the long-range ion-induced dipole forces peculiar to the reaction $Ar^+(D_2, D)ArD^+$

$$V(r_1, r_2, r_3) = V_{LEPS}(r_1, r_2, r_3) + \mu(r_1, r_2, r_3) \qquad (3)$$

where r_1, r_2 and r_3 are the AB, BC and CA internuclear distances, respectively. The ion-induced dipole potential, $\mu(r_1, r_2, r_3)$, is

$$\mu(r_1, r_2, r_3) = -\frac{e^2}{2}\left[\frac{\alpha(r_2)}{r_1^4} + \frac{\alpha(r_2)}{r_3^4}\right] + \beta\left[\frac{1}{r_1^n} + \frac{1}{r_3^n}\right] \qquad (4)$$

where $e^2 = 332 \cdot 11$ Å kcal/mol and $n = 9$. The constant β is chosen so the function $\mu(r_1, \alpha, \infty)$ has the same equilibrium distance as the corresponding one in the LEPS function

$$\beta = 2\alpha_\infty e^2 (r_1^0)^{n-4} \qquad (5)$$

r_1^0 is the ground state bond distance and

$$\alpha_\infty = \lim_{r_2 \to \infty} \alpha(r_2) \qquad (6)$$

In most surfaces $\alpha(r_2)$ was equal to the polarizability of the hydrogen atom.

The angular distributions realized from these calculations were forward peaked for all energies studied ($E = 5$ to 150 kcal). No evidence was obtained for complex formation. The authors also discuss their results in terms of a 'migration' mechanism,[37] i.e. in the reaction $A^+ + BC$, A^+ has the ability to switch from B to C and there is a definite probability of both products, AC^+ and AB^+, being formed. It will be noted that this mechanism is very different from another direct mechanism, the so-called spectator stripping mechanism in which A^+ picks up the atom B without any transfer of momentum to the atom C, the product ion moves strictly in the forward direction with the same momentum as possessed by the projectile ion. In such a mechanism conservation of energy and momentum give rise to a critical energy beyond which the product ion should not exist since it possesses an internal energy which exceeds its bond energy. This critical energy is given by

$$E_{crit} = (D - W)\frac{M + m}{m} \qquad (7)$$

where W is the heat of reaction, D is the dissociation energy of the product
ion, M is the mass of the projectile ion and m the mass of the transferred
atom.[38]

Most three atom ion–molecule reactions that exhibit direct mechanism
behaviour do not proceed solely by a spectator stripping mechanism since
product ions exist far beyond the critical energy. The migration mechanism is
an attractive candidate for the direct mechanism since it contributes to the
simultaneous occurrence of forward scattering, large momentum transfer
to the product atom and stability to the product ion far beyond the critical
energy of the spectator stripping model. It is, of course, clear that the actual
mechanism is a mixture of a number of direct processes.

Recently Tully[27] reported on some unpublished work of Chapman[39] and
Preston using the trajectory surface hopping method coupled with a 'dia
tomics-in-molecules' surface[40] which was constructed as a workable surface
for ion–molecule reactions of the rare gases with hydrogen by Kuntz and
Roach.[41] The cross sections and the centre of mass angular distribution
were calculated for $Ar^+ + H_2$ and again good agreement with experiment
was realized for both properties. It therefore appears that both theory and
experiment agree that in this reaction no persistent complex formation
should be expected and none is observed.

3. Reaction of Kr^+ with D_2

Another interesting three atom ion-molecule reaction involving a rare gas
is $Kr^+(D_2, D)KrD^+$. The product ion intensity of KrD^+ at two energies
are shown in Fig. 9.[42,43] This reaction has a cross section ten times smaller
than $Ar^+(D_2, D)ArD^+$. At the centre of mass energy of 2·7 eV the product ion

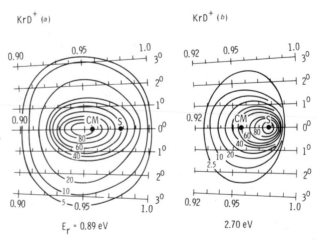

Fig. 9. Contour maps for KrD^+ for the reaction $Kr^+(D_2, D)KrD^+$
at 0·89 eV and 2·70 eV centre of mass collision energies.[42]

intensity peaks at the point S indicating a stripping or direct mechanism contribution. However, the 'hard sphere' contribution is of relatively greater importance here than the Ar^+ case (see Fig. 8). This is represented in Fig. 9b by the lower intensity contours which are concentric about the point CM. At the low collision energy of 0·89 eV the peak at S has disappeared and lies close to CM. Henglein concludes from these contour maps that in the Kr^+ reactions, collisions with strong repulsion between the products occur much more frequently than in the Ar^+ reaction. The near isotropic product distribution in the Kr^+ reaction, where apparently a direct mechanism is operative, illustrates the care observers must take in interpreting contour maps lest erroneous conclusions are drawn. Henglein[43] attributes the isotropic distribution to the scattering of the particles by the steep repulsive part of the potential energy curve which is reached in small impact parameter collisions. In such collisions a simplified view of the process may consist of interaction of the projectile with both atoms of the molecule followed by the pick up of one of the atoms and the result is 'impulsive isotropic' (hard sphere) scattering.

4. Reaction of N^+ with H_2

The reaction of N^+ in its ground electronic state (^3P) with H_2 is another interesting three atom ion–molecule system.[44] In the 2·5–8·1 eV range of initial projectile ion relative energies, the reaction exhibited predominant forward scattering indicative of a direct interaction mechanism. It is interesting that in this system a deep potential well is present as indicated by the thermodynamically stable NH_2^+ ion normally observed in the mass spectra of appropriate nitrogen compounds, however, the experimental results indicate that this well is not accessible to the reactants and no persistent complex is formed. A study of the translational exoergicity (difference in relative translational energy of products and reactants) shows that the forward scattered NH^+ ions are more internally excited than expected from the predictions of the spectator stripping model, indicating that there is coupling of all atoms even in the higher energy grazing collisions. The product internal energy was observed to decrease as the scattering angle increased. The variation of the most probable internal energy of the forward scattered products with initial relative energy of the projectile energy suggests that below 4 eV the principle product is $NH^+(a\ ^4\Sigma^-)$ while above 6·5 eV the dominant product is $NH^+(\tilde{X}\ ^2\Pi)$.

5. Reaction of N^+ with O_2

The $N^+ + O_2$ reaction, which is a very important upper atmosphere reaction, has also been studied by the crossed-beam method[45] over an initial barycentric energy range of 1·2 to 8·3 eV. The important features of the mechanism can be readily deduced from the experimental results. The NO^+ product ion distribution indicated a direct mechanism and at low initial

energies the Q of the reaction followed the Q expected from spectator stripping
At 5 eV the value levelled off to a plateau. One may conclude therefore tha
the reactive collisions did not involve the complex NO_2^+ in spite of it
thermodynamic stability. The position of the plateau was consistent only
with the production of $O(^3P)$ and a state of NO^+ which, with excess energy
dissociates to $N(^4S^0)$ and $O^+(^4S^0)$. Nine lowest energy channels for the
collision of N^+ with O_2 were considered. Five of these gave the NO^+ product
From a knowledge of the NO^+ electronic states, it was deduced that the NO^+
product must be formed in either the $\tilde{X}\ ^1\Sigma^+$ or $^3\Sigma^+$ state, neither of which
could be formed adiabatically from the initial channel. The $^1\Sigma^+$ state wa
eliminated since the reaction involving it would be highly exothermic
($-6\cdot6$ eV) and this was expected to lead to a product translational energy fa
greater than expected from spectator stripping. On the other hand if the
electronically excited $NO^+(^3\Sigma^+)$ product was formed, $6\cdot31$ eV of the exces
energy supplied by the exothermicity would be used up. The exothermicity
of the reaction would be $-0\cdot29$ eV and the product ion energy might b
expected to be close to the spectator stripping value. Further consideration
of the electronic correlation diagram of NO_2^+ led to the following probable
mechanism. At moderately large separation $N^+(^3P)$ and $O_2(^3\Sigma_g^+)$, originall
on a surface correlating with linear NO_2^+ undergoes an electron jump
leading to a surface correlating with $NO_2^+(^1\Sigma^-)$. Then as the reactant
approach closer an O atom is transferred leading adiabatically to $O(^3P)$ +
$NO^+(^3\Sigma^+)$.

6. Reaction of O^+ with N_2

The reaction $O^+(N_2, N)NO^+$ is of considerable importance in aeronomy
since it is a major source of NO^+ in the F layer of the ionosphere. It is also c
interest from the point of view of chemical kinetics since it is one of the few
exothermic ion molecule reactions that exhibits an activation energy. A
number of rate or cross section measurements have been made on thi
reaction.[46,47,48,49,50,51] In the thermal region 100–600°K the rate of reaction
falls with increasing temperature and in the eV region the measured cross
section rises and goes through a broad maximum at about 5 eV and then fall
again. No kinematic studies have been made of this reaction until recently
Leventhal[52] measured the kinetic energy distribution of the NO^+ ion at 0
to the beam and covering a projectile energy range of 3–12 eV. A plot of th
Q value vs projectile energy showed that the reaction followed a stripping
type collision process. More recently Smith and Cross[53] made a study of th
angular and energy distribution of the products over the laboratory energy
range of $1\cdot5$–$15\cdot8$ eV. The product contour maps are shown in Fig. 10 and i
is clear that a direct mechanism is indicated and no evidence was found fo
persistent complex formation. The variation of the translational exoergicit
(Q) with reactant relative translational energy also supported this conclusion

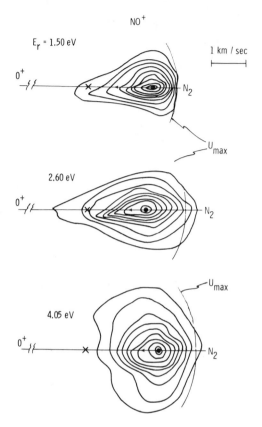

Fig. 10. Contour maps for NO^+ for the reaction
$O^+(N_2, N)NO^+$ at relative energies of 1.50 eV,
2.60 eV and 4.05 eV.[53]

In order to explain the unusual kinetics of this reaction Kaufman and
Koski[54] drew a set of schematic potential energy curves (Fig. 11) based on
the then known theoretical, optical and mass spectroscopic results. The spin
and symmetry rules indicated that both ground state reactants $O^+(^4S_u)$ +
$N_2(^1\Sigma_g^+)$ and ground state products $NO^+(^1\Sigma^+)$ + $N(^4S_u)$ are permitted to
combine uniquely on to a $^4\Sigma^-$ state of the intermediate N_2O^+. The $^4\Sigma^-$
repulsive curves cross the attractive potential curves of the ground and some
of the excited states of the intermediate N_2O^+. It is both spin and symmetry
forbidden to go directly from ground state reactants to the $^2\Pi_i$ ground state
of N_2O^+ and then in turn to ground state products. It is this forbiddenness
and the fact that the ground state products and ground state reactants can
combine uniquely only to a $^4\Sigma^-$ state that apparently account for the low
reaction efficiency in the thermal energy region reported by Schmeltekopf

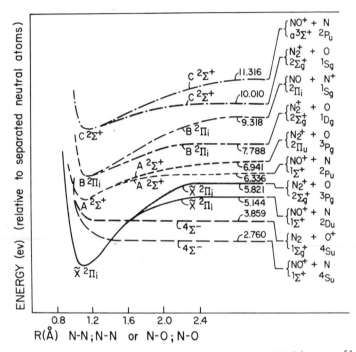

Fig. 11. Schematic potential energy diagram of the N–N–O$^+$ system.[54]

et al.[55] The absence of an activation energy in the thermal energy region was attributed by Kaufman and Koski[54] to the perturbation of the potential energy curve in the reactant channel by the ion-induced dipole force. The increase of the cross section with energy in the eV region was interpreted as arising from crossings of the ground state repulsive $^4\Sigma^-$ curve with the higher lying attractive states of the intermediate opening up additional channels for reaction. Smith and Cross[53] have expressed concern about this mechanism since the spin–orbit interaction which couples the $^4\Sigma^-$ and $^2\Pi$ state and causes the surface hopping might be expected to be small for light atoms and hence may predict a cross section that is too small for the reaction. They in turn prefer a similar mechanism but involving the $^4\Sigma^-$ and $^4\Pi$ surfaces. This explanation suffers from the disadvantage that the $^4\Pi$ may not be sufficiently low for the O$^+$ + N$_2$ system to jump from the $^4\Sigma^-$ to the $^4\Pi$ surface at low energies. The $^4\Pi$ state was not included in the schematic potential of Kaufman and Koski but is shown in the *ab initio* configuration interaction calculations of Pipano and Kaufman[56] which were carried out to verify the earlier qualitative arguments. Cuts through the calculated surfaces are shown in Figs 12 and 13. The $^4\Pi$ curve is a complicated one

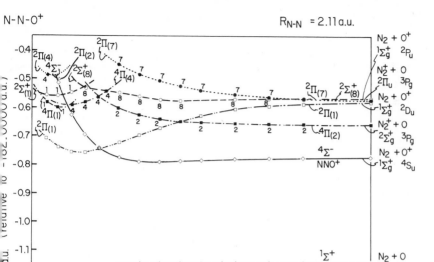

Fig. 12. Configuration interaction energies of NNO$^+$ vs R$_{N-O}$ a.u. [O^x approaching N$_2^y$ linearly, $x + y = 1$, R$_{N-N}$ = 2·11 a.u.].[56]

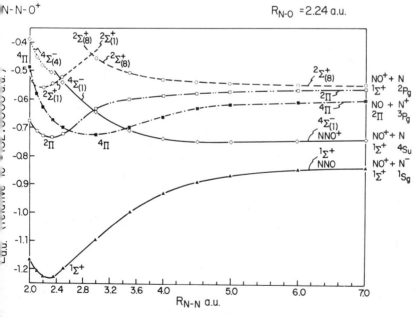

Fig. 13. Configuration interaction energies of NNO$^+$ vs R$_{N-N}$ a.u. [N^x receding from NO^y linearly, $(x + y = +1$, R$_{N-O}$ = 2·24 a.u.].[56]

with several loops and it appears to be located too high to realize a crossing of the $^4\Sigma^-$ curve with the $^4\Pi$ at low energies. The question therefore arises as to how accurately are the calculated surfaces located relative to each other since the calculations were made with a limited basis set. The authors (PK) feel that they are accurate to within 1 eV relative to one another. If this is correct the $^4\Sigma^- - ^4\Pi$ mechanism would not be feasible. It is clear that only additional work on the calculations of *ab initio* surfaces with a larger basis set and some reliable trajectory surface hopping calculations will resolve these mechanistic details. The important feature of such studies from the chemical dynamics point of view is not the narrow details but rather the broad implications that such investigations show the role of spin, symmetry, potential surface crossings and so forth in determining the course of a chemical reaction.

Before leaving this reaction it may be of interest to point out a contribution that this study makes to our criteria for persistent complex formation in ion–molecule reactions. It has been pointed out that in order to realize a persistent complex one must have a deep potential well and the system must have access to it. In the reaction $O^+(N_2, N)NO^+$ these requirements have been met yet experiment indicates a direct mechanism and no evidence of complex formation. Pipano and Kaufman[56] point out a rationalization for this behaviour. As the reactants come in on the $^4\Sigma^-$ surface and they cross into the $\tilde{X}\,^2\Pi_i$ state of the N_2O^+ the system is now in a deep well, however, as vibrations take place the system is oscillating through a lower crossing point (where the product $^4\Sigma^-$ curve crosses $\tilde{X}\,^2\Pi_i$ surface) where there is a certain probability of dissociating to the ground state of $NO^+ + N$ along its unique $^4\Sigma^-$ curve. Since there are a number of vibrations taking place for each rotational period, the lower crossing point will be traversed a number of times and if the transition probability has an appropriate value the N_2O^+ will dissociate into products, $NO^+ + N$, in a time less than several rotation periods and a direct mechanism will be observed experimentally. It would therefore appear that we should include in our criteria for persistent complex formation, not only the existence and the availability of a potential well of a suitable depth, but also the conditions for the system to reside in the well for a sufficiently long time.

7. Reaction of O^+ with D_2

The reaction of $O^+(^4S)$ with D_2 has been studied by Harris *et al.*[57] at a relative energy range of 0·76–3·58 eV using ion beam techniques. No angular scans were made so only the velocity spectra at $0°$ were reported. Excited electronic states of OD^+ are not accessible below about 1·3 eV, so below this energy the internal energy in the ground state ($\tilde{X}\,^3\Sigma$) of OD^+ was in the form of vibrational–rotational excitation. It was found that OD^+ was formed with internal energy over the entire projectile ion energy range. The Q vs projectile ion energy gave a linear relation which followed the spectator stripping value.

Gillen, Mahan and Winn[58] have studied the dynamics of the $O^+ - H_2$ reaction and have examined the product velocity vector distribution for the reaction $O^+(H_2, H)OH^+$ as well as corresponding reactions with HD and D_2 over a range of relative energies from 4·5 eV to 11·9 eV. It was found that the reaction proceeded by direct interaction in which the spectator stripping was prominently featured. At relative energies in excess of 8 eV evidence was found for the production of OH^+ in the $^1\Delta$ state. At lower energies the ground state reactants gave rise to ionic products predominantly in the ground electronic state.

8. Reaction of C^+ with D_2

The $C^+(^2P) + D_2$ is a three-atom endothermic ion–molecule reaction which provides an example of a reaction that proceeds at low energies by persistent complex formation and, as the projectile ion kinetic energy increases, the mechanism gradually changes to a direct mechanism. This type of transition from complex to direct mechanisms is not unusual in polyatomic systems but is rare in three atom systems. The energy and angular distributions of CD^+ formed by the reaction $C^+(D_2, D)CD^+$ were measured[59] for barycentric energies between 3·47 eV and 9·14 eV. The contour maps which are given by Iden et al.[59] showed a product ion distribution which was nearly isotropic and symmetrical within experimental error about the barycentric angle of $\pm 90°$. This symmetry was interpreted by the authors as indicative of persistent complex formation with a lifetime exceeding the rotational period of the system (10^{-12} sec). As the kinetic energy was increased stepwise up to 9·14 eV an asymmetry developed until finally the product intensity distribution was well forward of the centre of mass of the system. The sequence of contour maps showed the change from a complex mechanism to one that is completely direct. In addition, the plot of the translational exoergicity vs primary ion energy showed that Q fell with projectile ion energy along a curve expected for complex formation rather than stripping. An approximate plateau was reached when the bond dissociation energy of CD^+ was reached.

More recently Mahan and Sloane[60] have made a more detailed study of the $C^+ - H_2$ system using reactions of C^+ with HD, and D_2 as well as C^+ ion-reactive scattering to illucidate the dynamics. Their main conclusions are similar to those reached by Iden et al.[59] with one difference. In the 2–3 eV relative energy range the product velocity vector distributions from $C^+(H_2, H)CH^+$ and its isotopic variants showed a high degree of symmetry about the barycentric angle of $\pm 90°$, however, there was a small but measurable deviation from perfect symmetry. Mahan et al.[60] interpret their small deviation from perfect symmetry and the results of the non-reactive scattering measurements as being consistent with a strong interaction between all three atoms which lasts for approximately one rotational period of the $C^+ - H_2$ collision complex.

B. Four Atom Ion–Molecule Systems

In the above category we have attempted to review some interesting highlights in the studies of three atom ion–molecule reactions. Such relatively simple systems are more amenable to theoretical considerations and quantitative calculations, and as such may be more likely to contribute more to our understanding of the details of chemical dynamics, consequently they were stressed more. We now turn our attention to the four atom systems and again no attempt will be made to make an exhaustive coverage of the literature but rather to illustrate the interesting results of representative investigations in reactions of positive-ion reactive scattering. Comparatively few four atom ion–molecule reactions have been studied, however, several of those that have appeared in the literature have interesting features in them.

1. Reaction of H_2^+ with H_2

The simplest of four atom ion–molecule reactions is the $H_2^+ + H_2$ reaction and its stable isotopic variants which has been reviewed recently.[4] However, it is interesting to reexamine it in view of subsequent measurements. Early studies[64,65] indicated several direct channels at all energies and at low energy the H_3^+ product ion intensity contours had a symmetry around the centre of mass that was interpreted as arising from a persistent complex formation. Recent measurements by Henglein[43] on the elastic scattering of H_2^+ by H_2 enabled him to place the energy level of H_4^+ about midway between $H_2^+ + H_2$ and $H_3^+ + H$. Hence H_4^+ is unstable relative to $H_3^+ + H$ and since this energy level diagram indicated no potential well for H_4^+ he concluded that this reaction was not proceeding by a persistent complex and that the observed symmetry of the H_3^+ ion intensity around the centre of mass was due to impulsive isotropic (hard sphere) scattering as in the case of $Kr^+ + H_2$.

2. Reaction of O_2^+ with D_2

The reaction $O_2^+(D_2, D)O_2D^+$ has been studied in two laboratories. In Mahan's[61,62] group the reaction was studied with O_2^+ predominantly in the ground state. In this state the reaction is endothermic by 1.9 eV and at a relative energy of 3.86 eV the reaction appears to proceed by a persistent complex mechanism as indicated by the contour maps for O_2D^+. As the projectile energy is increased the mechanism gradually changes to a direct one. Henglein's group,[63,43] on the other hand, used O_2^+ in a mixture of the ground state ($^2\Pi_g$) and excited state. The reaction with the excited state is exothermic by -2.0 eV; at a relative energy of 1.0 eV the ground state will not react and the contour map in Fig. 14 indicates a direct mechanism. As the energy is increased both states react and the contour maps in Fig. 14 have been interpreted as arising from contributions from both direct and complex mechanisms. Henglein also measured the velocity spectra and obtained two

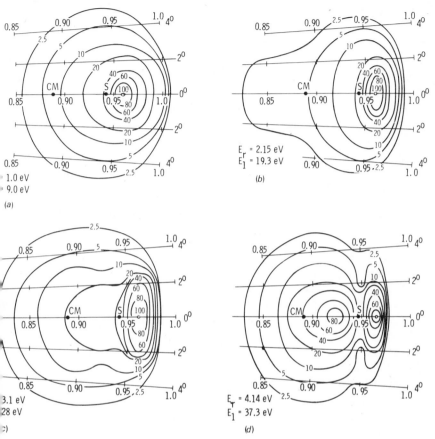

Fig. 14. Contour maps for O_2D^+ ions from the reaction $O_2^+(D_2, D)O_2D^+$ at various initial energies.[63]

peaks in the intensity vs energy plot. A stripping peak due to reaction with $O_2^+({}^4\Pi_u)$ and a complex peak resulting from the reaction $O_2^+({}^2\Pi_g)$ with HD were observed consistent with the contour maps. Application of Kassels' equation at 3 eV to the ground state reaction of O_2^+ gives a lifetime of about 10^{-10} sec. The small cross section (0·1 Å) and an absence of large angle scattering[61] implies that only small impact parameters lead to reaction and consequently the angular momenta are small and most of the collision energy ends up in vibrational modes in the complex.[63]

3. Reaction of N_2^+ with H_2

This reaction also has been previously reported by Herman et al.,[4] however, recent work reported by Henglein[43] on elastic scattering of N_2^+ in H_2 has a contribution to make to the interpretation of this reaction and is commented on at this point. The elastic scattering of N_2^+ with H_2 permitted the positioning of the energy of $N_2 - H_2^+$ between that of $N_2^+ + H_2$ and the products of the reaction $N_2^+(H_2, H)N_2H^+$. The scattering measurements indicated that no potential well was present for $N_2 - H_2^+$ system so it would be expected that the reaction should proceed by a direct mechanism in agreement with experiment. The fact that a stable ion $N_2H_2^+$ exists in the mass spectra of hydrazine would suggest a deep well for the complex if the system was proceeding by this path. However, the reaction apparently proceeds on a part of the potential hypersurface far away from the deep well. Presumably the complicated rearrangement of heavy particles to produce the intermediate $HNNH^+$ is unlikely and the more probable mode is an intermediate of the type $N-N-H_2^+$.

C. Polyatomic Ion–Molecule Systems

The reaction scattering in polyatomic ion–molecule systems has been studied in a number of cases and both direct and persistent complex mechanisms have been observed. Although these reactions are too complex to be amenable to the detailed theoretical analysis in the manner that three atom ion–molecule systems are, they can provide us with valuable information on internal energy equilibrium, uni-molecular reaction rates, life times of short-lived intermediates, and so on.

The reactions of $X^+(CD_4, CD_3)XD^+$ where X^+ is N_2^+ or Ar^+ have been studied by Henglein's group[66] and by Mahan;[67] a spectator stripping type of mechanism was observed. The energy balance as determined by the latter group indicated a considerable amount of excitation energy residing in the CD_3 group showing that the reaction is not proceeding by an ideal spectator stripping process. Likewise crossed-beam studies of[68] the reaction $CH_3^+(CH_4, H_2)C_2H_5^+$ showed an asymmetric distribution of products about the centre of mass confirming an earlier measurement[69] indicating a direct process. The reaction $N_2^+(D_2O, OD)N_2D^{+}$[70] has also been reported to proceed by an impulsive mechanism.

1. Reaction of $C_2H_4^+$ with C_2H_4

The reaction of $C_2H_4^+$ with C_2H_4 has been reviewed previously,[4] however, a more recent study of this reaction and its application to tests of models of uni-molecular decay make it worthy of additional comment. The reaction can be represented by the process[71]

$$C_2H_4^+ + C_2H_4 \rightarrow \begin{array}{c} [C_4H_8^+] ----- \\ \nearrow \qquad\qquad \searrow \\ CH_3 + C_3H_5^+ \rightarrow C_3H_3^+ + H_2 \\ \searrow \qquad\qquad \nearrow \\ ------- \end{array}$$

$C_3H_5^+$ and $C_3H_3^+$ are final products and $C_4H_8^+$ is a possible intermediate.

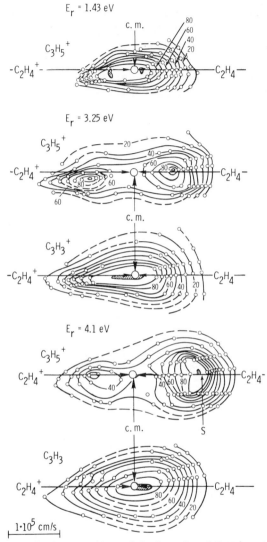

Fig. 15. Contour maps of the intensity of $C_3H_5^+$ and $C_3H_3^+$ from the reaction $C_2H_4^+ + C_2H_4$.[71]

The contour maps at several relative energies are given in Fig. 15. At the lower energies the distribution of the product is to within experimental error symmetric with respect to the plane passing through the centre of mass and normal to the collision axis indicating a persistent complex with a lifetime exceeding the rotational period of the system. As the projectile energy is increased an asymmetry begins to appear in the $C_3H_5^+$ distribution indicating a shortening of the lifetime of the complex, $C_4H_8^+$. The $C_3H_3^+$ product, which presumably results from the decomposition of a highly excited $C_3H_5^+$, has a distribution that exhibits considerable symmetry. The fact that the lifetime of the complex could be varied by varying the projectile energy led the group at Yale University[72] to use this reaction as a diagnostic tool to check the role that inernal energy equilibration plays in decomposition in a uni-molecular process. Each of the contour maps were integrated over three dimensional velocity space to yield $P(E_t')$, the probability of decay to products having a relative translational energy E_t'. $P(E_T')$ was then calculated from transition state theory[73] assuming that the only energy barrier to decomposition is the centrifugal barrier. This is frequently the case in exothermic ion–molecule reactions. The expression is for a $1/r^4$ potential

$$P(E_t') = N_{vr}(E' - E_t')E_t'^{1/2} \qquad (8)$$

where E' is the total energy of the products and N_{vr} the density of vibrational–rotational states in the critical configuration which is strongly dependent on s, the number of degrees of vibrational–rotational freedom. The semi-empirical approximation of Rabinovitch[74,75] was used to calculate the expression $N_{vr}(E' - E_t')$. The comparison between experimental points and the calculated values is given in Fig. 16. The dashed curves give the calculations when the number of oscillators s was the 'full' number of oscillators whereas the solid line gives calculated results when an 'effective' number of oscillators was used. The effective s was 16·9, 15·1 and 13·0 for the energies 1·43, 2·11 and 3·26 eV, respectively. It is evident that if an effective number of

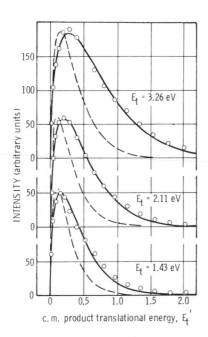

Fig. 16. Product relative translational energy distribution[72] $P(E_T')$; points are experimental; solid line: RRKM calculations using an effective $S(16·9, 15·1$ and $13·9$, respectively): dashed line: RRKM calculations using 'full' number of oscillators ($S = 27$).

oscillators was used excellent agreement could be realized between the experimental and calculated points. The authors concluded from these results that there is incomplete equilibration of energy in the complex and that the decomposition of the complex competes effectively with energy equilibration. Using RRKM theory[76,77] and the s eff obtained from the above theory the mean lives of the complex were calculated. These are all of the order 10^{-12} sec and decreased with energy. The fact that these lifetimes were only about an order of magnitude longer than vibrational periods is compatible with the conclusion that there was insufficient time to realize complete energy equilibration.

2. Reactions of $C_2H_2^+ + C_2H_4$ and $I_2^+ + C_2H_4$

The reaction

$$C_2H_2^+ + C_2H_4 \rightarrow [C_4H_6^+] \rightarrow C_3H_3^+ + CH_3$$

was studied in Herman's laboratory and a preliminary report has been made.[15] The contour maps at two energies are shown in Fig. 17. They

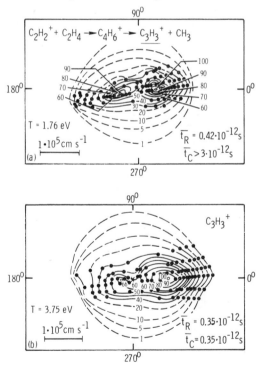

Fig. 17. Contour maps of the intensity of $C_3H_3^+$ from the reaction $C_2H_2^+ + C_2H_4 \rightarrow C_3H_3^+ + CH_3$. The indicated lifetime of the complex T_c and the mean rotational period, T_R, are estimated from the experiment.[15]

indicate a decreasing forward–backward symmetry over a barycentric energy range 1·7–3·5 eV. The distribution of relative translational energies of the products again indicates an incomplete equilibration of energy in the intermediate, $C_4H_6^+$.

A preliminary report has also been made[15] on the reaction $I_2^+ + C_2H_4 \rightarrow C_2H_3I^+ + HI$. The scattering map at barycentric energy of 0·25 eV has been reported as having a symmetrical distribution about the centre of mass indicating complex formation and, in addition, a forward asymmetric component due to a direct process was observed. The distribution of relative translational energy of the products associated with the complex mechanism appears to indicate involvement of a limited number of degrees of vibrational freedom, probably connected with the heavy atom skeleton of the complex.

D. Conditions for Complex Formation

The above examples of results on reactions proceeding by direct and persistent complex mechanisms show that if one obtains a contour map with the product ion intensity asymmetrically disposed about the $\pm 90°$ line one can with some certainty conclude that the reaction is proceeding by a direct mechanism. On the other hand, if one has an isotropic or symmetric distribution of the product ions one must interpret such curves with some care since such distributions may not unambiguously indicate complex formation judging from the results obtained on such reactions as the $Kr^+(D_2, D)KrD^+$ and $H_2^+(H_2, H)H_3^+$.

At this point it may be appropriate in concluding the paragraphs on the reactive scattering of positive ions to recall the conditions required for persistent complex formation.[78] The most obvious requirement for complex formation is that the system exhibit a potental energy curve of sufficient depth. Another requirement as pointed out by a number of authors is that the existing well must be accessible to the reactants, i.e. it is not prohibited by spin, symmetry, energy or other restrictions. The work on the $O^+(N_2,O)NO^+$ reactions suggests a third condition, namely that the system be able to reside in such a well for an appropriate time and not be able to escape because of the presence of a curve crossing, etc. Once the system is in such a well, one can then estimate the lifetime (t) of the complex from uni-molecular decomposition theory such as the RRKM[76,77] or more simply with the correspondingly rougher estimate by the RRK theory[79,80]

$$t \approx 10^{-13}[(E - E^*)/E]^{1-S} \tag{9}$$

where E = total energy of the complex, E^* = decomposition threshold energy and S = number of effective vibrational modes. Using such considerations Wolfgang[78] pointed out that a long life time will be favoured for the complex if (1) the total excitation energy is low, (2) high threshold for decomposition is present (endothermic reactions would be favoured) and (3) polyatomic products are formed since the internal energy can distribute itself over many vibrational modes. Such criteria are useful in rationalizing experimental results after the fact, however, they are rarely useful for predicting unambiguously that complex formation will occur since, short of a reliable calculation of a potential energy surface of the system, one cannot with certainty estimate the depth, availability of the well and other pertinent factors leading to a persistent complex.

Recently Mahan[81] has used molecular orbital correlations in ion–molecule reaction dynamics to explain several puzzling features in three atom ion–molecule and more complicated reactions and also to examine the method for predicting mechanisms of reactions. For example, stable ions of type H–C–H$^+$, H–N–H$^+$, H–O–H$^+$, exist in the mass spectra of appropriate compounds and consequently one might expect them to be intermediates in ion–molecule reactions of C$^+$, N$^+$ or O$^+$ with H$_2$.

If one considers a binary collision process, the molecular orbitals of the reactants interact and evolve into orbitals of the collision complex and then into those of the product molecules. One can, frequently with the help of Herzberg's[82] correlation tables, make a correlation diagram which shows the connection between molecular orbitals of products and reactants. One can then inquire whether the correlation diagrams permit the formation of an insertion complex such as H–X–H$^+$, linear or bent, since these are the structures of the corresponding mass spectral entities. In considering the cases of reaction of C$^+$, N$^+$, O$^+$ and F$^+$ with H$_2$, Mahan in applying his correlation diagrams concluded that in the absence of a_1–b_2 orbital interaction in no case can the ground electronic configuration of the intermediate HXH$^+$ be reached adiabatically from the reactants. All four reactants can give XH$^+$ by a direct interaction through a linear XHH$^+$ complex, consequently one might expect an asymmetric product ion distribution characteristic of a direct mechanism. These considerations indicate that the deep well represented by the collision complex HXH$^+$ is not accessible to the system and hence the reaction does not proceed on that portion of the surface. Indeed, the ground state reactions of N$^+$, O$^+$ and F$^+$ with H$_2$ do proceed by direct mechanisms, on the other hand the reaction of C$^+$ with H$_2$ apparently does not conform to these expectations. Preliminary *ab initio* configuration interaction calculations of the attack of C$^+$ on H$_2$ indicate that the system has a well of significant depth even for a linear approach.[83] Clearly some more research on this system should be fruitful. Mahan and Sloane[60] in a paper that appeared as this manuscript was completed report on an extension of their application of molecular orbital correlations to ion–molecule reactions by including electronic state correlations. The state correlations applied to the C$^+$ – H$_2$ system show that the orbital configurations do not need to be conserved[84] as the reactants approach and the relatively deep potential

well of the ground state of the symmetric CH_2^+ intermediate is indeed accessible to the reactants and can dissociate to the products $CH^+ + H$.

The answer to many questions that arise in binary collision processes lies
in a knowledge of detailed potential energy surfaces which for the three
atom systems are now attainable, however, at considerable effort and
expense.[85] Until such surfaces become available, we must rely on approximate or schematic potential energy curves constructed from empirical
calculations and experimental information to interpret reactive scattering
and rate data of ion–molecule reactions.

IV. INELASTIC SCATTERING OF POSITIVE IONS

In studies of collisional processes frequently questions associated with the
transfer of energy from translational into internal degrees of freedom or vice
versa arise. In this section we will not attempt to examine those inelastic
processes which involve chemical change, ionization or charge transfer. We
will, however, discuss those inelastic collisions which result in conversion
of energy from translational kinetic energy of the projectile ion into internal
degrees of freedom of the products. We can represent such processes by the
equation

$$A^+ + B \rightarrow A^+ + B^*$$

We will also discuss processes in which internal energy of the projectile ion
is converted into translation energy and can be represented by the equation

$$A^{+*} + B \rightarrow A^{+**} + B$$

where the product ion A^{+**} has more kinetic energy after collision than
the internally excited A^{+*} ion had before collision. This process is sometimes
called superelastic scattering. Obviously corresponding effects can take
place in charge transfer, reactive processes and in ionization processes, however, these will not be treated for lack of space. For a recent authoritative
review of this subject the reader is referred to Doering.[86]

A. Experimental

The apparatus used for studying energy loss spectra of projectile ions
interacting inelastically with target molecules is similar in principle to that
used for reactive scattering although more attention has to be given to such
factors as energy and angular spread and resolution of the ion beam, especially if one is attempting to study energy loss due to rotational excitation.
An example of such an instrument used by Toennies' group[87] at Gottingen
to study the energy loss of Li^+ ions passing through H_2 is shown in Fig. 18.
The Li^+ ion beam ($\sim 100 \, \mu A/cm^2$) is produced by a β-eucryptite surface
ionization source heated to about 1300°C. The resulting ions are then passed

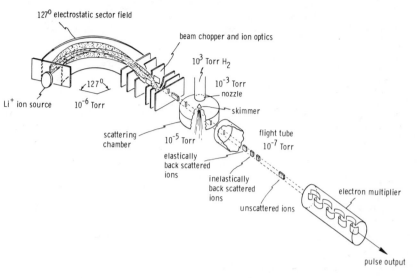

Fig. 18. Apparatus used to study $Li^+ + H_2$ vibrational and rotational excitation by a time of flight technique.[87]

through a 127° electrostatic analyser (resolution 0·5%, energy spread 40 meV). Then the ions are focused by an Einzel lens on to a 0·4 mm slit. Between the Einzel lens and the slit two 13 mm long deflection plates are provided for deflecting the beam. Short bursts of ions (50 nsec at 15 eV) are produced at a repetition rate of 10 KHz. The chopped ion beam is then focused on to the axis of the secondary H_2 beam which crosses the incident ion beam at 90°. The H_2 beam has an angular spread of $\pm 1\cdot 5°$ and an intensity of $\sim 5 \times 10^{20}$ particles/sterad. sec. The scattered ions then pass through a flight tube on to an electron multiplier. By measuring the ion flight times for two flight lengths, the difference of which can be accurately measured, the absolute energies of the incident and scattered ions can be determined. The time of flight spectra are measured by using a time–amplitude converter and a pulse height analyser. Several other instruments similar in principle but differing[88,89,90,91] markedly in design details have been built by others.

B. Electronic Excitation

There have been a number of investigations of the excitation of electronic states of atomic and molecular targets by positive ions in recent years. The earliest work using energy loss techniques in this area have been carried out by Park and his coworkers.[92,93,94] They worked in an energy range of 20–120 keV using H^+, H_2^+, He^+ and Ar^+ as projectile ions. From the inelastic energy-loss spectra of helium for proton impact they were able to obtain the total cross sections for the sum of $1'S \to 2'S$ and $1'S \to 2'P$ excitations and also the relative contributions of each were estimated.[92] The impact of H^+, H_2^+ and Ar^+ on N_2 gave peaks in the energy-loss spectra which were attributed to excitation of the Lyman–Birge–Hopfield system of N_2. A second energy-loss peak at $13 \cdot 8$ eV was believed to be primarily due to the excitation of the $b\ ^1\Pi_u$ band of the Worley–Jenkins series.[9] Significant differences were observed, when different ions were used as projectiles. With 20–110 keV protons on O_2 excitation of the Schumann–Runge dissociation continuum was realized and absolute cross sections were determined.[93] The energy resolution of the equipment was about 2 eV consequently no vibrational fine structure was observed.

Moore and Doering[88] have studied inelastic collisions of 150–500 eV H^+ and H_2^+ ions with N_2, CO, C_2H_2. Most of the observed energy losses were assigned to known electronic transitions in the targets. The cross sections of the processes were reported to be large, at times approaching 10^{-16} cm^2. Ion impact excitation in the cases studied appeared to be very specific because only certain types of transitions were excited in molecules which have a large number of available electronic states. In all target molecules studied, proton impact was found to excite only singlet–singlet transitions and in an isoelectronic series such as N_2, CO and C_2H_2 the same transition ($\tilde{X}\ ^1\Sigma \to\ ^1\Pi$) was excited. In the case of H_2^+ collisions both singlet–singlet and singlet–triplet transitions were observed with an intensity much greater than observed for proton or electron impact. It was suggested that the reason for the difference between the behaviour of H^+ and H_2^+ may be due to the operation of a process analogous to exchange scattering in electron impact.[86] Such a process would not occur with H^+ because of the absence of an electron in the projectile. On the other hand multiplicity-forbidden transitions such as the singlet–triplet transitions are produced with considerable intensity by H_2^+ impact. An example of this effect is shown in Fig. 19 where the results for H^+ impact and H_2^+ impact on C_2H_4 is shown.

Since the projectile ion energies were in excess of 150 eV and the observed scattering was strongly peaked in the forward direction, the formation of a long-lived complex between projectile ion and target was ruled out as an excitation mechanism. The authors suggest a much shorter lived association between projectile ion, such as H_2^+ or He^+, and the target molecule. During this association excitation of singlet–triplet transitions may occur by an electron exchange process with conservation of total spin angular momentum. The latter statement implies spin conservation and Moore[95,96] has made some recent measurements which have bearing on this point. He prepared a beam of N^+ containing the ground and two excited states. The ground state is a triplet and the two excited states singlets. The ^1D state is $1 \cdot 90$ eV and the ^1S state is $4 \cdot 05$ eV above the ground state. On collision of such a beam with a rare gas target, one has the possibility of superelastic scattering in which the

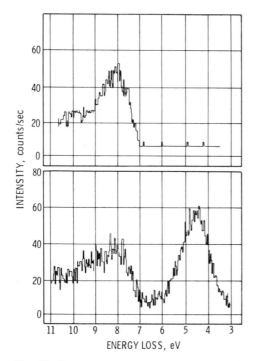

Fig. 19. Energy loss spectrum of 400 eV H^+ and H_2^+ ions.[88] The peak at 4·5 eV is due to a singlet–triplet transition to the lowest excited triplet state of ethylene. It will be noted that this peak is absent in the upper curve which represents the results for H^+ on C_2H_4. The lower curve represents the results of H_2^+ on C_2H_4.

energy of the excited electronic state is converted to kinetic energy of the target atom. Both the inelastic and superelastic collisions are observed when N^+ collides with He. Moore's results are illustrated in Fig. 20. The transitions are $^1S \rightarrow {}^1D$ and $^1S \leftarrow {}^1D$ (see Fig. 20b). Obviously spin conservation is realized in these transitions. It is clear on the other hand that the optical selection rules do not hold here since one is observing rather intense $^1S \leftrightarrow {}^1D$ transitions. Careful examination of the figure reveals weak peaks at $\pm 1·9$ eV corresponding to $^1D \leftrightarrow {}^3P$ transitions. These latter transitions violate the Wigner spin conservation rules but they are about 0·04 as intense as the singlet–singlet transitions. It appears therefore that in the case of N^+ collisions with He and other rare gases, the Wigner spin conservation rule is dominant.

Moore has also studied the collision of N^+ with O_2 which provides an interesting contrast with the $N^+ - He$ system, and which tends to put our

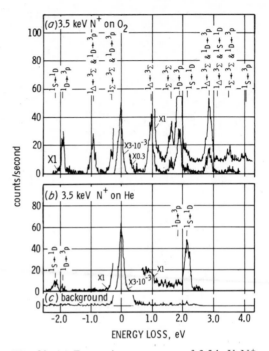

Fig. 20. (a) Energy-loss spectrum of 3·5-keV N⁺
scattered from O_2. The scattering angle was 0°
and the target gas pressure was 5·5 mtorr. (b) Energy
loss spectrum of 3·5-keV N⁺ scattered from helium
at 0° and 6·0 mtorr. (c) Energy spectrum of the
primary beam at 0° with no gas in collision
chamber.[96]

present knowledge of the phenomena into perspective. The results for oxygen
are given in the upper part of Fig. 20. The large number of transitions
observed arises from the large number of low lying excited states present in
O_2. The large intensity of the $^1D - ^3P$ transitions is probably due to the
large amount of spin–orbit coupling present.[86] Of all the transitions observed
in the $N^+ - O_2$ system only the weak peak at 3·13 eV (Fig. 20) which arises
from the reaction $N^+(^1D) + O_2(^3\Sigma_g^-) \rightarrow N^+(^1S) + O_2(^1\Delta_g)$ is due to a spin
non-conservative process. A study of the angular dependence of the scattered
N^+ ions shows that scattering is strongly peaked in the forward direction
indicating large impact parameter processes. No significant differences were
observed in the angular dependences of the inelastic and superelastic
processes.

It appears therefore that in some simple systems spin conservation is
realized, however, in the present state of our knowledge of these systems
generalizations should only be made with caution.

1. Vibrational and Rotational States in Electronic Transitions

In diatomic molecules in which the spacing between vibrational levels is greater than 0.1 eV, it would be expected that electronic transitions induced by inelastic scattering of positive ions should show evidence of individual vibrational transitions if the resolution of the equipment is sufficiently high. Indeed Moore and Doering[88] have reported some partially resolved vibration structure for a few transitions in N_2 and CO resulting from 500 eV H^+ bombardment. They compared the band envelope of their spectrum to the vibrational band predicted by the Franck–Condon factors of Benesch et al.[97], assuming all transitions to originate in the $v'' = 0$ level of N_2 ground state. They concluded that the process was due to a $\tilde{X}\,^1\Sigma_g^+ \rightarrow a\,^1\Pi_g$ transition. There was also a large amount of excess excitation to the higher vibrational states. It was concluded[86] that there was a breakdown of the Franck–Condon principle at low projectile energies as a consequence of the perturbation of the target molecule by the electric field of the incident ion.

A similar situation existed in the case of CO. The envelope of the observed transition agreed quite well with Nicholls'[98] Franck–Condon factors except that again the population of the higher vibrational states was larger than expected. More recently, Moore[99] has reexamined the N_2 and CO problem by exciting these molecules with 1–3 keV H^+, H_2^+ and N^+ ions. Again the distribution of the vibrational transitions in the electronic transition $\tilde{X}\,^1\Sigma_g^+ \rightarrow a\,^1\Pi_g$ was greater than predicted by the Franck–Condon principle. On the other hand vibrational transitions in other electronic transitions were well described by the Franck–Condon principle. Moore concluded that the anomalous intensity of the $\tilde{X}\,^1\Sigma_g^+ \rightarrow a\,^1\Pi_g$ transition was due to the presence of a second underlying transition. This situation is to be contrasted with the case in N_2 when charge transfer is involved. At projectile velocities below 5×10^7 cm/sec the intensities of the higher vibrational states become anomalously large.[24,86,100] However, this aspect of the subject will not be treated here.

As far as the author is aware there has been no reported resolution of individual rotational lines of vibrational bands in electronic transitions using inelastic scattering techniques. This development will have to await equipment with much better resolution. However, there have been some recent developments in scattering processes involving pure rotational transitions and this will be commented on below.

C. Vibration–Rotation Excitation

In this section we will consider vibration–rotation transitions within the ground electronic state. In the simplest molecule, i.e. H_2, the vibrational levels have a spacing of the order of 0.5 eV and in view of the fact that modern scattering spectrometers have resolutions less than 100 meV FWHM one would expect to see vibrational–rotational excitations in the inelastic scattering of positive ions. Moore and Doering[101] first reported such vibrational–rotational transitions in the $H^+ + H_2$ system. Using a few hundred eV protons they reported a cross section for the process in excess of 1 Å². This work was closely followed by the work of Moran and Cosby[102] on the vibrational excitation in the $Ar^+ + D_2$ system. Toennies'[103,104,87] group reported vibrational excitation of H_2 by Li^+ impact. Cheng et al.[105] and Cosby and Moran[106] have studied vibrational excitation in NO^+ and O_2^+ impacting on rare gases and Petty and Moran[107] have investigated the $CO^+ - Ar$ system. This experimental activity coupled with recent theoretical work on the $H^+ + H_2$ system,[108] the $Li^+ - H_2$ system[109,110,111] and on the $H^+ + HF$ system[112] indicate vigorous activity in the area and interesting progress can be expected in the near future.

1. Vibration–Rotation Excitation by Protons

The excitation of H_2 by proton impact is a particularly attractive system from a theoretical point of view since it involves a structureless particle and a molecule that can be put in a well-defined state interacting through an accurately known potential. Consequently the results from such a system should provide a significant test of theories for vibrational excitation.

Moore and Doering[101] first reported vibration–rotation excitation in the $H^+ + H_2$ system and more recently Herrero and Doering[113] have carried out a more complete study of this system. They covered an energy range from 25–1500 eV and determined the absolute cross sections for proton and deuteron impact excitation for the first four vibrational levels of the ground state of H_2. They used a high quality proton beam with an energy spread of 80 meV and an angular divergence $< \pm 1°$. Their cross sections as a function of ion energy for the various vibrational transitions are given in Fig. 21 and

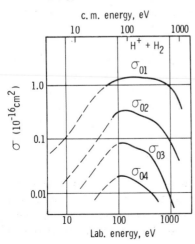

Fig. 21. Cross sections for transition from the $v' = 0$ vibrational state of H_2 to the first four excited vibrational states excited by protons of various energies.[113]

show that the vibrational excitation cross section in this system has a broad maximum in energy between 10 and 500 eV and then declines slowly. Essentially the same cross sections were obtained for D^+ collisions with H_2. It is interesting to note that although the instrumental resolution was not sufficient to resolve individual rotational transitions it was possible to conclude, from the shifts of the energy loss peaks, that the initial and final rotational states involved in the collision were the same at all energies involved in this experiment. Herrero and Doering were able qualitatively to explain features of their cross section vs energy curves by applying the theoretical considerations of Slawsky and coworkers[114,115] and also the work of Shin.[116] The work of Doering and coworkers has been confirmed and extended by Udseth et al.[117,118] The latter report the transition probabilities and differential cross sections for vibrational excitation in collisions of H^+ with H_2, HD and D_2 over a c.m. energy range of 4–21 eV and at c.m. angles between 6° and 36°. Again there seems to be an absence of significant rotational excitation.

2. Vibration–Rotation Excitation by H_2^+

In going from protons as projectiles to the simplest molecular ion, H_2^+, a considerable increase in complexity is reported as far as vibrational rotation excitation is concerned. This arises in part from the fact that the H_2^+ projectile can be in vibrationally excited states, consequently inelastic and superelastic processes can be expected. Herrero and Doering,[119,86] using a duoplasmatron ion source, produced 500 eV H_2^+ ions with a significant vibrational population up to $v' = 4$ and scattered such projectiles from the rare gases and H_2. They observed a number of inelastic and superelastic vibrational transitions. In going through the rare gases from Ne to Xe the cross sections for the excitation processes systematically increased suggesting that the target polarizability was playing a significant role in the process.

3. Vibration–Rotation Excitation by Other Ions

A number of inelastic scattering processes have been carried out which may be cited as examples of vibration–rotation excitation in which ions other than H^+ or H_2^+ were the projectiles.

Moran and Cosby[120] have reported the study of non-reactive scattering processes in low energy $Ar^+ - D_2$ collisions and observed the following processes.

$$Ar^+(^2P_{3/2}) + D_2(v = 0) \rightarrow Ar^+(^2P_{3/2}) + D_2(v = v')$$

$$\rightarrow Ar^+(^2P_{1/2}) + D_2(v = v')$$

where $v' = 0$, 1 and 2.

Energy loss spectra of H_3^+ passing through Ne with an energy of 16 eV were measured by Petty and Moran[121] at laboratory scattering angles of

$\theta = 0°$ and $10°$. The energy losses corresponded to those expected for vibrational excitation of H_3^+ and the frequencies obtained agreed with the quantum mechanical calculations.[122,123]

Toennies and his coworkers[103,104,87] have carried out a number of studies on scattering of Li^+ from H_2 and also have reported some preliminary results with N_2 and CO.[124] The use of Li^+ as a projectile is attractive since it is a closed shell system and by keeping the energy low one can easily eliminate complications from reactive scattering or electronic excitation. The reactions with N_2 and CO are interesting. The studies were carried out at $\simeq 7\,eV$ centre of mass energy and a barycentric angle of $37°$. Although these two systems are isoelectronic, have the same reduced masses and similar rotational and vibrational energy level spacings significant differences are observed in the time-of-flight spectra. The apparatus that was used was the one described in the experimental section above. In the case of CO two peaks, one for the elastic scattering ($\Delta v = 0$) and another for the vibrational excitation ($\Delta v = 1$), were clearly evident. In the case of N_2 the energy loss peaks were smeared out so that a single distorted peak was obtained. The authors attributed the distortion to a larger amount of rotation excitation in the N_2 case and they felt that the observed differences were related to differences in the shape of the potential hypersurface at small internuclear separations.

The $Li^+ - H_2$ system is particularly attractive for study since the potential energy surface[125] appears to be well known. Toennies and coworkers studied this system and were able to resolve clearly the elastic peak ($\Delta v = 0$) from the inelastic peak ($\Delta v = 1$). Some evidence was obtained for the presence of rotational excitation. The $\Delta v = 0$ cross section was measured and the value agreed to within experimental error with the calculated value using the spherically symmetric part of the Lester potential.[126] The $0 \rightarrow 1$ differential cross section was estimated to be about $0.13\,Å^2/sr$.

The situation is considerably more complicated if heavier ions, such as O_2^+, are used since the low lying electronic states are important even at low impact energies. This is illustrated by the study of Cosby and Moran.[106] In the system $O_2^+ - Ar$ a number of maxima were observed in the scattered ion intensity. Two states, $\tilde{X}\,^2\Pi_g$ and $a\,^4\Pi_u$, were present in the O_2^+ beam. Using the known spectroscopic constants for O_2^+ ions in the two states, satisfactory assignments to the various observed maxima were made. Measured vibrational transition probabilities were in reasonable agreement with Shin's[116] theoretical impact parameter model. It was found that multiquantum transitions were more important as the projectile ion kinetic energy increased from 10 to 20 eV. For a given projectile ion energy higher vibrational transitions predominated at larger scattering angles.

Moran and coworkers[107,127] have studied vibrationally inelastic collisions of CO^+ with Ar and satisfactory vibrational assignments were made.

In this study, analogous to the findings in the O_2^+ system, it was found that the probability of multi-quantum vibrational transitions increased with reactant ion energy and scattering angle. They also reported a large amount of rotational excitation. A comparison of the results obtained with the more complicated ions with those obtained from the simple ions such as H^+ and H_2^+ is of interest. For the heavier systems excitation cross sections reached a maximum at fairly low energies (20 eV) and multi-quantum transitions increased in importance with higher projectile kinetic energy. The question of rotational excitation is probably even more interesting. In the case of protons on H_2, Herrero and Doering[119] reported no rotational excitation and this was confirmed by Udseth et al.[117,118] in their work with H_2, HD and D_2. On the other hand Toennies and coworkers[124] and Petty and Moran[107,127] found evidence for rotational excitation in vibration–rotation transitions in heavier systems. No satisfactory explanation for this difference between the $H^+ - H_2$ system and $CO^+ - Ar$ system has yet been proposed. However, a recent study by Udseth et al. has some interesting implications in this connection which are referred to below.

D. Pure Rotational Excitation

There is a dearth of experimental information on pure rotational excitation by positive ion impact. This, of course, arises from the severe instrumental resolution requirements. Recently, however, experimentalists have been overcoming this obstacle.

In investigating the $CO^+ - Ar$ system Petty and Moran[107] have demonstrated unresolved rotational excitation for projectile energy below the threshold for vibrational excitation and at large scattering angles.

The first resolved individual rotational quantum transitions, produced by positive ion scattering, have been realized in Toennies'[128] laboratory. Using the equipment described in the experimental section above with Li^+ as projectile ions and H_2 as a target gas and working at a centre of mass energy of 0.6 eV and at centre of mass angles between 14° and 32°, they obtained time of flight spectra one of which is illustrated in Fig. 22. They were able to obtain the differential inelastic cross sections for $j = 0 \rightarrow 2$ and $j = 1 \rightarrow 3$ rotational transitions relative to the differential elastic cross sections ($j = 0 \rightarrow 0, j = 1 \rightarrow 1$). The ratio of these transitions, i.e. $(0 \rightarrow 0 + 1 \rightarrow 1)$: $(0 \rightarrow 2):(1 \rightarrow 3)$, was 0.84:0.10:0.16. Using the classical Monte Carlo calculations of La Budde and Bernstein[129] and Lesters' potential energy surface for $Li^+ + H_2$, they calculated an inelastic energy transfer $\langle \Delta E_{rot} \rangle_{Av}/E_{cm} \approx 0.02$ compared to the observed value of ≈ 0.03. Considering the uncertainties in theory and experiment the agreement is good.

In a recent theoretical study on the $Li^+ - H_2$ system by Schaffer and Lester,[109] coupled channel calculations of integral cross sections for rotational and vibrational excitation of H_2 at a centre of mass energy 1.2 eV

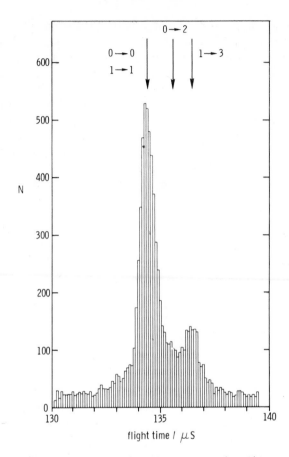

Fig. 22. Measured time of flight spectrum for Li^+–H_2 at $\theta_{lu} = 5°$, $E_{lab} = 2.7$ eV and $L = 115$. The arrows show the calculated location of the elastic and inelastic maxima corresponding to the rotational transitions $j = 0 \rightarrow 2$ and $j = 1 \rightarrow 3$.[128]

were carried out. Pure rotational excitation dominated the inelastic scattering at this energy. A very interesting feature reported is that preparation of H_2 in non-zero rotational states enhanced the $0 \rightarrow 1$ vibrational transition by about an order of magnitude. This is illustrated by the curves in Fig. 23 and Fig. 24.

Finally, the study of Udseth, Giese and Gentry[130] on the rotational excitation in the small angle scattering of protons from various diatomic molecules is an interesting contribution to this area. This group, as did Herrero and Doering, failed to find any rotational excitation associated with vibrational transitions induced by protons in H_2. Udseth et al.[131]

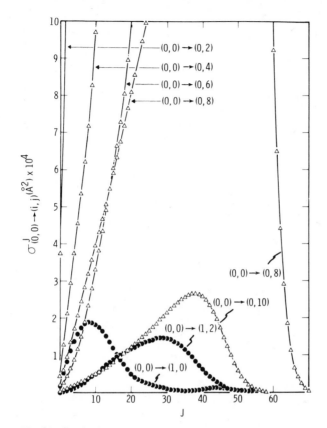

Fig. 23. Partial integral inelastic cross sections for excitation
from the ground $(0, 0)$ to selected final (v, j) states.[109]

point out that their data were taken at angles smaller than the rainbow
angle and consequently were most sensitive to the attractive part of the
potential energy surface which is very nearly spherical for H_3^+. Classical
trajectory calculations using an accurate H_3^+ potential also gave small
rotational excitation under their experimental conditions.

By contrast, if one considers the collision of an ion with a molecule possessing a permanent dipole the interaction potential is dominated at large
separations, r, by the strongly anisotropic charge-dipole term

$$V = \frac{e\mathbf{D} \cdot \mathbf{r}}{r^3} \tag{10}$$

where D is the dipole moment. As a result of this long range torque, one might
expect that the rotational inelasticity in ion–dipole collisions would be much
greater than in ion–homonuclear molecule collisions. For this reason they

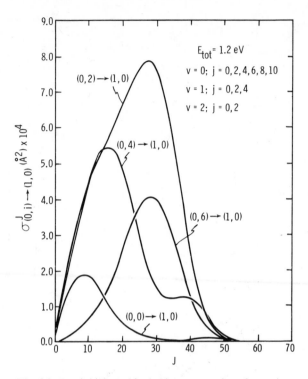

Fig. 24. Partial integral inelastic cross sections for excitation from selected initially excited rotational $(0, i)$ states to the $(1, 0)$ state.[109]

undertook the investigation of proton scattering from the polar molecules CO, HF and HCl. To illustrate the effect of a small dipole moment they studied proton scattering from N_2 and the isoelectronic CO which has a dipole moment of 0·1 Debye. The CM kinetic energy was 12 eV and the angle 15°. They found that for N_2 the energy distribution was symmetric about $Q = 0$ and was identical to the distribution for purely elastic $H^+ - H$ scattering. The data for CO, however, showed a distinct broadening of the energy distribution on the negative Q side, indicating a translational energy loss which they attributed to probable rotational excitation. Interestingly enough this seems to be diametrically opposite to the conclusions reached by Bottner, Ross and Toennies[124] in their study of $Li^+ - N_2$ and $Li^+ - CO$ systems. Obviously further work is necessary to resolve the reasons for this difference.

For $H^+ - HF$ scattering in Udseth *et al.*'s study[131] there appears to be no doubt that much of the observed excitation is rotational. The CM kinetic energy was 10 eV and the scattering angle 5° in these measurements. Since the instrumental resolution was 0·23 eV FWHM and the vibrational spacing in HF is 0·49 eV one might expect clear resolution of vibrational peaks, however, only a completely unresolved energy-loss spectrum was observed. Presumably because of the large amount of rotational excitation the vibrational peak was smeared out. The inelasticity was found to be greater for HF than for HCl which was consistent with the relative dipole moments and reduced masses of the two molecules. The average excitation energies for HF were in qualitative agreement with a classical perturbation calculation for pure rotational excitation.[132] For HCl the classical perturbation calculations gave a much lower rotational excitation energy than the average observed translational energy loss which was interpreted as indicating concurrent vibrational excitation which was not resolved in the presence of the rotational excitation.

One can briefly summarize the situation in inelastic collisions between ions and molecules in their ground states in cases where no reactive or charge transfer processes take place by the following:

1. At low projectile ion energies, in the region close to the elastic scattered peak one sees pure rotational excitation. At present it is only in the case of the $Li^+ - H_2$ where clean cut resolution has been realized.

2. In the energy region a few tenths of an eV, peaks corresponding to vibration–rotation transitions have been observed. Doering[86] points out that these are the most intense transitions in the inelastic scattering spectrum with cross sections exceeding 1 $Å^2$ and at times comparable in intensity to elastic scattering.

3. The third category of transitions observed is due to the electronic energy levels and it will be recalled that superelastic scattering has been observed in the electronic and vibrational rotation transitions.

One may conclude that although many questions remain concerning the mechanisms of energy transfer from one degree of freedom to another, our knowledge in the area is rapidly unfolding.

V. ELASTIC SCATTERING OF POSITIVE IONS

A knowledge of the potential energy curves is a basic requirement to understanding many chemical and physical phenomena. In the case of reactive scattering of ions we have seen that a knowledge of the potential energy surface is necessary in rationalizing the magnitude of the rate constants, life time of the complex, mechanism of reaction, and so forth. During the past decade low energy elastic scattering has been used to evaluate intermolecular forces and, in the case of elastic scattering of ions by rare gas atoms, potential energy curves for the corresponding diatomic rare gas hydride ions have been realized. More recently elastic scattering of atomic ions on H_2 and H_2^+ ions on rare gases have been used to estimate

average potential well depths and this information has been of use in drawing mechanistic conclusions on ion molecule reactions. Our comments of
this subject will be, of necessity, brief and the reader is referred to a recent
excellent review by Weise[133] for more details and additional references.

A. Classical Considerations

In the evaluation of elastic scattering data to obtain information about
the potential energy curve of the system, quantum mechanical methods are
used exclusively. However, it is useful to combine the classical and quantum
mechanical approaches to illustrate qualitatively the procedure used to
extract the pertinent information from the experimental data. Using the
principles of energy and angular momentum conservation of classical
mechanics, one can arrive at the following expression for the angle of deflection of a projectile impinging on a target.[134]

$$\theta = \pi - 2\beta \int_{\rho_0}^{\infty} \frac{d\rho}{\rho^2 \sqrt{1 - U(\rho)/K - \beta^2/\rho^2}} \tag{11}$$

ρ, the reduced distance $= r/r_m$, $r_m =$ equilibrium distance, $U_p = V(\rho)/\epsilon =$
reduced potential and $\epsilon =$ potential well depth, which is the dissociation
energy plus the zero point energy, θ is the centre of mass deflection angle,
$\beta = b/r_m$ the reduced impact parameter, ρ_0 is the distance of closest approach,
the reduced energy $(K) = E_c/\epsilon$ where E_c is the centre of mass collision energy.
For near head-on collisions (small impact parameter) the ions are back
scattered in the cm system and θ is positive, as the impact parameter increases
the scattering angle decreases and at a point where $\beta = \beta_0$ the deflection
function becomes zero and negative when $\beta > \beta_0$ and when $\beta \to \infty \to \theta \to 0$.
Some classical trajectories are illustrated in Fig. 25.

Fig. 25. Some classical trajectories corresponding to various impact
parameters.

In Fig. 25, β_2 and β_3 illustrate trajectories with comparatively large impact
parameters showing the influence of the positive portion of the potential
with negative deflection angles. β_1 in Fig. 25 illustrates a trajectory with a
smaller impact parameter which is influenced by the repulsive part of the
potential exhibiting a positive deflection angle. β_4 illustrates the situation for

backward scattering. It is clear that because of the cylindrical symmetry of the system, it is impossible to distinguish between positive θ's describing repulsive processes and negative θ's representing attractive scattering.

The classical differential cross section can easily be shown to be given by

$$I(\theta) = \left| \frac{b}{\sin\theta} \frac{db}{d\theta} \right| \tag{12}$$

A plot of the differential cross section vs θ decreases as the scattering angle increases however, with further increase of scattering angle $d\theta/db \to 0$, since the deflection function goes through a minimum and this leads to a singularity in the scattering intensity (rainbow). It is also evident that if $\theta = n\Pi$ ($n = 0, 1, \ldots$) other singularities (glories) can develop, however, they will not concern us here. The angle corresponding to the rainbow, θ_r, is called the rainbow angle. For small scattering angles θ_r is about equal to const./E_c/ϵ indicating that the position of the rainbow singularity is sensitive to potential well depth, an important parameter in reactive scattering considerations.

B. Quantum Mechanical Considerations

In the classical treatment of rainbow scattering the wave nature of the particles involved is completely neglected. In low energy collisions (eV) for heavy particles the wave length is much smaller than the range of the potential and the potential does not change significantly over the distance represented by one wave length. Under such conditions one expects interference effects in the differential cross section analogous to optical diffraction patterns. In the quantum mechanical case the classical trajectories are replaced by partial waves. The phases of these partial waves undergo shifts due to interaction with the potential. Interference arises between partial waves scattered in the same direction and this results in an oscillatory pattern in the differential cross section. The periods of the oscillations are given by

$$\Delta\theta_f = \frac{2\pi}{kr_m(\beta_1 + \beta_2)} \qquad \Delta\theta_s = \frac{2\pi}{kr_m(\beta_3 - \beta_2)} \tag{13}$$

$$k = \frac{2\pi}{\lambda} = \frac{1}{\hbar}\sqrt{2\mu E_c} \qquad \mu = \frac{m_1 m_2}{m_1 + m_2} \tag{14}$$

and m_1 is the projectile mass and m_2 the target mass. Consequently one may conclude that interference between partial waves associated with β_1 and β_2 gives rise to the higher frequency oscillations since $\beta_1 + \beta_2$ is larger, whereas interference between partial waves related to β_3 and β_2 give rise to the lower frequency oscillations since $\beta_3 - \beta_2$ is much smaller than the sum of $\beta_1 + \beta_2$. The lower frequency maximum in the extreme right of Fig. 26, is called the primary rainbow and the other low frequency maxima are the secondary

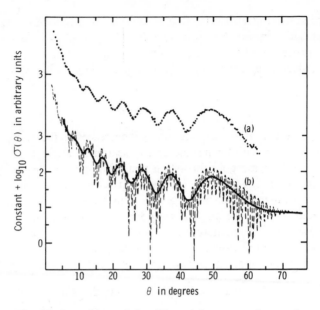

Fig. 26. Logarithm of the differential cross section vs the centre of mass angle for $H^+ + Ar$, $E = 6\,eV$ (a) Experimental observations, (b) dashed line are the results of JWKB calculation. The solid line is the convolution of the dashed line.[140]

rainbows. The dashed line in Fig. 26 shows the high frequency oscillations calculated by quantum mechanical methods. In the vicinity of the primary rainbow, $\beta_1 + \beta_2$ is approximately constant so that it is necessary to resolve the high frequency oscillations in order to obtain an accurate value of the equilibrium distance r_m. This point has been emphasized and illustrated by the work of Champion's group[135] on the $H^+ - Ne$ potential. The relation involving $\beta_3 - \beta_2$ depends on the shape of the potential well, and the angular position of the primary rainbow depends on the depth of the potential well. It is clear that high resolution of the scattering of monoatomic ions of atomic targets can yield the shape of the potential energy curve, its equilibrium distance and the well depth. If polyatomic targets are used, the information obtained from elastic scattering is reduced, however as will be seen some useful data can still be obtained.

It is clear that in order to analyse the high frequency oscillations it is necessary to resort to quantum mechanical methods. This involves the solution of the Schroedinger equation

$$\left[-\frac{\hbar^2}{2\mu}\nabla^2 + V(r)\right]\psi(r) = E_c\psi(r) \tag{15}$$

where $V(r)$ is the adiabatic interaction potential. The solution[136] can be expressed by the following

$$\phi(r, \theta, \phi) = \sum_{l=0}^{\infty} \sum_{m=-l}^{1} C_{lm} r^{-1} u_l(r) Y_{lm}(\theta, \phi) \tag{16}$$

where Y_{lm} is the orbital angular momentum wave function and $r^{-1}u_l(r)$ is the radial wave function which satisfies the differential equation.

$$\frac{\hbar^2}{2} \frac{d^2 u_l(r)}{dr^2} + \left[E - V(r) - \frac{l(l+1)\hbar^2}{2\mu r^2} \right] u_l(r) = 0 \tag{17}$$

This equation has the solution for infinite r of

$$r^{-1}u_l(r) = \frac{\sin(kr - l\pi/2 + \eta_l)}{kr} \tag{18}$$

where n_l is the phase shift.

The coefficients C_{lm} are determined from the boundary conditions which arises from the following considerations. A collimated beam of non-interacting particles may be represented by a plane wave $\psi(r) = e^{ikz}$ where z is the beam direction. If target molecules are present and some of the primary beam particles interact with them the scattered beam is represented by a radially outgoing wave, $f(\theta) e^{ikr}/r$ where the angular coefficient, $f(\theta)$, which is also a function of k, is the scattering amplitude. Since the distance of the particle detector from the point of interaction is effectively infinite compared to atomic dimensions the wave functions at the detector must be of the form

$$\psi(r) \underset{r \to \infty}{=} e^{ikz} + f(\theta) e^{ikr/r} \tag{19}$$

The square of the absolute value of the scattering amplitude $|f(\theta)|$ is the differential cross section. $f(\theta)$ is given by

$$f(\theta) = \frac{1}{2ik} \sum_{l=0}^{\infty} (2l+1)[e^{2i\eta_l} - 1]P_l(\cos\theta) \tag{20}$$

l is the orbital angular momentum quantum number, $l = 0, 1, 2\ldots$ and $P_l(\cos\theta)$ are spherical harmonics. If we can determine the phase shift and if the series in l converges reasonably rapidly one can compute $f(\theta)$ from the above expression and consequently one can arrive at a value of the differential cross section.

The phase shifts, which contain information about the potential, are determined from a solution of the radial equation for each value of l. In order to reduce the amount of computational time approximate methods are used.

One that is frequently used is the WKB-approximation which gives for η_l

$$\eta_l = kr_m \left[\int_{\rho_0}^{\infty} \left[1 - \frac{U(\rho)_\mathbb{c}}{E_c} - \frac{\beta_l^2}{\rho^2} \right]^{1/2} d\rho - \int_{\beta_l}^{\infty} \left[1 - \frac{\beta_l^2}{\rho^2} \right]^{1/2} d\rho \right] \quad (21)$$

where $\beta_l = (l + \frac{1}{2})/kr_m$.

1. Evaluation of the Potential Parameters

Two categories of procedure have been used to evaluate the potential parameters, iterative fitting and direct inversion. Probably the simplest method for extracting information is a fitting procedure in which an analytical form of the potential with suitable adjustable parameters is used. The differential cross section is calculated using the above quantum mechanical equation and the analytical form of the potential with assumed values for the adjustable parameters. On the basis of the quality of the fit between the calculated and experimental results, a new set of adjustable parameters is introduced and the procedure is repeated. This iterative procedure is continued until satisfactory agreement is obtained between the experimental rainbow and the high frequency oscillations and the corresponding calculated results. This gives one an analytical form of the potential energy curve. The group at the Hahn–Meitner Institute[137,138,139] used this fitting technique very successfully for proton–rare gas potentials. Good results were obtained using a modified Morse potential.

$$U = \exp\left[2G_1 G_2(1 - \rho)\right] - 2 \exp\left[G_1 G_2(1 - \rho)\right] \quad (22)$$

$$G_2 = 1 \quad \text{for } \rho < 1 \qquad G_2 \neq 1 \quad \text{for } \rho \geq 1$$

$$G_1 \text{ and } G_2 \text{ are adjustable parameters}$$

The use of this potential is attractive since one can vary the repulsive and attractive branches of the potential independently. The group also studied the use of other potentials.

Champion, Doverspike and coworkers[135,140,141] have also determined the potential energy curves of proton–rare gas systems and they preferred to use inversion procedures which permit a more direct determination of the potential from the differential scattering cross section. Their preference stems from the fact that if the experimental resolution is not sufficiently good one may not get a reliable intermolecular potential. This was their experience with the $H^+ + Ne$ case. The inversion technique that they used was the Remler–Regge[142,143] method. Buck and Pauly[144] have also developed an inversion technique which they applied successfully to the $Na - Hg$ system. No attempt will be made to outline these inversion techniques so the reader is referred to the review by Weise[133] or to the original references.

2. Proton–Rare Gas Systems

Most detailed work has been carried out on the proton–rare gas atom systems. The spherical symmetry of the potential permits an experimental observation not only of the primary and some secondary rainbows but also useful resolution of the high frequency oscillations in the region of the primary rainbow. This has resulted in good potential energy curves for $H^+ - X$ systems in which X is a rare gas. In general the procedure was to determine the potential energy curve from the experimental measurement of the rainbow and the high frequency oscillations using either the fitting procedure or the inversion technique and then to compare the experimental results with ab initio quantum chemical calculations. Following Weise[133] we summarize the various results in Table I.

TABLE I

A summary of potential parameters obtained from scattering measurements and theoretical calculations on the H^+–rare gas systems

System	ε(eV)	r_m (Å)	ε (eV)	r_m (Å)	ε (eV)	r_m (Å)
HeH$^+$	2·18	0·75	2·00	0·77	2·04	0·775[145]
NeH$^+$	2·27	1·01	2·28	0·99	2·21	0·97[146]
ArH$^+$	4·22	1·24	4·04	1·31	4·09	1·38[147]
KrH$^+$	4·60	1·53	4·45	1·47		
XeH$^+$			6.75	1.74		

The first four columns of numbers give the experimental results and the last two give the results of the ab initio calculations. The agreement between the shapes of the experimentally and theoretically determined potential energy curves was quite good. These systems are of particular interest from the point of view of reactive scattering because of the extensive studies that have been carried out on the corresponding ion–molecule reactions.

3. Polyatomic Systems

It was pointed out earlier that an important factor in persistent complex formation in reactive scattering systems is the depth of the potential well. In studying the elastic scattering of protons on molecules the Hahn–Meitner Institute group showed that even though one can only deduce average potentials in these molecular systems, such information can have significant applications in interpretation of ion–molecule reactions.

In the case of atomic ions scattered by neutral atoms, the potential is spherically symmetric and depends only on the distance between the colliding particles. For polyatomic systems the potential energy is a function of coordinates of all the atoms involved, consequently it is dependent on orientation of the molecule. Since the molecule is rotating and vibrating and the potential is anisotropic, the net result is that the high frequency oscillations are completely washed out and even the rainbow undulations are considerably damped.

The Hahn–Meitner Institute group[139] has reported results on the elastic scattering of protons from N_2, CO, CO_2, SF_6, H_2 and CH_4 and also the elastic scattering of H_2^+ on Ar and Kr. An illustrative result is given in Fig. 27

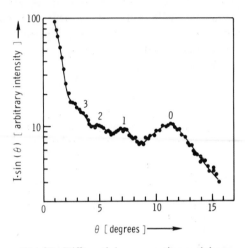

Fig. 27. Differential cross section vs laboratory scattering angle for the scattering of 20·8 eV protons on N_2.[139]

which represents the elastic scattering of protons on N_2 at a laboratory energy of 20.8 eV. Rainbow oscillations are clearly visible, however, it is clear that the high frequency oscillations are completely wiped out. In some cases such as $H^+ - CH_4$ only the primary rainbow was observed. In the cases of N_2, CO, CO_2, SF_6 and H_2 the differential cross sections were fitted using a simple Morse potential.

$$U = \exp[2G(1 - \rho)] - 2\exp[G(1 - \rho)] \qquad (23)$$

The equilibrium distances r_m were estimated from the relation $r_m = 0.72 r_g$ where r_g is the gas kinetic radius of the molecule. This relation was found to hold approximately in the H^+-noble gas systems. The case of $H^+ - CH_4$ was evaluated classically. The results are tabulated in Table II.

The uncertainties that the authors associate with these measurements range from ± 0.30 to ± 0.50 eV. It should be emphasized that because of the potential anisotropies that these values represent at best average values. The theoretical values for H_3^+ are those reported by Polanyi et al.[148] The

TABLE II

Potential well depths obtained from scattering
experiments and theoretical calculations for some
H^+-molecule interactions

System	Experimental $[\varepsilon (eV)]$	Theoretical
H^+-N_2	4·15	
H^+-CO	4·70	
H^+-CO_2	5·10	
H^+-SF_6	3·66	
H^+-H_2	4·04	4·56 2·87
H^+-CH_4	3·80	5·45

4·56 eV value corresponds to the more stable triangular configuration and 2·87 eV to the collinear configuration.

For the CH_5^+ system an *ab initio* calculation[149] gives 5·45 eV for the potential well depth. It is not clear that the experimental well depths should agree with the theoretical values in molecular systems and the authors rationalize the apparent disagreement in the CH_5^+ system by the non-adiabacity of the scattering process with respect to the movement of the H atoms in methane.

In the case of H_2^+ on Ar and Kr the rainbow undulations are damped even more severely and the well depths are reported as $1·3 \pm 0·2$ eV and $1·1 \pm 0·2$ eV, respectively.

It is clear that the polyatomic systems give less impressive and probably less reliable results, on the other hand they do seem to give consistent results when applied to rare gas reactions such as $Ar^+(H_2, H)ArH^+$, $H_2^+(Ar, H)ArH^+$.[43]

Finally one can cite a different use of non-reactive and elastic scattering which provides a very useful insight into the general nature of the potential energy surfaces and, when coupled with reactive scattering, gives one additional mechanistic information. This approach has been first used by Mahan's group and reference has been made to it earlier. Recently Mahan *et al.*[150] have used reactive and non-reactive scattering measurements to study the dynamics of the $O^+ + H_2$ reaction in the relative energy range 15–50 eV. Frequently, such three atom ion–molecule reactions are interpreted at low energies in terms of the spectator stripping model. A system obeying such a model has a projectile ion critical energy beyond which the product ion should not exist because its internal energy exceeds its bond energy. It frequently happens, however, that product ions are observed for projectile energies far in excess of the critical energy. A model that is frequently invoked to explain the presence of ionic products at high energy is the sequential impulse model proposed by Bates *et al.*[151]

If we consider a reaction such as $A^+(BC, C)AB^+$, the projectile A^+ is scattered off B which in turn is scattered by C, then A^+ and B can combine to form AB^+ if certain restrictive conditions on the relative energy between A^+ and B and the directions of their motions are satisfied. The product can then exist at energies considerably in excess of the critical energy expected for the spectator stripping model. Mahan and coworkers[150] found that at relatively high energies the sequential impulse model could explain various features of the reaction of O^+ with molecular hydrogen and its stable isotopic variants. Non-reactive and elastic scattering measurements played a significant part in elucidating the role of the sequential impulse model in this reaction.

An illustration of the contour maps obtained of O^+ scattered from D_2 at a relative energy of 20.1 eV and from HD at a relative energy of 27.6 eV is given in Figs. 28 and 29. In Fig. 28 it is clear that at angles 45° or less there is a well defined ridge of intensity which follows the circle labelled elastic $O^+(D_2, D_2)O^+$. This elastic component of the non-reactive scattering arises from collisions in which O^+ makes a grazing encounter with the D_2 molecule as a whole. At angles greater than 45° the scattering becomes inelastic. The large angle scattering falls well within the circle marked $Q = -4.5$ eV in which the scattering is sufficiently inelastic to dissociate D_2. Attention is called to the circle marked elastic $O^+(D, D)O^+$. This is the locus of elastic

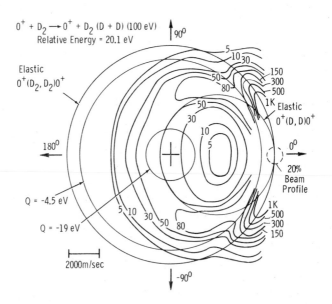

Fig. 28. Contour map of the intensity of O^+ scattered from D_2 at 20.1 eV relative initial energy.[150]

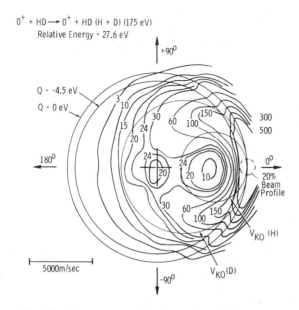

$O^+ + HD \longrightarrow O^+ + HD$ (H + D) (175 eV)
Relative Energy = 27.6 eV

Fig. 29. Contour map of the intensity of O^+ scattered from HD at 27·6 eV initial relative energy.[150]

scattering of O^+ by a free atom. The fact that the ridge of inelastically scattered O^+ falls close to this circle indicates that many of the collisions are such that O^+ interacts nearly impulsively with one atom of the D_2 molecule. A closely related contour map of O^+ scattering from HD is given in Fig. 29. If the non-reactive scattering was ideally impulsive one would expect two inelastic ridges, one corresponding to $O^+ - H$ collisions and another to $O^+ - D$ collisions. In Fig. 29 the loci of impulsive scattering of O^+ by H and by D are indicated by the circles labelled $V_{KO}(H)$ and $V_{KO}(D)$ respectively. The broad observed inelastic ridge is really two overlapping ridges which cover both circles over much of the angular range. However, it will be noted that in the region near the centre of mass origin the broad ridge does separate into two ridges so the observations are qualitatively consistent with the impulse model. At higher relative energies some of the features referred to above stand out more impressively since the conditions for impulsive scattering are more nearly met. The experimental results in this study are compared with the predictions of an impulsive model in which reactions occur as a result of a sequence of two body hard sphere collisions which lead to a final bound product. The general forms of the product angular distributions and their dependence on the isotopic composition of the target are fairly well represented by the model. The reader is referred to the original article for details.

The work of the author as reported in this chapter was supported by the U.S. Atomic Energy Commission and the U.S. Army Research Office, Durham.

References

1. L. Friedman and B. G. Reuben, in *Advances in Chemical Physics*, Vol. XIX (Ed. I. Prigogine and S. A. Rice), Wiley, New York, 1971.
2. J. H. Futrell and T. O. Tiernan, in *Ion-Molecule Reactions*, Vol. 2 (Ed. J. L. Franklin), Plenum Press, New York, 1972.
3. G. A. Gray, in *Advances in Chemical Physics*, Vol. XIX (Ed. I. Prigogine and S. A. Rice), Wiley, New York, 1971.
4. Z. Herman and R. Wolfgang, in *Ion-Molecule Reactions*, Vol. 2 (Ed. J. L. Franklin), Plenum Press, New York, 1972.
5. C. R. Iden, K. Wendell and W. S. Koski, to be published.
6. C. E. Kuyatt and J. A. Simpson, *Rev. Sci. Inst.*, **38**, 103 (1967); J. A. Simpson, *Rev. Sci. Inst.*, **35**, 1698 (1964).
7. K. Spangenberg and L. M. Field, *Elect. Comm.*, **21**, 194 (1943); F. H. Read, A. Adams and J. R. Soto-Monfiel, *J. Phys. E. Sci. Instrum.*, **4**, 625 (1971); S. Natali, D. DiChio, E. Uva and C. E. Kuyatt, *Rev. Sci. Inst.*, **43**, 80 (1972).
8. R. Wolfgang and J. R. Cross, Jr., *J. Phys. Chem.*, **73**, 743 (1969).
9. R. L. Champion, L. D. Doverspike and T. L. Bailey, *J. Chem. Phys.*, **45**, 4377 (1966).
10. W. R. Gentry, E. A. Gislason, Y. T. Lee, B. H. Mahan and C. W. Tsao, *Disc. Faraday Soc.*, **44**, 137 (1967).
11. W. R. Gentry, E. A. Gislason, B. H. Mahan and C. W. Tsao, *J. Chem. Phys.*, **49**, 3058 (1968).
12. B. Mahan, *Acc. Chem. Res.*, **1**, 217 (1968).
13. M. Chiang, E. A. Gislason, B. H. Mahan, C. W. Tsao and A. S. Werner, *J. Chem. Phys.*, **52**, 2698 (1970).
14. Z. Herman, J. Kerstetter, T. Rose and R. Wolfgang, *Disc. Faraday Soc.*, **44**, 123 (1967).
15. Z. Herman, K. Birkinshaw, *Ber. Bunsenges. Phys. Chem.*, **75**, 413 (1971).
16. Z. Herman, J. D. Kerstetter, T. L. Rose and R. Wolfgang, *Rev. Sci. Instr.*, **40**, 538 (1969).
17. E. Gustafsson and E. Lindholm, *Ark. Fys.*, **18**, 219 (1960).
18. B. R. Turner, J. A. Rutherford and D. M. J. Compton, *J. Chem. Phys.*, **48**, 1602 (1968).
19. E. Lindemann, R. W. Rozett and W. S. Koski, *J. Chem. Phys.*, **56**, 5490 (1972).
20. R. C. C. Lao, R. W. Rozett and W. S. Koski, *J. Chem. Phys.*, **49**, 4202 (1968).
21. R. J. Cotter and W. S. Koski, *J. Chem. Phys.*, **59**, 784 (1973).
22. K. C. Lin, R. J. Cotter and W. S. Koski, *J. Chem. Phys.*, **60**, 3412 (1974).
23. W. A. Chupka, M. E. Russell and K. Refaey, *J. Chem. Phys.*, **48**, 1518 (1968); *J. Chem. Phys.*, **48**, 1527 (1968).
24. D. Brandt, C. Ottinger and J. Simonis, *Ber. Bunsenges. phys. Chem.*, **77**, 648 (1973).
25. J. Krenos and R. Wolfgang, *J. Chem. Phys.*, **52**, 5961 (1968).
26. J. Krenos, R. K. Preston, R. Wolfgang and J. C. Tully, *J. Chem. Phys.*, **60**, 1634 (1974).
27. J. C. Tully, *Ber. Bunsenges. Phys. Chem.*, **77**, 557 (1973).
28. J. C. Tully and R. K. Preston, *J. Chem. Phys.*, **55**, 562 (1971).

29. D. L. Bunker, *Methods in Computational Physics*, **10**, 287 (1971), Ácademic Press, New York.
30. R. K. Preston and J. C. Tully, *J. Chem. Phys.*, **54**, 4297 (1971).
31. H. Conroy, *J. Chem. Phys.*, **51**, 3979 (1969).
32. Z. Herman, J. Kersetter, T. L. Rose and R. Wolfgang, *Disc. Farad. Soc.*, **44**, 123 (1966).
33. L. Friedman and B. G. Reuben, in *Advances in Chemical Physics*, Vol. XIX (Ed. I. Prigogine and S. A. Rice), Wiley-Interscience, New York, 1971, p. 33.
34. F. J. Kuntz and A. C. Roach, *J. Chem. Phys.*, **59**, 6299 (1973).
35. J. C. Polanyi and P. J. Kuntz, *Disc. Farad. Soc.*, **44**, 180 (1967).
36. J. C. Polanyi and W. H. Wong, *J. Chem. Phys.*, **51**, 1439 (1969).
37. P. J. Kuntz, M. H. Mok and J. C. Polanyi, *J. Chem. Phys.*, **50**, 4623 (1969).
38. A. Henglein, *Molecular Beam and Reaction Kinetics*, Academic Press, New York, 1970.
39. S. Chapman, *Ph.D. Thesis*, Yale Univ., New Haven, Conn., 1973.
40. F. O. Ellison, *J. Amer. Chem. Soc.*, **85**, 3540 (1963); G. V. Pfeiffer and F. O. Ellison, *J. Chem. Phys.*, **43**, 4305 (1965); A. A. Wu and F. O. Ellison, *J. Chem. Phys.*, **48**, 727 (1968).
41. P. J. Kuntz and A. C. Roach, *Faraday Soc. Trans. II*, **68**, 259 (1972).
42. G. Bone, A. Ding and A. Henglein, *Z. Naturforsch.*, **26a**, 932 (1971).
43. A. Henglein, *J. Phys. Chem.*, **76**, 3883 (1972).
44. E. A. Gislason, B. H. Mahan, C. W. Tsao and A. S. Werner, *J. Chem. Phys.*, **54**, 3897 (1971).
45. J. C. Tully, Z. Herman and R. Wolfgang, *J. Chem. Phys.*, **54**, 1730 (1971).
46. E. E. Fergusson, D. K. Bohme, F. C. Fesenfeld and D. B. Dinkin, *J. Chem. Phys.*, **50**, 5039 (1969).
47. P. Warneck, *Planet. Space Sci.*, **15**, 1349 (1967).
48. D. K. Bohme, P. P. Ong, J. B. Hastead and L. R. Megill, *Planet. Space Sci.*, **15**, 1777 (1967).
49. C. G. Geise, *Advan. Chem.*, **58**, 63 (1966).
50. R. F. Stebbings, B. R. Turner and J. A. Rutherford, *J. Geophys. Res.*, **71**, 771 (1966).
51. J. A. Rutherford and D. A. Vroom, *J. Chem. Phys.*, **55**, 5622 (1971).
52. J. J. Leventhal, *J. Chem. Phys.*, **54**, 5102 (1971).
53. G. P. K. Smith and R. J. Cross, Jr., *J. Chem. Phys.*, **60**, 2125 (1974).
54. Joyce J. Kaufman and W. S. Koski, *J. Chem. Phys.*, **50**, 1942 (1969).
55. A. L. Schmeltekopf, E. E. Ferguson and F. C. Fehsenfeld, *J. Chem. Phys.*, **48**, 2966 (1968).
56. A. Pipano and Joyce J. Kaufman, *J. Chem. Phys.*, **56**, 5258 (1972).
57. H. H. Harris and J. J. Leventhal, *J. Chem. Phys.*, **58**, 2333 (1973).
58. K. T. Gillen, B. H. Mahan and J. S. Winn, *J. Chem. Phys.*, **58**, 5373 (1973).
59. C. R. Iden, R. Liardon and W. S. Koski, *J. Chem. Phys.*, **56**, 851 (1972).
60. B. H. Mahan and T. H. Sloane, *J. Chem. Phys.*, **59**, 5661 (1973).
61. E. A. Gislason, B. H. Mahan, C. W. Tsao and A. S. Werner, *J. Chem. Phys.*, **50**, 5418 (1969).
62. M. H. Chang, B. H. Mahan, C. W. Tsao and A. S. Werner, *J. Chem. Phys.*, **53**, 3752 (1970).
63. G. Bosse, A. Ding and A. Henglein, *Ber. Bunsenges. Phys. Chem.*, **75**, 413 (1971).
64. M. T. Bowers, D. D. Elleman and J. King, *J. Chem. Phys.*, **50**, 4787 (1969).
65. L. D. Doverspike and R. L. Champion, *J. Chem. Phys.*, **46**, 4718 (1967).
66. A. Ding, H. Henglein, D. Hyatt and K. Lacmann, *Z. Naturforsch.*, **23a**, 2084 (1968).

67. E. A. Gislason, B. H. Mahan, D. W. Tsao and A. S. Werner, *J. Chem. Phys.*, **50**, 142 (1969).
68. Z. Herman, P. Hierl, A. Lee and R. Wolfgang, *J. Chem. Phys.*, **51**, 454 (1969).
69. A. Ding, A. Henglein and K. Lacmann, *Z. Naturforsch.*, **23a**, 780 (1968).
70. W. Felder, N. Sbar and J. Durbin, *Chem. Phys. Lett.*, **6**, 385 (1970).
71. Z. Herman, A. Lee and R. Wolfgang, *J. Chem. Phys.*, **51**, 452 (1969).
72. A. Lee, R. L. LeRoy, Z. Herman, R. Wolfgang and J. C. Tully, *Chem. Phys. Lett.*, **12**, 569 (1972).
73. S. A. Safron, N. D. Weinstein, D. R. Herschbach and J. C. Tully, *Chem. Phys. Lett.*, **12**, 564 (1972).
74. G. Z. Whitten and B. S. Ravinovitch, *J. Chem. Phys.*, **38**, 2466 (1963).
75. D. C. Tardy, B. S. Rabinovitch and G. Z. Whitten, *J. Chem. Phys.*, **48**, 1427 (1968).
76. R. A. Marcus, *J. Chem. Phys.*, **20**, 359 (1952).
77. G. M. Wieder and R. A. Marcus, *J. Chem. Phys.*, **37**, 1835 (1962).
78. R. Wolfgang, *Acc. Chem. Res.*, **3**, 48 (1970).
79. L. S. Kassel, *Kinetics of Homogeneous Gas Reactions*, The Chemical Catalog Co., New York, 1932.
80. H. S. Johnston, *Gas Phase Reaction Rate Theory*, Ronald Press, New York, 1966.
81. B. H. Mahan, *J. Chem. Phys.*, **55**, 1436 (1971).
82. G. Herzberg, *Electronic Spectra and Electronic Structure of Polyatomic Molecules*, Van Nostrand, Princeton, New Jersey, 1966.
83. Marie M. Heaton, A. Pipano, K. C. Lin and Joyce J. Kaufman, unpublished results.
84. L. Cusachs, M. Krieger and C. W. McCurdy, *Int. J. Quantum Chem.*, **53**, 67 (1969).
85. S. V. O'Neil, P. K. Pearson, H. F. Shaeffer and C. F. Bender, *J. Chem. Phys.*, **58**, 1126 (1973).
86. J. P. Doering, *Ber. Bunsenges. Physik. Chem.*, **77**, 593 (1973).
87. R. David, M. Faubel and J. P. Toennies, *Chem. Phys. Lett.*, **18**, 87 (1973).
88. J. H. Moore and J. P. Doering, *J. Chem. Phys.*, **52**, 1692 (1970).
89. P. C. Cosby and T. F. Moran, *J. Chem. Phys.*, **52**, 6157 (1970).
90. H. Udseth, C. F. Giese and W. R. Gentry, *Phys. Rev. A*, **8**, 2483 (1973).
91. J. H. Moore, *J. Chem. Phys.*, **55**, 2760 (1971).
92. J. T. Park and F. D. Schowengerdt, *Phys. Rev.*, **185**, 152 (1969).
93. F. D. Schowengerdt and J. T. Park, *Phys. Rev.*, **A1**, 848 (1970).
94. J. T. Park, F. D. Schowengerdt and D. R. Schoonover, *Phys. Rev.*, **A3**, 679 (1971).
95. J. H. Moore, 'Electronic and Atomic Collisions', in *Abstracts of Papers VIII*, *ICPEAC*, Vol. 2, 657, Beograd, Yugoslavia, 1973.
96. J. H. Moore, *Phys. Rev.*, **A8**, 2359 (1973).
97. W. Benesch, J. T. Vanderslice, S. G. Tilford and P. G. Williamson, *Astrophysics J.*, **143**, 236 (1966).
98. R. W. Nicholls, *J. Quant. Spectry. Radiative Transfer*, **2**, 433 (1962).
99. J. H. Moore, *Bull. Amer. Phys. Soc.*, **17**, 1142 (1972), *Phys. Rev. A*, **9**, 2043 (1974).
100. J. P. Doering, 'Physics of Electronic and Atomic Collisions', in *VII ICPEAC*, Amsterdam, 1971, p. 341.
101. J. H. Moore and J. P. Doering, *Phys. Rev. Lett.*, **23**, 564 (1969).
102. T. F. Moran and P. C. Cosby, *J. Chem. Phys.*, **51**, 5724 (1969).
103. R. David, M. Faubel, P. March and J. P. Toennies, 'Electronic and Atomic Collisions', in *VII ICPEAC*, North-Holland, Amsterdam, 1971, p. 252.
104. W. D. Held, J. Schottler and J. P. Toennies, *Chem. Phys. Lett.*, **6**, 304 (1970).

05. M. H. Cheng, E. A. Gislason, B. H. Mahan, C. W. Tsao and A. S. Werner, *J. Chem. Phys.*, **52**, 6150 (1970).
06. P. C. Cosby and T. F. Moran, *J. Chem. Phys.*, **52**, 6157 (1970).
07. F. Petty and T. F. Moran, *Phys. Rev.*, **A5**, 266 (1972).
08. B. Ritchie, *Phys. Rev.*, **A6**, 1902 (1972).
09. J. Schaeffer and W. A. Lester, *Chem. Phys. Lett.*, **20**, 575 (1973).
10. J. Schaeffer, W. A. Lester, D. Kouri and C. A. Wells, *IBM Research Report*, RJ 1300, 1973.
11. D. J. Kouri and C. A. Wells, *J. Chem. Phys.*, **60**, 2296 (1974).
12. W. R. Gentry, *J. Chem. Phys.*, **60**, 2547 (1974).
13. F. W. Herrero and J. P. Doering, *Phys. Rev.*, **A5**, 702 (1972).
14. F. A. De Wette and Z. I., Slawsky, *Phys.*, **20**, 1169 (1954).
15. J. Korobkin and Z. I. Slawsky, *J. Chem. Phys.*, **37**, 226 (1962).
16. H. K. Shin, *J. Phys. Chem.*, **73**, 4521 (1969).
17. H. Udseth, C. F. Giese and W. R. Gentry, *J. Chem. Phys.*, **54**, 3642 (1971).
18. H. K. Udseth, C. F. Giese and W. R. Gentry, *Phys. Rev.*, **A8**, 2483 (1973).
19. F. A. Herrero and J. P. Doering, *Phys. Rev. Lett.*, **29**, 609 (1972).
20. T. F. Moran and P. C. Cosby, *J. Chem. Phys.*, **51**, 5724 (1969).
21. F. Petty and T. F. Moran, *Chem. Phys. Lett.*, **5**, 64 (1970).
22. R. E. Christoffersen, *J. Chem. Phys.*, **41**, 960 (1964).
23. M. E. Schwartz and L. J. Schaad, *J. Chem. Phys.*, **47**, 5325 (1967).
24. R. Bottner, U. Ross and J. P. Toennies, Report 122, Max Planck Institut fur Stromungforschung, July 1973.
25. W. Kutzelnigg, V. Staemmler and C. Hoheisel, *Chem. Phys.*, **1**, 27 (1973).
26. W. A. Lester, *J. Chem. Phys.*, **53**, 1511 (1970); **54**, 3171 (1971).
27. T. F. Moran, F. Petty and G. S. Turner, *Chem. Phys. Lett.*, **9**, 379 (1971).
28. H. E. Van den Bergh, M. Faubel and J. P. Toennies, *Faraday Discussions* (1973).
29. R. A. La Budde and R. B. Bernstein, *J. Chem. Phys.*, **55**, 5499 (1971).
30. H. Udseth, C. F. Giese and R. W. Gentry, *J. Chem. Phys.*, **60**, 3051 (1974).
31. H. Udseth, C. F. Giese and R. W. Gentry, *J. Chem. Phys.*, in press.
32. W. R. Gentry, *J. Chem. Phys.*, in press.
33. H. P. Weise, *Ber. Bunsenges. Phys. Chem.*, **77**, 578 (1973).
34. J. B. Marion, *Classical Dynamics of Particles and Systems*, Academic Press, New York, 1967.
35. S. M. Bobbio, W. G. Rich, L. D. Doverspike and R. L. Champion, *Phys. Rev.*, **A4**, 957 (1971).
36. L. S. Rodberg and R. M. Thaler, *Introduction to the Quantum Theory of Scattering*, Academic Press, New York (1967).
37. H. U. Mittman, H. P. Weise, A. Ding and A. Henglein, *Z. Naturforsch.*, **26a**, 1112 (1971).
38. H. P. Weise, H. U. Mittman, A. Ding and A. Henglein, *Z. Naturforsch.*, **26a**, 1122 (1971).
39. H. U. Mittman, H. P. Weise, A. Ding and A. Henglein, *Z. Naturforsch.*, **26a**, 1282 (1971).
40. R. L. Champion, L. D. Doverspike, W. G. Rich and S. M. Bobbio, *Phys. Rev.*, **A2**, 2327 (1970).
41. W. G. Rich, S. M. Bobbio, R. L. Champion and L. D. Doverspike, *Phys. Rev.*, **A4**, 2253 (1971).
42. E. A. Remler, *Phys. Rev.*, **A3**, 1949 (1971).
43. I. Regge, *Nuovo Cimento*, **14**, 951 (1959).

144. U. Buck, *J. Chem. Phys.*, **54**, 1923 (1971); U. Buck and H. Pauly, *J. Chem. Phys.* **54**, 1929 (1971).
145. L. Wolniewicz, *J. Chem. Phys.*, **43**, 1087 (1965).
146. S. Peyerimhoff, *J. Chem. Phys.*, **43**, 998 (1965).
147. A. C. Roach and F. J. Kuntz, *Chem. Comm.*, 1336 (1970).
148. I. G. Csizmadia, R. E. Mari, J. C. Polanyi, A. C. Roach and M. A. Robb, *J. Chem Phys.*, **52**, 6205 (1970).
149. V. Dyczmons, V. Staemmler and W. Kutzelnigg, *Chem. Phys. Lett.*, **5**, 361 (1970)
150. K. T. Gillen, B. H. Mahan and J. S. Winn, *J. Chem. Phys.*, **59**, 6380 (1973).
151. D. R. Bates, C. J. Cook and F. J. Smith, *Proc. Phys. Soc. London*, **83**, 49 (1964)

REACTIVE SCATTERING

ROGER GRICE

Department of Theoretical Chemistry, Cambridge University,
Cambridge, CB2 1EW

CONTENTS

I. Alkali Atom Reactions 249
  A. Velocity Analysis 250
  B. Reaction at Superthermal Energies 256
  C. Rotational Polarization Analysis 258
  D. Molecular Beam Electric Resonance Spectra of Reaction Products . . . 258
  E. Reaction of Oriented Molecules 258
II. Alkali Dimer Reactions 259
  A. Reactions with Halogen Molecules 260
  B. Reactions with Polyhalide Molecules 265
  C. Reactions with Halogen Atoms 269
  D. Reactions with Hydrogen Atoms 270
  E. Reactions with Alkali Atoms 272
II. Alkaline Earth Atom Reactions 274
  A. Angular Distribution Measurements 275
  B. Chemiluminescence Measurements 276
  C. Laser-Induced Fluorescence 277
  D. Comparison with Alkali Dimer Reactions 280
V. Non-Metal Reactions 281
  A. Reactions of Halogen Atoms with Halogen Molecules 283
  B. Reactions of Hydrogen Atoms with Halogen Molecules. 285
  C. Reactions of Methyl Radicals with Halogen Molecules 289
  D. Reactions of Oxygen Atoms with Halogen Molecules 292
  E. Reactions of Halogen Atoms with Hydrogen, Hydrogen Halides and Methyl
    Halides 294
  F. Reactions of Hydrogen Atoms with Hydrogen Molecules 296
  G. Reactions of Halogen Atoms with Unsaturated Hydrocarbons . . . 296
  H. More Four-Centre Reactions 300
V. Theoretical Models 302
  A. Long-Lived Collision Complexes 302
  B. Intersection of Covalent and Ionic Potential Energy Surfaces . . . 304
Acknowledgements 307
References 307

'All is flux, nothing is stationary.'
Heracleitus (513 B.C.)

247

The last few years have seen a rapid advance in the power and scope c reactive scattering studies in molecular beams. The initial phase of work i this field followed on the first reactive scattering experiment[1] by Taylor an Datz, using a differential surface ionization detector to study the reactio $K + HBr \rightarrow KBr + H$. The reactions of alkali atoms with a wide range c halogen-containing molecules were intensively investigated. The experimenta techniques required for these studies were rendered relatively simple an straightforward by the high sensitivity and specificity of the surface ioniza tion detector and the condensable materials forming the reactant beam which could be pumped extremely rapidly by liquid nitrogen cooled surface Many of the basic methods for measurement of angular distribution velocity distributions and rotational excitation of reaction products wer developed. Quadrupole mass spectrometer detectors were successfull employed with surface ionization sources. The basic techniques also c kinematic transformation of laboratory measurements to the centre c mass frame were established. The results of these studies laid the foundation for understanding reaction dynamics and their relation to the electroni structure of the reaction potential energy surface. Thus, reactions of alka atoms with alkyl halides exhibited *rebound* dynamics, with halogen molecule *stripping* dynamics and with alkali halides *long-lived complex* dynamic These serve as archetypal limiting cases to which the diverse reactio mechanisms of other reactants are often conveniently related. This earl period is substantially covered by the review of Herschbach[2] and the 44t Discussion of the Faraday Society held in Toronto in September, 1967.

More recently, the burgeoning of more sophisticated experimenta techniques has rendered a vastly increased range of reactions accessible t reactive scattering studies. Most significant among these developments ha been the advent of molecular beam machines employing an electron bombarc ment mass spectrometer detector, and capable of *non-alkali* reactiv scattering measurements. This removes the principal constraint on th application of molecular beam methods to the study of chemical kinetic The investigation of a range of atom and free radical reactions, important i gas kinetics, is now underway in several laboratories. The increased use c supersonic nozzle beam sources, in place of the earlier effusive beam source has increased the intensity and decreased the velocity spread of reactar beams. Consequently, experiments of higher resolution, probing mor deeply into the details of differential reaction cross sections are being unde taken. Novel reactants (e.g. dimer molecules) may be prepared by cor densations in nozzle expansions. Increased collision energies may b achieved by accelerating heavy reactant molecules 'seeded' in the nozz expansion of a large excess of a light 'driver' gas. Most recently, the applic tion of lasers to the production of vibrationally excited reactants and to th measurement of quantum states of product molecules has added further t the armoury of techniques at the beamist's disposal. Concomitant advances in our theoretical understanding of reaction dynamics in relation to the electronic structure of the potential energy surface are being made at a qualitative and semi-empirical level. However, the calculation of accurate *ab initio* potential energy surfaces has been slower to respond to the stimulus of experimental observations.

Stages in the progress of this rapid advance have been evidenced in reviews of the last few years, by Herschbach,[3] Lee[4] and Kinsey.[5] It is perhaps most strongly apparent in the 55th Faraday Discussion held in London in April, 1973. Thus, the field of molecular beam reactive scattering is now 'coming of age', after a protracted adolescence and should yield insights at a fundamental level into many of the processes of chemical kinetics. This review will endeavour to cover those areas of reactive scattering where experimental progress and the improvement in our understanding of reaction dynamics seem presently most promising. Particular attention will be given to those developments which have emerged since the review of Kinsey.[5]

'It is the customary fate of new truths to begin as heresies and to end as superstitions.'

T. H. Huxley (1825–1895).

I. ALKALI ATOM REACTIONS

The study of alkali atom reactions with halogen-containing molecules comprises much of the history of reactive scattering in molecular beams. The broad features of the reaction dynamics and their relation to the electronic structure of the potential energy surface are well understood.[2] The reaction is initiated by an electron jump transition in which the valence electron of the alkali atom M is transferred to the halogen-containing molecule RX. Subsequent interaction of the alkali ion and the molecule anion, in the exit valley of the potential surface, leads to an alkali halide product molecule MX.

$$M + RX \rightarrow M^+ + RX^- \rightarrow M^+X^- + R \tag{1}$$

However, many of the detailed features have remained obscure till recently. The interactions involved in the electron jump transition were not quantitatively understood. The exit valley interaction remained qualitatively ambiguous even in the case when RX is a diatomic halogen molecule and still more obscure for more complicated polyhalide molecules. Moreover, alkali atom reactions continue to provide an attractive proving ground for the development of more sophisticated techniques of reactant preparation and product analysis, due to the otherwise tractable experimental situation.

Thus, alkali atom reactions are still presenting interesting problems even though they no longer hold the unchallenged centre of the stage.

A. Velocity Analysis

Early experiments[2] involved the measurement of angular distributions $I_{lab}(\Theta)$ of reactive scattering as a function of laboratory scattering angle Θ. Information about the centre of mass differential cross section $I_{cm}(\theta, u)$, where θ and u are the centre of mass scattering angle and velocity for the observed reaction product, may only be inferred by invoking assumptions, which amount to requiring that the differential cross section be factorizable.

$$I_{cm}(\theta, u) = T(\theta)U(u) \tag{2}$$

A vastly more comprehensive and reliable view of the centre of mass differential cross section is gained, if the distribution over laboratory velocity v is measured for the observed product. This may be performed[2] by augmenting the apparatus used for angular distribution measurements with a slotted disc velocity selector placed between the scattering zone and the detector. Early velocity selectors[2] were large and involve a long path length which makes them cumbersome in use and displaces the detector well back from the scattering zone, with a consequent reduction in signal. Thus, early velocity analysis studies[2] gave information only at a limited number of laboratory scattering angles where the reactive scattering is most intense. The introduction of more compact velocity selectors[6,7] has reduced these problems and permitted velocity analysis measurements over the full range of physically accessible laboratory scattering angles. The apparatus[7] used in one sequence of such velocity analysis studies is shown in Fig. 1.

The stripping reactions $K, Cs + Br_2, ICl$ have been studied[8] with this apparatus and the results are shown as contour maps of alkali halide flux in both laboratory and centre of mass coordinates in Figs 2 and 3. These maps illustrate the rather satisfying level of detail in the differential cross section which is revealed. Such detailed information offers a much more demanding target for theoretical models to seek to explain. These contour maps do confirm the broad conclusions of sharply forward ($\theta = 0°$) peaked differential cross sections and low product translational energies E', which were inferred from angular distribution measurements.[2] Indeed, the angular variation of the differential cross section is very close to that deduced from laboratory angular distribution measurements. The distributions of product translational energy $P(E')$ peak at very low energy, $E' \simeq 1$–2 kcal mol^{-1}, compared with the reaction exoergicities, $\Delta D_0 \sim 40$–50 kcal mol^{-1}, again in accord with conclusions from angular distribution measurements. However, the product translational energy distributions also exhibit a long tail extending to much higher energies, $E' \sim \Delta D_0$. These features were

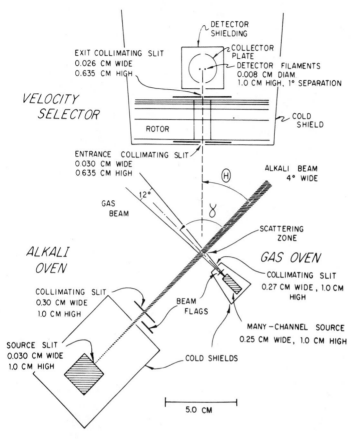

Fig. 1. Alkali atom reactive scattering apparatus incorporating a compact selector for product velocity analysis.

lso apparent in the velocity analysis study of Gillen et al.[9] of the reaction K + I_2, which was discussed in a previous review,[5] though in that study the se of a velocity selected K beam and a large velocity selector for product nalysis, confined information essentially to the forward hemisphere $\theta < 90°$) scattering. The velocity analysis measurements of Figs 2 and 3 overing the complete angular range, clearly reveal a subsidiary backward eak ($\theta = 180°$) in addition to the forward peak ($\theta = 0°$). Moreover, the ontour maps also show a considerable variation in the product translational nergy distribution with scattering angle. The peak velocity in the forward nd backward directions is higher by a factor ~ 1.5 than the peak velocity n the sideways scattering ($\theta = 90°$). This was unobserved in previous tudies and is explicitly assumed to be absent in the analysis of angular istribution measurements.

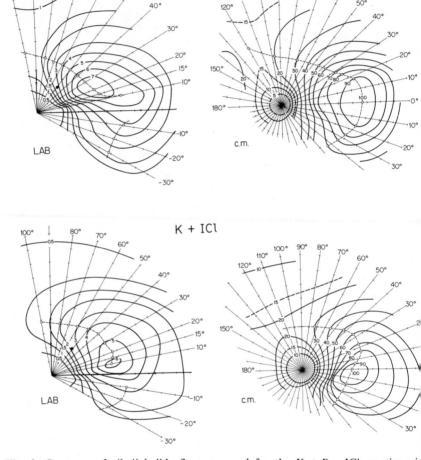

Fig. 2. Contours of alkali halide flux measured for the $K + Br_2$, ICl reactions in laboratory and centre-of-mass coordinates.

The rebound reactions K, Cs + CH_3I have also been the subject[10] of velocity analysis measurements, which confirm the strong backward peaking. Once again, the angular variation of the differential cross section agrees closely with that deduced from angular distribution measurements. The product translational energy distribution is now established[10] as being rather sharply peaked at ~60% of the total available energy, and independent of scattering angle.

Velocity analysis measurements[11] over a wide angular range have established that the reactions K, Rb, Cs + $SnCl_4$, SF_6 involve a long-lived

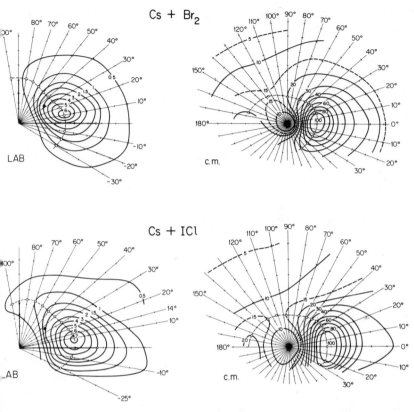

ig. 3. Contours of alkali halide flux measured for the Cs + Br$_2$, ICl reactions in laboratory and centre-of-mass coordinates.

ollision complex. In the SnCl$_4$ case the complex dissociates by two reaction)aths yielding alternatively the alkali chloride or the alkali chlorostanite.

$$M + SnCl_4 \rightarrow MCl + SnCl_3$$
$$\rightarrow MSnCl_3 + Cl \tag{3}$$

[his situation was not suspected from early angular distribution measurements[12] using a thermal alkali beam, but was deduced prior to the velocity _nalysis paper from angular distribution measurements[13] using a supersonic ʿ atom beam. The narrow velocity distribution of the K beam greatly mproves the resolution of the angular distribution measurements. The ›bservation of long-lived collision complexes for the M + SnCl$_4$, SF$_6$ ·eactions are interesting since the exoergicities of these reactions are substantial, $\Delta D_0 \sim 25$–40 kcal mol^{-1}. The long lifetimes of the ionic complexes M$^+$SnCl$_4^-$ and M$^+$SF$_6^-$) arise from their large number of degrees of freedom and the high bond energies of the $SnCl_4^-$, SF_6^- anions. The observation o two competing modes of dissociation for the $M^+SnCl_4^-$ complex require that both reaction paths have closely similar exoergicities. Long-lived complexes have also been observed in recent velocity analysis studies[14] o K atoms with $ZnCl_2$, ZnI_2 and CdI_2. However, short-lived osculating complexes are observed[14] for K + $HgBr_2$, HgI_2. These short-lived comple dynamics are confirmed by angular distribution measurements[15] using a supersonic K atom beam, which indicate that the lifetime may be appreciably lower at supersonic initial translation energy, $E \simeq 5.5 \, \text{kcal mol}^{-1}$, compare with the thermal energies $E \simeq 2 \, \text{kcal mol}^{-1}$ of the velocity analysis study.[14]

Perversely, velocity analysis measurements[16] of alkali atom reaction with CCl_4 indicate strong repulsion disposing energy into product transla tion, as for CH_3I. Angular distribution measurements, both with a therma alkali beam[12] for M + CCl_4 and with a supersonic K beam[17] for K + CCl_4, CBr_4, indicate that the differential cross sections are all sideway peaked with the preferred direction depending on the particular reactants.

This wide variation in reaction dynamics finds a ready explanation[11,1 in terms of the electronic structure of the molecule anion formed in th electron jump, RX^- of (1). When the transferred electron enters a non-bonding σ orbital or weakly antibonding Π^* orbital which does not interpose a node into the halogen bond to be broken, the molecule anion is stable and resistan to dissociation. This is the case for $SnCl_4$, SF_6, ZnX_2, CdI_2 and HgX_2 and a long-lived complex, or for HgX_2 an osculating complex, results. Alternatively when the electron enters an antibonding σ^* orbital which interposes a node in the halogen bond to be broken the molecule anion dissociates rapidly This is the case for Br_2, ICl, CH_3I, CCl_4, CBr_4 and results in direct reaction dynamics. The preferred direction of recoil depends on the strength of th repulsion in the halogen bond, the extent to which it is offset by longer rang attraction, the range of impact parameters over which reaction occurs an any orientation dependence of the molecule for reaction. Thus, for haloger molecules (Br_2, ICl) the repulsion is considerably[18] offset by long-rang attraction and reaction occurs at large impact parameters, $b < 7$ Å, with the collinear orientation of the molecule preferred; stripping dynamic are observed with the alkali halide recoiling in the forward direction. Fo CH_3I the repulsion is very strong, reaction occurs only at small impac parameters, $b < 3$ Å, in collisions with the iodine end[19] of the molecule rebound dynamics are observed with the alkali halide recoiling in th backward direction. For CCl_4, CBr_4 the situation is intermediate fo repulsive forces[20] and impact parameters, and there is no preferred orienta tion for reaction; the alkali halide product is observed to recoil sideways The variation is graphically illustrated by comparing the potential energy curves of the molecules and their anions shown in Fig. 4 (Br_2, $HgCl_2$)[15] and Fig. 5 (CCl_4 and CH_3I).[17] The intermediate situation is also found for othe

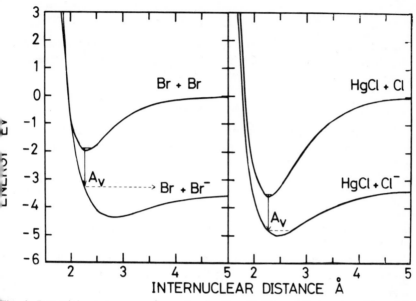

Fig. 4. Potential energy curves for molecules Br_2, $HgCl_2$ and their anions formed in the electron jump mechanism (from D. R. Hardin *et al.*[15] by permission of Taylor and Francis Ltd).

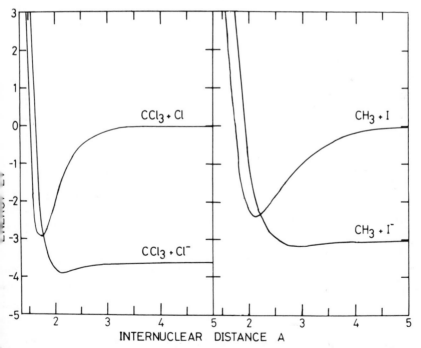

Fig. 5. Potential energy curves for molecules CCl_4, CH_3I and their anions formed in the electron jump mechanism.

halomethane molecules in reactive scattering studies with supersonic[17] and thermal[21] beams, but long-lived complexes are observed[22] in reactions of unsaturated halogen hydrocarbons.

B. Reaction at Superthermal Energies

Mention has been made in the previous section of reactive scattering measurements[13,15,17] with a supersonic alkali atom beam, which gives a modest increase in collision energy, $E \sim 5\text{--}6\,\mathrm{kcal\,mol}^{-1}$, compared with a thermal alkali beam, $E \sim 1\text{--}2\,\mathrm{kcal\,mol}^{-1}$. The change observed in reaction dynamics is confined, at most, to a modest increase in forward scattering relative to backward scattering. The change is naturally small, since the increase in reactant translational energy is minor compared with reaction exoergicity.

However, a much more significant range of collision energy may be explored by seeding[23] the heavy halogen-containing molecules in a nozzle beam of a light driver gas. Using this method, the total reaction cross section for $K + CH_3I \rightarrow KI + CH_3$ has been measured[24] over the energy range $E \simeq 2\text{--}22\,\mathrm{kcal\,mol}^{-1}$, compared with the reaction exoergicity, $\Delta D_0 = 23\,\mathrm{kcal\,mol}^{-1}$. The reaction cross section, shown in Fig. 6, initially increases

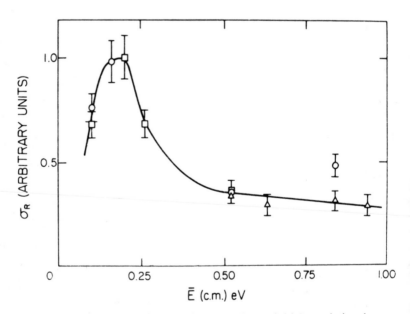

Fig. 6. Dependence of total reaction cross section on initial translational energy for the $K + Ch_3I$ reaction (from M. E. Gersh et al.[24] by permission of the American Institute of Physics).

to a peak at $E \simeq 4 \, \text{kcal mol}^{-1}$, and then declines rapidly with energy to $E \sim 12 \, \text{kcal mol}^{-1}$, thereafter remaining almost constant with further energy increase. Comparison with cross sections for reactive plus inelastic scattering, from optical model analysis[25] of elastic scattering, reveals that the drop in reaction cross section, above $E \simeq 4 \, \text{kcal mol}^{-1}$, is due to the onset of inelastic scattering.

Two models have been proposed to account for this behaviour. In the first,[26] classical trajectory calculations were performed on a potential surface constructed according to the electron jump model (1). The ionic potential surface $(K^+ + CH_3I^-)$ intersects the covalent surface $(K + CH_3I)$ in the region of covalent repulsion $(>2 \, \text{kcal mol}^{-1})$ with internuclear distance $R \gtrsim 3 \, \text{Å}$. The electron is transferred only when the trajectory crosses the line of intersection, having overcome the covalent repulsion. At low energy $(E < 4 \, \text{kcal mol}^{-1})$ all trajectories following the electron jump lead to reaction and the total reaction cross section increases with energy. However, at higher collision energies $(E > 4 \, \text{kcal mol}^{-1})$ trajectories which have crossed the line of intersection may be reflected by short-range repulsive interaction on the ionic surface and traverse the intersection for a second time, returning to the covalent surface. Trajectories crossing the intersection an odd number of times give reaction and an even number of times give inelastic scattering. Thus, the reaction cross section decreases by diversion of trajectories into inelastic scattering, though the sum of reactive and inelastic cross sections continues to increase with energy. The second model[27] also postulates a potential surface with a barrier at internuclear distance $R \sim 4 \cdot 5 \, \text{Å}$ and an attractive well at smaller internuclear distance. Again the requirement of overcoming the repulsive barrier causes the reaction cross section to increase with energy at low energies. However, the decline of the surface behind the barrier is much slower than in the first model. Trajectories which have surmounted the barrier may still fail to give reaction at higher energies due to centrifugal repulsion on the more weakly attractive surface. This broad barrier, as proposed in the second model, is in better accord with the expectations of electronic structure calculations (Section V.B). The covalent–ionic interaction matrix element is expected to be large for an intersection at this small internuclear distance. Indeed, just such a broad region of covalent–ionic interaction has been suggested, without a barrier, in a previous model[28] of the $K + CH_3I$ reaction.

The non-reactive scattering of $K + SF_6, CCl_4$ and $SnCl_4$ has been studied[29] over a range of initial translational energies, $E = 2-5 \, \text{kcal mol}^{-1}$, and as a function of internal energy for molecules in the temperature range 300–600°K. The molecular reactivity, as estimated by the optical potential model,[25] was found to increase with internal energy for SF_6, to a lesser extent for CCl_4, but not for $SnCl_4$. No effect of initial translational energy was observed.

C. Rotational Polarization Analysis

The measurement of angular and velocity distributions of reactive scattering may go a long way to specifying the dynamics and potential surface of a reaction. Measurement of mean product rotational energies,[30] or better, rotational state distributions,[31] taking account of the enhanced dipole moments[32] of vibrationally excited alkali halides, permit still closer specification. However, even this level of information may be inadequate to determine unambiguously the full nature of the reaction dynamics. The recent development[33] of electric deflection fields capable of determining the polarization (angle χ) of the angular momentum vector J' of the product alkali halide relative to the initial relative velocity vector v, provides an important new tool for reactive scattering studies. Measurements[33] of the reactions K, Cs + HBr, HI reveal that J' is strongly polarized perpendicular ($\chi = 90°$) to v. This is expected from kinematics since the light hydrogen atom is unable to carry off appreciable angular momentum. Thus, the product orbital angular momentum L' is small; as is the reactant molecule angular momentum J in these experiments. Conservation of total angular momentum

$$\mathcal{J} = L + J = L' + J' \tag{4}$$

then requires that the initial orbital angular momentum L be efficiently converted to product angular momentum $J' \simeq L$. However, L is necessarily polarized perpendicular to v and hence J' is likewise polarized. Polarization is also substantial for CH_3I, where the kinematics are also favourable, and for Br_2 and CF_3I due to the weak exit valley interactions in these reactions. However, polarization is undetectably small for CCl_4, $SnCl_4$ and SF_6, indicating that the distribution of J' is essentially isotropic for these reactions.

D. Molecular Beam Electric Resonance Spectra of Reaction Products

Molecular beam electric resonance spectroscopy[34] has been used to measure directly the vibrational state distribution of CsF product from the Cs + SF_6 reaction. The results were in accord with equipartition of energy in a long-lived $CsSF_6$ complex. Recent measurements[35] for LiF from Li + SF_6 indicate a vibrational temperature which is only half of that corresponding to equipartition of energy. Thus in contrast to Cs + SF_6, the Li + SF_6 reaction does not proceed via a long-lived complex, due to the increased exoergicity of the Li reaction. A similar effect is observed in the exchange reactions of alkali atoms with alkali halides, which proceed via a long-lived complex except [36] in the case of Li atoms reacting with alkali fluorides, when the reaction complexes become short-lived.

E. Reaction of Oriented Molecules

The dependence of chemical reaction on the orientation of the reactant molecule with respect to the initial relative velocity vector may be studied by measuring[19] the intensity of reactive scattering of a beam of oriented symmetric top molecules as a function of orientation. Early work on the K, Rb + CH_3I reactions[19] showed that reaction occurred preferentially in alkali atom collisions with the iodine end of the molecule. In contrast, reaction is preferred[37] for K + CF_3I in collisions with the CF_3 end of the molecule, even though reaction yields KI product. These experiments have now been extended[38] to full angular distribution measurements of reactive scattering from oriented K + CF_3I. The reaction of K atoms with the CF_3 end of the molecule gives rise to KI recoiling in the forward hemisphere, while reaction with the I end gives KI rebounding into the backward hemisphere. These results are nicely in accord with the electron jump model. Since reaction with the I end can occur in small impact parameter collisions, repulsive interaction with CF_3 causes the KI to rebound backward. In reaction with the CF_3 end of imperfectly aligned CF_3I, only K atom collisions at large impact parameters can view the I atom and the KI is free to recoil forward. Angular distribution measurements[39] with oriented CH_3I and t-BuI show similar results for reaction at either end but with higher intensity for collisions with the I end. Angular distributions for oriented K + $CHCl_3$ are independent of orientation despite the high degree of alignment attainable with oblate tops. Evidently the H atom is too diminutive to exert any observable steric hindrance.

II. ALKALI DIMER REACTIONS

Alkali dimer beams may be conveniently generated[40,41] from a supersonic nozzle expansion of alkali vapour with Mach number $M \gtrsim 10$. The conversion of random thermal motion in the alkali vapour into bulk forward translation during the expansion causes condensation of alkali atoms to alkali dimers (typically ~ 25 mol%) and a narrow velocity distribution ($\sim 15\%$ full width at half maximum) for both atoms and dimers. Alkali atoms may be removed from the beam by magnetic deflection. Velocity analysis measurements[40,41] indicated that alkali dimers might be formed in vibrationally excited states, though for K_2 zero vibrational excitation[41] was within the experimental uncertainty. More direct laser-induced fluorescence studies[42] for Na_2 and K_2 beams show the dimers to be concentrated in the lowest vibrational and rotational states. More recent work[43] indicates that the alkali dimer rotation is polarized with the angular momentum vector perpendicular to the beam velocity vector.

Alkali dimers are attractive reactants for molecular beam studies since they provide an unusual example of highly labile ground state diatomic molecules. The dimer bond is very weak ($D_0(Li_2) = 26$ kcal mol^{-1}, $D_0(Cs_2) = 10$ kcal mol^{-1}) but each of the alkali atoms can form strong ionic bonds with electronegative atoms. Hence, many alkali dimer reactions have exceptionally high reaction exoergicities. The last few years have witnessed a considerable range of alkali dimer reactive scattering studies which testify to their versatility as reagents.

A. Reactions with Halogen Molecules

The reactions[44,45] of alkali dimers with halogen molecules offer an opportunity to study the dynamics of four-centre reactions. Three-centre reactions of an atom plus diatomic molecule have been extensively studied and also four-centre reactions[46] of an atom plus triatomic molecule. However, there are few examples known[47] of reactive four-centre diatomic plus diatomic systems.

Reactive scattering of $K_2 + Br_2$, IBr, ICl and BrCN[44,45] and $K_2 + Cl_2$[47] has been studied by measuring the angular distribution of alkali halide (cyanide) product. The differential cross sections peak in the forward direction with large total reaction cross sections, $Q > 100 \text{ Å}^2$. This lack of symmetry about $\theta = 90°$ for $K_2 + Br_2$ indicates immediately that in the forward scattering the reaction must form an alkali halide, alkali atom and halogen atom products (reaction path 1) to an appreciable extent.

$$K_2 + XY \rightarrow K + KX + Y \qquad \text{path 1}$$
$$\rightarrow KX + KY \qquad \text{path 2} \tag{5}$$

For, if the reaction followed path 2 forming (in the case $X = Y$) two equivalent alkali halide molecules, then conservation of linear momentum in the centre of mass system would require symmetry about $\theta = 90°$ independent of dynamical mechanism. The angular distributions of the reactively scattered K atoms have also been determined[48,49] approximately by comparison of the sum of the K atom and K_2 dimer scattering with the elastic scattering of a supersonic K atom beam, which in centre of mass coordinates closely approximates the form of the elastic scattering of K_2 dimers. The differential cross sections obtained for alkali halide (cyanide) KX and reactively scattered K atoms for $K_2 + Br_2$ and $K_2 + BrCN$ are shown in Figs 7 and 8. Both differential cross sections peak in the forward direction with closely similar flux densities at $\theta = 0°$, indicating that path 1 is essentially the only reaction path contributing to the forward scattering. However, the differential cross section for reactively scattered K atoms falls much more rapidly at wide angles than that for KX, indicating that reaction path 2 becomes significant in the wide angle scattering. The ratio of total reaction cross sections for path 1 relative to path 2 have been estimated[49] as ~ 3.0 for $K_2 + Br_2$ and ~ 1.8 for $K_2 + BrCN$.

For collisions at large impact parameters, these observations have been interpreted[44,45] in terms of reaction dynamics which are initiated by an electron jump in the entrance valley of the potential surface. A mutual ion dissociation reaction follows, in which the halogen molecule anion dissociates

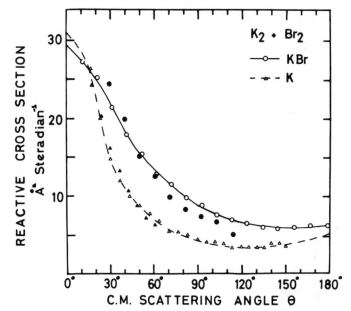

Fig. 7. Differential cross sections for KBr and K products from the $K_2 + Br_2$ reaction (from J. C. Whitehead *et al.*[49] by permission of Taylor and Francis Ltd).

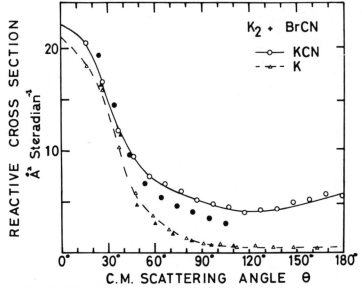

Fig. 8. Differential cross sections for KCN and K products from the $K_2 + BrCN$ reaction (from J. C. Whitehead *et al.*[49] by permission of Taylor and Francis Ltd).

more rapidly than the alkali dimer ion. The K_2X radical thus formed recoils in the forward direction just as in an alkali atom stripping reaction and then dissociates to $K + KX$. This may be summarized schematically by

$$K_2 + XY \rightarrow K_2^+ + XY^- \rightarrow K_2X + Y \rightarrow K + KX + Y \qquad (6)$$

where, for $K_2 + Br_2$, calculation of the likely K_2Br lifetime indicates that this should be viewed as a concerted mechanism but for the less exoergic $K_2 + BrCN$, it should proceed more by separate well-defined steps.

More recent measurements[50] of the $K_2 + I_2$ reaction, shown in Fig. 9, provide further insight into these four-centre reaction dynamics. While very similar in general form to the previous results (Figs 7, 8), the flux of reactive K atoms in the forward direction from $K_2 + I_2$ appears to be greater than

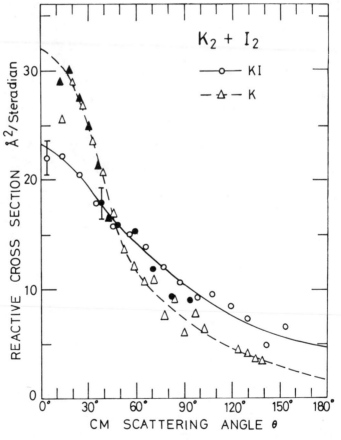

Fig. 9. Differential cross sections for KI and K products from the $K_2 + I_2$ reaction.

that for KI, though it falls rapidly below the KI flux at wider angles. This suggests that the K_2^+ dissociation occurs in a time comparable to (or less than) that of the I_2^- dissociation, resulting in the K atom stripping in the forward direction, while residual I_2^- interaction deflects the KI to wider angles.

Chemiluminescence studies[51] of the $K_2 + Cl_2$ reaction have detected electronically excited K atoms formed by the reaction

$$K_2 + XY \rightarrow K^* + KX + Y \qquad \text{path 3} \qquad (7)$$

with a smaller total cross section $Q < 3 \text{ Å}^2$. This is also consistent with the mechanism outlined above, since vibrationally excited K_2X radicals which attain the KXK configuration may undergo a non-adiabatic transition[52] to an electronically excited state of K_2X which dissociates to $K^* + KX$.

Chemi-ionization studies[53,54] of the $K_2 + XY$ reactions have revealed surprisingly large cross sections, $Q \sim 3-10 \text{ Å}^2$, for reaction leading to ions. Three chemi-ionization pathways are energetically accessible,

$$K_2 + XY \rightarrow K^+ + KX + Y^- \qquad \text{path 4}$$
$$\rightarrow K_2X^+ + Y^- \qquad \text{path 5} \qquad (8)$$
$$\rightarrow K^+ + KXY^- \qquad \text{path 6}$$

The experiment did not identify the ions formed but the total chemi-ionization cross section appeared to correlate most closely with the exoergicity of path 4.

This proliferation of reaction paths may be subdivided into those which involve the transfer of only one electron (paths 1 and 3) and those in which a second electron jump also occurs (paths 2, 4, 5 and 6). Molecular orbital symmetry favours[45] the first electron jump in collisions where the halogen molecule has a roughly collinear orientation with respect to the K_2 dimer, but places no requirement on the orientation of the K_2 dimer itself. Examination[54] of the electronic structure of the incipient $K_2^+ XY^-$ complex, suggests that a second electron jump is inhibited when the orientation of the K_2 dimer is broadside to the halogen molecule axis and enhanced when the K_2 axis is collinear. In particular, chemi-ionization seems most favoured when the KKXY complex is most nearly in a linear configuration. Thus, the reaction dynamics may be summarized as illustrated in Fig. 10.

Recent chemiluminescence studies[55,56] have revealed yet another reaction path leading to alkali halide molecules in the lowest electronically excited state

$$K_2 + XY \rightarrow KX + KY^* \qquad \text{path 7} \qquad (9)$$

This straddles both categories since the reaction initially following the dynamics of Equation (6) to yield $K + KX + Y$ does not involve a second electron jump. Rather[55] the KX molecule acts as a chaperone to bring

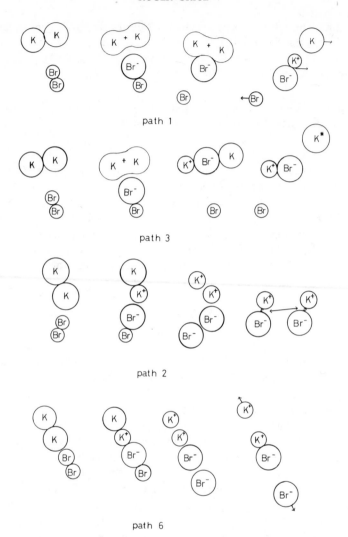

Fig. 10. Schematic representation of reaction dynamics for the $K_2 + Br_2$ reaction (from S. M. Lin *et al.*[54] by permission of Taylor and Francis Ltd).

together the K and Y atoms to form the alkali halide KY* molecule in a homopolar excited state. The molecular electron charge distribution is then rearranged to the ionic ground electronic state by emission of a photon.

$$KY^* \rightarrow K^+Y^- + h\nu \tag{10}$$

As Fig. 10 indicates, the alkali dimer plus halogen molecule reactions exemplify highly energetic motion over the K_2XY potential hypersurface. A prominent feature of this surface is a deep potential well corresponding to the quadrupolar alkali halide dimer[57] $\begin{smallmatrix} K^+X^- \\ Y^-K^+ \end{smallmatrix}$ which is bound by ~ 40–50 kcal mol^{-1} with respect to the separated alkali halides, $KX + KY$. Much less energetic motion confined to this area of the potential hypersurface has been explored in a study[58] of the alkali halide exchange reactions of CsCl with KCl and KI, which also provide a second example[47] of a facile four-center reaction. Since these reactions are thermo-neutral and the initial translational energy $E \simeq 4$ kcal mol^{-1} is small compared with the potential well depth, they exhibit very long-lived reaction complexes. Large reaction cross sections $Q \sim 200$ Å^2 are observed due to the dipole–dipole attraction, which provides an ideal situation for application of the theory[59,60] of long-lived complexes governed by angular momentum conservation (see Section V.A). The observed differential cross sections are in excellent accord with theory, except that the branching ratio for decomposition of the complex to products or back to reactants is not statistical. Recent trajectory calculations[61] on a semi-empirical potential surface show this discrepancy to be due to subsidiary minima in the potential surface corresponding to linear configurations $K^+X^-K^+Y^-$ and $K^+Y^-K^+X^-$. Collisions with very high angular momenta[58] may sample only these extended configurations and subsequently dissociate back to reactants.

A similar, though much shallower, subsidiary minimum[52] in the M_2X potential surface corresponding to a linear MMX configuration, accounts[62] for a discrepancy in the branching ratio for alkali atom–alkali halide exchange reactions. These reactions also proceed[59] by a long-lived complex and the differential cross sections[63] are otherwise in excellent agreement with theory.

B. Reactions with Polyhalide Molecules

Extension of alkali dimer reactive scattering studies to reactions with polyhalide molecules[64,65,50] increases our insight further into the dynamics of diatomic reactants. Measurements attempting[49] to find reactively scattered K atoms from the $K_2 + SnCl_4$ reaction found that both K atoms of the K_2 dimer invariably react to form alkali halide molecules in sharp contrast to the reactions with halogen molecules. Angular distributions of alkali halide reactive scattering have been measured for a wide range[64,65,50] of polyhalide molecules. In the majority of cases the differential cross sections appear to peak in the forward direction but to have significant backward peaks indicative of short-lived reaction complexes with lifetimes comparable to the rotational period of the complex. Typical examples are shown by the reactions $K_2 + SnCl_4$,[64] HgI_2[65] and CBr_4[50] in Figs 11–13. In all the polyhalide reactions, where the approximate method[49] of determining reactively scattered K atoms can be applied (i.e. reactions with large

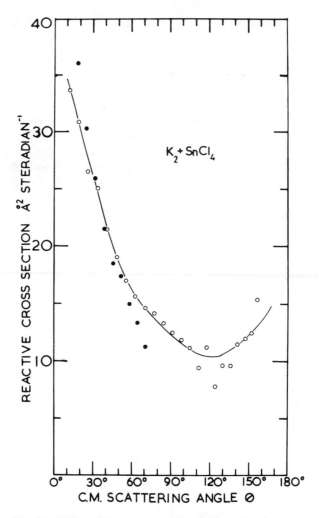

Fig. 11. Differential cross section for KCl product from the $K_2 + SnCl_4$ reaction (from P. B. Foreman *et al.*[64] by permission of Taylor and Francis Ltd).

total reaction cross sections, $Q \gtrsim 70$ Å^2), the yield of reactively scattered K atoms is found to be absent or very low. This rough uniformity of reaction dynamics in the alkali dimer reactions contrasts vividly with the wide variation in differential cross section exhibited by the K atom reactions with polyhalide molecules (see Section I.A). This is in spite of the very similar total reaction cross sections for the atom and dimer reactions,[64,65,50] which arise since both are determined by an electron jump transition in the entrance valley of the potential surface.

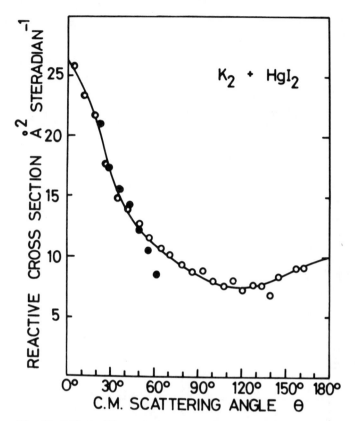

Fig. 12. Differential cross section for KI product from the K_2 + HgI_2 reaction (from D. R. Hardin *et al.*[65] by permission of Taylor and Francis Ltd).

Capture of both atoms of the alkali dimer as alkali halide molecules implies the transfer of both alkali valence electrons in these reactions. Thus, following the first electron jump which initiates reaction, a second electron jump must occur in all reactive collisions during the lifetime of the short-lived reaction complex. This could occur in some collisions, following the dissociation of the K_2^+ ion to K^+ + K, from the released K atom to the polyhalide radical. However, the alkali dimer–halogen molecule reactions suggest that the dissociation of K_2^+ may occur rather slowly in collisions at large impact parameters. Thus an alternative mechanism for the second electron jump has been suggested,[64–66] which does not require prior K_2^+ dissociation. Extended Rittner calculations[67] of the potential energy surfaces for M_2X^+ ions show them to be bound by $\sim 40\,\text{kcal mol}^{-1}$ relative to M^+ + MX, in agreement with mass spectrometric[68] estimates. Comparison with semi-empirical potential energy surfaces[52] for the corresponding

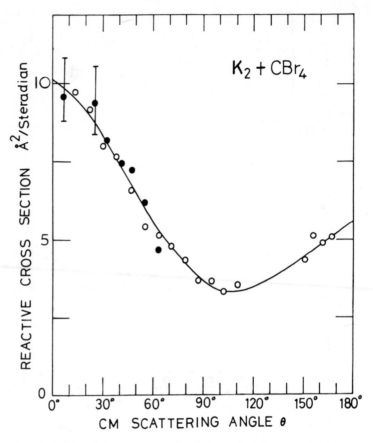

Fig. 13. Differential cross section for KBr product from the $K_2 + CBr_4$
reaction.

M_2X radicals which are more weakly bound, indicate surprisingly low
vertical ionization potentials for M_2X radicals (comparable to the alkali
atom M) over a wide range of bent geometry. Thus, as the K_2^+ ion begins
to polarize a halogen ion X^- from the polyhalide anion following the first
electron jump, a second electron jump can occur from the incipient K_2X
fragment to the polyhalide fragment. The reaction dynamics have been
summarized schematically as

$$K_2 + AX_n \rightarrow K_2^+ + AX_n^-$$
$$\rightarrow K_2^+X^- \dots AX_{n-1} \rightarrow K_2X^+ \dots AX_{n-1}^- \qquad (11)$$

with branches to $2KX + AX_{n-2}$ and $KX + KAX_{n-1}$

This provides a rapid mechanism for the second electron jump, without requiring prior dissociation of the K_2^+ ion. It was first suggested[64] to explain the $K_2 + SnCl_4$ reaction, but the argument for its necessity is rendered equivocal by the durability (see Section I.A) of the $SnCl_4^-$ ion, which could provoke K_2^+ dissociation. However, efficient capture of both atoms of the K_2 dimer has since been observed in the reactions $K_2 + HgX_2$[65] and $K_2 + CH_2I_2$, CHI_3, CBr_4[50] where dissociation of the corresponding anions is expected to be facile. The reactions of K_2 with alkyl and allyl halides[69] have also been interpreted in terms of a second electron jump from the K_2X fragment to the alkyl or allyl radical. The time scale for the second electron jump in this case is particularly stringent since the alkyl halide anion is formed in a strongly repulsive state and dissociates in $\sim 10^{-13}$ sec.

A second electron jump fails to occur by this mechanism in the K_2 plus halogen molecule reactions in large impact parameter collisions due to the high, single-valued electron affinity of the remaining halogen atom A(Y). This is comparable to the ionization potential of the M_2X fragment, $I(M_2X) \sim A(Y)$. Thus the crossing radius between covalent and ionic states occurs at such large inter-nuclear distance, $R_c > 20$ Å, that their interaction is insufficient to effect the second electron jump. The electron affinity of the polyhalide radicals AX_{n-1} is lower and varies with geometry as the radical vibrates. Hence, the second electron jump occurs at some much smaller crossing radius, $R_c < 10$ Å.

Chemi-ionization has been observed[54] in the reactions of K_2 with HgX_2 and SnX_4, but with much lower cross sections for ion formation, $Q_i \sim 0.1$ Å^2, than in the reactions with halogen molecules, $Q_i \sim 3$–10 Å^2. Other polyhalide molecules studied did not yield any observable chemi-ionization giving an upper bound to their chemi-ionization cross sections, $Q_i < 0.001$ Å^2.

C. Reactions with Halogen Atoms

Chemiluminescence studies[51,55] of alkali dimer (Na_2, K_2, Rb_2, Cs_2) reactions with halogen atoms (Cl, Br, I) have discovered large cross sections, $Q \sim 10$–100 Å^2, for formation of alkali atoms in the lowest excited state.

$$M_2 + X \rightarrow MX + M^* \qquad (12)$$

The reaction yield is at least comparable to that of the analogous reaction giving ground state alkali atoms. Higher electronically excited states of the alkali atom M^* are also formed[55,70] up to the limit of the reaction exoergicity. Chemiluminescence was predicted for these reactions long ago by Magee,[71] who argued that collisions with the vacant halogen atom p orbital perpendicular to the plane of the three atoms must correlate with an excited ^2P state of the alkali atom product.

D. Reactions with Hydrogen Atoms

In the reactions of alkali atoms and dimers with halogen-containing molecules reactively scattered alkali halide may be identified[1] by differential surface ionization. This method is also applicable to alkali cyanides,[46] alkali chlorostannites[11,13] and perhaps other alkali salt molecules, but it is by no means the general rule. In the reactions[72] of alkali dimers with hydrogen or deuterium atoms, alkali hydride (or deuteride) product cannot be detected by differential surface ionization. However, elastic collisions with light H or D atoms deflect the heavy alkali dimers only to a small extent in laboratory coordinates. Thus the alkali dimer elastic scattering is confined to a narrow range of laboratory scattering angles close to the dimer beam. In contrast, reactive collisions separate the heavy alkali atoms of the dimer and a much more extensive range of laboratory scattering angles becomes accessible to reactive scattering by virtue of the even disposition of mass in centre-of-mass coordinates. Thus kinematic identification of reactive scattering can be made unambiguously for intensity observed with a surface ionization detector outside the range of laboratory scattering angles permitted to elastic scattering.

Angular distributions of reactive scattering of H and D atoms with K_2, Rb_2 and Cs_2, determined in this way, are shown in Fig. 14. The reaction path

$$H + M_2 \rightarrow M + MH \qquad (13)$$

is the path most consistent with the data, with total reaction cross sections $Q < 50 \text{ Å}^2$ for H, $Q < 100 \text{ Å}^2$ for D, but the analogous reaction path producing alkali atoms in the lowest electronically excited state is also energetically accessible. The pronounced disparity of reactant masses required for kinematic identification of reactive scattering also has a profound influence on the reaction dynamics. The motion of the light H atom is extremely rapid compared with the sluggish internal motion of the alkali dimer; e.g. the incident H atom travels 1 Å in $\sim 10^{-14}$ sec, in which time the K_2 molecule will have rotated $\sim 0.6°$ and executed $\sim 3\%$ of a vibrational period. For an encounter of hard spheres this would imply very inefficient momentum transfer to the diatomic molecule and most collisions would be non-reactive. However, in the $H + M_2$ reaction an electron jump transition may occur to the $H^- + M_2^+$ ionic state. At small internuclear distances this has the structure of a positive core enveloped by a diffuse electron cloud $(M_2H^+)^-$, so that extensive charge migration retards the departure of the H atom. Reaction then occurs with the H atom gyrating wildly about the slowly dissociating alkali diatomic, without imparting significant momentum. Thus, the differential reactive cross section mirrors the dependence of reaction on the initial orientation of the alkali dimer relative to the initial relative velocity vector. The available energy is disposed almost completely into vibrational and rotational excitation of the MH product molecule.

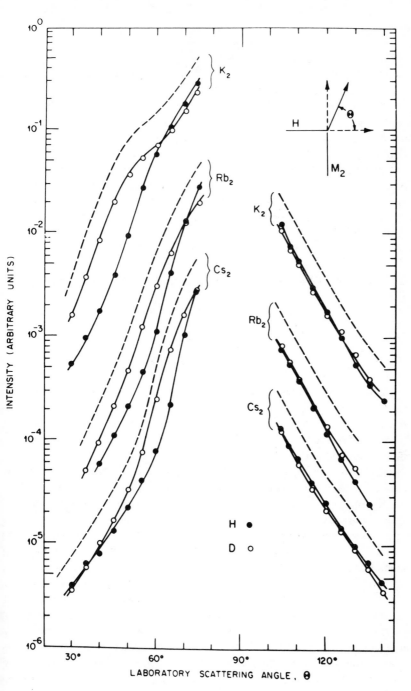

Fig. 14. Laboratory angular distributions of reactive scattering from $H + M_2$ reactions identified by kinematic discrimination (from Y. T. Lee *et al.*[72] by permission of the American Institute of Physics).

The data suggest that the differential cross section peaks sideways at $\theta = 90°$, implying optimum reaction of M_2 in the broadside orientation. A detailed examination of the electronic structure of the potential surface[73] confirms that reaction is preferred in this orientation due to configuration interaction involving excited states of M_2 and M_2^+. The $M_2H(^2A_1)$ potential surface correlating with $M_2(^1\Sigma_g^+) + H$, which is the lower surface in the triangular configuration at long range, is intersected by the $M_2H(^2B_2)$ surface, which correlates with excited $M_2(^3\Sigma_u^+) + H$ and becomes the lower surface in the triangular configuration at close range. Thus the adiabatic potential surface E^- correlating with reaction products is strongly attractive in broadside collisions, but probably lies sufficiently high to inhibit reaction in collinear collisions.

Similar light atom effects are also postulated[3,74] for the $H + KX \rightarrow HX + K$ reaction which shows a shift from sideways to forward scattering as X is changed $F \rightarrow Br$, despite the small reaction cross section $Q \sim 1\,Å^2$.

E. Reactions with Alkali Atoms

The alkali atom–dimer exchange reactions

$$M' + M_2 \rightarrow M'M + M \tag{14}$$

offer a novel class of alkali dimer reactions, since they involve little change in the character of the chemical bonding and thus have low reaction exoergicities, $\Delta D_0 < 3\,kcal\,mol^{-1}$. They involve largely covalent bonding but the atoms are highly polarizable and hence large reaction cross sections might be anticipated.

Reactive scattering measurements have recently[75] been made on the alkali atom–dimer exchange reactions of Na and K atoms with Cs_2, Rb_2 and K_2 dimers, using a quadrupole mass spectrometer detector with a surface ionization source. This detector cannot distinguish alkali atoms from dimers since both are ionized to alkali ions, M^+, which are then mass analysed by the mass spectrometer. Reactive scattering is identified by kinematic separation just as in the H atom reactions discussed in Section II.D. The laboratory angular distributions of M^+ signal from the reactions $Na + Cs_2$, Rb_2, K_2 and $K + Rb_2$ are shown in Fig. 15, together with the Newton diagrams which indicate the spectrum of velocity vectors (inner sphere) accessible to elastically scattered M_2 dimers. The limiting laboratory scattering angles for M_2 elastic scattering are indicated by arrows and it is apparent that there is substantial intensity of reactive scattering outside these limits. Although the shape of the differential cross sections could not be clearly resolved, the total reaction cross sections were estimated to be large, $Q \sim 100\,Å^2$. Calculation of capture cross sections for the long-range van der Waals attraction in these reactions gives $Q_{cap} \sim 150$–$180\,Å^2$. Thus a

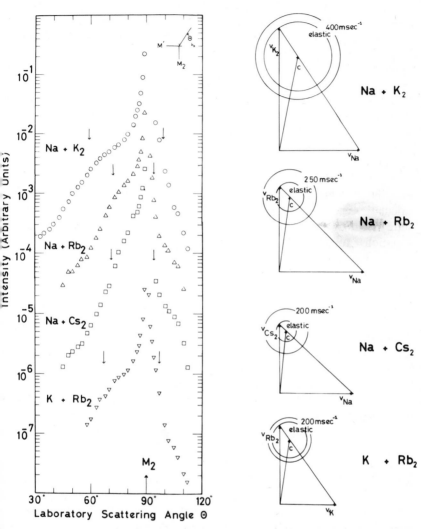

Fig. 15. Laboratory angular distributions of reactive scattering from $M' + M_2$ reactions identified (outside the arrows) by kinematic discrimination (from J. C. Whitehead *et al.*[75] by permission of the Chemical Society).

major fraction of collisions drawn into small internuclear distance undergo reaction. The energy disposed into product translation is low ($\sim 20\%$ of the total available energy) and hence the MM' dimer products are vibrationally excited to $\sim 30\%$ of the MM' bond energy.

Semi-empirical calculations[76,77] of potential energy surfaces for Na and Li exchange reactions indicate a potential well at small internuclear distances extending into the entrance and exit valleys without activation energy. The most stable configurations of the lowest state are nominally linear, with the lighter alkali atom occupying the central position. However, the energy is found to be very flat over a considerable range of bending. Rough calculations of the lifetimes of reaction complexes in the potential energy wells for $Na + Cs_2, Rb_2, K_2$ suggest only short-lived complexes, but a longer lifetime is indicated for $K + Rb_2$. Just as for the $H + M_2$ potential surfaces, configuration interaction between the $M'M_2(^2\Sigma^+, {}^2A_1)$ state and the $M'M_2(^2\Sigma^+, {}^2B_2)$ might be expected to influence the form of the lowest adiabatic surface in the triangular configuration. Again, the energy of the $M'M_2(^2\Sigma^+, {}^2B_2)$ state is strongly dependent on orientation having its lowest energy in the triangular configuration at small internuclear distance. However, the effect may be much milder for $M'M_2$ than HM_2 since semi-empirical calculations[77] indicate that the $^2B_2$ state minimum remains above the $^2A_1$ state minimum for $M'M_2$ in the triangular configuration in contrast to the situation[72] for HM_2. Preliminary results[78] of *ab initio* calculations on Li_3 show some support for this picture.

III. ALKALINE EARTH ATOM REACTIONS

The alkaline earth atoms Ca, Sr, Ba have low ionization potentials, $I \sim 5$–6 eV [however $I(Mg) = 7\cdot5$ eV], which do not greatly exceed those of the alkali atoms. Indeed, the ionization potential of barium, $I(Ba) = 5\cdot2$ eV, is less than that of lithium, $I(Li) = 5\cdot4$ eV. Thus we might anticipate that the reactions of alkaline earth atoms would resemble those of alkali atoms, particularly in exhibiting electron jump transitions. However, the alkaline earth atoms are divalent and two valence electrons can potentially be transferred. This should introduce interesting new features into the reaction dynamics. A particularly close analogy might thus be expected with the reactions of alkali dimers discussed in Section II.

The ionization potentials of alkaline earth atoms and those of their salts[79] do not permit the use of differential surface ionization detection. Consequently reactive scattering must be detected by electron bombardment mass spectrometry. The low ionization potentials of the alkaline earth atoms and many of their salts[79] result in electron bombardment ionization cross sections considerably larger than those of more typical ambient gas molecules. Moreover, the alkaline earth atoms and their salts are notably involatile; thus every effectively pumped by liquid nitrogen-cooled surfaces. Hence, the detection of alkaline earth atom reactive scattering by electron bombardment mass spectrometry presents less severe problems than the study of 'non-metal' reactions, which will be the subject of Section IV. Therefore, less sophisticated apparatus design suffices for the study of alkaline earth atom reactions and discussion of designing reactive scattering apparatus employing electron bombardment detection will be delayed until Section IV. The alkaline earth salts have low-lying electronic states, which render them eminently suitable for detection by laser-induced fluorescence, within the range of wavelengths accessible to present tuneable dye lasers.

A. Angular Distribution Measurements

Angular distribution measurements of reactive scattering of alkaline earth atoms A by halogen molecules X_2 have been reported[80] for Ba + Cl_2 and more extensively[81,82] for Ba, Sr, Ca, Mg with Cl_2, Br_2, ICl, BrCN. The reactions form monohalide products AX, thus following primarily the reaction path

$$A + X_2 \to A^+ + X_2^- \to A^+X^+ + X \qquad (15)$$

The differential cross sections peak strongly in the forward direction for ACl, ABr, ACN, with increasing sharpness along the sequence A = Mg, Ca, Sr, Ba in each case. Only a small fraction, ~ 10–20%, of the reaction exoergicity is disposed into product translation. Thus, the reactions follow a stripping mechanism[2] similar to the alkali atom–halogen molecule reactions. A careful search failed to detect any dihalide product AX_2 indicating that it constitutes $<5\%$ of the reaction yield. The reactions with ICl favour ACl product over AI by a factor ~ 2 and with BrCN, favour ACN over ABr by ~ 10. The same qualitative preferences in reaction product are exhibited by the corresponding[18,46] alkali atom reactions. The differential cross sections for AI product peak in the forward direction but are much broader than the other reactions and more energy is disposed into product translation. The close similarity of these reactions to the corresponding alkali atom reactions indicates that they also proceed by an electron jump mechanism (compare (1) and (15)).

Angular distributions of alkaline earth atom reactive scattering with halomethanes[83] and inorganic polyhalides[84] are also closely analogous to the corresponding alkali atom reactions. Thus, in the reactions of Ba, Sr and Ca with CH_3I, the AI product rebounds backward with substantial recoil energy $E' \sim 10$–15 kcal mol^{-1}. In reactions with CH_2I_2, CF_3I and CCl_4 product is scattered preferentially in the forward hemisphere, but the identity of the reaction product remains in doubt. Only monohalide ions AX^+ were observed, which may have arisen either from AX product or from AX_2 product molecules fragmented during electron bombardment ionization. Circumstantial evidence suggests that AI and AI_2 are formed from A + CH_2I_2 and BaI and BaIF from Ba + CF_3I. The Ba + SF_6 reaction proceeds by a long-lived complex mechanism while the product distributions for Ba, Sr + PCl_3 peak in the forward direction, indicating a direct mechanism; again analogous to alkali atom reactions, though here also the identity of the product remains ambiguous. The reactions[80,84] with NO_2 and $(CH_3)_2CHNO_2$ both yield AO as the predominant reaction product scattering in the forward direction. The reactions with $SnCl_4$ give rise only to ACl or

ACl_2 product but not to $ASnCl_3$ in contrast to the alkali atom reactions which yield both MCl and $MSnCl_3$ (see Section I.A).

Angular distribution measurements[85] of AI from Ca, Sr and Ba with HI yield no information concerning the centre of mass differential cross sections due to kinematic constraint of AI to the centroid distribution. However, advantage was taken of this constraint to estimate the threshold translational energies required for reaction of 5, 4 and 2·5 kcal mol^{-1}, respectively. These values establish lower bounds to the AI bond dissociation energies.

Following early misleading results,[86] the Ba + O_2 reaction is now established[87] as proceeding via a long-lived $Ba^+O_2^-$ complex, at collision energies up to ~ 18 kcal mol^{-1} and despite a reaction exoergicity of ~ 20 kcal mol^{-1}. Unimolecular decay theory requires a $Ba^+O_2^-$ potential energy well of depth > 150 kcal mol^{-1}. Molecular orbital theory would suggest that the $Ba^+O_2^-$ complex is a singlet electronic state, since there are an even number of electrons and non-degenerate orbitals in the bent configuration. However, the reactants and products both lie on a triplet potential energy surface. Thus, the reaction may undergo two singlet–triplet electronic transitions and this, rather than the well depth, may determine the lifetime of the complex. The total reaction cross section[88] has been roughly estimated $Q \sim 1\,\text{Å}^2$.

B. Chemiluminescence Measurements

Chemiluminescence has been observed[89] in beam-scattering gas experiments for the Ba + Cl_2 reaction from radiative two-body recombination

$$A + X_2 \rightarrow AX_2^* \rightarrow AX_2 + h\nu \tag{16}$$

and, to a smaller extent, from the formation of electronically excited BaCl* molecules.

$$A + X_2 \rightarrow AX^* + X \tag{17}$$

Recent *ab initio* calculations[90] on CaF_2 support the assignment of AX_2^* to the excited $Ba^+Cl_2^-(^1B_2)$ state radiating to the ground $Cl^-Ba^{2+}Cl^-(^1A_1)$ state in (16).

Analysis of higher resolution measurements[91] on BaCl* in flow tube experiments indicates a statistical distribution of vibrational states and it was proposed that the reaction of (17) proceeds via a long-lived AX_2^* complex as in (16). The analogous reactions of the other alkaline earths and halogen molecules are presently the subject of further study.[92] The radiative lifetime of $BaCl_2^*$ has been estimated[93] as $\tau_R > 100$ μsec from quenching experiments in a flow system and a remarkably long lifetime ~ 1 μsec deduced for the $BaCl_2^*$ complex. However, it should be borne in mind that, due to the extreme sensitivity of chemiluminescence measurements, products from secondary reactions may be difficult to distinguish from minor pathways of primary reactions. The problem is much more severe in flow experiments than beam experiments; ideally two cross-beams should be employed.

Chemiluminescence has also been observed in beam-scattering gas experiments[94] for the reactions Ba, Ca + N_2O, NO_2. The spectrum for Ba + NO_2 is well resolved and attributed to $BaO(A\ ^1\Sigma)$ with high vibrational excitation but low rotational excitation. The dense spectrum for Ba + N_2O has now been attributed[92] to $BaO(A\ ^2\Sigma)$ with extremely high rotational excitation. A lower bound was calculated to the BaO bond dissociation energy. A lower bound to the AlO dissociation energy has similarly been determined[95,96] from Al + O_3, O_2 chemiluminescence.

Associative electron detachment has been observed[97,98] in the reaction of uranium atoms with oxygen atoms and molecules. The electron detachment cross section is ~ 100 times greater with O atoms than O_2 molecules since only $\sim 1\%$ of U + O_2 collisions detach an electron, while $\sim 99\%$ give neutral products UO + O, with a total reaction cross section $Q \sim 20\ \text{Å}^2$.

C. Laser-Induced Fluorescence

The technique of laser-induced fluorescence spectroscopy[42] has been used to measure vibrational and rotational state distributions in products from the reactions Ba + O_2[99] and Ba + HF, HCl, HBr and HI.[100] The experimental configuration, shown in Fig. 16, consists of a barium atom beam crossed at right-angles by a gas beam. The scattering zone is intersected by the light beam from a tuneable dye laser and is viewed by a photomultiplier looking perpendicular to the laser beam. The apparatus is equipped with extensive blackened baffles to reduce scattering of laser light onto the photomultiplier. The wavelength of the laser is scanned over the absorption spectrum of product molecules for transitions to a bound excited electronic state. When the laser wavelength coincides with an absorption line, product molecules ($\sim 80\%$) are excited to the higher state. The fluorescence emitted on their return to the ground electronic state is detected by the photomultiplier. After correction for Franck–Condon factors, the intensities of fluorescence lines measure the relative number densities of vibrational and rotational states of product molecules in the scattering zone. Calculation of relative total cross sections for product states requires additional knowledge of the product laboratory velocities in order to convert number density to flux.

The results for the Ba + O_2 reactions,[99] which employed a beam-scattering gas arrangement, show a statistical distribution over vibrational states corresponding to a temperature $\sim 2500°K$. The distribution over rotational states was also statistical, corresponding to a temperature of only $\sim 500°K$, but higher gas pressures were required to resolve rotation and some rotational relaxation may have occurred. The results are in accord with the conclusion,

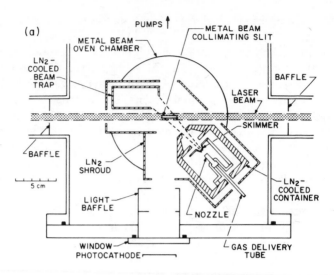

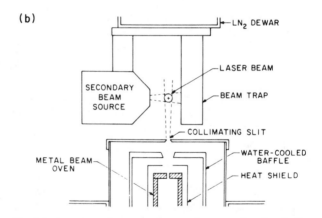

Fig. 16. Apparatus for measuring product rotational and vibrational state distributions by laser-induced fluorescence (from H. W. Cruse *et al.*[100] by permission of the Chemical Society).

drawn from reactive scattering measurements,[87] that the Ba + O_2 reaction proceeds via a long-lived complex.

The Ba + HX experiments[100] employed the cross beam arrangement and kinematics circumvent the requirement of knowing laboratory product velocities; all BaX product has essentially the velocity of the centroid vector. Thus relative total cross sections for individual product quantum states were obtained. The distributions over product vibrational states are shown in Fig. 17. Results indicate that the available energy is disposed

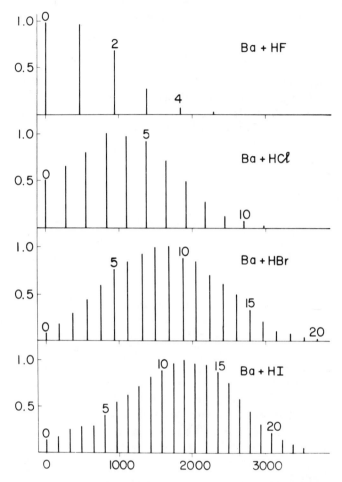

Fig. 17. Distributions of BaX vibrational states from the Ba + HX reactions determined by laser-induced fluorescence (from H. W. Cruse *et al.*[100] by permission of the Chemical Society).

primarily into translation; a fraction 0·75 for HF, 0·5 for HCl. The fractions disposed into vibration varies from 0·12 for HF to 0·36 for HBr, while the fraction into rotation was 0·13 for HF and 0·18 for HCl.

The laser-induced fluorescence method is capable of greater efficiency in detecting molecules distributed over a small number of quantum states than electron bombardment detection (see Section IV). Using current laser technology, it should be possible to measure angular distributions of molecules in individual quantum states. At present the technique is limited to molecules with low-lying excited states, but as laser technology extends the range of wavelengths available from tuneable lasers, this restriction should be reduced.

D. Comparison with Alkali Dimer Reactions

In the preceding discussion of alkaline earth atom reactions, attention has been drawn to analogies with the corresponding alkali atom reactions particularly in Section III.A. However, a still more revealing comparison might be anticipated with alkali dimer reactions, since both alkaline earth atoms A and alkali dimers M_2 have two valence electrons in a totally symmetric orbital giving a singlet state of low ionization potential.

The alkaline earth atom reactions with halogen molecules parallel those of alkali dimers discussed in Section II.A. In both cases, reaction is initiated by transfer of one valence electron. The second valence electron can be transferred only at smaller internuclear distance because (a) the singly positive ions A^+ and M_2^+ have much higher ionization potentials than A and M_2, and (b) the halogen molecule ion X_2^- has a negative electron affinity. The crossing distances in the entrance valley for Ba + Cl_2 have been calculated[90] as $R_c \sim 5$ Å for the first electron jump and $R_c \sim 2$–3 Å for the second electron jump with broadside orientation of Cl_2. However, symmetry arguments (Section V.B) indicate that the first electron jump is preferred in the near-collinear orientation of the halogen molecule, where the crossing for the second electron jump will be at a still smaller internuclear distance. Thus, for reactive collisions with large impact parameters the crossing distance for a second electron jump may not be reached in the entrance valley of the potential surface. Moreover, dissociation of the halogen anion X_2^- is likely to be rapid compared with approach of A^+ or M_2^+. In the exit valley of the potential surface, a second electron jump becomes possible for both reactions, due to the low ionization potentials of the AX and M_2X radicals[79,68] which can transfer a weakly-bonding electron to the departing halogen atom. In both reactions, such a second electron jump leads to a very stable, doubly ionic state: the alkaline earth dihalide $X^-A^{2+}X^-$ and the alkali halide dimer $\frac{M^+X^-}{X^-M^+}$, respectively. However, the preferred paths in the forward scattering for both reactions (15) and (6) involve only the first electron jump. In the alkali dimer reactions, the lack of a second electron jump was attributed (see Section II.A) to the requisite crossing occurring at large internuclear distance ($R_c > 20$ Å) in the exit valley. This argument might also be invoked for some of the alkaline earth atom reactions, e.g. Ba + Cl_2 where $I(BaCl) = 5.0$ eV, $A(Cl) = 3.6$ eV, hence $R_c \simeq 10$ Å. However, it is clearly inappropriate for reactions with lower ionization potentials and higher electron affinities, e.g. Mg + ClI where $I(MgCl) = 7.5$ eV, $A(I) = 3.1$ eV, hence $R_c \simeq 3.5$ Å. Yet all the alkaline earth atom plus halogen molecule reactions[82] show very similar reaction dynamics, as revealed by angular distribution measurements

Probably, most reactive trajectories in the alkaline earth atom reactions sample the avoided intersection (see Section V.B) between the singly ionic and doubly ionic potential surfaces. However, for collisions with large impact parameter, centrifugal repulsion due to the high orbital angular momentum will favour a return to the singly ionic surface, yielding reaction products AX + X (15). This will be enhanced for near-collinear configurations where the second electron jump provokes electrostatic repulsion between the halogen atoms $A^{2+}-X^--X^-$. A complete transition to the doubly ionic surface will be most favoured in broadside collisions $X^-A^{2+}X^-$ and may be followed[90] by radiative recombination (16).

The second electron jump occurs with very high probability in the reactions of alkali dimers with polyhalide molecules (see Section II.B), resulting in the capture of both alkali atoms as bound alkali halides MX. Thus, it would be very interesting to see whether the dihalides AX_2 become prominent reaction products in the alkaline earth atom reactions with polyhalide molecules. Unfortunately, angular distribution measurements are presently equivocal on this point (see Section III.A).

IV. NON-METAL REACTIONS

The problem of detecting reactive scattering of non-metal reactions poses considerable technical difficulties, due to the low efficiency of electron bombardment ionization sources ($\sim 0.1\%$) and the high level of noise relative to signal, arising from ionization of residual gas molecules even at low pressures. The first difficulty requires intense reactant beams in order to provide a high intensity of reactive scattering, while the second requires extensive differential pumping of both sources and detector in order to reduce the background partial pressure in the ion source to $\gtrsim 10^{-14}$ Torr at the product mass peak. The requisite conditions were first attained by Lee, McDonald, LeBreton and Herschbach[101] and several other machines[4,102] have been constructed following this basic design. A diagram of one of these[102] is shown in Fig. 18. Other apparatus designs,[103,104] relying on extensive liquid helium cryotrapping, have also proved successful but are considerably more expensive to operate. With the advent of these 'universal' machines, the 'mainstream' reactions of chemical kinetics have become amenable to the method of molecular beam reactive scattering and the study of an increasing range of atom and free radical reactions is being vigorously pursued.

Except for particular reactions,[4] a reactive scattering apparatus, using electron bombardment detection, is not well suited to the direct observation of product vibrational and rotational states. The measurement of infrared chemiluminescence has, for some time,[105] provided information on vibrational and rotational state distributions of diatomic hydride reaction

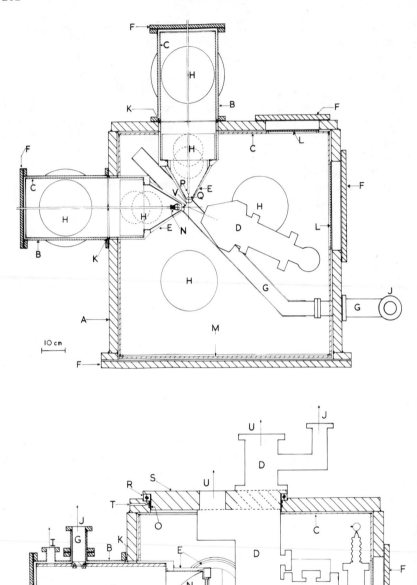

10 cm

10 cm

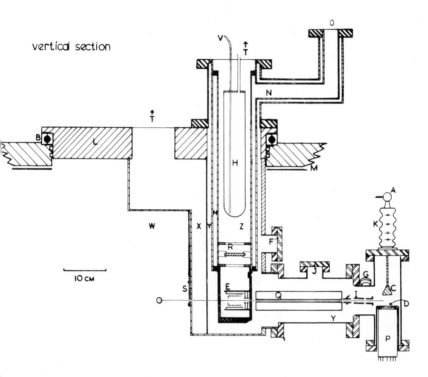

vertical section

10 CM

Fig. 18. Diagram of reactive scattering apparatus for the study of non-metal reactions: A, scattering chamber; B, source chambers; C, liquid nitrogen cooled cold shield; D, detector; E, source bulkheads; G, liquid nitrogen trap; H, oil diffusion pumps; N, free radical source; P, nozzle source; Q, skimmer; E, ion source; H, liquid He trap; , ion lenses; P, photomultiplier; Q, quadrupole rods; R, light baffle; S, slide valve; , radial electric field pumps (from C. F. Carter et al.[102] by permission of the Chemical Society).

products. The experimental technique of these studies has now developed[106] to the point where measurements are made under essentially beam conditions. We may also anticipate that the more sensitive laser-induced fluorescence method, described in Section III.C, will be extended to the measurement of product internal state distributions from non-metal reactions. This technique may even permit the determination of angular distributions of reactive scattering for specific product quantum states.

A. Reactions of Halogen Atoms with Halogen Molecules

Angular distribution measurements[107–109] of reactive scattering of halogen atoms Cl and Br by halogen molecules Br_2, I_2, IBr, ICl marked the advent of successful molecular beam studies of non-metal reactions, and were subsequently augmented[110–113] by velocity analysis measurements. These halogen exchange reactions exhibit forward peaked differential cross sections despite having total reaction cross sections, $Q \sim 1$–10 Å^2, which are significantly smaller than the hard-sphere cross sections. This is attributed to short-range attractive forces arising from a stable trihalogen complex. Molecular orbital theory[114] predicts near linear complexes (interbond angle $\alpha > 140°$) with their stability determined primarily by the nature of the central atom. Thus, the stability is greatest when iodine is the central atom and angular distribution measurements[109] indicate an osculating complex for the reactions Cl + II, Br + II and Cl + IBr. Stability is reduced when bromine is the central atom and angular distribution measurements[109] show a more strongly forward peaked stripping distribution for Cl + BrBr and Cl + BrI. The trihalogen complex becomes least stable when chlorine is the central atom and angular distribution measurements[109] show a backward peaked, rebound distribution for Br + ClI, with a cross section lower by a factor ~ 5 and an activation energy $E_a \sim 5$ kcal mol^{-1}. Velocity analysis measurements,[113] using a slotted disc velocity selector, indicate that the backward peaking attributed to an osculating complex may have been over estimated by the angular distribution measurements, but confirm a trend of increasing forward peaking along the sequence Cl + IBr, Cl + II, Cl + BrBr and Cl + BrI. The product translational energies were found to depend primarily on the identity of the newly formed bond. When this is Cl − I, $\sim 50\%$ of the total available energy is disposed into internal energy but when Cl–Br is the bond formed, $\sim 80\%$ is disposed into internal energy. In all the reactions, the behaviour is characterized by the end of the molecule which is attacked, indicating that migration of the attacking atom from one end of the molecule to the other is insignificant in the reaction dynamics.

Recently studies of halogen exchange reactions have been extended to angular distribution measurements[115,116] of fluorine atom reactions with halogen molecules. These should exhibit the effects of increased electronegativity of the attacking atom. Preliminary results for the F + I$_2$ reaction have been compared for an F atom beam produced[115] from a discharge source ($T \simeq 300°$K) and from[116] a thermal dissociation source ($T \simeq 1000°$K). Preliminary kinematic analysis indicates a broad angular distribution with energy being disposed primarily into product internal excitation. The differential cross section appears to be unsymmetrical about $\theta = 90°$, indicating that the reaction complex is not very long-lived. However, velocity analysis data are required for an unambiguous interpretation. Angular distribution measurements[117] for the F + IBr and F + ICl reactions may be susceptible to a clearer interpretation, since they agree more closely with the predictions[6] for a long-lived collision complex, as illustrated in Fig. 19 for F + ICl. Particular stability is expected for the FICl trihalogen complex due to the extreme difference in the electronegativity of the peripheral atoms compared

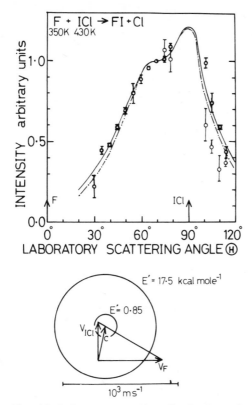

Fig. 19. Laboratory angular distribution of reactively scattered Fl from F + ICl, showing the predicted distributions for a long-lived collision complex.

with the central atom, which maximizes charge transfer stabilization. When the electronegativity of the terminal atom is decreased, as in FII, the stability of the trihalogen complex will be reduced by the increased exoergicity of the product exit valley.

B. Reactions of Hydrogen Atoms with Halogen Molecules

Initially, angular distribution measurements[118,119] of reactive scattering of H and D atoms by halogen molecules were measured in an out-of-plane configuration using an uncollimated halogen beam. Subsequently, a very comprehensive angular and velocity distribution study[120] of D atoms with Cl_2, Br_2, I_2, ICl, IBr has been reported using time-of-flight analysis of the in-plane scattering from narrow ($\sim 5°$) crossed beams. The D atom beam was generated by thermal dissociation of D_2 in a tungsten oven and the halogen beam from a supersonic nozzle beam source, giving an initial translational

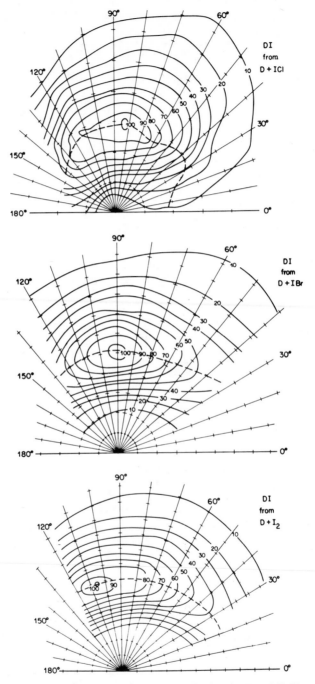

Fig. 20. Contours of DI flux measured for the D + ICl, IBr and I₂ reactions in centre-of-mass coordinates (from J. D. McDonald *et al.*[120] by permission of the American Institute of Physics).

energy $E \sim 10\,\text{kcal mol}^{-1}$. The contour maps in centre-of-mass coordinates of DI flux from D + ICl, IBr, I_2 are shown in Fig. 20. The preferred direction of product DX recoil, in the D + XY reaction depends primarily on the identity of the transferred atom X. Thus for D + Cl_2, the DCl rebounds strongly backward, peaking at $\theta > 150°$; for D + Br_2, the DBr peaks further forward at $\theta \sim 120°–140°$; for D + I_2, the DI peaks sideways at $\theta \sim 100°–110°$. The identity of the departing atom Y has a smaller subsidiary effect on this general trend in the opposite sense. Thus, for D + IBr the DI peaks at $\theta \sim 90°$ shifted slightly forward relative to D + I_2, while for D + ICl the DI peaks at $\theta \sim 80°$, which is shifted slightly further forward. Likewise, the direction of DBr recoil from D + BrI is similar to that from D + Br_2 and DCl from D + ClI similar to that from D + Cl_2. All the reactions dispose a substantial fraction of the available energy into product translation, the fraction being 0.44 for D + Cl_2 (the highest case) and ~ 0.3 for the other reactions, except 0.24 for D + ICl (the lowest case). In the mixed halogen molecule reactions the yields of DI and DBr from D + IBr are comparable, but the yield of DI from D + ICl exceeds the DCl yield by a factor $\sim 3–4$ despite the greater exoergicity ($\sim 30\,\text{kcal mol}^{-1}$ higher) for DCl product. Measurements were also made for the D + I_2 reaction using a D atom beam generated from a microwave discharge source, giving a lower collision energy, $E \sim 1\,\text{kcal mol}^{-1}$. The peak of the differential cross section was observed to shift backward by $\sim 15°$ and the DI recoil velocity decreased, compared with the results at higher collision energy. However, the fraction of available energy disposed into translation remained about constant. The reaction yield from D + Br_2, Cl_2 was too low for distribution measurements, since the collision energy was below the activation energy for these reactions.

These reactions again exemplify the light atom effect, which was discussed in Section II.D, for the reactions of H and D atoms with alkali dimers. The motion of the light D atom is rapid compared with the ponderous halogen molecule, and the differential cross section for reactive scattering mirrors the dependence of reaction on the orientation of the halogen molecule relative to the initial velocity vector. However, the reactions of D atoms with halogen molecules have small total reaction cross sections, $Q \sim 5\,\text{Å}^2$, so that reaction occurs only when the D atom interacts with the halogen molecule at close range and invokes repulsion in the halogen bond. Dissociation of the halogen bond with 'repulsive energy release' thus produces substantial product translational energy. This view is also in accord with triatomic molecular orbital theory,[121] since a linear DClCl complex with 15 valence electrons places an electron in the anti-bonding $3\sigma^*$ orbital (see Fig. 22 of Section IV.C). The $3\sigma^*$ orbital is formed from bonding overlap of the D atom $1s$ orbital with the strongly antibonding $\sigma^*(Cl_2)$ orbital, which is occupied in Cl_2 photodissociation. Indeed, the velocity distribution of DCl from D + Cl_2 correlates with the continuous absorption spectrum of

Cl_2. The contribution to the $3\sigma^*$ orbital of a DXY complex of the anti-bonding $\sigma^*(XY)$ halogen molecule orbital is expected to decrease for Br_2 and I_2 in accord with the decreased fraction of energy disposed into product translation for $D + Br_2, I_2$. The geometry of a 15 electron DXY complex according to Walsh theory[121] may have an interbond angle lying anywhere between $\alpha = 90-180°$, but the molecule is expected to become increasingly bent as the electronegativity of the central X atom decreases. This is in accord with the transition from backward to sideways peaking, which is observed as the central atom changes $Cl \rightarrow Br \rightarrow I$. Molecular orbital theory also predicts, in accord with experiment, that the mixed halogen molecules ICl, IBr will be more reactive at the I end, since the $\sigma^*(IY)$ orbital places electron density predominantly on the I atom. The configuration of the DXY complex with the I atom in the central position will then be more favourable, since this maximizes bonding overlap with the D atom $1s$ orbital.

Matrix isolation experiments[122] have detected a stable bichloride radical Cl–H–Cl, implying a substantial well in the potential energy surface for this configuration. Clearly such a well plays no part in the reaction dynamics for $H + Cl_2$, suggesting that the Cl–H–Cl configuration may be rendered inaccessible by a potential barrier. A simple valence bond argument[3] for the existence of such a barrier follows from noting that insertion of an H atom in the Cl_2 bond cannot pair the H atom electron spin with both Cl atom electron spins Cl↓H↑Cl↑, but reaction with one end of the Cl_2 can satisfy spin pairing H↓Cl↑Cl↓ for adjacent atoms. Molecular orbital symmetry correlations[123] lead to a similar conclusion.

The distributions[124] over vibrational and rotational states of product molecules for $H + Cl_2$, Br_2 and $D + Cl_2$ have been observed directly in infra-red chemiluminescence experiments, using H and D atoms from a discharge source. The fractions of available energy disposed into vibration plus rotation were $0.39 + 0.07$ for $H + Cl_2$; $0.39 + 0.10$ for $D + Cl_2$ and $0.55 + 0.04$ for $H + Br_2$, which imply substantial fractions of the available energy being disposed into product translation; 0.53 for $H + Cl_2$, 0.51 for $D + Cl_2$ and 0.41 for $H + Br_2$. These values are slightly higher than the fractions estimated for translation from velocity analysis of reactive scattering; 0.44 for $D + Cl_2$ and 0.31 for $D + Br_2$. The rotational energy distributions show little variation with vibrational state; thus lower vibrational states are formed with higher translational energies. The effect of varying initial translational energy and vibrational energy of the diatomic molecule for the $H + Cl_2$ reaction has been studied[106] in chemiluminescence experiments by forming the H atom beam by thermal dissociation of H_2 in a tungsten oven and by heating the Cl_2 gas in a graphite oven. It was found that increasing initial translational energy increases the total reaction cross section more effectively than increasing initial vibration. Additional initial translational energy enhances primarily product translation and to a lesser extent rotation. Additional initial vibrational energy enhances primarily HCl product vibration, and results in a bimodal distribution with the principle peak at vibrational quantum number $v = 5$ but a subsidiary peak at $v = 1$. The double peaked distribution arises[125] from reaction of H with Cl_2 at the extremities of its vibrational motion; the extended Cl_2 enhances HCl product vibration and the compressed Cl_2 diminishes HCl vibration. The peak at high vibrational excitation of HCl is dominant because the vibrationally excited Cl_2 spends a longer time at the outer turning point than the inner. The $H + F_2$ reaction has also been studied[126] by infra-red chemiluminescence and the fractions of total energy disposed into product vibration and rotation were 0·53 and 0·03; the fraction into translation 0·44 is again substantial. The reaction of H with vibrationally excited F_2 gives[127] a more pronounced bimodal distribution of HF vibrational excitation than in the $H + Cl_2$ reaction.

Angular distribution measurements[128] of DF reactive scattering from $D + ClF_3$ show backward scattering.

C. Reactions of Methyl Radicals with Halogen Molecules

Methyl radical beams cannot be generated in the same way as atom beams, by thermal dissociation in an oven or electric discharge of a diatomic molecule. Methyl radicals react extremely rapidly by recombination reactions or by reaction with precursor molecules and are not present in significant concentrations for equilibrium dissociation. The problem may be solved[129] by pyrolysing a suitable precursor gas, flowing through a short heated tantalum tube and then issuing directly into the vacuum. The methyl radical reactions following on pyrolysis in the tube are halted in the collision-free conditions of the molecular beam. Using sources of this design, angular distributions of methyl radical reactive scattering with halogen molecules have been measured using azomethane precursor[130,102,131] and using $Zn(CH_3)_2$, $Cd(CH_3)_2$ precursors.[132,104] The laboratory angular distribution[102,131] of CH_3I from $CH_3 + I_2$ is shown in Fig. 21. The methyl radical plus halogen molecule reactions, $CH_3 + XY$, exhibit striking similarities to the D atom reactions discussed in Section IV.B. The CH_3Cl product from $CH_3 + Cl_2$ rebounds in the backward direction; the preferred direction of recoil moves forward for $CH_3 + Br_2$ and $CH_3 + I_2$. The preferred direction of recoil depends primarily on the identity of the transferred atom, but the identity of the leaving atom exerts a smaller subsidiary effect in the opposite sense. The CH_3I product from $CH_3 + IBr$ peaks slightly further forward than $CH_3 + I_2$ and $CH_3 + ICl$ is further forward still. Although the precise directions of recoil for the $CH_3 + XY$ reactions can only be determined once velocity analysis measurements have been obtained, nominal angles determined from the angular distributions indicate a close parallel between $CH_3 + XY$ and $D + XY$, with the CH_3 reactions being pitched further

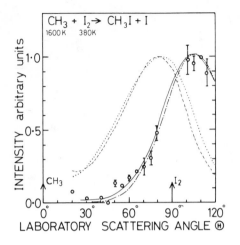

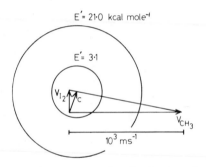

Fig. 21. Laboratory angular distribution of
reactively scattered CH_3I from $CH_3 + I_2$.
The broken curves show the predicted distribution for a long-lived complex.

backward (by $\sim 20°$) than the corresponding D atom reactions. Clearly, the
$CH_3 + XY$ reactions are in distinct contrast to the halogen atom–molecule
exchange reactions (discussed in Section IV.A) which peak strongly in the
forward direction.

It has been suggested,[102] that the $CH_3 + XY$ reactions involve repulsive
energy release in the halogen bond and that the CH_3 radical is sufficiently
light compared with the heavy halogen molecule that the differential reactive
cross section mirrors the dependence of reaction on the orientation of the
halogen molecule, just as in the $D + XY$ reactions. Monte Carlo trajectory
calculations[133] using a potential energy surface appropriate to $D + I_2$
have been performed using a point mass of 15 a.m.u. appropriate to CH_3.
These confirm that the CH_3 exhibits a light atom effect and find that the direction of CH_3I recoil is shifted backward by $\sim 10–20°$ with respect to $D + I_2$.

Thus the backward shift in the CH_3 reaction may be partly a mass effect. Qualitative molecular orbital theory is able to account[102] for a similarity in the potential surfaces of H + XY and CH_3 + XY and their contrast to the halogen atom plus halogen molecule, Z + XY, potential surface. The molecular orbitals are illustrated for the linear reaction complexes in Fig. 22.

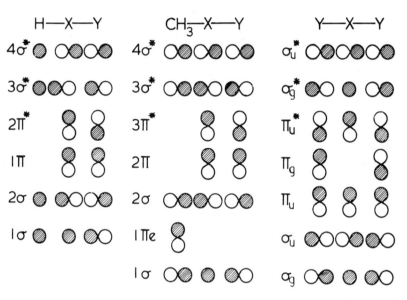

Fig. 22. Schematic diagram of molecular orbitals in the linear configuration for triatomic reaction complexes H + XY, CH_3 + XY and Y + XY, where X, Y denote halogen atoms (from C. F. Carter et al.[102] by permission of the Chemical Society).

The CH_3XY complex with 21 valence electrons is formally equivalent to the trihalogen YXY (or ZXY) complex which also has 21 valence electrons, while the HXY complex has 15 valence electrons. The molecular orbital configurations are

$$HXY(\sigma s_Y)^2(1\sigma)^2(2\sigma)^2(1\pi)^4(2\pi^*)^4(3\sigma^*)$$

$$YXY(s_Y)^2(s_Y)^2(\sigma_g)^2(\sigma_u)^2(\pi_u)^4(\pi_g)^4(\pi_u^*)^4(\sigma_g^*) \tag{18}$$

$$CH_3XY(sa_1)^2(\sigma s_Y)^2(1\sigma)^2(1\pi e)^4(2\sigma)^2(2\pi)^4(3\pi^*)^4(3\sigma^*)$$

However, the orbitals with π symmetry ($1\pi e$) on the CH_3 group are the main CH bonding orbitals and have much higher ionization energies than the halogen π orbitals. The CH_3 ($1\pi e$) orbitals are essentially core orbitals and do not mix significantly with the π orbitals of the XY molecule. Thus, the form of the 2π and $3\pi^*$ molecular orbitals of CH_3XY correspond much more closely to those of DXY than to those of ZXY. Similarly, the $3\sigma^*$ molecular orbital is expected to be CH_3X bonding and XY anti-bonding, hence provoking repulsive energy release, similar to the $3\sigma^*$ orbital of HXY but contrasting with the $3\sigma^*$ orbital of YXY which is symmetrically anti-bonding in both XY bonds. Finally, the H and CH_3 groups with low electron affinities are ineffective in participating in charge transfer stabilization which is important in the YXY(ZXY) complex. Examination[102] of the Walsh diagrams[114,121] for these reaction complexes suggests that the similarities in the molecular orbitals for HXY and CH_3XY also extends to bent configurations. Thus, qualitatively similar trends in the interbond angles α for corresponding HXY and CH_3XY complexes are expected, in accord with the preferred directions of recoil in the experimental differential cross sections. The CH_3 + ICl reaction may give a greater yield of CH_3I product than CH_3Cl despite the higher reaction exoergicity for CH_3Cl (by ~ 27 kcal mol^{-1}), indicating the absence of any migration of the CH_3 group between the I and Cl atoms. Just as in the D + ICl reaction, this is attributable to more effective bonding overlap in the $3\sigma^*$ orbital for CH_3 with the I end of the ICl molecule.

D. Reactions of Oxygen Atoms with Halogen Molecules

The reactions of oxygen atoms with halogen molecules have been studied independently by two groups[87,115,134,135] using a microwave discharge through O_2 as the source of O atoms. In the study by the Cambridge group[115] the discharge is ~ 60 cm from the O atom source slit, so that the source temperature ($\sim 350°K$) gives an initial translational energy, $E \simeq 0.8$ kcal mol^{-1}. Angular distributions of reactive scattering have been measured for O + I_2, Br_2, ICl, and velocity distributions for O + I_2, ICl by time-of-flight analysis. The data[115] for O + I_2 as shown in Fig. 23 gives excellent agreement with the predictions[60] for a long-lived reaction complex: as also does the O + ICl, Br_2 data. The O + ICl reaction yields predominantly OI product by a nominally thermo-neutral path despite the greater exoergicity (~ 13 kcal mol^{-1}) for the formation of OCl product. In the study by the Harvard group,[87,134,135] the discharge extends up to the exit slit of the O atom source, so that the source temperature is higher ($\sim 1000°K$) and thus also the initial translational energy, $E \simeq 2.5$ kcal mol^{-1}. Angular distribution measurements were made for O + Cl_2, Br_2, I_2, ICl, IBr and velocity analysis measurements[135] for O + Br_2, which show clearly that the reactions proceed via a long-lived complex. The predominant reaction product is OI from both O + IBr and O + ICl.

The $O(^3P) + X_2(^1\Sigma_g^+)$ reactions are of particular interest because the reactants approach on a triplet potential surface. However, the lowest potential surface is a singlet corresponding to the bent symmetrical XOX configuration, which corresponds to[136] chemically stable molecules for

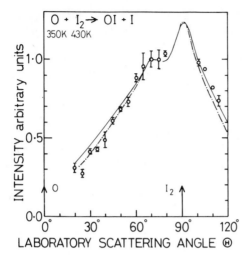

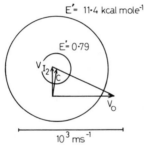

Fig. 23. Laboratory angular distribution of reactively scattered OI from $O + I_2$, showing the predicted distributions for a long-lived collision complex.

OCl_2 and OBr_2. Thus the question arises[115] as to whether the $O + X_2$ reaction involves a triplet to singlet transition with the long-lived reaction complex being formed in the strongly attractive singlet potential. However, this can be ruled out by the $O + ICl, IBr$ reactions which, if they passed through the symmetrical singlet states (IOCl, IOBr), would yield the more exoergic reaction products OCl, OBr, in contradiction to experiment. Fig. 22 would predict the lowest triplet state to have the molecular orbital configuration

$$OXX(1\sigma)^2(2\sigma)^2(1\pi)^4(2\pi)^4(3\pi^*)^3(3\sigma^*) \tag{19}$$

which, in view of the occupied $3\sigma^*$ orbital, might not be expected to be sufficiently attractive to support a long-lived complex, particularly for $O + I_2$ with exoergicity $\Delta D_0 \sim 10.6 \text{ kcal mol}^{-1}$. However, it has been suggested[87,135] that in fact the $3\sigma^*$ orbital lies lower than the $3\pi^*$ orbital in the near-linear configuration. Thus, the lowest electronic state for the linear configuration is a triplet

$$OXX(1\sigma)^2(2\sigma)^2(1\pi)^4(2\pi)^4(3\sigma^*)^2(3\pi^*)^2 \qquad (20)$$

This[87,135] is strongly supported by observation[137] of an unsymmetrical OClCl bound state of the OCl_2 molecule in matrix isolation experiments. Thus, the triplet potential surface should be sufficiently attractive (~ 13 kcal mol^{-1} for $O + Br_2$) in the near-linear configuration to sustain a long-lived complex. This conclusion is reinforced by the case of $O + F_2$, which would place the least electronegative atom in a peripheral position for the OFF configuration, thus rendering it unstable. Indeed, flow tube experiments[138] fail to detect the $O + F_2$ reaction under conditions where the $O + Br_2$ reaction is readily observed, despite the greater exoergicity of the $O + F_2$ reaction.

The reaction of oxygen atoms with CS_2 has been observed[139] with the OS product distribution peaking in the forward direction.

E. Reactions of Halogen Atoms with Hydrogen, Hydrogen Halides and Methyl Halides

When the product of a reaction is a diatomic hydride with widely spaced vibrational energy levels, the reaction exoergicity may be sufficient to populate only a few vibrational states. If the 'excess' energy for low vibrational states appears as product translational energy rather than rotation, structure may be resolved in the angular distribution of reactive scattering. The vibrational state distribution of DF product from the $F + D_2$ reaction has been observed[4,140] in this way. The results are in essential agreement with the structured DF translational energy distribution,[141] calculated from infrared chemiluminescence measurements of the DF vibrational and rotational states. Angular distributions for the $F + D_2$ reaction have now been measured[4] for a range of initial translational energies. At the lowest energy $E = 0.8$ kcal mol^{-1}, which is comparable with the activation energy $E_a \simeq 1.0$ kcal mol^{-1}, the FD product recoils in the backward direction with only the highest accessible vibrational states $v = 4$ and 3 populated. As the initial translational energy is increased, a forward shift in the differential cross section for the $v = 4$ state is observed, though there is relatively little change for the $v = 3$ state. This specific coupling of initial translational energy with dynamics for the $v = 4$ state is a consequence of it being only just energetically accessible. Thus, at the lowest initial translational energy, little energy is available for translation or rotation of the DF ($v = 4$) product and it is necessarily formed with low orbital and molecular angular momentum. Hence, conservation of angular momentum permits the formation of DF($v = 4$) only from small impact parameter collisions with low initial orbital angular momentum. As the initial translational energy is raised, higher rotational states of DF ($v = 4$) become accessible from collisions with larger impact parameters, with a concommitant increase in the forward scattering.

Infrared chemiluminescence measurements[142] have determined the vibrational and rotational state distributions of HF and DF from the $F + H_2, D_2$ reactions, and find substantial fractions of vibrational (0·66) and rotational (0·08) excitation. The potential surface[143] appears to be predominantly repulsive but the heavy mass of the attacking atom relative to the light transferred atom channels the energy of repulsion into product vibration. This is termed mixed energy release on a repulsive potential surface.

The reactions of halogen atoms with hydrogen halides have been extensively investigated[106,144,145] by infrared chemiluminescence. Channeling of energy into vibrational and rotational excitation is even more pronounced in these reactions, e.g. for $Cl + HI$, DI, the fractions are vibration 0·61 and rotation 0·13. Once again the potential surface is predominantly repulsive but mixed energy release is particularly effective in channeling energy into internal excitation when a light atom is transferred between two relatively heavy atoms. Experiments[106,145] for $F + HCl$ at higher initial translational energy find the extra energy disposed primarily into product translation and rotation, but reaction of vibrationally excited HCl transfers the extra energy into product vibration. The same correlation was found for the $H + X_2$ reactions, discussed in Section IV.B. Specific relative reaction rates for the $HCl(v = 0, 1, 2)$ states were also estimated[106] and found to increase with vibrational quantum number. The dependence of the reaction rate on rotational quantum number J for the $HCl(v = 1)$ state has also been determined by observing the depletion of the infrared fluorescence of these levels by reaction with F atoms. The rate of the endoergic reaction $Br + HCl \rightarrow HBr + Cl$, $\Delta D_0 = -16·5 \text{ kcal mol}^{-1}$, has been measured[146] for the vibrationally excited reactant states $HCl(v = 1-4)$ by fluorescence depletion. The reaction is energetically accessible for the $v = 2$ state, but the rate increases by a factor ~ 7 for $v = 3$ and ~ 10 for $v = 4$ relative to $v = 2$. Thus, reaction is strongly promoted by vibrational excitation.

In contrast, reactive scattering from the endoergic reaction $I + CH_3Br$ has been observed[4] when CH_3Br, seeded in a supersonic H_2 nozzle beam, is crossed with an I atom beam formed from thermal dissociation of I_2 in a nickel oven. At initial translational energies $E > 30 \text{ kcal mol}^{-1}$, HI and IBr products are observed which are both endoergic by 25 kcal mol^{-1} but CH_3I product which is endoergic by only $12·5 \text{ kcal mol}^{-1}$ is not detected. Such chemical preference for H atom or halogen atom transfer rather than transfer of a CH_3 group is well known in exoergic reactions. The HI product peaks in the forward direction while the IBr product peaks backward.

F. Reactions of Hydrogen Atoms with Hydrogen Molecules

The exchange reaction of hydrogen atoms with hydrogen molecules is the simplest chemical reaction of neutral species and offers a tractable system for direct comparison of experiment with *ab initio* theory. The angular distribution of HD reactive scattering from the $D + H_2$ reaction has been measured[103,147] in a crossed beam experiment employing extensive liquid helium cryotrapping. The laboratory distribution is bimodal due to the 'edge effect' in the Jacobian transforming from laboratory to centre-of-mass coordinates. Normally this effect is smoothed out by the distribution of product velocities in centre-of-mass coordinates. However, the $D + H_2$ reaction populates only the HD ground vibrational state and low rotational states, so that the HD velocity distribution is very narrow. The differential cross section in centre-of-mass coordinates peaks backward at $\theta = 180°$ with an angular dependence $I(\theta) = \cos^2[1.35(180°-\theta)]$, in general accord with classical trajectory calculations[148] using a valence bond H_3 potential energy surface.[149] Detailed comparison[150] of classical calculations with the experimental laboratory distribution confirms this general agreement but indicates a rather more pronounced bimodal distribution than is observed. Agreement is also found with a quantum calculation[151] of the angular distribution of reactive scattering for $\theta > 120°$ but significant discrepancy appears at smaller scattering angles.

Very recently[152] the exchange reaction of H atoms with tritium molecules has been observed by trapping the T atom products on MoO_3 detector buttons placed around the scattering centre and measuring the radioactivity of the buttons after removal from the vacuum. The laboratory angular distributions of T atom scattering also indicate a differential reaction cross section peaking backward for HT with respect to the incident H atom. However, the differential cross section obtained is significantly broader than that deduced[103,147] for $D + H_2$. Reference to quantum DWBA calculations[151] indicates that, in experiments with a broad distribution of initial translational energy, the contribution of high energy collisions favouring forward scattering may significantly broaden the observed differential cross section.

G. Reactions of Halogen Atoms with Unsaturated Hydrocarbons

The substitution reactions of fluorine atoms with unsaturated hydrocarbons have been investigated in a comprehensive range of experiments.[153–159] A velocity selected beam of F atoms, formed by thermal dissociation of F_2 in a nickel oven, is intersected by a hydrocarbon beam from a supersonic nozzle beam source. Angular[153] and velocity[154] distributions of C_2H_3F product were measured for the displacement of a hydrogen atom from ethylene

$$F + C_2H_4 \rightarrow C_2H_3F + H \tag{21}$$

$$F + C_2H_4 \longrightarrow C_2H_3F + H$$

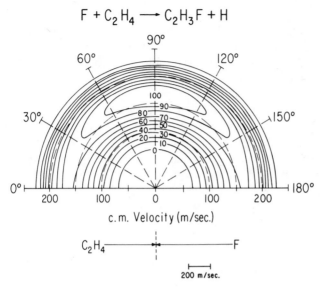

Fig. 24. Contours of C_2H_3F flux measured for the $F + C_2H_4$ reaction in centre-of-mass coordinates (from J. M. Parson *et al.*[154] by permission of the Chemical Society).

The differential cross section, shown in Fig. 24, is symmetrical about $\theta = 90°$, indicative of a long-lived reaction complex and exhibits a mild sideways peak at $\theta = 90°$. This may be attributed to conservation of angular momentum in the long-lived complex as illustrated in Fig. 25. The F atom interacts with the π electrons of the C_2H_4 double bond in collisions where its direction of approach is roughly perpendicular to the C_2H_4 plane. Due to rotational relaxation in the supersonic nozzle expansion, the C_2H_4 occupies only low rotational states. The total angular momentum $\mathscr{J}$ of the C_2H_4F complex is essentially given by the initial orbital angular momentum of the

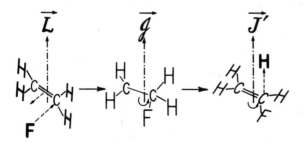

Fig. 25. Schematic mechanism for the $F + C_2H_4 \rightarrow C_2H_3F + H$ reaction (from J. M. Parson *et al.*[154] by permission of the Chemical Society).

collision, $L \simeq \mathcal{J}$ and is perpendicular to the initial relative velocity vector $\mathbf{v}$. However, $\mathcal{J}$ is also perpendicular to the plane of the heavy atoms C–C–F. The H atom is displaced from the complex also by interaction with the π electrons of the C_2H_3F protomolecule and leaves in a direction perpendicular to the C_2H_3F plane. Thus, the final relative velocity vector $\mathbf{v}'$ is either parallel or antiparallel to L and perpendicular to $\mathbf{v}$, i.e. the products recoil sideways. Emission of a light hydrogen atom can accommodate only a small product orbital angular momentum so that most of the total angular momentum $\mathcal{J}$ appears as rotational angular momentum $\mathbf{J}'$ of the C_2H_3F product.

The distribution of product translational energy $P(E')$ determined from experiment does not agree with either RRKM theory or phase space theory if the full 12 oscillators of C_2H_4F are included, but agreement is closer if only 5 oscillators are included, corresponding to the heavy atoms. For large molecules, phase space theory and RRKM theory, including angular momentum (RRKM–AM),[60] give essentially equivalent[160] predictions.

The reaction of F atoms with butenes[154,155] can proceed either by H atom displacement

$$F + C_4H_8 \rightarrow C_4H_7F + H \tag{22}$$

or by methyl radical displacement

$$F + C_4H_8 \rightarrow C_3H_5F + CH_3 \tag{23}$$

The differential cross sections are now found to be isotropic as would be anticipated from the mechanism of Fig. 25, since the C_4H_8F complex now has more than three heavy atoms; the orientation of the total angular momentum is no longer related to the double bond and the attacking F atom. The product translational energy for both H atom and CH_3 displacement is again in better accord with statistical theory if the H atoms of the C_4H_8F complex are ignored in assigning the number of oscillators. However, the ratio of total reaction cross sections for H atom versus CH_3 displacement are found to be in agreement with the RRKM prediction including all atoms.

The reactions of F atoms with substituted olefins, cyclic olefins and dienes have been studied[156] with conclusions similar to those for the butene reactions. The D atom displacement reaction of F atoms with perdeutero benzene[157] shows mild sideways peaking. The mechanism advanced for C_2H_4, Fig. 25, again applies since the rigid benzene ring plays the same role as the carbon double bond. Once again, the product translational energy distribution is found to be in poor accord with the predictions of statistical theory. The H atom and CH_3 displacement reactions of F atoms with other aromatic and heterocyclic molecules[158] yield similar discrepancies with statistical theory. However Cl atom displacement from chlorobenzene[158] and dichloroethylene[159] gives product translational energy distributions in agreement with statistical theory.

The Br atom displacement reactions[160] of Cl atoms with vinyl bromide, 1-bromopropene and 2-bromopropene also proceed by a long-lived complex mechanism with larger reaction cross sections, $Q \sim 20\text{--}35 \text{ Å}^2$. The product translational energy distributions are found to be in excellent agreement with RRKM–AM theory. The distributions using the approximate quantum level density in the RRKM–AM theory were found to be very similar (for C_4H_8F) to the distribution calculated using the classical level density but omitting the H atoms. Thus, the quantum weighting of the light hydrogen atoms makes their effect small even when they are included in the randomization of energy. The Br atom displacement reaction[160] of Cl atoms with allyl bromide proceeds by a short-lived complex as indicated by a forward peak which is three times more intense than the backward peak. The product translational energy distribution is displaced to higher energy than that predicted by the statistical model due to the short lifetime. However, agreement with experiment is achieved by a statistical distribution corresponding to a three-atom complex rather than five. The anisotropy of the differential cross section suggests that the transition state places both Cl and Br atoms on the same C atom forming a roughly linear configuration.

The interpretation of the apparent discrepancy between experiment and statistical theory in the F atom displacement of H and CH_3 is controversial.[134,161,162] The exit valley of the potential surface for these reactions involves an energy barrier. The authors of the experimental work suggest[162] that, in displacement of a light H or CH_3, the barrier energy is converted efficiently into product translational energy. They support the assertion[163] with model calculations of vibrational–translational energy transfer between the retreating H or CH_3 and the remaining polyatomic molecule. They then ascribe the discrepancy of theory and experiment to energy randomization in the long-lived complex remaining incomplete, saying it occurs more slowly than reaction. However, it has been questioned[134,161] whether measurements of product translational energy distributions can in fact test energy randomization in the reaction complex. Product energy distributions are also affected by possibly complicated exit valley interactions as a tight activated complex dissociates from a strained configuration. These effects are solved for a loose complex by the RRKM–AM theory (see Section V.A) but for a complex which is not loose, the problem may involve more than just one dimensional vibrational–translational energy interaction. Only a modest energy exchange ($\pm 2 \text{ kcal mol}^{-1}$) is required[134] as the products descend from the transition state in order to make the data compatible with energy randomization in the reaction complex for F plus isobutene.

Recently, the vibrational energy distribution of cyclooctanone formed by addition of an oxygen atom to cyclooctene has been observed[164] directly by infrared chemiluminescence. The measured populations of different vibrational modes were in accord with statistical randomization of internal energy. However, the vibrational energy distribution observed corresponds to excited cyclloctanone with a lifetime $\sim 10^{-3}$ sec in the field of view of the spectrometer. This is very much longer than the typical lifetimes ($\sim 10^{-11}$–10^{-9} sec) of long-lived collision complexes observed in reactive scattering.

H. More Four-Centre Reactions

The investigations of two classes of facile four-centre reactions, alkali dimers plus halogen molecules and alkali halide–alkali halide exchange, have been discussed in Section II.A. However, facile four-centre reactions are more notable for their scarcity than their abundance. Shock tube and laser Raman studies[165] of the reaction $H_2 + D_2 \rightarrow 2HD$ show that it requires vibrational excitation and has an activation energy, $E_a \sim 40$ kcal mol^{-1}. The textbook reaction $H_2 + I_2 \rightarrow 2HI$ has been shown[166] to have a negligible rate compared with the termolecular reaction $I + H_2 + I$. Indeed, molecular orbital correlations[167] predict these reactions to be 'thermally forbidden' with activation energies comparable to the reactant bond energy.

The halogen molecule exchange reaction

$$Cl_2 + Br_2 \rightarrow 2BrCl \tag{24}$$

is also thermally forbidden but inclusion[168] of the in plane p orbitals in the correlation diagram suggests a lower activation energy, $E_a \sim 20$–30 kcal mol^{-1}. The reaction was sought by crossing a Br_2 nozzle beam with a nozzle beam of Cl_2 seeded in helium, to obtain enhanced translational energy. A significant fraction of the reactant molecules were also vibrationally excited. However, no BrCl product attributable to halogen molecule exchange could be detected for translational energies $E \gtrsim 30$ kcal mol^{-1}, placing an upper limit on the reaction cross section of $Q \gtrsim 0.01$ Å^2.

However, one further class of facile four-centre reactions has been discovered:[47] the alkali halide plus halogen molecule reactions. For the reaction

$$CsI + Cl_2 \rightarrow CsCl + ICl \tag{25}$$

angular and velocity distributions of ICl product show forward and backward peaking, symmetrical about $\theta = 90°$ and indicative of a long-lived collision complex. The product translational energy distribution shows excellent agreement with the predictions of RRKM–AM theory[60] with a preference for the 'tight' complex model. However, for the reaction

$$CsBr + ICl \rightarrow CsCl + IBr \tag{26}$$

angular distribution measurements of IBr product show a prominent backward peak relative to incident ICl, which is more than twice as intense as the forward peak. Thus, roughly half of the reaction complexes have lifetimes less than the rotational period and are non-statistical. The endoergic reaction products CsI + BrCl are not observed even for translational energies above the threshold, obtained by seeding the ICl beam in helium.

The previous examples (Section II.A) of facile four-centre reactions appear to be enabled by the intrusion of ionic interactions into the electronic structure of the potential energy surfaces. The alkali halide–alkali halide exchange reactions involve purely electrostatic interactions. The alkali dimer–halogen molecule reactions involve an electron jump in the entrance valley of the potential surface and thereafter proceed as ion recombination reactions. The present alkali halide–halogen molecule reactions also appear to be permitted by virtue of ionic interactions. The reaction complex, as illustrated in Fig. 26, has the structure of an alkali trihalide molecule. The observed

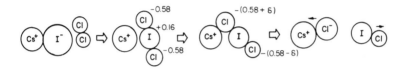

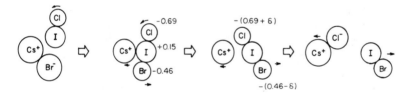

Fig. 26. Schematic mechanisms for the CsI + Cl_2 and CsBr + ICl reactions (from D. L. King et al.[47] by permission of the Chemical Society).

features of the reaction dynamics find ready interpretation in terms of the structure[169] of the linear trihalogen anion. Thus, the CsI + Cl_2 reaction forms the most stable $(ClICl)^-$ configuration with the least electronegative atom I in the central position, by insertion of I^- into the Cl_2 bond. This insertion is allowed by valence bond theory because the I^- has no unpaired spin, in contrast to the halogen atom–halogen molecule reactions (Section IV.A). The Cl^- ion may be detached from $(ClICl)^-$ by the Cs^+ ion only when a sufficiently collinear configuration of $Cs^+(ClICl)^-$ polarizes the $(ClICl)^-$ ion strongly. However, the CsBr + ICl reaction forms the most stable $(ClIBr)^-$ configuration by attachment of Br^- to the end of ICl. The negative charge in the unsymmetrical $(ClIBr)^-$ resides primarily on the Cl atom, which is readily detached by the Cs^+ ion, resulting in a short complex

lifetime. The I atom is the least negatively charged atom in (ClIBr)⁻ and is
not transferred to Cs⁺ even when this is energetically allowed.

V. THEORETICAL MODELS

The quantum theory of reactive scattering and the calculation of potential
energy surfaces are reviewed elsewhere in this volume. Here attention will
be confined to two recent theoretical models which are both simple and
should be widely applicable. Thus, they are appropriate to the analysis of
experimental data at an early stage, in order to help discern the nature of the
reaction dynamics.

A. Long-Lived Collision Complexes

The angular dependence $T(\theta)$ of the differential reaction cross section
for a long-lived collision complex is governed[59,170] by conservation of
the total angular momentum $\mathscr{J}$ (see (4), where L, L' denote the reactant and
product orbital angular momenta and J, J' denote the angular momenta of
reactant and product molecules). The initial orbital angular momentum L
is perpendicular to the initial relative velocity vector $\mathbf{v}$, distributed uniformly
over aximuthal orientations, while the reactant molecule angular momentum
J is isotropically distributed. Thus, the distribution of $\mathscr{J}$ is cylindrically
symmetric about $\mathbf{v}$ and its projection M on $\mathbf{v}$ is conserved. Since the lifetime
of the collision complex is much longer than its rotational period, the distribution of product relative velocity vectors $\mathbf{v}'$ must be symmetrically distributed
about $\mathscr{J}$ with a projection M' of $\mathscr{J}$ on $\mathbf{v}'$. The probability distribution for the
scattering angle θ between $\mathbf{v}$ and $\mathbf{v}'$ is given by the geometrical transformation
of uniform precession of $\mathbf{v}'$ about $\mathscr{J}$, followed by uniform precession of $\mathscr{J}$
about $\mathbf{v}$. The resulting quantum form factors are thus the squares of the
Wigner matrices[171] for the transformation

$$|d_{MM'}^{\mathscr{J}}(\cos\theta)|^2 + |d_{M-M'}^{\mathscr{J}}(\cos\theta)|^2 \tag{27}$$

including both terms differing by the sign of M'. Special cases and the corresponding classical expressions are given in Table 5 of Reference 59. The
observed angular dependence $T(\theta)$ is found by averaging the appropriate
form factors over the distributions of $\mathscr{J}$, M and M'. The distribution of impact
parameters b may be taken as uniform in b^2 up to a maximum value. Hence

$$P(L) = 2L/L_m^2, \qquad L \gtrsim L_m = (2\mu EQ_c/\pi)^{1/2} \tag{28}$$

where Q_c is the total cross section for complex formation. The distribution
over M is uniform between $-J \leqslant M \leqslant J$ and the distribution over M' taken
to be Boltzmann at the transition state of the dissociating complex.

$$P(M') = \exp\left(\pm\tfrac{1}{2}M'^2/M_a'^2\right) \tag{29}$$

The resulting angular dependence $T(\theta)$ depends only on $X = L_m/M_a'$ in the frequent case $L_m \gg J_{mp}$, and is shown in Fig. 10 of Ref. 59. When the transition state corresponds to a prolate rotor the minus sign applies in (29) and $T(\theta)$ peaks forward and backward at $\theta = 0°, 180°$. When the transition state corresponds to an oblate rotor, the plus sign applies in (29) and $T(\theta)$ peaks sideways at $\theta = 90°$. In both cases the sharpness of peaking increases with X and $T(\theta)$ is isotropic for $X = 0$.

The product translational energy distribution $P(E')$ may also be determined[60] for the case $L_m \gg J_{mp}$ in the form

$$P(E') = N_{vr}^{\dagger}(E_{tot} - E')A(E') \tag{30}$$

where $E_{tot} = E + E_{int} + \Delta D_0$ denotes the total energy available to products. The energy level density $N_{vr}^{\dagger}(E_{tot} - E')$ of active vibrations and rotations at the transition state as given by RRKM theory[172] is multiplied by a factor $A(E')$ which accounts for the centrifugal barrier to dissociation. The product translational energy is given by the sum of the translational energy along the dissociation axis $\varepsilon^{\dagger}$ in the transition state at the top of the centrifugal barrier plus the barrier height B'. The distribution of product translational energy, given by averaging over the distribution of barrier heights B', reduces to (30) with

$$A(E') = \int_0^{E'} P(B')\, dB' \tag{31}$$

This factor may be evaluated for the case $L \simeq \mathcal{J} \simeq L'$ when the reactants and products are governed by a long-range attractive potential of the form $V(r) = -C/r^n$.

$$A(E') \simeq (E'/B_m')^{(n-2)/n} \quad \text{for } E' < B_m'$$
$$\simeq 1 \quad \text{for } E' > B_m \tag{32}$$

where

$$B_m = (\mu/\mu')^{n/(n-2)}(C/C')^{2/(n-2)}E \tag{33}$$

For the case of a steep exit potential, the factor $n/(n-2)$ is replaced by unity in (32) and (33) and $C = C'$ in (33).

The energy level density may be replaced by its classical limit for most complexes (see, however, the comments of Section IV.G when H atoms are present)

$$N_{vr}^{\dagger}(E_{tot} - E') \simeq (E_{tot} - E')^{s+r/2-2} \tag{34}$$

where s and r are the number of active vibrations and rotations, see Table 1. Thus at this level of approximation

$$P(E') = (f/f_B)^p(1 - f)^q \quad f < f_B$$
$$= (1 - f)^q \quad f > f_B \tag{35}$$

TABLE I
Numbers of active vibrations, s, and rotations, r, for 3 atom and 4 atom complexes

Complex	3 atom		4 atom	
	s	r	s	r
Tight, linear	4	0	7	0
Tight, bent	3	1	6	1
Loose	2	2	3	4

where $f = E'/E_{tot}$, $f_B = B'_m/E_{tot}$ and p denotes the exponent $[n/(n-2)$ or 1$]$ in (32), and q the exponent $[s + r/2 - 2]$ in (34). The distribution peaks at $f = p/(p+q)$ or f_B whichever is the less.

The opposite limit of direct reaction with repulsive energy release also invites interpretation in terms of simple models. The situation has recently been reviewed by Herschbach[134] who shows that the appropriate repulsive potential may be determined from photodissociation spectra of the diatomic reactant in the case of the canonical reactions $K + CH_3I$ and $H + Cl_2$.

B. Intersection of Covalent and Ionic Potential Energy Surfaces

The electron jump mechanism, (see 1), has long been invoked[71,2] in a highly simplified manner to explain the stripping dynamics of alkali atom–halogen molecule reactions $M + X_2$. Electron transfer occurs in the entrance valley of the covalent $M + X_2$ potential surface near the intersection with the ionic $M^+ + X_2^-$ potential surface. The $M \rightarrow X_2$ internuclear distance R at the intersection R_c is roughly estimated

$$R_c = e^2/[I(M) - A_v(X_2)] \tag{36}$$

where $I(M)$ is the ionization potential of M and $A_v(X_2)$ the vertical electron affinity of X_2. The use of the vertical electron affinity being justified[2] only on heuristic grounds. Typical values of $R_c \sim 4$–7 Å are given by this formula.

Theoretical investigation[173] of the intersection proceeds from the two-state interaction of a covalent state ψ_c with an ionic state ψ_i to yield two adiabatic states ψ_+ and ψ_-. The separation between the upper adiabatic surface V_+ and the lower adiabatic surface V_- at the crossing radius R_c is given by

$$\Delta V(R_c) = 2(H_{ii}S - H_{ic})/(1 - S^2) \tag{37}$$

where $S = \langle \psi_i | \psi_c \rangle$ and H is the Hamiltonian for the system. The topology of the intersection may be revealed[45,173] by examining the symmetries of the molecular states involved. The alkali atom $M(^2S)$, ion $M^+(^1S)$ and halogen molecule $X_2(^1\Sigma_g^+)$ are all totally symmetric. However, the halogen anion $X_2^-(^2\Sigma_u^+)$ is odd under inversion since the electron enters an antibonding σ_u^* orbital. In the linear $M + X_2$ configuration ($C_{\infty v}$ symmetry) both covalent and ionic states belong to the representation $^2\Sigma^+$; the interaction matrix elements S and H_{ic} are non-zero. However, in the isosceles triangular configuration (C_{2v} symmetry), the covalent state belongs to the representation $^2A_1$, while the ionic state belongs to $^2B_2$; the interaction matrix elements vanish. Accordingly, the adiabatic potential surfaces V_+ and V_- have a conical intersection,[174] meeting at the point $R = R_c, \eta = 90°$, for a given X_2 internuclear distance (η denotes the angle between $\mathbf{R}$ and the X_2 axis). The intersection is avoided for all other orientations (C_s symmetry) where both covalent and ionic states belong to the representation $^2A'$. The splitting $\Delta V(R_c)$ is maximal in the collinear configuration ($\eta = 0°$) and varies[175] roughly as $\cos \eta$. Thus calculation of $\Delta V(R_c)$ in the collinear configuration is adequate to determine the quantitative form of the potential surface intersection.

Recently a theoretical model[173] for long-range covalent–ionic interaction has been proposed, based on asymptotic pseudopotential wavefunctions. Apart from R_c, the only parameters required to determine $\Delta V(R_c)$ are $I(M)$, $A_v(X_2)$ and the atomic radius[176] of the halogen atom, $r_{at}(X)$. The results obtained for alkali halides and alkali hydrides, spanning a similar range of R_c to the $M + X_2$ reactions, are in good agreement with experimental estimates. Thus, calculations for $M + X_2$ reactions should be qualitatively reliable. Fig. 27 shows the lower adiabatic surface $V_-(R)$ calculated for collinear approach of K to Br_2, fixed at the equilibrium bond length, for two values of the electron affinity, $A_v(Br_2) = 1\cdot2\ eV$[177] and the improbably high value $A_v(Br_2) = 2\cdot1\ eV$. The potential surface is seen to be lowered significantly for thermal collisions at internuclear distances well outside the crossing radius, $R > R_c$. The splitting $\Delta V(R_c)$ considerably exceeds thermal collision energies. Thus trajectories will be largely confined to the lower adiabatic surface, except in broadside collisions. The total reaction cross section Q will be determined[175] by the orbiting criterion on this attractive surface. As illustrated by the effective potential curves in Fig. 27, reaction occurs for all collisions with impact parameters less than the orbiting impact parameter b_0 for a given energy. Hence

$$Q = \pi b_0^2 \tag{38}$$

rather than the more primitive formula

$$Q = \pi R_c^2 \tag{39}$$

and b_0 significantly exceeds R_c. At the radius R_0 of the maximum in the effective potential for $b = b_0$ which is $R_0 \sim R_c + 2\ Å$, the ionic contribution

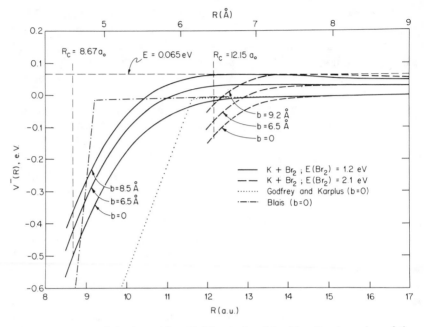

Fig. 27. Lower adiabatic surfaces $V_-(R)$ calculated for K + Br_2 in region of the intersection of covalent and ionic potential surfaces (from R. Grice *et al.*[173] by permission of Taylor and Francis Ltd).

to the adiabatic wavefunction is only $\sim 10\%$. Thus the Br_2 bond length will not be greatly disturbed and the effective electron affinity determining Q will be close to the vertical electron affinity. Interestingly, the total reaction cross section is rather insensitive to the halogen electron affinity. When $A_v(Br_2)$ is increased to 2·1 eV in Fig. 27, R_c increases substantially but the concommittant decrease in $\Delta V(R_c)$ leaves the adiabatic surface little altered in the region of R_0. Thus the increase in b_0 is much less than that in R_c. This is in line with experimental cross section estimates.[178]

According to Fig. 27, the electron jump transition will only be half completed (50% ionic) when the internuclear distance has decreased by ~ 2 Å from R_0 to R_c. This rather overstates the case, since the ionic contribution, increasing as R decreases, will expand the Br_2 internuclear distance. Thus, the Br_2 electron affinity and hence R_c both increase. Moreover, the distance between R_0 and R_c will be smaller for nonlinear configurations. Even so, the calculations show that the nuclei will move a distance ~ 1 Å during the electron jump transition, which is therefore a Franck–Condon process only in the sense that the vertical electron affinity governs the orbiting impact parameter.

The role of X_2 vibrational quantization in the $M + X_2$ electron jump mechanism has also received recent theoretical attention. It has been suggested[179] that this may be treated classically to a good approximation and criteria for the validity of this have been determined[180] from semi-classical theory. Potential energy surfaces have been calculated semi-empirically[181] for $K + Cl_2$ and *ab initio*[182] for $Li + F_2$.

Finally, mention should be made of the development of information theoretic methods for characterizing product energy distributions. Surprisal analysis[183] may offer a means of compacting and parametrizing distributions for a wide range of reactions, by comparing with the statistically expected distribution in each case.

Acknowledgements

Generous support by the Science Research Council of the experimental work by the Cambridge Molecular Beams Group, reported here *inter alia*, is gratefully acknowledged. It is a great pleasure to acknowledge the immense contributions made to this work by members of the group over the past five years: P. B. Foreman, G. M. Kendall, J. C. Whitehead, D. R. Hardin, M. R. Levy, C. F. Carter, K. B. Woodall, D. St. A. G. Radlein, S. M. Lin and D. J. Mascord.

References

1. E. H. Taylor and S. Datz, *J. Chem. Phys.*, **23**, 1711 (1955).
2. D. R. Herschbach, *Advan. Chem. Phys.*, **10**, 319 (1966).
3. D. R. Herschbach, *Proc. Conf. on Potential Energy Surfaces in Chemistry* (Ed. W. A. Lester), IBM, San Jose, 44, 1970.
4. Y. T. Lee, 'Physics of Electronic and Atomic Collisions', *VII ICPEAC*, Amsterdam, North-Holland, 357, 1971.
5. J. L. Kinsey, *MTP International Review of Science*, Physical Chemistry, Butterworths, **9**, 173, 1972.
6. A. E. Grosser, *Rev. Sci. Instr.*, **38**, 257 (1967).
7. R. Grice, *Ph.D. Thesis*, Harvard University, 1967.
8. Ch. Ottinger, R. Grice and D. R. Herschbach, *J. Chem. Phys.*, to be published. Unfolding of these intensity maps has been investigated by P. E. Siska, *J. Chem. Phys.*, **59**, 6052 (1973).
9. K. T. Gillen, A. M. Rulis and R. B. Bernstein, *J. Chem. Phys.*, **54**, 2831 (1971).
10. A. M. Rulis and R. B. Bernstein, *J. Chem. Phys.*, **57**, 5497 (1972); R. B. Bernstein and A. M. Rulis, *J.C.S. Faraday Disc.*, **55**, 293 (1973); Ch. Ottinger, P. M. Strudler and D. R. Herschbach, *J. Chem. Phys.*, to be published.
11. S. J. Riley and D. R. Herschbach, *J. Chem. Phys.*, **58**, 27 (1973).
12. K. R. Wilson and D. R. Herschbach, *J. Chem. Phys.*, **49**, 2676 (1968).
13. J. C. Whitehead, D. R. Hardin and R. Grice, *Molec. Phys.*, **23**, 787 (1972).
14. M. K. Bullitt, C. H. Fisher and J. L. Kinsey, *J. Chem. Phys.*, **60**, 478 (1974).
15. D. R. Hardin, K. B. Woodall and R. Grice, *Molec. Phys.*, **26**, 1073 (1973).
16. D. R. Herschbach, private communication, 1973.
17. S. M. Lin, D. J. Mascord and R. Grice, *Molec. Phys.*, **28**, 975 (1974).
18. G. H. Kwei and D. R. Herschbach, *J. Chem. Phys.*, **51**, 1742 (1969).

19. P. R. Brooks and E. M. Jones, *J. Chem. Phys.*, **45**, 3449 (1966); R. J. Beuhler, R. B. Bernstein and K. H. Kramer, *J. Amer. Chem. Soc.*, **88**, 5331 (1966); R. J. Beuhler and R. B. Bernstein, *Chem. Phys. Lett.*, **2**, 166 (1968); *J. Chem. Phys.*, **51**, 5305 (1969).
20. W. E. Wentworth, R. George and H. Keith, *J. Chem. Phys.*, **51**, 1791 (1969).
21. E. A. Entemann, *J. Chem. Phys.*, **55**, 4872 (1971).
22. E. A. Entemann and G. H. Kwei, *J. Chem. Phys.*, **55**, 4879 (1971).
23. J. B. Anderson, R. P. Andres and J. B. Fenn, *Advan. Chem. Phys.*, **10**, 275 (1966).
24. M. E. Gersh and R. B. Bernstein, *J. Chem. Phys.*, **55**, 4661 (1971); *J. Chem. Phys.*, **56**, 6131 (1972).
25. R. M. Harris and J. F. Wilson, *J. Chem. Phys.*, **54**, 2088 (1971).
26. R. A. La Budde, P. J. Kuntz, R. B. Bernstein and R. D. Levine, *Chem. Phys. Lett.*, **19**, 7 (1973); *J. Chem. Phys.*, **59**, 6286 (1973).
27. R. M. Harris and D. R. Herschbach, *J.C.S. Faraday Disc.*, **55**, 121 (1973).
28. R. Grice and D. R. Hardin, *Molec. Phys.*, **21**, 805 (1971).
29. T. M. Sloane, S. Y. Yang and J. Ross, *J. Chem. Phys.*, **57**, 2745 (1972).
30. R. R. Herm and D. R. Herschbach, *J. Chem. Phys.*, **43**, 2139 (1965); C. Maltz and D. R. Herschbach, *Disc. Faraday Soc.*, **44**, 176 (1967).
31. R. Grice, J. E. Mosch, S. A. Safron and J. P. Toennies, *J. Chem. Phys.*, **53**, 3376 (1970).
32. C. Maltz, *Chem. Phys. Lett.*, **3**, 707 (1969); R. Grice, *Molec. Phys.*, **18**, 545 (1970).
33. C. Maltz, N. D. Weinstein and D. R. Herschbach, *Molec. Phys.*, **24**, 133 (1972); D. S. Y. Hsu and D. R. Herschbach, *J.C.S. Faraday Disc.*, **55**, 116 (1973).
34. S. Freund, G. A. Fisk, D. R. Herschbach and W. Klemperer, *J. Chem. Phys.*, **54**, 2510 (1971); H. G. Bennewitz, R. Haertern and G. Müller, *Chem. Phys. Lett.*, **12**, 335 (1971).
35. R. P. Mariella, D. R. Herschbach and W. Klemperer, *J. Chem. Phys.*, **58**, 3785 (1973).
36. A. B. Lees and G. H. Kwei, *J. Chem. Phys.*, **58**, 1710 (1973).
37. P. R. Brooks, *J. Chem. Phys.*, **50**, 5031 (1969).
38. P. R. Brooks, *J.C.S. Faraday Disc.*, **55**, 299 (1973).
39. G. Marcelin and P. R. Brooks, *J.C.S. Faraday Disc.*, **55**, 318 (1973); *J. Amer. Chem. Soc.*, **95**, 7885 (1973).
40. R. J. Gordon, Y. T. Lee and D. R. Herschbach, *J. Chem. Phys.*, **54**, 2393 (1971).
41. P. B. Foreman, G. M. Kendall and R. Grice, *Molec. Phys.*, **23**, 117 (1972).
42. M. P. Sinha, A. Schultz and R. N. Zare, *J. Chem. Phys.*, **58**, 549 (1973).
43. M. P. Sinha and R. N. Zare, private communication, 1973.
44. G. M. Kendall, P. B. Foreman and R. Grice, 'Electronic and Atomic Collisions', (Ed. L. M. Branscomb *et al.*), in *VII ICPEAC*, North-Holland, 1971, p. 23.
45. P. B. Foreman, G. M. Kendall and R. Grice, *Molec. Phys.*, **23**, 127 (1972).
46. R. Grice, M. R. Cosandey and D. R. Herschbach, *Ber. Bunsenges. Phys. Chem.*, **72**, 975 (1968).
47. D. L. King and D. R. Herschbach, *J.C.S. Faraday Disc.*, **55**, 331 (1973).
48. J. C. Whitehead, D. R. Hardin and R. Grice, *Chem. Phys. Lett.*, **13**, 319 (1972).
49. J. C. Whitehead, D. R. Hardin and R. Grice, *Molec. Phys.*, **25**, 515 (1973).
50. S. M. Lin, D. J. Mascord and R. Grice, *Molec. Phys.*, **28**, 957 (1974).
51. W. S. Struve, T. Kitagawa and D. R. Herschbach, *J. Chem. Phys.*, **54**, 2759 (1971).
52. A. C. Roach and M. S. Child, *Molec. Phys.*, **14**, 1 (1968); W. S. Struve, *Molec. Phys.*, **25**, 777 (1973); P. K. Pearson, W. J. Hunt, C. F. Bender and H. F. Schaefer, *J. Chem. Phys.*, **58**, 5358 (1973).
53. S. M. Lin and R. Grice, *J.C.S. Faraday Disc.*, **55**, 370 (1973).
54. S. M. Lin, J. C. Whitehead and R. Grice, *Molec. Phys.*, **27**, 741 (1974).

55. W. S. Struve, J. R. Krenos, D. L. McFadden and D. R. Herschbach, *J.C.S. Faraday Disc.*, **55**, 314 (1973).
56. P. J. Dagdigian reporting work by R. C. Oldenborg, J. L. Gole and R. N. Zare, *J.C.S. Faraday Disc.*, **55**, 311 (1973).
57. L. Brewer and E. Bracket, *Chem. Rev.*, **61**, 425 (1961); S. H. Bauer and R. I. Porter, in *Molten Salt Chemistry* (Ed. M. Blanden), Interscience, 1964, p. 607.
58. W. B. Miller, S. A. Safron and D. R. Herschbach, *J. Chem. Phys.*, **56**, 3581 (1972).
59. W. B. Miller, S. A. Safron and D. R. Herschbach, *Disc. Faraday Soc.*, **44**, 108 (1967).
60. S. A. Safron, N. D. Weinstein, D. R. Herschbach and J. C. Tully, *Chem. Phys. Lett.*, **12**, 564 (1972).
61. P. Brumer and M. Karplus, *J.C.S. Faraday Disc.*, **55**, 80 (1973).
62. D. R. Herschbach, *J.C.S. Faraday Disc.*, **55**, 113 (1973).
63. S. A. Safron, *Ph.D. Thesis*, Harvard University, 1969.
64. P. B. Foreman, G. M. Kendall and R. Grice, *Molec. Phys.*, **25**, 529 (1973).
65. D. R. Hardin, K. B. Woodall and R. Grice, *Molec. Phys.*, **26**, 1057 (1973).
66. S. M. Lin and R. Grice, *J.C.S. Faraday Disc.*, **55**, 113 (1973).
67. S. M. Lin, J. G. Wharton and R. Grice, *Molec. Phys.*, **26**, 317 (1973).
68. W. A. Chupka, *J. Chem. Phys.*, **30**, 459 (1959).
69. P. B. Foreman, G. M. Kendall and R. Grice, *Molec. Phys.*, **25**, 551 (1973).
70. D. O. Ham, *J.C.S. Faraday Disc.*, **55**, 313 (1973).
71. J. L. Magee, *J. Chem. Phys.*, **8**, 687 (1940).
72. Y. T. Lee, R. J. Gordon and D. R. Herschbach, *J. Chem. Phys.*, **54**, 2410 (1971).
73. A. L. Companion, *J. Chem. Phys.*, **48**, 1186 (1968).
74. P. E. Siska, *Ph.D. Thesis*, Harvard University, 1970.
75. J. C. Whitehead and R. Grice, *J.C.S. Faraday Disc.*, **55**, 320, 374 (1973).
76. A. L. Companion, D. J. Steible and A. J. Starshak, *J. Chem. Phys.*, **49**, 3637 (1968).
77. J. C. Whitehead and R. Grice, *Molec. Phys.*, **26**, 267 (1973).
78. D. W. Davies and G. del Conde, *J.C.S. Faraday Disc.*, **55**, 369 (1973).
79. D. L. Hildenbrand, *J. Chem. Phys.*, **48**, 3657 (1968); *J. Chem. Phys.*, **52**, 5751 (1970).
80. J. A. Haberman, K. G. Anlauf, R. B. Bernstein and F. J. van Itallie, *Chem. Phys. Lett.*, **16**, 442 (1972).
81. S. M. Lin, C. A. Mims and R. R. Herm, *J. Chem. Phys.*, **58**, 327 (1973).
82. C. A. Mims, S. M. Lin and R. R. Herm, *J. Chem. Phys.*, **58**, 1983 (1973).
83. S. M. Lin, C. A. Mims and R. R. Herm, *J. Phys. Chem.*, **77**, 569 (1973).
84. R. R. Herm, S. M. Lin and C. A. Mims, *J. Phys. Chem.*, **77**, 2931 (1973).
85. C. A. Mims, S. M. Lin and R. R. Herm, *J. Phys. Chem.*, **57**, 3099 (1972).
86. C. Batalli-Cosmovici and K. W. Michel, *Chem. Phys. Lett.*, **11**, 245 (1971); *Chem. Phys. Lett.*, **16**, 77 (1972).
87. D. A. Dixon, D. D. Parrish and D. R. Herschbach, *J.C.S. Faraday Disc.*, **55**, 385 (1973) reporting work by H. J. Loesch and D. R. Herschbach.
88. J. Fricke, B. Kim and W. L. Fite, 'Electronic and Atomic Collisions', in *VII ICPEAC*, Amsterdam, North-Holland, 1971, p. 37.
89. C. D. Johan and R. N. Zare, *Chem. Phys. Lett.*, **9**, 65 (1971).
90. D. R. Yarkony, W. J. Hunt and H. F. Schaefer, *Molec. Phys.*, **26**, 941 (1973).
91. M. Menzinger and D. J. Wren, *Chem. Phys. Lett.*, **18**, 431 (1973).
92. J. L. Gole and R. N. Zare, private communication.
93. D. J. Wren and M. Menzinger, *Chem. Phys. Lett.*, **20**, 471 (1973); *J.C.S. Faraday Disc.*, **55**, 312 (1973).
94. C. D. Jonah, R. N. Zare and Ch. Ottinger, *J. Chem. Phys.*, **56**, 263 (1972).

95. J. L. Gole and R. N. Zare, *J. Chem. Phys.*, **57**, 5331 (1972).
96. R. N. Zare, *Ber. Bunsenges. Phys. Chem.*, **78**, 153 (1974).
97. W. L. Fite and P. Irving, *J. Chem. Phys.*, **56**, 4227 (1972).
98. W. L. Fite, H. H. Lo and P. Irving, *J. Chem. Phys.*, **60**, 1236 (1974).
99. A. Schultz, H. W. Cruse and R. N. Zare, *J. Chem. Phys.*, **57**, 1354 (1972).
100. H. W. Cruse, P. J. Dagdigian and R. N. Zare, *J.C.S. Faraday Disc.*, **55**, 277 (1973).
101. Y. T. Lee, J. D. McDonald, P. R. Le Breton and D. R. Herschbach, *Rev. Sci. Instr.*, **40**, 1402 (1969).
102. C. F. Carter, M. R. Levy and R. Grice, *J.C.S. Faraday Disc.*, **55**, 357 (1973).
103. J. Geddes, H. F. Krause and W. L. Fite, *J. Chem. Phys.*, **56**, 3298 (1972).
104. D. L. McFadden, E. A. McCullough, F. Kalos and J. Ross, *J. Chem. Phys.*, **59**, 121 (1973).
105. T. Carrington and J. C. Polanyi, *MTP International Review of Science*, Physical Chemistry, Butterworths, **9**, 135 (1973).
106. A. M. G. Ding, L. J. Kirsch, D. S. Perry, J. C. Polanyi and J. L. Schreiber, *J.C.S. Faraday Disc.*, **55**, 252 (1973).
107. Y. T. Lee, J. D. McDonald, P. R. LeBreton and D. R. Herschbach, *J. Chem. Phys.*, **49**, 2447 (1968).
108. D. Beck, F. Engelke and H. J. Loesch, *Ber. Bunsenges. Phys. Chem.*, **72**, 1105 (1968).
109. Y. T. Lee, P. R. LeBreton, J. D. McDonald and D. R. Herschbach, *J. Chem. Phys.*, **51**, 455 (1969).
110. J. B. Cross and N. C. Blais, *J. Chem. Phys.*, **50**, 4108 (1969).
111. N. C. Blais and J. B. Cross, *J. Chem. Phys.*, **52**, 3580 (1970).
112. J. B. Cross and N. C. Blais, *J. Chem. Phys.*, **55**, 3970 (1971).
113. H. J. Loesch and D. Beck, *Ber. Bunsenges. Phys. Chem.*, **75**, 736 (1971).
114. A. D. Walsh, *J. Chem. Soc.*, 2266 (1953).
115. C. F. Carter, M. R. Levy, K. B. Woodall and R. Grice, *J.C.S. Faraday Disc.*, **55**, 381, 385 (1973); D. St. A. G. Radkin, J. C. Whitehead and R. Grice, to be published.
116. Y. C. Wong and Y. T. Lee, *J.C.S. Faraday Disc.*, **55**, 383 (1973).
117. C. F. Carter, M. R. Levy, K. B. Woodall and R. Grice, to be published.
118. J. Grosser and H. Haberland, *Chem. Phys. Lett.*, **7**, 442 (1970); *Chem. Phys.*, **2**, 342 (1973).
119. H. Haberland and J. Grosser, 'Electronic and Atomic Collisions', (Ed. L. M. Branscomb *et al.*), *VII ICPEAC*, North-Holland, 1971, p. 27.
120. J. D. McDonald, P. R. LeBreton, Y. T. Lee and D. R. Herschbach, *J. Chem. Phys.*, **56**, 769 (1972).
121. A. D. Walsh, *J. Chem. Soc.*, 2288 (1953); for an *ab initio* calculation of $H + F_2$ see S. V. O'Neil, P. K. Pearson, H. F. Schaefer and C. F. Bender, *J. Chem. Phys.*, **58**, 1126 (1973).
122. P. N. Noble and G. C. Pimentel, *J. Chem. Phys.*, **49**, 3165 (1968); V. Bondybey, G. C. Pimentel and P. N. Noble, *J. Chem. Phys.*, **55**, 540 (1971).
123. C. Maltz, *Chem. Phys. Lett.*, **9**, 251 (1971).
124. K. G. Anlauf, D. S. Horne, R. G. McDonald, J. C. Polanyi and K. B. Woodall, *J. Chem. Phys.*, **57**, 1561 (1972).
125. C. A. Parr, J. C. Polanyi, W. H. Wong and D. C. Tardy, *J.C.S. Faraday Disc.*, **55**, 308 (1973).
126. J. C. Polanyi and J. J. Sloan, *J. Chem. Phys.*, **57**, 4988 (1972).
127. J. C. Polanyi, J. L. Schreiber and J. J. Sloan, *J.C.S. Faraday Disc.*, **55**, 124 (1973).
128. J. B. Cross, *J. Chem. Phys.*, **59**, 966 (1973).
129. F. Kalos and A. E. Grosser, *Rev. Sci. Instr.*, **40**, 804 (1969).

30. C. F. Carter, M. R. Levy and R. Grice, *Chem. Phys. Lett.*, **17**, 414 (1972).
31. C. F. Carter, *Ph.D. Thesis*, Cambridge University, 1973; M. R. Levy, *Ph.D. Thesis*, Cambridge University, 1973.
32. D. L. McFadden, E. A. McCullough, F. Kalos, W. R. Gentry and J. Ross, *J. Chem. Phys.*, **57**, 1351 (1972).
33. J. C. Polanyi and J. L. Schreiber, *J.C.S. Faraday Disc.*, **55**, 372 (1973).
34. D. R. Herschbach, *J.C.S. Faraday Disc.*, **55**, 233 (1973).
35. D. D. Parrish and D. R. Herschbach, *J. Amer. Chem. Soc.*, **95**, 6133 (1973).
36. F. A. Cotton and G. Wilkinson, *Advanced Inorganic Chemistry*, Wiley-Interscience, 1962; C. Campbell, J. P. M. Jones and J. J. Turner, *Chem. Comm.*, 888 (1968).
37. M. M. Rochind and G. C. Pimentel, *J. Chem. Phys.*, **46**, 4481 (1967).
38. M. Kaufman and C. E. Holb, *Chem. Instr.*, **3**, 175 (1971).
39. P. L. Moore, P. N. Clough and J. Geddes, *Chem. Phys. Lett.*, **17**, 608 (1972).
40. T. P. Schafer, P. E. Siska, J. M. Parson, F. P. Tully, Y. C. Wong and Y. T. Lee, *J. Chem. Phys.*, **53**, 3385 (1970).
41. K. G. Anlauf, P. E. Charters, D. S. Horne, R. G. McDonald, D. H. Maylotte, J. C. Polanyi, W. J. Skrlac, D. C. Tardy and K. B. Woodall, *J. Chem. Phys.*, **53**, 4091 (1970).
42. J. C. Polanyi and K. B. Woodall, *J. Chem. Phys.*, **57**, 1574 (1972).
43. C. F. Bender, P. K. Pearson, S. V. O'Neil and H. F. Schaefer, *J. Chem. Phys.*, **56**, 4626 (1972); C. F. Bender, S. V. O'Neil, P. K. Pearson and H. F. Schaefer, *Science*, **176**, 1412 (1972).
44. L. T. Cowley, D. S. Horne and J. C. Polanyi, *Chem. Phys. Lett.*, **12**, 144 (1971); D. H. Maylotte, J. C. Polanyi and K. B. Woodall, *J. Chem. Phys.*, **57**, 1547 (1972).
45. L. J. Kirsch and J. C. Polanyi, *J. Chem. Phys.*, **57**, 4498 (1972).
46. D. J. Douglas, J. C. Polanyi and J. J. Sloan, *J. Chem. Phys.*, **59**, 6679 (1973).
47. J. Geddes, H. F. Krause and W. L. Fite, *J. Chem. Phys.*, **52**, 3296 (1970); *J. Chem. Phys.*, **59**, 566 (1973).
48. M. Karplus, R. N. Porter and R. D. Sharma, *J. Chem. Phys.*, **43**, 3259 (1965).
49. R. N. Porter and M. Karplus, *J. Chem. Phys.*, **40**, 1105 (1964).
50. P. Brumer and M. Karplus, *J. Chem. Phys.*, **54**, 4955 (1971).
51. D. A. Micha, *Ark. Fysik.*, **30**, 437, 425 (1965); K. T. Tang and M. Karplus, *Phys. Rev.*, **A4**, 1844 (1971).
52. G. H. Kwei, V. W. S. Lo and E. A. Entemann, *J. Chem. Phys.*, **59**, 3421 (1973).
53. J. M. Parson and Y. T. Lee, *J. Chem. Phys.*, **56**, 4658 (1972).
54. J. M. Parson, K. Shobatake, Y. T. Lee and S. A. Rice, *J.C.S. Faraday Disc.*, **55**, 344 (1973).
55. J. M. Parson, K. Shobatake, Y. T. Lee and S. A. Rice, *J. Chem. Phys.*, **59**, 1402 (1973).
56. K. Shobatake, J. M. Parson, Y. T. Lee and S. A. Rice, *J. Chem. Phys.*, **59**, 1416 (1973).
57. K. Shobatake, J. M. Parson, Y. T. Lee and S. A. Rice, *J. Chem. Phys.*, **59**, 1427 (1973).
58. K. Shobatake, Y. T. Lee and S. A. Rice, *J. Chem. Phys.*, **59**, 1435 (1973).
59. K. Shobatake, Y. T. Lee and S. A. Rice, *J. Chem. Phys.*, **59**, 6104 (1973); Y. T. Lee, *Ber. Bunsenges. phys. Chem.*, **78**, 135 (1974).
60. J. T. Cheung, J. D. McDonald and D. R. Herschbach, *J/C.S. Faraday Disc.*, **55**, 377 (1973); *J. Amer. Chem. Soc.*, **95**, 7889 (1973).
61. R. A. Marcus, *J.C.S. Faraday Disc.*, **55**, 379, 381 (1973).
62. J. M. Parson, K. Shobatake, Y. T. Lee and S. A. Rice, *J.C.S. Faraday Disc.*, **55**, 380 (1973).

163. K. Shobatake, Y. T. Lee and S. A. Rice, *J. Chem. Phys.*, **59**, 2483 (1973).
164. J. G. Moehlmann and J. D. McDonald, *J. Chem. Phys.*, **59**, 6683 (1973).
165. S. H. Bauer, D. M. Ledermann, E. L. Kesler and E. R. Fisher, *Int. J. Chem Kinetics*, **5**, 93 (1973).
166. J. H. Sullivan, *J. Chem. Phys.*, **46**, 73 (1967).
167. R. Hoffman, *J. Chem. Phys.*, **49**, 3739 (1968); L. C. Cusachs, M. Krieger and C. W. McCurdy, *J. Chem. Phys.*, **49**, 3740 (1968).
168. D. A. Dixon, D. L. King and D. R. Herschbach, *J.C.S. Faraday Disc.*, **55**, 37! (1973).
169. E. H. Wiebenga, E. E. Havinga and K. H. Boswijk, *Advances in Inorganic and Radiochemistry* (Ed. H. J. Emeleus and A. G. Sharp), Academic Press, **3**, 13. (1961). A. I. Popov, *MTP International Review of Science: Inorganic Chemistry* (Ed. V. Gutman), Butterworths, **3**, 53 (1972).
170. A. Bohr, *Proc. Int. Conf. on Peaceful Uses of Atomic Energy*, Geneva, **2** 151 (1956); I. Halpern, *Ann. Rev. Nucl. Sci.*, **9**, 245 (1959).
171. M. E. Rose, *Elementary Theory of Angular Momentum*, Wiley, 1957.
172. R. A. Marcus, *J. Chem. Phys.*, **43**, 2658 (1965).
173. R. Grice and D. R. Herschbach, *Molec. Phys.*, **27**, 159 (1974).
174. G. Herzberg and H. C. Longuet-Higgins, *Disc. Faraday Soc.*, **35**, 77 (1963).
175. R. W. Anderson, *Ph.D. Thesis*, Harvard University, 1968.
176. J. C. Slater, *J. Chem. Phys.*, **41**, 3199 (1964); J. T. Waber and D. L. Cromer, *J Chem. Phys.*, **42**, 4116 (1965).
177. W. B. Person, *J. Chem. Phys.*, **38**, 109 (1963).
178. S. A. Edelstein and P. Davidovits, *J. Chem. Phys.*, **55**, 5164 (1973); J. Maya and P. Davidovits, *J. Chem. Phys.*, **59**, 3143 (1973).
179. G. M. Kendall and R. Grice, *Molec. Phys.*, **24**, 1373 (1972).
180. M. S. Child, *J.C.S. Faraday Disc.*, **55**, 30 (1973).
181. C. Nyeland and J. Ross, *J. Chem. Phys.*, **54**, 1665 (1971).
182. G. G. Balint-Kurti and M. Karplus, *Chem. Phys. Lett.*, **11**, 203 (1971); G. G Balint-Kurti, *Molec. Phys.*, **25**, 393 (1973).
183. R. D. Levine and R. B. Bernstein, *J.C.S. Faraday Disc.*, **55**, 100 (1973); R. D Levine, *J.C.S. Faraday Disc.*, **55**, 125 (1973).

ELASTIC SCATTERING

U. BUCK

Max-Planck-Institut für Strömungsforschung, Göttingen, Germany

CONTENTS

I. Introduction 313
II. Experimental Techniques 314
   A. Basic Scattering Apparatus 314
   B. Experimental Components 317
III. Theoretical Description 319
   A. Uniform Approximation in Semiclassical Scattering 320
   B. Summary of Cross Section Features 326
   C. Perturbation of the Elastic Cross Section 331
IV. Methods for the Determination of the Potential 334
   A. Fitting Procedures 334
   B. Inversion Procedures 336
     1. General Remarks 336
     2. Semiclassical Procedures for the Determination of the Potential . . . 338
     3. Practical Methods 341
     4. Discussion 352
V. Results 353
   A. Diatomic Systems (One Potential) 356
     1. Alkali–Mercury Systems 356
     2. Rare Gas Systems 359
     3. Interactions with H-atoms 367
   B. Diatomic Systems (more than one potential) 371
     1. Alkali Halides 372
     2. Metastable Atoms 373
   C. Polyatomic Systems 374
     1. Alkali–Molecule Systems 374
     2. Rare Gas–Molecule Systems 377
     3. Hydrogen–Molecule Systems 378
Acknowledgements 382
References 382

I INTRODUCTION

Elastic scattering of two atoms or molecules provides a direct method to determine the intermolecular potential between the colliding particles. It cannot compete with spectroscopy concerning the precision of the potential determined. However, the beam scattering method is in principle more universal than spectroscopy and covers the entire energy range. All other methods which are usually applied for the determination of the potential suffer from the drawback that the observed values are averaged over many interactions and thus are often not very sensitive to important details of the potential. But it should be noted that the principle advantage of molecular beam experiments can only be fully utilized if high resolution measurements are performed. Otherwise similar problems to those in the interpretation of bulk properties of the matter arise.

The field of elastic scattering has been extensively developed within the last 15 years and most of the important phenomena have been observed and understood (Pauly and Toennies, 1965; Bernstein, 1966; Pauly and Toennies, 1968; Bernstein and Muckermann, 1967; Schlier, 1969; Beck, 1970; Toennies, 1974; Pauly, 1974). As a result, elastic scattering has emerged as an almost routine laboratory tool (but not a simple one) for measuring potential curves. Nevertheless, there have been some new developments in the last three years which make it worthwhile to review this field not only from the aspect of an inventory of all systems measured. These are: (1) The possibility to measure high resolution cross section for nearly all species of the periodic table. For instance, the molecular beam results for the rare gas system help appreciably to find a potential which is able to fit all data available including beam scattering, spectroscopy, bulk properties of gaseous, liquid and solid matter, and transport phenomena (Maitland and Smith, 1973a). (2) The development of new data evaluation schemes, the inversion techniques, which allow the determination of the potential without assuming potential models (Buck, 1974).

The scope of this chapter will be on these recent developments of the last three years, 1971 to 1973. We restrict ourselves to the interaction of neutral partners, with no restriction to the energy or state of the species. The article starts with a brief review of basic scattering laws and experimental techniques used in the experiments (Section II). Then the theory of elastic scattering is reviewed, the different cross section features are summarized and their sensitivity to the potential is discussed. Perturbations of these results caused from anisotropic potentials, inelasticities or potential crossings are described (Section III). The next Section IV contains a survey of methods of how to determine the potential from the data. The emphasis will be on the inversion procedures. Finally, the results are summarized in Section V where, besides the atom–atom interactions in the ground states, new results for metastable species and molecules are included.

II. EXPERIMENTAL TECHNIQUES

A. Basic Scattering Apparatus

The quantities usually measured when a beam of particles is scattered by a target gas or by a secondary beam are the differential cross section σ and the total cross section Q. The two-particle collision problem is formally reduced to a one-particle problem with the reduced mass μ and the relative energy E by the transformation to the centre-of-mass system. In centre-of-mass coordinates the differential cross section is defined by the ratio of the number of particles scattered into the solid angle $d\omega$ per unit time in a direction making an angle ϑ with the incident beam to the incident flux density j_0. Thus, in classical mechanics, the number of particles, which are observed in a solid angle $d\omega = \sin \vartheta \, d\vartheta \, d\varphi$ and which have originated from the area $b \, db \, d\varphi$ is $j_0 b \, db \, d\varphi$. The definition of the cross section gives

$$\sigma_E(\vartheta) = b \, db/\sin \vartheta \, d\vartheta \tag{1}$$

In the quantum mechanical treatment of stationary scattering theory the wave function $\psi_k(\mathbf{r})$ corresponding to the wave number $k^2 = 2\mu E/\hbar^2$ must obey the boundary condition

$$\psi_k(\mathbf{r}) \xrightarrow[r \to \infty]{} e^{i(\mathbf{kr})} + \frac{e^{ikr}}{r} f(\vartheta, E) \tag{2}$$

The differential cross section is obtained by calculating the incident and scattered particle flux j_0 and j_s, which yields, with $j_0 = \hbar k/\mu$ and $j_s = \hbar k |f(\vartheta_1 E)|^2/(\mu r^2)$

$$\sigma(\vartheta_1 E) = |f(\vartheta, E)|^2 \tag{3}$$

The total cross section is obtained by integrating the differential cross section over the solid angle

$$Q(E) = \int d\omega \, \sigma(\vartheta, E) \tag{4}$$

A schematic diagram of a typical experimental setup used to measure these quantities is shown in Fig. 1. Two beams selected for their direction and velocity are crossed. The collisions occur in the intersections volume. A measurement of the intensity of the scattered primary beam with the movable detector as a function of the laboratory (L) scattering angle θ yields the quantity $I(\theta)$, which is proportional to the differential cross section. The two other detectors serve as monitors for checking the beam intensities. In order to compare the measured values with theoretical calculations they have to be transformed into the centre-of-mass (CM) system. Accurate data evaluation techniques have to account for the velocity and angular distributions of the two beams as well as the finite dimensions of the scattering volume and the detector. These steps may be illustrated by the following scheme (see, for instance, Helbing, 1968, 1969b; Creaser et al., 1973)

$$I(\theta) \xrightarrow{\text{L} \to \text{CM}} I(\vartheta) \quad \text{to be compared with } \sigma(\vartheta)$$

or

$$\sigma(\vartheta) \xrightarrow{\text{CM} \to \text{L}} \sigma(\theta) \xrightarrow{\text{apparatus}} \sigma_{\text{eff}}(\theta) \quad \text{to be compared with } I(\theta)$$

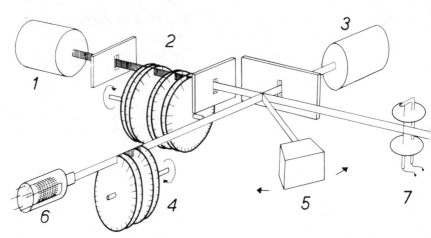

Fig. 1. Schematic diagram of a crossed molecular beam apparatus. (1) primary beam oven, (2) velocity selector, (3) secondary beam oven, (4) velocity analyser for the secondary beam, (5) detector for measuring the differential cross section, (6) monitor detector, (7) detector for measuring the total cross section.

where I is the measured and σ the calculated differential cross section. The total cross section is usually measured by monitoring the intensities without and with scattering particles in the forward direction using Beer's law

$$I = I_0 \exp\left(-nlQ_{\text{eff}}\right) \tag{5}$$

where n is the particle density of the secondary beam or target gas and l the scattering length. Usually only an effective cross section is measured due to the motion of the target particles. Another important correction has to be applied for the finite angular resolving power. All these corrections are

TABLE I
Notation of quantities

θ	Laboratory scattering angle
ϑ	Centre-of-mass scattering angle
I	Measured differential cross section
σ	Calculated differential cross section
Q	Total cross section
ε	Well depth of potential minimum
r_{m}	Position of the potential minimum
μ	Reduced mass
$A = kr_{\text{m}}$	Reduced wave number
$K = E/\varepsilon$	Reduced energy
$B = 2\mu\varepsilon r_{\text{m}}^2/\hbar^2$	Quantum parameter, potential well capacity
Θ	Deflection function
η	Phase shifts
$v_{\text{c}} = 2\varepsilon r_{\text{m}}/\hbar\mu$	Characteristic velocity

treated in detail by Pauly and Toennies (1965). Table I gives a survey of the notation of the most important quantities used in this article.

B. Experimental Components

The different components which are used in a beam scattering experiment are described in detail in several review articles (Pauly and Toennies, 1965; Pauly and Toennies, 1968; Toennies, 1974). Thus we shall only give a short outline of new developments with emphasis on techniques originally used in other fields.

Sources: Nozzle sources have proved to be by far the best molecular beam sources for high resolution experiments (for a review see Anderson *et al.*, 1966). They combine narrow velocity distributions and narrow angular spreads with high intensities. Carefully designed skimmers, very small nozzle diameters (between 50 and 100 μ) and large pumping speeds (3000 l/s to 6000 l/s) generate rare gas and hydrogen beams with Mach numbers between 20–30 for room temperature sources and more than 50 for sources at liquid nitrogen temperature (see, for instance, Buck *et al.*, 1973b). This corresponds to intensities of 10^{20} particles/sr . s and a velocity distribution of about 3% FWHM. The temperatures of the translational motion are reduced to about 1°K at a source temperature of 300°K. The same occurs, to a less extent, to the internal degrees of freedom of molecules. A rotational temperature of 200°K has been found for H_2, whereas for N_2 a value of 8°K has been reported which clearly shows the different behaviour of these molecules (Buck *et al.*, 1973b; Gallagher and Fenn, 1974). In summary, these beams are more intensive than effusive beams by several orders of magnitude. For many experiments additional state and velocity selection may not be needed, resulting in a further intensity gain over the conventional sources used together with velocity and state selectors. The beam energy can be varied by simply changing the temperature of the source or by using a mixture of a heavy and a light gas. Because of the high density during the expansion, the velocities in the emerging beam are the same, but the energies are determined by the different molecular weights. The energy for such a 'seeded' beam is given by

$$E = m\bar{C}_p T_0 / \bar{m} \tag{6}$$

where $\bar{m} = Xm + X_t m_t$ is the average molecular weight (X is the mole fraction, the index t means carrier gas), T_0 the source temperature, and $\bar{C}_p$ the average molar heat capacity. Thus, depending on whether an excess of the heavier or the lighter component is added, the solute beam energy is greatly decreased or increased, respectively. These ideal values are only reached in 'good' expansions. Otherwise, velocity slip effects reduce the energy (Abuaf *et al.*, 1967). Energies up to 15 eV for 1% Xe seeded in 99% H_2 have been reported (Haberland *et al.*, 1973a). Again, very small velocity spreads are found. The intensities of such beams are enlarged by about six orders of magnitude compared to beams produced by charge exchange. It should be noted that the large expense for the pumping equipment of these sources is often a great disadvantage.

Beams of unstable species like H, O or N are usually produced by thermal dissociation, in radio frequency (RF) discharges (100 MHz) or microwave (MW) discharges (2·45 GHz), where the RF source is, in general, simpler to construct and to employ but of larger dimensions. Dissociation degrees of nearly 90 % at pressures below 1 torr have been obtained for RF sources (100 W power) and MW sources (1 KW power). The application of these sources is restricted by the low pressure and the considerably high effective temperatures (measured by means of the velocity distribution of the beam) which could only be reduced by cooling to 300–800°K (Wilsch, 1972; Miller, 1971b; Aquilanti *et al.*, 1972a). Two new developments to overcome these limitations should be mentioned. Welz *et al.* (1974) constructed a RF source which is cooled down to an effective temperature of 40°K thus extending the velocity range of molecular beam experiments with H-atoms down to 280 m/s. A high pressure MW source has been developed by Lagomarsino *et al.* (1973) for which 40 % dissociation is obtained at 10 torr. Another possibility is to combine the technique of seeded beams with a RF-discharge source, as has been done by Miller and Patch (1969) for the production of O atoms.

The production of excited states is usually performed by electron bombardment. Although only a small fraction of the ground state atoms are excited (10^{10} atoms/sr . s) this intensity proved sufficient to perform experiment with metastable Hg*($^3P_2$) (Davidson *et al.*, 1973) and He*(2^1S) (Haberland *et al.*, 1973b) in the thermal energy range. Metastable He*(2^3S) atoms between 5 and 10 eV have been produced by near resonant charge exchange of He$^+$-ions with Cs (Morgenstern *et al.*, 1973). This source produces nearly exclusively He*(2^3S) metastables since the charge exchange reaction goes mainly to the He(2^3P) and He(2^1P) states and He(2^1P) decays rapidly to the ground state whereas He(2^3P) gives the desired He*(2^3S). A new possibility to excite states of short lifetime is opened by the application of tunable cw dye lasers. This technique has successfully been applied to the deflection of a sodium beam (Schieder *et al.*, 1972). The spectral radiation density available with an argon-ion laser pumped dye-laser is sufficient to saturate the Na($3^2P_{3/2}$) state. Because of depopulation effects the laser has to be tuned to a special hyperfine structure transition $2S_{1/2}F = 2$ to $^2P_{3/2}F = 3$. In a scattering experiment the laser beam is directed into the scattering volume where the atom undergoes about 100 optical pumping cycles (Hertel and Stoll, 1974). The laser frequency must be tunable to better than the remaining Doppler width which is approximately 30 MHz. The wavelength from

4200–7000 Å for which tunable dye-lasers are available should make this very promising method applicable also to other beams.

Velocity Selectors: The velocity selection is generally achieved by a rotating slotted disc selector or a time-of-flight (TOF) device. The disadvantage of both methods is the decrease in intensity of the beam when the resolution is improved. In high resolution experiments only $5 \cdot 10^{-2}$ of the incident particles are velocity-selected. This number can be increased to 0·5 if a special design of the chopper is used in a TOF experiment. The beam is chopped by a pseudo-random sequence and the resulting detector signal is cross-correlated with the modulating sequence to produce the TOF-spectrum (Sköld, 1968). The chopper is open half the time independent of the length of the sequence, resulting in a beam intensity which is independent of the resolution. This method is especially suited for measurements where the signal to background ratio is small or the TOF-spectrum consists of a small number of peaks of nearly equal intensity (Gläser and Gompf, 1969). Otherwise the results are worse than those obtained by the single pulse TOF method.

Another possibility for a velocity selection without losing too much intensity is the use of hexapole magnets. Besides the velocity selection, paramagnetic atoms such as H (Barcellona *et al.*, 1973) and Li (Spindler *et al.*, 1973) are focused and polarized.

Detectors: No fundamentally new ideas can be reported on this subject. The most widely applicable method is the ionization of the beam by electron bombardment and subsequent mass selection of the ion beams. Typically only one in a thousand beam atoms are ionized (Bickes and Bernstein, 1970 and references cited therein). The background current from the ionization of the residual gas is greatly reduced by using ultrahigh vacuum techniques with pressures down to 10^{-11} torr (Lee *et al.*, 1970). Further improvement of the signal to noise ratio is generally achieved by applying counting techniques and increasing the overall measuring time.

Low temperature bolometers have also been used for detecting neutral particles in high resolution scattering experiments (Cavallini *et al.*, 1971c). Especially the detection of H-atoms proved to be very efficient due to recombination effects on the surface (Marenco *et al.*, 1972).

III. THEORETICAL DESCRIPTION

The theoretical description of the elastic scattering of two colliding atoms or molecules is well understood, at least as far as the interaction is spherically symmetric. For most atomic species the wavelength of the relative motion is small compared to a characteristic distance of the potential over which it varies appreciably. Thus classical mechanics are a reasonably good approximation for describing the dynamics of the collision. However, the results of molecular beam measurements show a number of important features of the cross section which require a quantum mechanical treatment. Therefore most of the calculations are performed using the exact quantum formulae of the partial wave analysis for the cross section where under the above condition the phase shifts are obtained from the JWKB-approximation. Since such a series converges very slowly under semiclassical conditions and does not show any insight into the phenomena involved, semiclassical procedures prove to be a good compromise between an exact description of the process and the sight of the physical picture (Ford and Wheeler, 1959). In semiclassical theory the classical particles are replaced by wave packets which are able to interfere but which follow the classical paths.

Semiclassical procedures which give results in quantitative agreement with the exact quantum results over the whole angular range have been published only recently (Berry, 1966; Miller, 1968; Berry, 1969; Mullen and Thomas, 1973; Mount, 1973). Therefore we start with a description of such a uniform semiclassical approximation (Section III.A). Next, all cross section features are summarized. Their dependence on the reduced energy and reduced wave number as well as their sensitivity to the potential are discussed in Section III.B. Finally, the influence of other than spherically symmetric interactions such as anisotropic potentials occurring in atom–molecule scattering or any kind of inelasticities and chemical reactions are studied (Section III.C).

A. Uniform Approximation in Semiclassical Scattering

The differential cross section for a spherical symmetric potential is given in quantum mechanics by (3) where the scattering amplitude $f(\vartheta)$ is related to the phase shifts η_l by

$$f(\vartheta) = (2ik)^{-1} \sum_{l=0}^{\infty} (2l + 1)P_l(\cos \vartheta)[\exp(2i\eta_l) - 1] \qquad (7)$$

where l is the orbital angular momentum quantum number and k the wave number of the relative motion as defined in Section II.A. P_l denotes the Legendre polynomials.

The semiclassical approximation is reached in three different steps (Berry and Mount, 1972). The first stage is to transform the summation over the discrete values of l into an integration over the continuous variable $\lambda = (l + \frac{1}{2})$. The proper way to do this is to use the Poisson summation formula which gives

$$f(\vartheta) = (ik)^{-1} \sum_{m=-\infty}^{\infty} \exp(-im\pi) \int_0^{\infty} d\lambda \, \lambda P_{\lambda - 1/2}(\cos \vartheta)$$
$$\times [\exp(2i\eta_{\lambda - 1/2}) - 1] \exp[2im\pi] \qquad (8)$$

where a suitable interpolation for η_l and P_l has to be found. In a second stage the quantities in the integrand are replaced by their asymptotic forms for large l. The phase shifts are given in the JWKB-approximation by

$$\eta_l = k\left[\int_{r_0}^{\infty} dr \left(1 - \frac{V(r)}{E} - \frac{\lambda^2}{k^2 r^2}\right)^{1/2} - \int_{\lambda/k}^{\infty} dr \left(1 - \frac{\lambda^2}{k^2 r^2}\right)^{1/2}\right] \tag{9}$$

The Legendre polynomials are replaced by

$$P_l(\cos \vartheta) \simeq (\pi\lambda \sin \vartheta)^{-1/2} 2^{1/2} \sin \left(\lambda\vartheta + \frac{\pi}{4}\right) \tag{10}$$

for $\lambda^{-1} < \vartheta < \pi - \lambda^{-1}$. An exact treatment of the cross section near $\vartheta = 0$ and $\vartheta = \pi$ requires different asymptotic forms which contain Bessel functions (Berry and Mount, 1972). The scattering amplitude can now be written as (where $\sum (2l + 1)P_l(\cos \vartheta) = 0$ for $\vartheta \neq 0$ is used)

$$f(\vartheta) = (ik)^{-1}(2\pi \sin \vartheta)^{-1/2} \sum_{m=-\infty}^{\infty} \exp(-im\pi)[I_m^+ - I_m^-] \tag{11}$$

where

$$I_m^+ = \int_0^{\infty} d\lambda \, \lambda^{1/2} \exp\left[i\left(2\eta(\lambda) + 2m\pi \pm \lambda\vartheta \pm \frac{\pi}{4}\right)\right] \tag{12}$$

The third stage of the approximation is the evaluation of the integrals $I_m^{\pm}$. The integrands of $I_m^{\pm}$ are rapidly oscillating functions and the dominant contribution to the integrals arise from stationary points in the exponents of the integrand which satisfy

$$2\frac{d\eta}{d\lambda}(\lambda_i) = \mp \vartheta - 2m\pi \tag{13}$$

Since $2 \, d\eta/d\lambda(\lambda_i)$ is equal to the classical deflection function $\Theta(\lambda_i)$ the above result can be interpreted as follows. The contributions to the scattering amplitude $f(\vartheta)$ are given by the classical paths which lead to the same deflection angle ϑ where the points of stationary phase correspond classically to the impact parameter $b = \lambda/k$ which corresponds to that angle ϑ. Since $\Theta(\lambda)$ is always less or equal to π the integrals $I_m^{\pm}$ have no stationary point for $m < 0$. Thus $I_m^-(I_m^+)$ accounts for classical paths which have undergone net repulsion (attraction) and which circle around the origin m times. If the stationary points are well separated from each other the integral may be evaluated by the methods of stationary phase expanding the exponent in the integrands to second order in $\lambda - \lambda_i$.

The resulting scattering amplitude is

$$f(\vartheta) = \sum_i \lambda_i^{1/2}(k^2 \sin \vartheta|\Theta|(\lambda_i)|)^{-1/2} \exp\left[iA_i(\vartheta)/\hbar\right] \tag{14}$$

where $A_i/\hbar = 2\eta(\lambda_i) + \lambda_i\vartheta$ is the classical action along the contributing path.

The amplitude of a single contribution is just the classical cross section, so that the classical result is obtained for the cross section. If more than one point of stationary phase exists the cross section displays characteristic interference oscillations which arise from the different classical paths:

$$\sigma(\vartheta) = \sum_{ij} (\sigma_i \sigma_j)^{1/2} \exp \{ih^{-1}[A_i(\vartheta) - A_j(\vartheta)]\} \tag{15}$$

This 'crude' semiclassical method which demonstrates the interfering effects cannot be applied to several typical cases which occur in atomic physics due to the special behaviour of the potential and thus of the points of stationary phase. Fig. 2 displays these typical trajectories and the corresponding deflection function where λ is replaced by the impact parameter b. There are:

1. The rainbow effect, at which $\Theta(b)$ has an extremum so that two points of stationary phases coalesce (b_2 and b_3 in Fig. 2).
2. The glory effect which arises when $\Theta(b)$ passes through zero (forward) or π (backward) or more general through $-2m\pi$ or $-(2m-1)\pi$ for $m = 0, 1, 2, \ldots$. Again we have the coalescence of two points of stationary phase, e.g. b_1 and b_2 for the forward glory.
3. The forward diffraction peak which is caused by contributions from very large b. Here the integrand is not a rapidly varying function of b so that the method described breaks down.
4. The orbiting effect which arises from collisions where the particle is kept spiralling around the other particle due to the potential barrier from the centrifugal part of the potential. This is a resonance behaviour and the deflection function displays a logarithmic singularity.

The 'crude' semiclassical approximation gives infinite cross sections for all these cases. The next steps are the so-called 'transitional' approximations which have been developed for these phenomena. They are valid in a small angular range near the critical angle, but they do not merge smoothly into the forms valid for the other angles so that they are not able to give cross sections which are in quantitative agreement with exact results. Most of the transitional approximation are treated in the pioneering paper of Ford and Wheeler (1959). For other applications concerning the higher order glories see Helbing (1969a), and the forward diffraction peak see Helbing and Pauly (1964) and Mason et al. (1964). In addition many reviews and books contain chapters on these topics.

We shall continue with an example of the method of uniform approximation which is the correct theoretical method for calculating the scattering amplitude, including the interference as well as the rainbow, the glory or the forward diffraction contribution. Only the orbiting has to be described by other methods due to the quite different nature of this phenomenon (see, for instance, Berry and Mount, 1972).

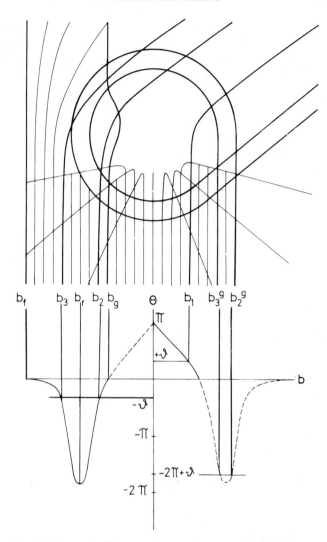

Fig. 2. Schematic diagram of classical trajectories and the corresponding deflection function for a realistic interatomic potential. Special trajectories which lead to forward (b_f), rainbow (b_r) and glory (b_g) scattering are marked. In addition the paths contributing to scattering at an angle of observation ϑ are drawn.

The problems in the 'crude' or the 'transitional' approximation arise from the fact that the stationary points of the integrals $I_m^{\pm}$ are treated as isolated or as common points, respectively. The uniform approximation tries to map the exact behaviour into a simpler integral which is hoped to be understood in terms of known standard functions. The mapping function is determined by the restriction that the structure of the stationary points must be the same. This method developed by Chester *et al.* (1957) was applied to the rainbow scattering by Berry (1966). The integrals could be solved by Airy functions and their derivatives. Glory scattering and the forward diffraction peak were treated by Berry (1969) where in the first case the Bessel functions J_0 and J_1 are the solution. The latter problem could not be solved by known functions, so that numerical methods have to be applied for evaluating the integrals. Miller (1968) applied a somewhat different method, treating glory and rainbow scattering in a single formula.

We shall discuss the rainbow cross section as an example in detail since this is the most important feature of the differential cross section at intermediate energies. The result for the cross section is when the contribution of the third impact parameter, b_1 of Fig. 2, is included (Boyle, 1971; Mullen and Thomas, 1973):

$$\sigma(\vartheta) = \sigma_1(\vartheta) + G^2 A_i^2(-z) + H^2 A_i'^2(-z) + 2\sigma_1^{1/2}(\vartheta)\left[G \cos\left(\alpha_r - \alpha_1 - \frac{\pi}{4}\right)\right.$$

$$\left. \times A_i(-z) + H \cos\left(\alpha_1 - \alpha_r - \frac{\pi}{4}\right) A_i'(-z)\right] \tag{16}$$

A_i and A_i' are the Airy functions and their derivatives, α_1, α_r and z are defined by

$$\alpha_1 = 2\eta(b_1) - kb_1\vartheta \tag{17}$$

$$2\alpha_r = [2\eta(b_2) + 2\eta(b_3) + k(b_2 + b_3)\vartheta] \tag{18}$$

$$z = \{0.75[2\eta(b_2) - 2\eta(b_3) + k(b_2 - b_3)\vartheta]\}^{2/3} \tag{19}$$

where

$$G = \pi^{1/2}z^{1/4}(\sigma_2^{1/2} + \sigma_3^{1/2}) \quad \text{and} \quad H = \pi^{1/2}z^{1/4}(\sigma_2^{1/2} - \sigma_3^{1/2})$$

σ_i ($i = 1, 2, 3$) are the classical cross sections, corresponding to the three interfering impact parameters. Since H is, in general, smaller than G, the cross section is mainly determined by two groups of oscillations. One is produced by $Ai^2(-z)$, coming from the interference of b_2 and b_3: the supernumerary rainbow oscillations. The other one is generated by the interference of the innermost impact parameter b_1 with the other two, determined by the phase factor $\alpha_r - \alpha_1 - \pi/4$: the rapid oscillations. For $2\eta(b_2) - 2\eta(b_3)$ we get

$$2\eta(b_2) - 2\eta(b_3) = k \int_{b_3}^{b_2} \Theta(b)\, db \tag{20}$$

If we substitute this result into z, it is seen that z is a measure of the shaded area in Fig. 3. If we make similar considerations for the phase factor of the

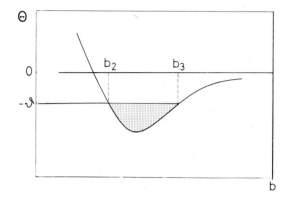

Fig. 3. Part of the deflection function which is responsible for rainbow scattering. The shaded area determines the positions of the rainbow oscillations which
are caused by the interference of trajectories around
b_2 and b_3.

rapid oscillations, we see that these oscillations are sensitive to

$$2(\alpha_r - \alpha_1) = 2[\eta(b_2) - \eta(b_1)] + 2[\eta(b_3) - \eta(b_1)] + k\vartheta(b_2 + b_3 + 2b_1)$$

$$= k\left[\int_{b_1}^{b_2} \Theta(b)\,db + \int_{b_1}^{b_3} \Theta(b)\,db + \vartheta(b_2 + b_3 + 2b_1)\right] \qquad (21)$$

Represented by the areas of Fig. 4 the result is

$$\alpha_r - \alpha_1 = k(F1 + F2 + F3 - \tfrac{1}{2}F4) \qquad (22)$$

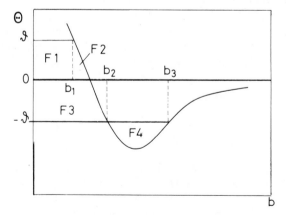

Fig. 4. Areas of the deflection function which determine the positions of the rapid oscillations due to the
interference of b_1 with b_2 and b_3.

Thus the rapid oscillations are very sensitive to the region around b_g the zero point of the deflection function if the rainbow oscillations are assumed to be known.

Numerical calculations show that the cross section (Equation 6) is in excellent agreement with calculations using the partial wave sum for $A \geq 50$ and for scattering angles $\vartheta > 2\pi/A$ where $A = kr_m$ is the reduced wave number (Berry and Mount, 1972; Mullen and Thomas, 1973). Even for $A = 20$ and $K = 2$ deviations are found only for the amplitudes (about 30%) whereas the angular positions of the extrema still agree to within a few tenths of a degree. Nearly the same results have been obtained for numerical calculations using the uniform approximation for the glory scattering and the forward diffraction peak which is valid over the whole angular region up to angles very close to the rainbow angle (Mount, 1973). When known hypergeometric functions are involved (rainbow and glory scattering) the computational time is greatly reduced especially for high A. Thus these formulae should be considered for the numerical calculation of differential cross sections in the semiclassical limit. For an application of the uniform approximation to rainbows in the framework of the optical model see Harris and Wilson (1971).

B. Summary of Cross Section Features

In the preceding Section the four main contributions to the *differential cross section* were classified. Except for the orbiting collisions the other features are well described by stationary phase methods in the semiclassical approximation. For energies $E > E_{orb}$ the cross section is governed by the forward diffraction peak and the rainbow effect. The forward glory is buried by the forward diffraction peak whereas backward glories can only be observed for energies very near to or smaller than the orbiting energy. In addition the cross section shows a characteristic interference structure which always occurs when more than one classical trajectory leads to the same deflection angle. The angular spacing of such an interference oscillation is given by differentiating

$$2\pi N = \hbar^{-1}[A(b_i) - A(b_k)] = k\vartheta(b_i - b_k) + 2[\eta(b_i) - \eta(b_k)] \qquad (23)$$

where $N = 1, 2, 3, \ldots$ is valid (see Equations 14 and 15). Usually there are three classical paths which contribute to the scattering amplitude as can be seen in Fig. 2. This behaviour leads to two different sets of interferences: widely spaced oscillations with an angular spacing inversely proportional to $b_2 - b_3$ and rapid oscillations with an angular spacing inversely proportional to $b_1 + b_{2,3}$.

The former referred to as supernumerary rainbows are sensitive to the attractive part of the potential. The latter usually called rapid oscillations, determine b_g and therefore the absolute scale of the potential. For angles ϑ greater than the rainbow angle there is only one contribution to the classical path so that the cross section is monotonic. The repulsive part of the deflection function and therefore of the potential is probed. These features are summarized for energies E greater than the orbiting energy E_{orb}:

1. Backward scattering $\vartheta > \vartheta_r$, monotonic cross section sensitive to the repulsive part of the potential.
2. Rainbow angle ϑ_r dependent on the energy E and the potential well depth ε.
3. The classical cross section in the forward direction which is proportional to $\sigma \sim \vartheta^{-(s+2)/s}$ for a potential of the asymptotic form $V = -Cr^{-s}$.
4. Forward diffraction peak, sensitive to the long range attractive part of the potential.
5. Supernumerary rainbows, interference between b_2 and b_3, very sensitive to the form of the attractive part of the potential near the inflection point.
6. Quantum oscillations, interference between b_1 and b_2 or b_3, sensitive to the glory impact parameter and the distance of the potential minimum r_m.

The energy dependence of the cross section is mainly determined by the rainbow angle which is inversely proportional to the reduced energy $\vartheta_r = a/K = a\varepsilon/E$ where the constant a is potential-dependent and varies, e.g. for a Lennard-Jones-potential from 1·79 for $n = 6$ to 2·15 for $n = 20$. For small K the rainbow structure moves to large angles whereas it disappears at small angles for large K. Fig. 5 displays these results. The angular spacing of the interference oscillation is mainly given by the reduced wave number. If this structure is nearly independent of the potential form, as is the case for the quantum oscillations, the angular spacing may be estimated by $\Delta\vartheta = \pi/A$. This means that these oscillations are only 'rapid' for A greater than 80. For $A = 10$ to 50 they are widely spaced and are often the only oscillatory contribution to the cross section since for such systems (small μ, ε and r_m) the rainbow is located at small angles. Thus, the amplitudes of these oscillations are sensitive both to the attractive and repulsive part of the potential. The occurrence of these oscillations, in spite of the fact that there are classically no impact parameters that interfere, results from the behaviour of the Airy function which tends to zero with $A^{-2/3}$, so that for small A there is a large range of impact parameters larger than b_r which are able to interfere.

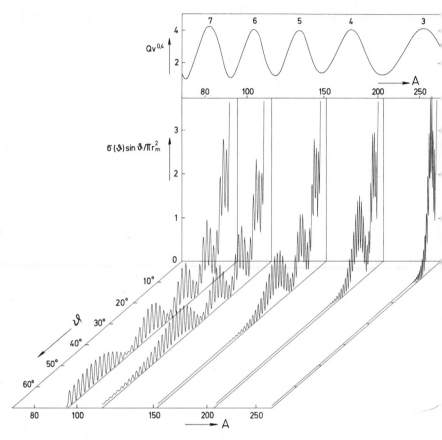

Fig. 5. Differential cross section (weighted with $\sin \vartheta$) as a function of the deflection angle ϑ and the reduced parameter $A = kr_m$ which is proportional to the velocity. The calculation was performed for a LJ 12–6 potential with $B = 5000$. In the upper part the total cross section multiplied by $v^{0.4}$ is plotted versus A. The close connection between the number of supernumerary rainbows and the number of the glory undulations is clearly demonstrated. Note that another rainbow oscillation is buried under the forward diffraction peak and not shown in the figure.

For energies smaller than the orbiting energy the cross section shows:

1. backward glories which means the rising of the cross section at $\vartheta = \pi$;
2. oscillatory behaviour over the whole angular range. Besides the known two interference structures additional interference occurs from the backward glories (see Fig. 2);
3. resonance behaviour in the energy dependence of the differential cross section; and
4. the forward diffraction peak.

Several special interference features should be mentioned which occur for identical particles, exchange cross sections and potential crossings. They can be treated in the same way as indicated by (23). The identical particle oscillations arise from the fact that one cannot distinguish between the scattering in the angle ϑ or $\pi - \vartheta$. The cross section is now symmetric about $\pi/2$ which leads to an interference between forward and recoil scattering. For energies greater than the orbiting energy these oscillations will mainly appear in the backward direction. The cross section oscillates with the phase

$$\gamma = 2\eta(b_1) - 2\eta(b_4) + kb_1\vartheta - kb_4(\pi - \vartheta) \tag{24}$$

where b_4 is the impact parameter which corresponds to recoil scattering. For a detailed description see Smith (1969).

The other two oscillations occur if two different trajectories are available from two different potential curves. The exchange oscillations are due to two potential curves of the gerade (g)–ungerade (u)-symmetry. In the case of a potential crossing there are two trajectories in the elastic channel, one corresponding to the basic potential and the other arising from a transition say to the excited state on the inward passage and a transition to the ground state on the outward passage. The interference pattern will be seen to be a perturbation on elastic scattering. For details see Stueckelberg (1932), Smith et al. (1967), Coffey et al. (1969), Smith et al. (1970), Olson and Smith (1971), Delvigne and Los (1973).

The theoretical description of the *total cross section* is easier in the semi-classical limit since the 'angular range' is indeed restricted to $\vartheta = 0$ so that the transition approximation gives results which are in quantitative agreement with those calculated by quantum mechanics. Using the parabola approximation for the phase shifts in the maximum or the straight line approximation for the deflection function at the zero point we have

$$Q = f(s)[Ck/E]^{2/(s-1)} - 4b_g[2\pi/k(-\Theta'_g)]^{1/2} \cos\left(2\eta_g - \frac{\pi}{4}\right) \tag{25}$$

where an asymptotic potential $V = C \cdot r^{-s}$ is assumed. Θ'_g is the derivative of the deflection function at b_g, f is a known function (Pauly and Toennies, 1965; Bernstein, 1966). In order to separate that part of the cross section which is sensitive to the repulsive part of the potential from that which is sensitive to the attractive part, we define a characteristic velocity $v_c = 2\varepsilon r_m/\hbar\mu$. Then we have for energies $E > E_{orb}$

1. for $v > v_c$: monotonic behaviour, sensitive to the repulsive part of the potential,
2. for $v < v_c$: oscillatory behaviour, interference between the forward glory contribution and the forward diffraction peak, sensitive to the attractive part of the potential, in particular to the region of the well

glory contribution) and the long range attractive part (forward contribution). The velocity spacing of these oscillations is sensitive to the maximum phase shift and the product $\varepsilon \cdot r_m$. For a detailed discussion see (Bernstein and La Budde, 1973; Greene and Mason, 1972 and 1973).

For $E < E_{orb}$ the orbiting resonances are found due to a short-lived complex of shape resonances formed behind the centrifugal barrier.

For identical particles, oscillations show up in the total cross section which are sensitive to the repulsive part of the potential. They are produced by head-on collisions for which $l = 0$ is valid. Expanding the phase shift near this point, one gets for small l

$$\eta(l) = \eta(0) + \tfrac{1}{2}\pi l - \varkappa l^2 \tag{26}$$

Because of the statistics—let us assume Bose statistics—only scattering phase shifts of even angular momentum contribute to the cross section. That means that the phase shifts have the separation π for all following l except for a small correction $\varkappa l^2$. In other words, all phase shifts make the same contribution to the cross section as the s-wave phase shift. The integral cross section therefore oscillates (Helbing, 1969a):

$$Q \propto \cos\left[2\eta(0) - \tfrac{1}{2}\pi - \varphi\right]$$

with $\tag{27}$

$$\varphi = \arctan\left[1 + 2(\pi\varkappa)^{-1/2}\right]^{-1}$$

Whether a system shows an interference structure or not depends only on the quantity $B = A^2/K = 2\mu\varepsilon r_m^2/\hbar^2$ which depends on the reduced mass, the strength and the distance parameter of the potential. The total number of bound vibrational states of a system is proportional to $B^{1/2}$ and equal to the number of glory oscillations for $v \to 0$ (Bernstein, 1963). The number of possible rainbow oscillations is then determined by the number of the glory oscillation at the energy where the differential cross section is measured Buck and Pauly, 1968). Figs. 5 and 6 display these connections. Fig. 5 shows the differential cross section as a function of ϑ and A calculated for a Lennard-Jones (12–6) potential with $B = 5000$ together with the total cross section. The number of rainbow oscillations is clearly correlated with the number of the glory undulation of the same energy. Fig. 6 illustrates the close relationship of possible glory undulation and the parameter B. The total cross section multiplied by $v^{2/5}$ in order to remove the monotonic part is plotted as a function of A/B, which is proportional to the velocity v and B. If the glory maxima $N = 1, 2, 3, \ldots$ are connected by lines, the extrapolation of these lines to $v = 0$ will give the B-values for which the potential just contains N vibrational states. In the region left to the dashed line $K = 1$ orbiting occurs. Thus these figures easily allow one to decide which structure will be displayed in the cross section for a chosen system.

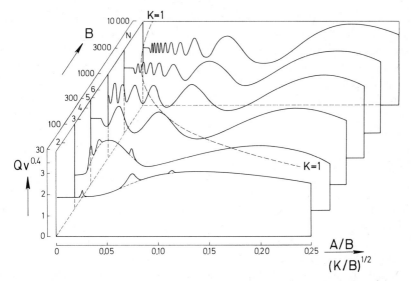

Fig. 6. The total cross sections calculated for a LJ 12–6 and multiplied by $v^{0.4}$ is plotted versus the potential capacity $B = 2\,\mu\varepsilon r_m^2/\hbar$ and the quantity $2\hbar v/\varepsilon r_m$ where v is the velocity. Note that the B scale is logarithmic. The close connection between the number of glory oscillation and B is demonstrated. Lines which connect the Nth glory maxima give for $v = 0$ the B values for which the potential just contains N bound states. On the left side of the dotted line, $K = E/\varepsilon = 1$, orbiting occurs. The number of glory oscillations (and, via Fig. 5, the rainbow oscillations) which can be expected for a system can easily be estimated from this graph.

C. Perturbation of the Elastic Cross Section

The description of the cross section given above is only valid for spherically symmetric interactions. The perturbation of these cross sections by other than spherically symmetric interactions, inelastic or rerrangement channels and the possibility to extract information out of these measurements will be treated in this Section. The cases which may occur are easily summarized:

1. The potential is angular dependent (anisotropic) and rotational transitions may occur.
2. The potential is non-stationary due to vibrating collision partners and vibrational inelastic transitions may happen.
3. Rearrangement, for instance chemical reaction, may happen.
4. Electronic transitions and curve crossing processes take place.

The influence of electronic excitation and curve crossing on elastic scattering is very important. In general, it is typical for atom–atom or atom–ion scattering at high energies and is discussed in Section III.B and IV.B.3. Thus we are left with three contributions which may occur in thermal collisions with molecules. These effects tend to destroy the coherence of the scattering process and therefore the interface oscillations will be damped or 'quenched'. Furthermore the cross section itself may be changed. A theoretical description of these effects only partially exists. Solutions of many coupled equations are necessary in an exact quantum treatment of this problem. Thus only a few model calculations are available, mainly for very light systems. Most of the results have been obtained in the high energy approximation (HEA) which unfortunately has proved to give no quantitative description of molecular collision processes at thermal energies (about 10% error in phase and 30% in amplitudes for a first order treatment). Recently, however, several attempts have been made to decouple the set of equations for rotational transitions. Rabitz (1972) presented an effective potential approach, McGuire and Kouri (1974) proposed the j_z-conserving coupled state approach, and Tsien et al. (1973) and Guerin and Eu (1974) studied the infinite order sudden approximation. The results obtained are very promising and calculations for heavy systems within these approximations will certainly be performed in the near future. Quite another approach describing inelastic and reactive collisions in the elastic channel is the optical potential model which has often been applied to these problems. We will give a short outline of the results of such calculations.

(1) The anisotropic part of the potential in the rigid rotator approximation is generally taken to be of the form

$$V(r) = V_0(r) + V_1(r)P_1(\cos \vartheta) + V_2(r)P_2(\cos \vartheta) + \ldots \qquad (28)$$

where $V_0(r)$ accounts for the spherically averaged potential. The angular dependent contributions are often approximated by $V_i(r) = V_{0\,\text{att}}a_i + V_{0\,\text{rep}}p_i$ where a_i and p_i are constant values which account for the attractive and the repulsive part of the potential. Explicit calculations for the quenching of the glory undulations in the total cross section have been reported by Olson and Bernstein (1968, 1969), Cross (1968) and Miller (1969a). The authors start from the distorted wave Born approximation (DWBA) or semiclassical procedures, but the results are essentially obtained in the high energy approximation. The differential cross section has been treated in the same approximation by Cross (1970). The results for the cross sections are:

Total cross section:

(a) The monotonic part of the cross section is not affected by the anisotropic potential.

(b) The positions of the undulations are shifted to lower velocities. The higher order extrema are more severely affected ($a_2 > 0.3$).

(c) The amplitudes are strongly quenched. The extent of the quenching increases rapidly as the asymmetry is increased. The degree of the quenching is roughly proportional to v^{-2}.

Differential cross section:

(a) The classical differential cross sections, including the position of the classical rainbow angle, are only slightly affected by the anisotropy.
(b) The rainbow oscillations and the rapid oscillations are severely quenched and shifted in phase to lower angles.

The reliability of these results especially for low energies and for the differential cross section (the HEA is indeed a poor approximation for large angles) can only be proved by exact quantum mechanical calculations or at least by better approximations. Unfortunately such a calculation has not been performed for heavy particle collisions. However, there are calculations available for the scattering of hydrogen molecules, H_2, and the isotopes D_2 and HT, on atoms (Allison and Dalgarno, 1967; Hayes et $al.$, 1971; Wolken et $al.$, 1972; Eastes and Secrest, 1972; McGuire and Micha, 1972). In particular Heukels and van de Ree (1972) studied the influence of the anisotropic potential and inelastic channels on the cross section. They found that the anisotropy in the D_2He potential has no influence both on the total and the differential cross section, at least for the small values of $a_2 = 0.12$ and $p_2 = 0.30$ they used. However, for the heteronuclear system HT–He in which the asymmetric mass distribution is properly accounted for in the potential, small but significant deviations occur. Thus, one has to be careful in interpreting experimental results of asymmetric molecules with a homonuclear potential with constant anisotropy parameters. The magnitude of the inelastic cross section for rotational excitation is generally only a few per cent of the elastic cross section.

(2) Vibrational quenching has been found unimportant in molecular collisions (Le Breton and Kramer, 1969; Helbing, 1969c). Furthermore the inelastic cross sections for vibrational excitation are very small in comparison to the elastic ones (McGuire and Micha, 1972).

(3) The influence of chemical reactions on elastic scattering has been extensively studied in the past. Nearly all treatments are based on the optical model (for a review see Ross and Green, 1970). Both the imaginary part of the potential (here assumed to be local) and the opacity function have been parameterized (Mariott and Micha, 1969; Harris and Wilson, 1971 and references cited therein). For a study of the total cross section see Düren et $al.$ (1972). A semiclassical study of a bimolecular exchange reaction where the three atoms are constrained to move on a straight line but the whole system is free to rotate in three dimensions, predicts a new kind of rainbow (Connor and Child, 1970).

All these effects may be used to derive information on the anisotropy, the inelastic processes and the reaction cross section, but not without difficulty. First, one has to decide which amount of quenching originates from which effect. Often this can only be done by knowledge from other sources. The different energy dependence of the quenching of the glory amplitudes may help (Herman *et al.*, 1973). Second, one would need to know accurately the spherically averaged potential to determine the extent of quenching or shifting. Third, the information on several anisotropy parameters has often to be determined from one measured quantity (see the explicit formulae of Olson and Bernstein, 1968), so that detailed knowledge of the form of these parameters is necessary. However, for high anisotropy parameters a new structure appears in the total differential cross section which seems to simplify the determination of these parameters (see Section V.C). Thus, in general, detailed information of these effects is better obtained from measurements with state selected beams (see Reuss, this book) and/or a direct measurement of the inelastic and reactive processes. On the other hand, one has to be very careful in evaluating measurements by a spherically symmetric potential where molecules are involved. The rule that the integral total cross section (the sum of the elastic and all of the inelastic cross sections) is identical to the elastic cross section calculated for $V_0(r)$ (Levine, 1972) has to be examined from case to case. Essentially it depends on which impact parameters are affected by the inelastic processes and which role they play in determining the elastic cross section (McGuire, 1974).

IV. METHODS FOR THE DETERMINATION OF THE POTENTIAL

A. Fitting Procedures

The main aim of elastic scattering experiments is to determine the forces of the interaction between the colliding particles. The most widely used method is to assume for the interaction a specific functional form and then to calculate the scattering phase shifts or the deflection function and the cross section. These calculations are compared with measured properties. The actual potentials are derived by varying parameters inserted in the potential by trial and error so as to obtain the best fit to the measured cross section. Although straightforward in concept this method suffers from the drawback that the parameters determined in this way are dependent on the potential model. Thus the most important task in applying such a method is to find a model potential which is flexible enough to fit the data and to have a method which is able to derive quantitative values for the reliability of the fit. The evaluation of the high resolution measurements of differential and total cross section now available for molecular systems shows that the commonly used potential models are not sufficient for a good description of the potential. These are the Lennard-Jones $(n, 6)$ potential and the potentials of the Buckingham type. The next step was to construct more flexible models by a superposition or distortion of the simple models (Buck and

Pauly, 1968; Düren et al., 1968; Mittmann et al., 1971). Another possibility is to use special functions only over a restricted range for which they are assumed to be the right description. As an example we consider the potential which is valid for neutral–neutral interaction

$$f(x) = V(r)/\varepsilon \qquad x = r/r_m$$

$$f(x) = A \exp\left[-\alpha(x - 1)\right] \qquad\qquad 0 < x < x_1$$

$$f(x) = \exp\left[-2\beta(x - 1)\right] - 2 \exp\left[\beta(x - 1)\right] \qquad x_2 < x < x_3 \qquad (29)$$

$$f(x) = -C_6 x^{-6} - C_8 x^{-8} \qquad\qquad x_4 < x < \infty$$

The short range repulsive part is represented by a Born–Mayer exponential (E) type, the region of the well by a Morse potential (M) and the long range attractive part by a dispersion potential with a dipole–dipole and a dipole–quadrupole term. These parts are connected by cubic spline functions (S)

$$\sum_{k=0}^{3} a_k[(x - x_j)/(x_i - x_j)]^k \qquad x_i < x < x_j \qquad (30)$$

which provide a continuous connection with a continuous derivative at the connection points. In this way the different parts are decoupled and can be varied independently in the minimization procedure. If necessary further regions can be introduced by another analytical expression or single points which are connected by the spline functions. This type of potential has been used by fitting cross sections of rare gases (Siska et al., 1971; Gegenbach et al., 1972; Buck et al., 1973a). Since all these potentials contain at least five free parameters the computational effort to find the best-fit parameters is large. The way this is usually done is by the χ^2 procedure where the expression

$$\chi^2 = \sum_{k=1}^{n} \sum_{i=1}^{N_k} [(I_{ik}^{ob} - I_{ik}^{cal})/\Delta I_{ik}]^2 \qquad (31)$$

is minimized. I_{ik}^{ob} is the measured and I_{ik}^{cal} the calculated ith data point of the kth data set. N_k is the number of data points of the kth data set and n the number of data sets. ΔI_{ik} is the statistical error of the measured point. It should be noted that the significance of χ^2 as a statistical measure of the compatibility between a model and a data set is only valid if either the data represent a complete statistical distribution or the model fully describes the physics. Then χ^2 is equal to its expectation value f, that is the number of degrees of freedom of the system (equal to the number of data points minus the number of free parameters) and it is possible to extract errors of the best fit parameters (Arndt and MacGregor, 1966; Düren et al., 1968). Unfortunately in almost all physical applications neither of these assumptions is correct. Experimental measurements often have a systematic error that is common to a number of data points. Also, the data set may be incomplete, which can give a misleading value of the error of the parameters. Neverthe-less, the χ^2-procedure is the only method which provides the possibility of distinguishing quantitatively between different models and of deriving the error bound for the parameters by renormalizing the experimental χ^2-value to the expectation value.

The number of parameters which can be determined from such a minimiza-tion procedure is restricted by the information content of the measurement. The easiest way to estimate this information content is to calculate the number of partial waves which contribute significantly to the scattering amplitude. For very low energies this information content reduces to one parameter so that such a measurement does not contain very much informa-tion on the potential (Farrar and Lee, 1972). Another method has been proposed by Ury and Wharton (1972) for the total cross section. An upper limit of the number of parameters is estimated from the order of a power series necessary to describe the experimental curve to that degree of experi-mental accuracy established when the fractional χ^2 of the polynomial with respect to the data equals the fractional variance of the experiment. In spite of the merits of the χ^2-procedure the disadvantages of a fit procedure may be summarized as follows:

(a) A large amount of computer time is required,
(b) errors and uniqueness of the potential are difficult to determine, and
(c) the range of the potential to which the measurements are sensitive is not given by the method. It can only be obtained by indirect guesses (see, for example, Ury and Wharton, 1972, and Bickes *et al.*, 1973a).

This last point should not be overlooked. Most of the discrepancies in the literature on interatomic potentials arise from this fact. Many authors over-estimate the range of the potential to which their measurement is sensitive.

B. Inversion Procedures

1. General Remarks

All the difficulties of a fit procedure mentioned at the end of the preceding Section can be avoided if an inversion procedure is applied. Starting from the observed angular dependence of the differential cross section or the energy dependence of the total cross sections, a set of phase shifts as a function of the angular momentum or of the energy is obtained and then the potential is deduced from the phase shifts. Thus the problem is naturally divided into two steps:

1. the determination of the phase shifts from the cross section,
2. the determination of the potential from the phase shifts.

Such a formulation is only valid for spherically symmetric potentials. A general solution in the framework of quantum mechanics exists for both steps. Since this subject was reviewed several times (Newton, 1966; Newton, 1972; Sabatier, 1972a; Buck, 1974) we shall only give a short summary of the results. The first step, the determination of the phase shift from the differential cross section can be performed if the cross section is smooth and small enough (Newton, 1968; Martin, 1969; Gerber and Karplus, 1970). The solution of the second step is in general not unique.

If the phase shifts are given as a function of the energy a unique solution can only be obtained if, in addition, the bound state energies and the normalization constants of the bound state wave function are known (Gelfand and Levitan, 1951). The explicit construction of the potential leads to a complicated solution of a Fredholm integral equation. Other procedures (Agranovich and Marchenko, 1963; Hylleraas, 1963) may be more convenient for a practical application (see Benn and Scharf, 1967, O'Brien and Bernstein, 1969) but up to now there has been no practical application of such a procedure in molecular physics.

If the phase shifts are given as a function of the angular momentum the uniqueness of the solution depends largely on the special choice of the interpolation of the phase shift for the values between the real angular momenta (Newton, 1962; Sabatier, 1966; Loeffel, 1968; Sabatier, 1972b). For practical purposes only the method of Sabatier and Qugen Van Phy (1971) seems to be applicable since interpolations or analytical continuations in the complex plane are avoided. Again a Fredholm integral equation has to be solved, but uniqueness of the potential can only be achieved for a restricted class of potentials which produce small phase shifts.

Because of these difficulties we turn to inversion procedures which are valid in the semiclassical limit since this approximation has proved to be applicable for most of the atomic and molecular collisions. Solutions of the second step, the determination of the potential, are treated in Section IV.B.2. In general, the input information will be the phase shifts or the deflection function. Only in the high energy approximation can the potential be derived directly from the cross section. For a detailed discussion of these procedures see Buck (1974). The possibilities of determining the phase shifts or the deflection function from the cross section are treated in Section IV.B.3. The advantage of such procedures and the general requirements on the data are discussed in Section IV.B.4. The emphasis will be on procedures which have been applied to real data. Extensions to non-central or optical interaction potentials are available. Most of them, however, are still in a 'formal' state, so that a direct application to molecular physics is not obvious. Two exceptions should be mentioned. One is a special inversion procedure for optical potentials derived by a perturbation formalism (Roberts and Ross, 1970). The other is a special attempt for a particular class of non-spherical potentials done by Gerber (1973). For potentials, the Fourier transform of which is a product of two functions which only depend on the magnitude and the direction of the transform space vector k, an iterative inversion procedure is established (Prosser, 1969) which converges in a unique way if the Born series converges for the k chosen. The input data required are the backward scattering amplitudes for all directions of the incident momentum at fixed energy and the value of the spherically averaged potential.

2. Semiclassical Procedures for the Determination of the Potential

The angular momentum dependence: Starting with the expression of the JWKB-phase shift (9) and using the transformation (Sabatier, 1965; Vollmer and Krüger, 1968; Vollmer, 1969)

$$s^2 = r^2(1 - V(r)/E) \tag{32}$$

and $\lambda/k = b$ one obtains after integration by parts

$$\eta(E, b) = -k \int_b^\infty ds\, I(s, E)s(s^2 - b^2)^{1/2} \tag{33}$$

where $\lim_{r \to \infty} rV(r) = 0$ is used. Now the phase function is a simple integral of the form of the Born integral and it is linear in

$$I(s, E) = \ln [r(s)/s] \tag{34}$$

Equation (33) has the form of an Abelian integral equation which is solved by

$$I(s, E) = \frac{2}{\pi} \int_s^\infty db\, \frac{d\eta}{db}(E, b)\frac{1}{k(b^2 - s^2)^{1/2}} \tag{35}$$

Now the potential is given through inverting (34) and (32) by

$$V(r) = E\{1 - \exp[-2I(s, E)]\} \tag{36}$$

With the semiclassical relationship $2(d\eta/db) = k\Theta(b)$ one obtains at once

$$I(s, E) = \pi^{-1} \int_s^\infty db\, \Theta(b, E)(b^2 - s^2)^{-1/2} \tag{37}$$

which is just the classical result (Firsov, 1953; Miller, 1969b). A suitable form of this integral for numerical application has been given by Buck (1971) and Klingbeil (1972). This procedure can be carried out under the following considerations:

(a) Since it is a semiclassical procedure the potential can only be determined up to the classical turning point r_0. If $\Theta(b)$ is not known over the whole range of impact parameters from zero up to ∞ then it is possible to perform the procedure up to a certain value of r, as can be seen from the nature of the integration. For the integration to infinity $\Theta(b)$ has to be extrapolated. This can often be done due to the known long range forces (see Section IV.B.3).

(b) In order to determine the potential in a unique manner Equation (33) must have a unique solution, this means $s(r)$ must be a monotonic function, otherwise the inverse function would not exist. This condition implies that the energy E must be greater than the energy E_{orb} at which orbiting occurs (Vollmer, 1969; Miller, 1969b).

(c) The phase shift or the deflection function have to be interpolated from discrete values of l so as to give a continuous function of b.

The integration of (35) can be performed numerically, but for a large number of functions the integration can be done explicitly. Vollmer (1969) gives a list of these functions. Special solutions are available for Gaussian shaped exponentials and negative power forms. Since $I(s, E)$ is linear, it can be constructed from a sum of single terms with great variety. Higher order correction terms of the WKB-phase shifts (Rosen and Yennie, 1964; Sabatier, 1965) can be treated in the same way as the usual WKB-solution (Vollmer, 1969) so that $I(s, E)$ can be obtained by successive approximation steps.

The energy dependence: For treating this case the WKB-phase shift is written as follows

$$\eta(E, \lambda) = m \left\{ \int_{r_0}^{\infty} dr \, [E - V(r) - \lambda^2 (mr)^{-2}]^{1/2} - \int_{r_1}^{\infty} dr \, [E - \lambda^2 (mr)^{-2}]^{1/2} \right\} \tag{38}$$

where m is given by $(2\mu/\hbar^2)^{1/2}$, and r_0 and r_1 are the zeros of the integrands. In order to apply the theory of Abelian integral equations we break the effective potential $U = V(r) + \lambda^2 (mr)^{-2}$, which usually shows an attractive minimum in molecular physics, into two branches:

$$r = r_1(u) \qquad r_0 \leq r \leq r_m$$
$$r = r_2(u) \qquad r_m \leq r \leq \infty \tag{39}$$

where $r_1(U)$ and $r_2(U)$ are the corresponding inverse functions. Here it is assumed that two branches are sufficient, which is fulfilled if either $l = 0$ or $l > l_{orb}$, where $2l_{orb}^2 = m \cdot \max [r^3 V'(r)]$. Then differentiating (38) with respect to E and introducing the quantities $y = U(r)$ and $z = Z(\rho) = \lambda^2 (m\rho)^{-2}$ as new variables of integration we have

$$\frac{d\eta}{dE} = \frac{m}{2} \left[\int_E^0 dy \, \frac{r_1'(y)}{(E - y)^{1/2}} + \int_0^{-\varepsilon} dy \, \frac{r_1'(y) - r_2'(y)}{(E - y)^{1/2}} - \int_E^0 dz \, \frac{\rho'(z)}{(E - Z)^{1/2}} \right] \tag{40}$$

where a prime denotes the derivative with respect to the variable in parenthesis. Applying the theory of Abelian integral equations we arrive at

$$\frac{2}{\pi} \frac{1}{m} \int_0^\alpha \frac{d\eta}{dE} \frac{dE}{(\alpha - E)^{1/2}} = -r_1(\alpha) + r_1(0) + \frac{\lambda}{k} - \rho(0) + I \tag{41}$$

where the remaining integral I can be expressed by the number $n(E)$ of bound states:

$$I = R(-\varepsilon) - R(0) + \frac{2}{m} \int_0^{-\varepsilon} \frac{dn}{dE} \frac{dE}{(\alpha - E)^{1/2}} \tag{42}$$

where the abbreviation $R(y) = r_1(y) - r_2(y)$ is used and the number of bound states is given by

$$(n + \tfrac{1}{2})\pi = m \int_{r_1}^{r_2} dr \, [E_n - U(r)]^{1/2} \tag{43}$$

(Miller, 1969b; Miller, 1971a; Buck, 1974). Since we have $R(-\varepsilon) = r_1(-\varepsilon) - r_2(-\varepsilon) = 0$, $R(0) = r_1(0) - r_2(0)$, and $r_2(0) = \rho(0)$, insertion of (42) in (41) yields, by setting $\alpha = U$,

$$r_1(U) = \frac{2}{\pi} \frac{1}{m} \int_0^U \frac{d\eta}{dE} \frac{dE}{(U - E)^{1/2}} + \frac{2}{m} \int_0^{-\varepsilon} \frac{dn}{dE} \frac{dE}{(U - E)^{1/2}} + \frac{\lambda}{k} \tag{44}$$

and the value of the potential is obtained from

$$V(r) = U(r) - \lambda^2 (mr)^{-2} \tag{45}$$

A more convenient form is obtained by changing the integration variables in both integrals (Miller, 1969b; Feltgen et al., 1973)

$$r_1(U) = -2(\pi m)^{-1} \int_{\eta(0)}^{\eta(U)} d\eta \, [U - E(\eta)]^{-1/2}$$

$$- 2m^{-1} \int_{-1/2}^{n(0)} dn \, [U - E(n)]^{-1/2} + \frac{\lambda}{k} \tag{46}$$

where $E(\eta)$ is the inverse function of the energy dependence of the phase shifts and $E(n)$ is the inverse function of the WKB bound state eigenvalue function. The potential for $U > 0$ is given by this expression if the phase shifts are known as a function of E for $l = $ const. In addition some knowledge of the attractive part of the potential which can be expressed by the bound states of the system is necessary (see integral I, Equation (42)), just as in the quantum mechanical case. The following restrictions have to be imposed on (46):

(a) the value of l must be such that the effective potential $U(r)$ does not possess a local maximum,

(b) $\eta(E)$ has to be known as a smooth and differentiable function,

(c) $\eta(E)$ has to be known in the limit $E \to 0$. Suitable guesses can be obtained from a knowledge of the attractive part for $E < 0$ which has to be known in any case for such an inversion. A restriction of the data to large E leads to a restricted range of the potential.

3. Practical Methods

The preceding Section shows that the problem of constructing the potential from the knowledge of the phase shifts or the deflection is comparably easy to solve in the semiclassical approximation. Now we are left with the problem of deriving the phase shifts or the deflection function from the measured cross section. In general no straightforward solution exists since the connection between these quantities is either very complicated (e.g. the partial wave sum of Equation 7) or it contains the phase shifts, the deflection angle ϑ and the multivalued inverse function $b(\vartheta)$ (e.g. Equation 19). For special cases a direct determination is possible. If the differential cross section is monotonic $b(\vartheta)$ is a single-valued function and it is obtained by a simple integration over the cross section (Firsov, 1953)

$$b^2(\vartheta) = 2 \int_\vartheta^\pi \sigma(\vartheta') \sin \vartheta' \, d\vartheta' \tag{47}$$

Another possibility is to analyse the phase differences corresponding to interference data (see Equation 23) which gives directly the difference between the two interacting impact parameters (Smith *et al.*, 1966; Pritchard, 1972). Information on the energy dependence of the phase shift curve from the total cross section can only be obtained for identical particles in the semiclassical limit. This case will be treated below in detail. For the differential cross section in practice two different methods have been applied for deriving information on $\Theta(b)$ or $\eta(b)$:

1. comparing semiclassical features of the cross section with calculated values via a parametrization of the deflection function (Buck, 1971),
2. deriving unknown coefficients of a parametrized phase shift curve or S-matrix through a minimization procedure on measured and calculated data (Vollmer, 1969; Remler, 1971; Klingbeil, 1972).

Now we start with a description of these methods where procedures which have been applied will be treated in more detail than those which have only been suggested.

Differential cross section: Deflection function. First we describe methods which take advantage of the close relationship between semi-classical cross sections and deflection function as outlined in Section III. A procedure which uses nearly all measurable quantities has been proposed and applied by Buck (1971). In order to unfold the multivalued character of $b(\vartheta)$, the deflection function is separated into monotonic functions $g_i(b)$ such that $\Theta(b) = \sum_i g_i(b)$ and $b = g_i^{-1}(\vartheta)$. The g_i are represented by the usual functional approximations made in the semiclassical scattering theory:

(i) a parabola in the minimum

$$\Theta(b) = -\vartheta_r + q(b - b_r)^2 \tag{48}$$

(ii) a straight line in the vicinity of the zero point

$$\Theta(b) = -a_1(b - b_g) \tag{49}$$

(iii) an inverse power law for the asymptotic region

$$\Theta(b) = -c_1 b^{-c_2} \tag{50}$$

The measurable quantities are then calculated with the help of these functions. For instance, the rainbow scattering is given by Equation (16). Because of the behaviour of A_i^2 and the smaller amplitude of the second term, the positions of the supernumerary rainbows z_N are given by the zeros of $A_i'(z_N)$ (maxima) and the zeros of $A_i(z_N)$ (minima). Now z_N is calculated using (19) and the function (48) with the result

(1) $$z_N = k^{2/3}q^{-1/3}(\vartheta_r - \vartheta_N) \tag{51}$$

for the region of the minimum, and using (49) and (50) with the result

(2) $$z_N = 0{\cdot}75^{2/3}[2\eta_g + kb_g(\vartheta_N) + 0{\cdot}5ka_1^{-1}\vartheta_N^2$$
$$- kc_1^{1/c_2}(1 - c_2^{-1})^{-1}\vartheta_N^{(1 - 1/c_2)}] \tag{52}$$

for the region next to the minimum, where η_g is the maximum phase shift. With the help of these equations the correct functional behaviour is tested (e.g. a straight line for region i) and the unknown quantities q, ϑ_r, b_r, b_g, η_g, a_1, c_1, c_2 are determined by a direct comparison with the measured z_N. The requirement of continuity at the ends of the region (i) reduce the number of unknowns by two. The maximum phase shifts can be determined either by an extrapolation of the z_N vs ϑ_N curves for $\vartheta = 0$ or, more precisely, from the velocity dependence of the glory oscillations of the total cross section (see Equation 25). Thus we are left with five coefficients which in principle can be determined if the number of measured rainbow extrema exceeds the number of unknown quantities in the corresponding regions. It should be noted that such a determination is not unique since the rainbow oscillations are only sensitive to certain areas of the deflection function (Boyle, 1971, see Section III; Pritchard, 1972; Farrar et al., 1973a). However, additional information on the rapid oscillations, which determine b_g, the amplitudes and the monotonic scattering cross section at large angles reduce the number of unknowns in a natural way and give a unique solution. In practice, the ratio of the amplitude of the first rainbow maximum to the monotonic cross section proved to be the best quantity for the additional information. Several other quantities, such as the ratio of the maximum to the minimum of the amplitudes were found to be affected too much by the finite energy and angular resolution even for high resolution experiments.

This procedure has been applied to the scattering of alkali atoms on mercury (Buck and Pauly, 1971; Buck et al., 1972; Buck et al., 1974; Barwig et al., 1975). These systems ($B = 3000$ to $B = 50\,000$) display a marked rainbow structure in the thermal energy range. Fig. 7 shows as an

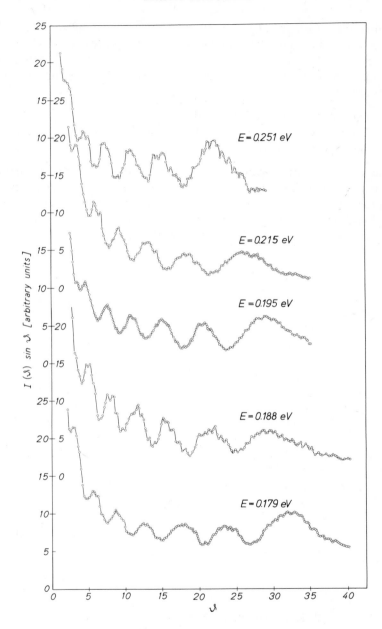

Fig. 7. Measured differential cross sections for NaHg of five different energies in the centre of mass system (Buck and Pauly, 1971). Supernumerary rainbows and rapid oscillation (at $E = 0.25$ eV and 0.19 eV) are well resolved.

example the measurement of Na–Hg performed at five different energies. Up to seven supernumerary rainbows with superimposed rapid oscillations are clearly resolved. From this structure the attractive part of the deflection function is determined as described above where the following data are used:

(1) The position of the rainbow oscillations.
(2) The separation of the rapid oscillations.
(3) The positions of the glory oscillations in the velocity dependence of the total cross section.

The repulsive part is built up from the monotonic backward scattering at one energy not shown in Fig. 7. Now the potential is calculated applying the procedure of Section IV.B.2. For the integration to infinity the deflection function is extrapolated with help of the theoretical van der Waal constant C (Stwalley and Kramer, 1968). Fig. 8 shows the final result. Each energy is

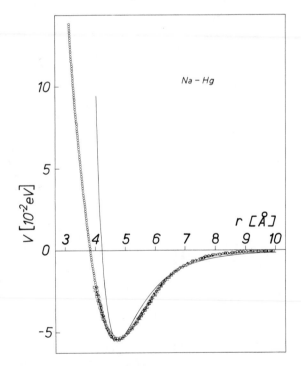

Fig. 8. The potential of NaHg obtained by the inversion of cross section data at five different energies which are denoted by different symbols: (○ for $E = 0.18\,\text{eV}$; ◇ for $E = 0.19\,\text{eV}$; + for $E = 0.20\,\text{eV}$; △ for $E = 0.22\,\text{eV}$; and ☐ for $E = 0.25\,\text{eV}$). The solid line is a LJ 12–6 potential fitted to the minimum (Buck and Pauly, 1971).

ndicated by a different symbol. As can be seen, the potential resulting from
each of the five energies is the same to within the experimental error. The
repulsive part is only determined by the data at one energy. The solid line is a
L.J. 12–6 potential fitted to the minimum which displays the great difference
of the inverted points to the commonly used analytic potential forms. The
error of the potential in the minimum is $\Delta r/r = 0.5\%$ and $\Delta V/V = 3.0\%$,
n the vicinity of 6·5 Å $\Delta r/r = 1.0\%$ and $\Delta V/V = 12\%$. In order to test the
validity of the potential derived from such an inversion procedure the
differential cross sections have been recalculated and compared with the
measured values. Fig. 9 illustrates such a comparison for the system LiHg
where the rapid oscillations are better resolved than for NaHg and the

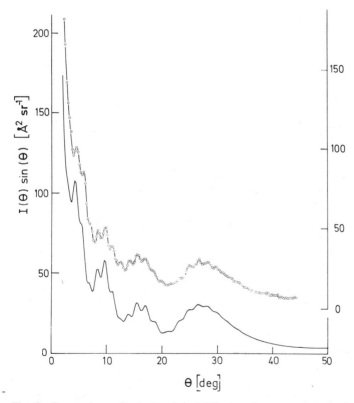

Fig. 9. Comparison of calculated (solid line) and measured (points)
differential cross sections for LiHg in the laboratory system. The
calculation has been performed with the potential obtained by the
inversion of the measured data (Buck *et al.*, 1974). The scale is
calibrated in Å²/sr.

ratio of the amplitude of the first rainbow maximum to the backward
scattering is used as additional input information (Buck et al., 1973b). The
upper curve shows the measured values whereas the lower curve is calculated
The finite energy and angular resolution of the apparatus has been taken into
account. Outside of a small distortion at 10° the agreement is good especially
in the region of the first rainbow maximum and the backward scattering
Both the positions and the amplitude of the cross sections are well reproduced.

The advantage of this inversion procedure is that only data which can be
measured without being influenced severely by the averaging processes is
used. In addition nearly all measurable quantities can be taken into account
On the other hand, restrictions are imposed by the set of simple functions
which reduce the number of unknowns for the deflection function and by the
well resolved rainbow structure of the cross section which is only possible
for certain values of the reduced parameter B and A (see Fig. 6). It should
be noted that, in principle, such a procedure is not restricted to the rainbow
structure. Any other interference structure can be treated in this way. Other
procedures similar to the one described have been proposed by Miller (1969b)
Kennedy and Smith (1969), Boyle (1971) and Pritchard (1972). To overcome
the multivalued character of the inverse of the deflection function $b(\vartheta)$
Miller introduces monotonic functions $p(x)$ and their inverses p^{-1} which are
parameterized by suitable expansions of simple power series. Boyle calculates
the deflection function first from an assumed potential. This $\Theta(b)$ is then
adjusted in a way such that the special features of the cross section measured
(say the rainbow and the rapid oscillations) are well represented. The
potential is calculated by use of the inversion techniques of Section IV.B.2
The procedure is then repeated until there is agreement between measured
and calculated quantities. The potential for Na–Xe has been determined in
this way. Kennedy and Smith start from the identical particle oscillations in
the differential cross section for He^+–He. The spacings of these oscillations
are a measure of the sum of the two impact parameters $b_1 + b_4$ (see Equation
24). This result can be used for constructing the $b(\vartheta)$-curve by an iterative
procedure. Pritchard starts from (23) which allows the direct determination
of the difference of two impact parameters. With additional information
similar to that mentioned above, the rainbow cross sections, the classical
cross section and the envelopes of the amplitudes, the deflection function is
constructed in a unique way. He suggests, in addition, the usefulness of the
reduced variables, introduced by Smith et al. (1966) for a small angle or high
energy approximation, for data taken at large angles and thermal energies
The advantage of such a procedure is obvious. Instead of (23) the reduced
action difference becomes

$$a(E, \tau) = (E/2\mu)^{1/2}\alpha(E_1\tau) = [N(\tau, E) - N_0](E/2\mu)^{1/2}2\pi\hbar \qquad (53$$

with

$$a(E, \tau) = a_0(\tau) + E^{-1}a_1(\tau) + \ldots \tag{54}$$

τ is given by the first member $\tau_0\,(b) = E\,.\,\vartheta$ of an expansion in E^{-1}, and it is only a function of the impact parameter. The quantity N_0 describes a quantal contribution to the relative phase. Thus the data taken at different energies can be combined into a *single* curve with greater accuracy than would be possible for the *single* curves. In addition this method helps to clarify and extend the range of data. The limits of applicability of these reduced variables for scattering at thermal energies have not yet been studied by a rigorous approach. Numerical calculations show that this relation is fulfilled to within 2% if the energies do not differ by more than 25%. Such reduced variables are indispensable for studying and identifying unknown features of the cross section.

The interference pattern due to the potential crossing can also be used for constructing the deflection function. Three significant types of information are available from these elastic perturbations: the reduced angle of the starting point of the oscillations τ_c, the oscillatory spacings and the amplitudes of the oscillations. By using the reduced angle τ rather than the angle itself to describe the threshold behaviour, it is easy to estimate the crossing distance from such τ_c values. Since τ is only a function of the impact parameter b, with some knowledge of the potential it can be related to the corresponding r values (for details see Smith et al., 1967, Coffey et al., 1969). The spacings are analysed in terms of the reduced action difference (see 53 and 54). The quantity N_0 which describes a quantal contribution to the relative phase may be determined either theoretically or by adjusting the experimental data so that the results from different energies fall into a common pattern. A plot of $E^{1/2}(N - N_0)$ versus τ is used to determine the difference of impact parameters involved which can then be used for a determination of the potentials if one of the potentials is known. The amplitudes of the oscillations associated with crossing contain information about the energy associated with the coupling between the states.

Differential cross section: phase shifts. Quite another procedure has been proposed by Remler (1971). First the S-matrix is decomposed into a repulsive and an attractive part with $S = \exp\,[2i(\eta_a + \eta_r)] = S_a S_r$. Then the S-matrix for the attractive part is parameterized by

$$S_a = \prod_{p=1}^{N} \left(\frac{\lambda^2 - \lambda_p^{*2}}{\lambda^2 - \lambda_p^2} \right) \tag{55}$$

where λ is the real angular momentum $l + \frac{1}{2}$ and λ_p is the position of the pth pole in the complex angular momentum plane. The scattering amplitude can now be computed using the Regge–Watson–Sommerfeld transformation

(De Alfaro and Regge, 1965)

$$f_a(\vartheta) = \frac{1}{2ik} \sum_{p=1}^{N} \frac{\pi}{\cos \pi\lambda} P_{\lambda-\frac{1}{2}}(-\cos \vartheta) \operatorname*{Res}_{p} [\lambda S_a(\lambda, k) - 1]$$

$$+ \oint_C d\lambda \frac{\lambda}{2k} \frac{S_a(\lambda, k) - 1}{\cos \pi\lambda} P_{\lambda-\frac{1}{2}}(-\cos \vartheta) \qquad (56)$$

where Res_p is the residue of the function at pole p. The integral has to be taken over the imaginary axis and the half circle including the complex plane for $\lambda > 0$. Since the function S is a quadratic function of λ, the integral over the imaginary axis vanishes and the remaining part is assumed to be zero. Therefore one gets

$$f_a(\vartheta) = \frac{1}{2ik} \sum_{p=1}^{N} \frac{\pi}{\cos \pi\lambda} P_{\lambda-\frac{1}{2}}(-\cos \vartheta)(\lambda_p^2 - \lambda_p^{*2}) \prod_{\substack{i=1 \\ p \neq i}}^{N} \left(\frac{\lambda_p^2 - \lambda_i^{*2}}{\lambda_p^2 - \lambda_i^2} \right) \qquad (57)$$

The partial wave sum is now reduced to a sum over few pole contributions in the complex plane of λ. The contribution of a single pole to the phase shift function and the deflection function can be obtained from the parameterization (55). Fig. 10 illustrates the result. $\Theta_p(l)$ is essentially a pulse centred at $l = \operatorname{Re}(\lambda_p - \frac{1}{2})$ with the depth $2/\operatorname{Im} \lambda_p$ and the width $2 \operatorname{Im} \lambda_p$. Now one proceeds as follows. Starting with N poles, which are placed on a small circle centred at λ_p in the complex λ-plane, the number of these poles (N) and the real and imaginary part of the central pole (λ_p) are derived from semiclassical quantities. The rainbow angle is given by $\vartheta_r = 2N/\operatorname{Im} \lambda_p$,

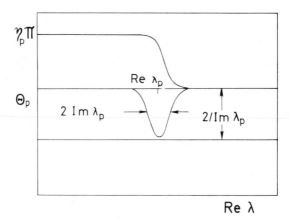

Fig. 10. Phase shift η_p and deflection function Θ_p of a single pole λ_p in the complex angular momentum plane according to a special parametrization of the S-matrix (Remler, 1971).

the rainbow angular momentum by $l_r = \text{Re } \lambda_p$ and the width of the deflection function at $\vartheta = \vartheta_r/2$ by $\Gamma_{1/2} \simeq 2 \text{ Im } \lambda_p$. Equation (57) is then summed, the results compared with the experimental data and the input information varied until a satisfactory agreement is achieved between the experimental and calculated cross sections. The number of poles which are sufficient for attaining good agreement with the experiment has been found to lie between 5 to 16 (Rich et al., 1971) when applying this procedure to proton–rare gas scattering. The repulsive part of the scattering amplitude f_r cannot be treated in this way, since here the large number of repulsive poles contributing to the cross section is cumbersome. The chosen parameterization of the S-matrix leads to the wrong asymptotic behaviour of the phase shift which should be important for systems where weak but long-range forces are involved. Nevertheless the results obtained by this method are in good agreement with other methods (Buck, 1974).

The easiest way to determine the phase shifts from the cross section is to assume a specific functional form for the phase shifts which represents the behaviour of the real phase function: negative phase shifts at small b, a maximum for positive values at larger b and an asymptotic behaviour of the phase shift at large b of the form

$$\eta \propto C b^{-(n-1)} \tag{58}$$

If the potential has the asymptotic form $V(r) = -Cr^{-n}$ (Pauly and Toennies, 1965; Bernstein and Muckerman, 1967; Pauly, 1973). Then the unknown coefficients of the phase function are determined by a non-linear least squares algorithm from the measured cross section.

An interesting parameterization is presented by Vollmer (1969, 1971). The phase shifts are described by a superposition of functions for which (35) can be solved analytically. Thus, once the unknown coefficients are known the potential is known. Klingbeil (1972) proposed a method where all phase shifts were treated as free parameters, except those for large b, where the expression (58) holds. The procedure does not seem well suited for atom–atom scattering processes, where more than hundreds of angular momenta l are involved. However, if a good estimate of the starting parameters is available (about 10 % deviation), and the number of l is not too high (small A), a rapid convergence of the minimization process is achieved (e.g. 20 iterations for 126 phase shifts). As an example, the differential cross section of Li and Ar measured by Detz and Wharton (1970) has been inverted by Klingbeil (1973). Fig. 11 shows the differential cross section multiplied by $\vartheta^{4/3}$ in order to remove the steep angular dependence at small angles. This system ($B = 360$, $A = 104$, $K = 30$) displays no rainbow structure but well separated rapid oscillations. Thus it is sensitive to the repulsive part of the potential. The phase shifts of the Buckingham-potential of Ury and Wharton (1972) were used as starting values. Then the phase shifts for $0 \le l \le 200$ are extracted

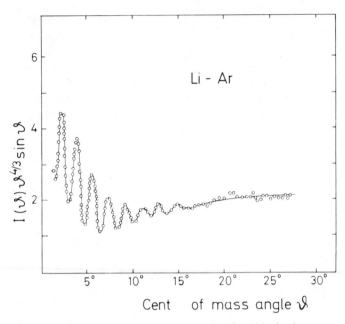

Fig. 11. Measured differential cross section for LiAr in the centre of mass system (Detz and Wharton, 1971). The cross section is multiplied by $\vartheta^{4/3}$ in order to remove the steep angular dependence at small angles. The rapid oscillations are clearly resolved.

from the measurement by the least squares procedure and the potential is obtained as described in Section IV.B.2. This method requires a well-behaved starting potential. It works without any assumption and is therefore well suited for systems which display well-resolved rapid oscillations without a rainbow structure. Such a case cannot be solved either by the method of Firsov or Buck or by the procedure of Remler.

Total cross section: identical particles. An example where the phase shifts for one l as a function of the energy can be extracted from the measurement is the total cross section for identical particles. Fig. 12 shows the cross section for ^{4}He on ^{4}He, ^{3}He on ^{3}He and ^{4}He on ^{3}He atoms measured by Feltgen et al. (1973 and 1974). The symmetry oscillations are clearly resolved for the identical systems ^{4}He^{4}He and ^{3}He^{3}He. Because of the different statistics involved the cross sections are different. The cross section for ^{4}He^{3}He displays no structure. Up to now only the ^{4}He^{4}He curve has been inverted to the potential so that we restrict ourselves to the discussion of this system. Since the potential well is very small most of the structure is sensitive to the repulsive part of the potential. Only the first shoulder is sensitive to the attractive part due to the special behaviour of the s-wave phase shift (analogous to the Ramsauer effect in electron scattering). The analysis of this

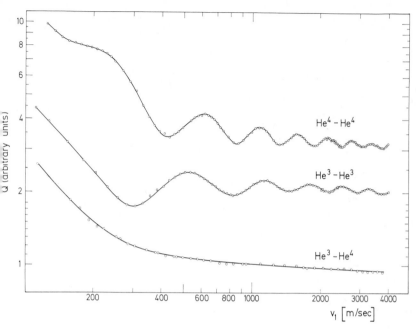

Fig. 12. Measured total cross section for ^{4}He–^{4}He, ^{3}He–^{3}He and ^{3}He–^{4}He as a function of the primary beam velocity (Feltgen *et al.*, 1973 and 1974). Only the identical particle systems display well-resolved symmetry oscillations which occur at different positions because of the different statistics involved for ^{4}He–^{4}He and ^{3}He–^{3}He.

cross section shows (see Section III.B) that all phase shifts are coupled to the -wave phase shift and that therefore these phase shifts can be deduced from the measured extrema of the cross section via (see Equation 27)

$$\eta(E_1 0) = [N(E) - \tfrac{1}{4}]\pi - \varphi(E)/2 \qquad (59)$$

The expression is especially valid at higher energies. In order to apply the procedure of Section IV.B.2 the potential well has to be known (bound state energies). In addition the phase shifts have to be extrapolated to $E \to 0$ for the integration. Both steps are solved by assuming several forms for the potential, performing the inversion and comparing the results with the measurements. The results of such an inversion (Feltgen *et al.*, 1973) are displayed in Fig. 13 along with other potentials derived from molecular beam experiments and gaseous properties. It should be noted that this inversion is in any case better than the usual fit procedure since a part of the cross section curve is correctly reproduced in any interation step, regardless of any assumed potential well.

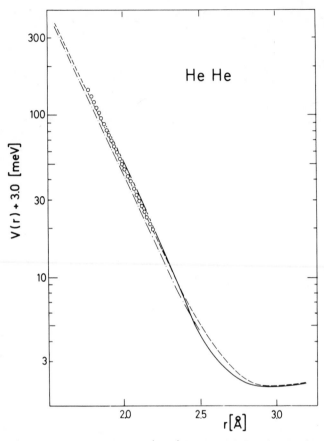

Fig. 13. The potential for ⁴He–⁴He. The circles denote the potential obtained by inversion (Feltgen *et al.*, 1973). The other potentials are determined by molecular beam scattering experiments: solid line (Farrar and Lee, 1972), dashed line below 30 meV (Bennewitz *et al.*, 1972a), dashed line above 30 meV (Gengenbach *et al.*, 1973a), and gaseous properties: dashed dotted line (Beck, 1968). The potential of Cantini *et al.* (1972) not shown in the figure coincides with the curves.

4. Discussion

The examples of the preceding Section show that inversion procedures are indeed a powerful tool for deducing potentials from measured cross sections. Several of the advantages should be pointed out. (1) The potential is automatically determined only for the region which is probed by the measurement. (2) The computation time is significantly reduced, being on the average a factor of about 5 to 10 smaller than for a fit procedure with χ^2-techniques. (3) Finally, due to the clear structure of the different steps involved in any inversion problem, the question of errors and uniqueness can be solved in a straightforward manner.

In the semiclassical approximation the step from the phase shifts or the deflection function to the potential is achieved by solving an Abelian integral equation. If the input information is known only as a function of energy, additional information about the potential well (bound states) is necessary for a unique solution; whereas for the angular momentum dependence of the input information, a unique solution is achieved if the interpolation from the angular momentum l to the impact parameter b is unique. The step from the cross section to the phase shifts is easily achieved if the data, the differential and the integral cross sections, are of absolute accuracy over the full range of the variable. In practice we are obliged to deal with a limited range of data with limited accuracy, e.g. for the differential cross section. The results of Section IV.B.3 show that the solution is unique if the whole interference structure is known together with some additional information from large angle scattering, the amplitude or the average cross section. The methods for this step which have been applied to real data are summarized in Table II in order of decreasing computational effort but increasing approximations. Which one of the methods is better suited for a special set of data depends largely on the cross section features which are resolved experimentally.

The limited angular (energy) range can in practice be overcome. At small angles (or low energies) an appropriate interpolation must be found, while a lack of large angle data (or high energy data) limits the range of the deduced potential. Another possibility is to vary the complementary, fixed parameter, e.g. the energy (the angular momentum). The different sets of phase shifts should then give the same potential. A more elegant possibility is the reduced variable formalism originally derived by Smith et al. (1966) as a small angle or high energy approximation. This procedure leads to an extension of the domain of available data. A severe restriction to all data is the finite resolution of energy and angles, which cannot be avoided in any real experiment. Since most of the procedures which have been applied need, as main input information, the positions and the amplitudes of the inteference oscillations of the cross sections, they can always be performed if the resolution is good enough so that these two quantities are not affected by a large amount of averaging.

V. RESULTS

This section summarizes the kinds of data available on systems studied by molecular beam scattering and the potentials derived from these measurements. The emphasis will be on developments of the last three years. First diatomic systems are treated which interact by way of only one potential curve (V.A). The systems are classified by the precision in the determination of the potential (see Table III). Very accurate potentials are now obtainable

TABLE II

Comparison of inversion procedures[a]

Author	Quantity	Unknowns	Number of unknowns	Restrictions	Cross section formula	Mathematical method	Best suited for
Klingbeil	$\eta(l)$	All phase shifts	10–200	None	Partial wave sum	Minimalization	Quantum oscillations, small A
Remler	$S(l)$	Poles in the complex angular momentum plane	5–16	Only attractive part	Regge–Watson sum	Minimalization	Rainbows
Buck	$\Theta(b)$	Semiclassical quantities	5	Interference oscillations	Semiclassical	Minimalization	Rainbows, large B
Firsov	$\Theta(b)$	—	—	Monotonic cross section	Classical	Direct integration	Backward scattering at high energies

[a] The table contains only methods for the step from the differential cross section to the phase shifts, S-matrix or deflection function.

TABLE III
Precision of the potentials derived from scattering experiments

Category	Type of data	Method	Potential model	Examples[b]	Experimental data[c]
1	High resolution data	Inversion	None	A–Hg, Li–Ar	$I(\vartheta)$
			None	He–He	$Q(v)$
2	Combined data sets	Best fit	Multiparameter	R–R	$I(\vartheta)$, $B(T)$
		Best fit	Multiparameter	A–R	$Q(v)$, $I(\vartheta)$
		RKR[a]	None	H–Hg	$Q(v)$, $G(\omega)$
3	One data set	Best fit	3 parameter	H–R, A–M, H–M, H_2–M, R–M,	$Q(v)$
				A–A, A–M, R–M, H_2–M	$I(\vartheta)$

[a] RKR: inversion of spectra.

[b] A, alkali atom; R, rare gas atom; M, molecule.

[c] $I(\vartheta)$, differential cross section; $Q(v)$, total cross section; $B(T)$, second virial coefficient; $G(\omega)$, vibrational energies.

for the alkali–mercury systems which all have been studied by direct inversion techniques. Rare gas–rare gas potentials are in the next category (2). These potentials have been evaluated by multiparameter potential models described in Section IV where more than one group of data of different origin (also other than beam data) has been used in the analysis. The same category should contain the alkali–rare gas systems which are not treated here since the main developments go back to earlier years (see Schlier, 1969; Buck and Pauly, 1968; Düren et al., 1968). For recent findings see Auerbach et al. (1971), Ury and Wharton (1972), Klingbeil (1973), and Averbach (1974). Also the system H–Hg for which the first measurement of orbiting resonances have been reported belongs to this category, since spectroscopic data have been used for the determination of the potential. The next category (3) deals with potentials produced by the interaction of hydrogen atoms with rare gases. Here only one set of data has been used (up to now the total cross section) to determine a three parameter potential. Next, we proceed to diatomic systems which interact by more than one potential curve (V.B). Here we have new developments for alkali halides, metastable helium and metastable mercury systems. For most cases the evaluation of the data is in a preliminary stage so that accurate potentials have not yet been determined. The $^1\Sigma$ and $^3\Sigma$ potentials of the alkali–alkali systems belong to this section but will be left out in this article. For details see Schlier (1969) and the references cited therein, Pritchard et al. (1970) and Kanes et al. (1971). The precision of the potentials derived is of the third category. Finally, results for polyatomic systems are given (V.C). Most of these results are available for the scattering of certain atoms or molecules on other molecules. Thus we classify the experiments according to the colliding particle which is common for several systems. Apart from experiments with alkali atoms and molecules containing alkalis, results for rare gas atoms, hydrogen atoms and hydrogen molecules with a large number of other molecules are presented. We will restrict ourselves to the evaluation of the spherically averaged potential which usually results from elastic scattering with unorientated molecules (see Section III.C). The anisotropic part of the potential can be better deduced from inelastic scattering experiments (see Toennies, 1974) or experiments with state selected and thus oriented beams. This type of experiments will be covered in another article in this book by Reuss. The precision of the potentials obtained for all these systems is of the third category.

A. Diatomic Systems (One Potential)

1. Alkali–Mercury Systems

These systems have been carefully studied in a series of high resolution experiments at thermal energies. A velocity resolution of about 2 % (FWHM) and an angular resolution of about 0·3° provided the possibility to resolve nearly all of the interference structure of the differential cross section. Results have been obtained for NaHg (Buck and Pauly, 1971), KHg and CsHg (Buck *et al.*, 1972) and for LiHg (Buck *et al.*, 1974). Examples of the results are shown in Figs 7 and 9. In addition, measurements of the total cross section are available for these systems (Beck and Loesch, 1966; Rothe and Veneklasen, 1967; Buck *et al.*, 1971). Since all these systems display a pronounced rainbow structure with several supernumerary rainbows and superimposed rapid oscillations they are well suited for applying the direct inversion procedure of Buck (1971), as described in detail in Section IV. In this way the attractive well and the repulsive part of the potential could be determined in a unique manner. Measurements of the differential cross section for KHG at energies from 30 to 250 eV (Barwig *et al.*, 1973) allow the range of this potential to extend up to more than 200 eV. Once again the potential has been obtained by direct inversion (Barwig *et al.*, 1973; Buck, 1974). These potentials should be regarded as the best known ones derived from scattering processes. The results are shown in Table IV and Fig. 14.

TABLE IV

Parameters for alkali–mercury potentials determined by inversion

System	ε [meV]	r_m [Å]	Reference
LiHg	$108 \pm 3{\cdot}0$	$3{\cdot}00 \pm 0{\cdot}02$	Buck *et al.* (1974)
NaHg	$55 \pm 1{\cdot}6$	$4{\cdot}72 \pm 0{\cdot}03$	Buck and Pauly (1971)
KHg	$52 \pm 1{\cdot}8$	$4{\cdot}91 \pm 0{\cdot}03$	Buck *et al.* (1972)
CsHg	$50 \pm 1{\cdot}5$	$5{\cdot}09 \pm 0{\cdot}02$	Buck *et al.* (1972)

Table IV gives the size parameters whereas Fig. 14 displays the reduced potentials in order to compare the different shapes. The shape of the potentials of the three heavier systems agrees well over the whole attractive part of the potential, with only small differences in the repulsive part. Also the size parameters are nearly the same. LiHg shows a quite different behaviour. The shape as well as the size parameters differ appreciably from the other three systems. A less severe restriction on the potential form than the reduced potential hypothesis, which can be traced back to the law of corresponding states, is given by the similar potential hypothesis introduced by Stwalley (1971). It says that proportionality constants relating observables to potential parameters are roughly equal in chemically similar systems. But even this hypothesis is violated when comparing the LiHg data with the other systems (Buck *et al.*, 1974). A possible explanation for this behaviour is a change in the interacting forces. The heavier systems are mainly governed by van der Waals forces leading to a comparably large r_m value of about 5 Å. For LiHg another mechanism is responsible for the small r_m value of 3 Å, a distortion

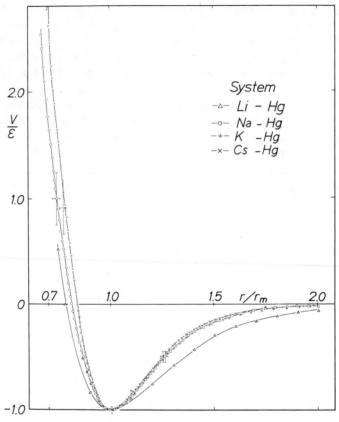

Fig. 14. Reduced form of alkali–mercury potentials derived by inversion.

of the $^2\Sigma^+$ van der Waals ground state by the strongly binding excited state of the same symmetry produced by the $Li(^2S_{1/2})$ and $Hg(^2P_1)$ states of the separated atoms (Herzberg, 1950).

It should be noted that the general shape of the potentials is quite different from a Lennard–Jones (LJ) (12–6) potential. The repulsive part is much softer whereas the minimum is wider and approaches zero more rapidly as r approaches infinity (see Fig. 8).

Another interesting point is the contribution of higher order terms to the asymptotic $V = -C_6/r^6$-behaviour. The potentials obtained by inversion show at their largest reduced separation $r/r_m \approx 2$ that, using the best theoretical estimate for C_6 (Stwalley and Kramer, 1968) one calculates that the $C_6 r^{-6}$ term accounts for only slightly more than half the total potential of NaHg, KHg and CsHg and nearly 80 % of the potential of LiHg at the same separation. Once again we notice a characteristic difference between LiHg and the other systems even at these large separations, which underlines the different character of the forces. There is great evidence to attribute the lacking part of the potential to a $-C_8 r^{-8}$ term, but any other attractive contribution cannot be excluded by the experimental findings alone. For a detailed discussion also of the earlier contributions to this problem see Schlier (1969) and Bernstein and La Budde (1973).

2. Rare Gas Systems

The interaction of two rare gas atoms is of special interest since for these systems data from many possible sources of information is available. Traditional sources are the gaseous bulk properties of the matter, the second virial coefficient $B(T)$ and the transport phenomena (particularly the viscosity) of the dilute gas. These only provide information on the potential in an integral form which is often deeply buried under several average processes. For a compilation of data see Dymond and Smith (1969) for $B(T)$ and Maitland and Smith (1972) for viscosities. However, in combination with spectroscopic data reliable potentials can be determined (Maitland and Smith, 1973a, b). More specific information can be extracted from data measured in the solid state. Lattice energy, the nearest neighbour distances, heat capacity, isothermal compressibility as well as elastic constants may be used to determine the potential. The main difficulty with this method is that one has to assume which part of the total energy is pairwise additive and which part comes from manybody interactions (Bobetic and Barker, 1970; Barker et al., 1970). On the other hand, if a very accurate two-body potential is known from other sources it should be possible to determine the contribution of the many-body forces. Information from theoretical calculations is available for the long range force constants (Starkschall and Gordon, 1972, and references cited therein) and short range interactions (Gordon and Kim, 1972) for most of the rare gas interactions. Ab initio calculations in the intermediate range including the van der Waal minimum are only available recently for the light systems (Bertoncini and Wahl, 1970; Schaefer et al., 1970). Besides the important results of u.v. absorption spectroscopy (Tanaka and Yoshino, 1970) molecular beam scattering experiments provide the only experimental method for a direct determination of the potential. Let us discuss the conditions for an accurate determination from the point of view of the experimental set up and of the information content of the data. As we have seen in Section III an accurate determination of the potential is only possible if at least one of the interference structures is resolved. For a unique determination all features have to be known. Thus the first measurements of total cross section (Düren et al., 1965) and low resolution differential cross section (Penta et al., 1967) proved not to be very profitable for the exact determination of the potential. A step forward was reached by improving the experimental techniques mainly by using supersonic nozzle beams and sophisticated pumping systems for the detection system (see Section 2). The higher intensities and narrower velocity distributions of the nozzle beams improved the resolution by a large amount and the first high resolution measurements of the differential cross section were reported (Cavallini *et al.*, 1970a; Cavallini *et al.*, 1970b; Siska *et al.*, 1970). An example of a recent high resolution measurement of the total cross section is given in Fig. 12. Measurements of the differential cross section for all rare gas dimers are displayed in

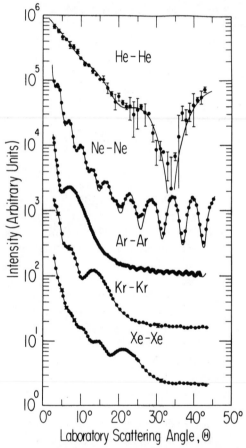

Fig. 15. Measured differential cross sections for rare gas pairs in the laboratory system (Farrar *et al.*, 1973a). The solid line is calculated from best fit potentials. The light systems mainly display symmetry oscillations whereas the heavier systems are governed by rainbow oscillations.

Fig. 15. They are obtained by Lee and coworkers (Farrar *et al.*, 1973a) by crossing two supersonic beams at various temperatures with velocity spreads of 8 % (FWHM) and angular spreads of 0·5° and 2°. The parameters B which determine the features of the differential cross section are listed in Table V. The systems with B smaller than 300 (see Figs 5 and 6) are not

TABLE V
Potentials for rare gas dimers

System	ε [meV]	r_m [Å]	Potential form	Experimental data[a]	Reference
He$_2$	0·948	2·963	ESMSV	$I(\vartheta), B(T)$	Farrar and Lee (1972)
$B = 8·2$	0·911	2·965	MLJV	$Q(v)$	Bennewitz *et al.* (1973)
	0·934	2·965	MLJV	$Q(v)$	Feltgen *et al.* (1974)
	1·03	2·963	None	CI	Schaefer *et al.* (1970)
	0·93	2·963	None	CI	Bertoncini and Wahl (1973)
	0·908	3·02	None	CI	Liu and McLean (1973)
Ne$_2$	3·62	3·102	ESMSV	$I(\vartheta), B(T)$, Sol	Farrar *et al.* (1973b)
$B = 175$	3·41	3·07	MV	$B(T), \eta(T), G(\omega)$	Maitland (1973)
	3·38	3·08	None	CI	Stevens *et al.* (1974)
Ar$_2$	12·2	3·761	Compound	Sol, $B(T)$, Liq	Barker *et al.* (1971)
$B = 1880$	12·3	3·76	Compound	$B(T), G(\omega), \eta(T)$	Maitland and Smith (1971)
	12·1	3·76	MSV	$I(\vartheta), B(T)$	Parson *et al.* (1972)
Kr$_2$	17·2	4·02	LJ$(n, \gamma, 6)$	$B(T), G(\omega), \eta(T)$	Gough *et al.* (1973)
$B = 5890$	17·4	4·006	Compound	Sol, $B(T), \eta(T)$ $G(\omega), I(\vartheta)$	Barker *et al.* (1974)
	17·2	4·03	MSV	$I(\vartheta), B(T)$	Buck *et al.* (1973a)
Xe$_2$	24·2	4·361	Compound	Sol, $B(T), \eta(T)$, $G(\omega), I(\vartheta)$	Barker *et al.* (1974)
$B = 15196$	23·0	4·42	LJ$(n, \gamma, 6)$	$B(T), G(\omega), \eta(T)$	Maitland and Smith (1973b)

[a] $I(\vartheta)$: differential cross section; $Q(v)$: total cross section; $B(T)$: virial coefficient; $\eta(T)$: viscosity; $G(\omega)$: spectroscopy; Sol: solid state data; Liq: liquid state data; CI: configuration interaction calculations.

expected to show a pronounced rainbow structure. They are mainly sensitive to the repulsive part of the potential. Because of the identity of the particles we would expect symmetry oscillations. As B increases from He$_2$ to Xe$_2$ (and therefore A) the angular spacing and the amplitude of the identical particle oscillations decreases and finally disappears because of the finite resolution of the apparatus whereas the rainbow structure appears in the cross sections. Ne$_2$ is the only system where both rapid oscillation and symmetry oscillations are resolved, whereas Ar$_2$ shows both a rainbow and symmetry oscillations.

Thus we conclude that the light systems mainly determine the repulsive wall of the potential whereas the heavy systems are mainly sensitive to the attractive well. We shall discuss the results for these systems in order of decreasing experimental material.

He_2: This system has probably been studied more than any other by molecular beam techniques. The measurements cover a range of more than four orders of magnitude. At very high energies measurements of the incomplete total cross section have been performed by Amdur and coworkers (Amdur et al., 1961; Jordan and Amdur, 1967). The next region is explored by a measurement of the total cross section in the energy range between 10 meV and 2 eV of Gengenbach et al. (1973a). Due to the temperature of the scattering chamber of 135°K no symmetry oscillations could be resolved. High resolution measurements of the total cross section are reported by Cantini et al. (1972), Bennewitz et al. (1972a) and Feltgen et al. (1973, 1974). The results of the last group for $^4He^4He$, $^3He^3He$ and $^4He^3He$ are shown in Fig. 12. They cover the energy range of 0·2 meV to 200 meV and are sensitive to the intermediate range of the potential as well as to the region of the well. The two other measurements fall inside this range. The results of Cantini et al. (1972) are more sensitive to the intermediate part whereas the results of Bennewitz et al. (1972a) are more sensitive to the low energy part. In addition, differential cross sections have been measured at energies of 60 meV, 23 meV and 5 meV by Siska et al. (1971) and Farrar and Lee (1972). The potentials derived from these measurements are shown in Fig. 13, except that of Cantini et al. (1972) which coincides with the other curves. The sensitivity of all these measurements is best demonstrated by the points derived by inversion (see Section IV). Therefore this region can be regarded as best explored. The region of the well is better probed by the measurement of the total than of the differential cross section since the energy reached is much lower. However, the results suffer from the drawback that at these low-energies only a very few partial waves contribute to the cross section and the complicated shape of well cannot be determined with sufficient accuracy. In addition the average effect of the target gas even at these very low temperatures ($T = $ 1·57°K), buries part of the structure. Nevertheless, the $^4He^4He$ potential is very near to the true one. The best values for the minimum parameters are given in Table V together with the results of ab initio calculations. The first value is based on configuration interaction (CI) calculations where only inter-atomic correlations have been used (Bertoncini and Wahl, 1970; Schaefer et al., 1970). The next value results from a calculation where intra-atomic correlations have been included (Bertoncini and Wahl, 1973). Finally, the most recent result (Liu and McLean, 1973) is listed which also accounts for the coupling between these correlations (which up to now have been estimated to be very small). Despite these discrepancies the comparison between the He-potential derived from molecular beam data and ab initio calculation gives excellent agreement in the well explored intermediate range as displayed in Fig. 16. At high energies the CI calculation of Phillipson (1962) and the results from the Statistical Thomas–Fermi–Dirac (TFD) model of Abrahamson (1963) differs from the molecular beam results. The

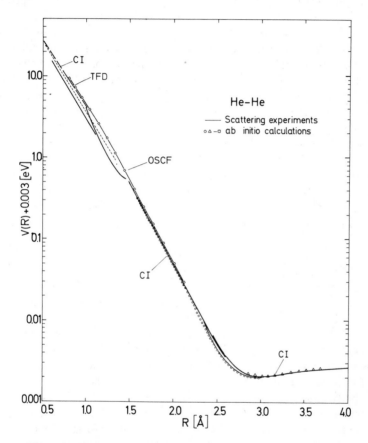

Fig. 16. Comparison of the He$_2$ potentials obtained from molecular beam scattering data (solid line, see Fig. 13) and *ab initio* calculations: dashed line, CI (Phillipson, 1962); dotted line, TFD (Abrahamson, 1963); line interrupted by quadrangles, OSCF (Gilbert and Wahl, 1967); circles, CI (Schaefer *et al.*, 1970); triangles CI (Bertoncini and Wahl, 1970).

optimized self-consistent field (OSCF) method of Gilbert and Wahl (1967) gives good results down to 0·05 eV. In this region the CI calculation of Schaefer *et al.* (1970) is also in good agreement with the experimental findings down to the well region. It should be noted that all these results rely on data for $^4He^4He$ measurements. Recent reports on differences between this potential and the $^3He^3He$ potential (Bennewitz et al., 1972b) are open to discussion. Up to now it is not clear if it is an artifact of the data evaluation method (Bennewitz et al., 1973) or a real discrepancy which may be traced back to mass polarization effects (Mittleman and Tai, 1973). It is to be hoped that the further evaluation of the data of Fig. 12 will clarify the situation.

Ar_2 and Kr_2: As mentioned above these systems are of particular interest due to the numerous experimental data on gaseous, liquid and solid bulk properties. High resolution molecular beam measurements are available for the differential (Cavallini et al., 1970b; Cavallini et al., 1971b; Parson et al., 1972; Schafer et al., 1971) and the total cross section (Bredewout, 1973) which show a well resolved rainbow structure or glory undulations, respectively (see Fig. 15). The differential cross sections of Ar_2 have been analysed in terms of a multiparameter potential described in Section IV (Parson et al., 1972). The resulting potential is in good agreement with second virial coefficients and spectroscopic data. Two other very reliable potentials are available for Ar_2. One is derived from spectroscopic, equilibrium and transport data (Maitland and Smith, 1971), the other from a careful analysis of equilibrium data of gaseous and solid (Bobetic and Barker, 1970) and liquid (Barker et al., 1971) argon. Especially the last one proved to give an excellent description of several properties of the solid state (Barker et al., 1972; Klein et al., 1973). The total cross section calculated from this potential is also in agreement with the measurements (Bredewout, 1973). A comparison of the size parameters of the Ar_2 potentials is shown in Table V. Also the shape of these potentials is very similar (see Maitland and Smith, 1973a). That means that nearly all measurable quantities can be explained by a single potential function. The crystal properties were reproduced under the assumption that the Axilrod–Teller–Muto (ATM) forces which account only for an attractive triple-dipole term were the only contributions from many-body forces.

A similar situation exists for Kr_2 although a smaller number of experimental properties has been studied. Reliable potentials from other than beam sources have been derived by Barker et al. (1972) using an approach similar to that for argon based on a careful analysis of the properties of solid krypton and by Gough et al. (1974) based on equilibrium and transport properties of the gas as well as spectroscopic results. Potentials which are based on differential cross section measurements and virial coefficients have been obtained by Docken and Schafer (1973) and Buck et al. (1973a). The potential of Buck et al. (1973a) is in agreement with spectroscopic data and several solid state data. Again, the result is consistent with the hypothesis, that the ATM force is the only one important for the lattice energy, and that at least 50% of it contributes to the pressure and possibly slightly less to the bulk modulus. The potential of Docken and Schafer (1973) proved inadequate when applied to solid state data (Buck *et al.*, 1973a), whereas the potential of Barker *et al.* (1972) was not consistent with differential cross sections and a recent calculation of the lattice constant of solid Kr at high temperature (Klein *et al.*, 1973). The size parameters of these potentials are also listed in Table V. It should be noted that preliminary calculations of the total cross section with these potentials show discrepancies with the measurements (Bredewout, 1973). It remains to be seen whether such discrepancies are due to the analysis of the experimental data or the potential model. Again, the analysis is very near to the true potential which is able to describe all measurable quantities. A new potential derived by Barker *et al.* (1974) on the basis of the virial coefficients, gas transport properties, solid state data, spectroscopic information on dimers and differential scattering cross sections confirms the results of Buck *et al.* (1973) and Gough *et al.* (1974).

Ne_2 *and* Xe_2 : Considerable progress has also been made in the analysis of the other two rare gas dimers. A potential for Ne_2 based on differential cross section data, second virial coefficients and solid state properties has been derived by Farrar *et al.* (1973b). This is compared in Table V with a potential given by Maitland (1973) working on the same line as for Ar and Kr. A very recent *ab initio* calculation of Stevens *et al.* (1974) is in excellent agreement with the size parameters of the other two potentials. For Xe_2 reliable potentials are published by Farrar *et al.* (1973a) and Maitland and Smith (1973b). A recent determination of this potential by Barker *et al.* (1974) using the same input information as for Kr_2 is in agreement with nearly all measured quantities available for this system.

Discussion: We conclude that considerable advancement has been made in the understanding of the intermolecular potential of the rare gas pairs. Both the shape and the size parameters appear to be defined to within narrow limits. This development has been made possible by using the results of quite different fields which often provide complementary information for the different parts of the potential function. Specifically, bulk properties of dilute gases give mainly integral information on the potential. The area of the potential is probed by the vibrational levels of the molecule obtained from spectroscopy. Properties of the solid at low temperatures are mainly sensitive to the attractive well and to the minimum distance r_m whereas specific scattering features determine very accurately specific parts of the potential, e.g. the rainbow structure determines the region near the inflection point (see Section III).

It should be noted that despite the similarity between the different estimates, there is still no perfect agreement with all experimental data. The total cross section, the low energy virial coefficient, the elastic constant and the Debye-temperature of the solid are examples of such deviations. Furthermore the question of the three-body forces is not solved completely. The fact that contributions of higher-order non-additive multipole forces were found to be negligibly small may be due to a fortuitous cancellation, so that general conclusions cannot be drawn with the present accuracy of the experimental results.

Nevertheless the rare gas pair potentials are sufficiently well established that a comparison of the reduced form is possible. Fig. 17 displays the results

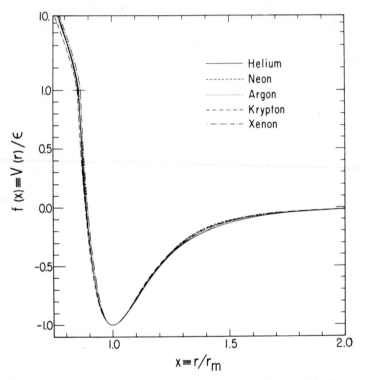

Fig. 17. Reduced form of the symmetric rare gas pair potentials (Farrar *et al.*, 1973a).

where the potential model of Farrar *et al.* (1973a) is used. The shape is the same to within small deviations for Ne_2 which shows the steepest attractive outer well and for Xe_2 which has the steepest repulsion. Lee and coworkers have not been able to find a single reduced potential curve which could fit all data available. Similar results are obtained by Maitland and Smith (1973b) when comparing their potential forms. Thus we may conclude that the shape is the same for the Ar_2, Kr_2 systems whereas small but significant deviations occur for the Ne_2 and the He_2 systems as well as for Xe_2.

Another interesting point is the behaviour of the unlike rare gas potentials and their relation to the results of the like molecules. An analysis which takes into account data of more than one specific origin has only been performed for He on Ne, Ar, Kr, Xe by Chen et al. (1973). Differential cross sections which show well resolved quantum oscillations and which are mainly sensitive to the repulsive wall and outer attractive well are used along with the second virial coefficients. The results are striking. The shapes are quite different from each other and from that of HeHe. The well depth of HeAr, HeKr and HeXe were found to be essentially identical as was the case for NeAr, NeKr, NeXe (Parson et al., 1970a; Helbing et al., 1968, Ng et al. 1974). A similar phenomenon has also been observed in the rare gas–alkali atom systems (Buck and Pauly, 1968). For systems with the same rare gas atom and different alkali atoms, the well depths are nearly the same, but there is substantial difference in the well depths for systems with the same alkali atom and different rare gas atoms. The situation for r_m is just reversed. It seems that for asymmetric systems the less polarizable atom plays a more important role in determining the well depth, whereas the reverse statement is valid for r_m. Table VI shows that this rule is fulfilled at best for ε and the three heavier systems Ar, Kr, Xe. The similar well depths in these cases also suggest that the commonly used geometric combining rule for the estimation of unlike pairs is unreliable. This question is discussed in detail using other combination rules by Chen et al. (1973). In summary, the likelihood of the existence of a simple universally valid combination rule is highly questionable, and the law of corresponding states can only be considered as an approximation for very similar systems.

3. Interactions with H-Atoms

Despite the fact that the H-atoms is the simplest of atoms, its interaction potentials with other atoms are largely unknown. Molecular beam experiments are not easily performed due to the difficulty of producing intensive beams of this labile atom. Notwithstanding these difficulties, measurements on the velocity dependence of the total cross section for H on Ar, Kr and Xe (Aquilanti et al., 1972b), for H on He (Gengenbach et al., 1973b), for H on He, Ne, Ar, Kr and Xe (Bickes et al., 1973a), and for H on Hg (Schutte et al., 1972) have recently been reported. Very recently high resolution differential cross sections for H on Ar and Kr have also been measured (Bassi et al., 1974). The last two experiments have been performed with a low temperature bolometer detector which proved to be very sensitive to atomic hydrogen (see Section II). In addition, the differential cross section experiment has been carried out with a magnetic hexapolar magnet for both velocity selection and focusing of the atoms. Up to now only the total cross section measurements have been analysed in terms of an interacting potential. The measured cross section for HHg is displayed in Fig. 18. A velocity selected (FWHM 3·5% and 8·5%) primary beam produced by a RF discharge source is crossed with a multichannel secondary beam of Hg which is inclined at about 73° relative

TABLE VI

The potential parameters ε [meV] and r_m [Å] for asymmetric rare gas systems

Systems	Ne		Ar		Kr		Xe		Potential	Reference
	ε	r_m	ε	r_m	ε	r_m	ε	r_m		
He	1·23	3·21	2·09	3·54	2·13	3·75	2·17	4·15	ESV	Chen et al. (1973)
Ne	—	—	6·20	3·48	6·74	3·58	6·46	3·75	ESMSV	Ng et al. (1974)
Ar	—	—	—		14·9	3·80	16·4	4·10	Mod. LJ	Parson et al. (1970b)

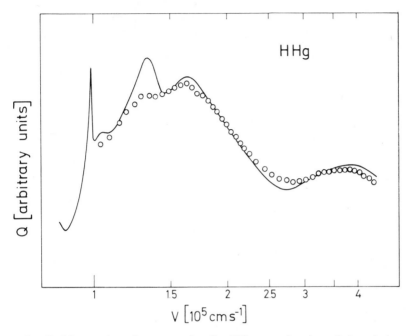

Fig. 18. Measured total cross section for HHg as a function of the relative velocity (Schutte *et al.*, 1972). The solid line is based on a calculation without accounting for the finite experimental resolution. Glory structure and orbiting resonances show up.

to the primary beam in order to improve the velocity resolution. Besides the glory structure two small bumps show up in the cross section which are attributed to orbiting resonances. This conclusion is confirmed by a calculation using an interaction potential which Stwalley (1972) obtained on combining theoretical C_6 and C_8 dispersion coefficients with an isotopically combined RKR-curve. Thus the observed structure at 1400 m/s is due to either the ($v = 4$, $l = 9$) or ($v = 3$, $l = 10$) quasibound level of HgH or a combination of the two. The solid line in Fig. 18 shows this calculation which, however, does not account for the finite velocity and angular resolution of the apparatus. A recent evaluation of the same data which accounts for this averaging effect yields an excellent fit when, in addition, the potential is modified slightly (Schutte *et al.*, 1975).

The study of the H-rare gas interaction potentials is mainly based on measurements of the total cross section. Fig. 19 shows results obtained by Bickes *et al.* (1973a). Because of the small *B*-values ($B = 2$–50) and the velocity range of the H-beam source we are in the transition region where the information content on the potential is small. Furthermore, a cancellation of attractive and repulsive forces makes it rather difficult to ascertain which

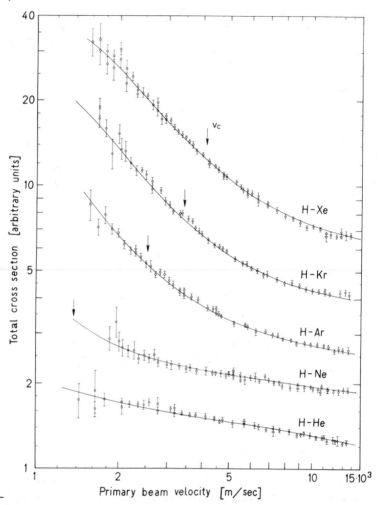

Fig. 19. Measured total cross sections as a function of the primary beam velocity for the H–rare gas systems (Bickes *et al.*, 1973a). The solid line is the best fit calculated curve for a LJ 12–6 potential. The velocity v_c marks the transition between attractive and repulsive forces.

part of the potential is really probed by which part of the measurement. At the present stage of the experiments it is only possible to determine ε and r_m for an assumed potential form, even if the absolute cross section is taken into account (Bickes *et al.*, 1973a). The potential parameters derived in this way for an exp-6 potential are given in Table VII. Unfortunately they do not agree with those deduced from other experiments (Aquilanti *et al.*, 1972b). However, a recent *ab initio* MCSCF-calculation for HAr by Das *et al.* (1974)

TABLE VII
Potential parameters for hydrogen atom–rare gas systems

System	ε [meV]	r_m [Å]	Potential form	Experimental data[a]	Reference
HHe	0·46	3·60	ESV	$Q(v)$	b, c
	0·46	3·72	None	MCSCF	e
HNe	2·82	3·18	exp $(\alpha, 6)$	$Q(v)$	f
HAr	4·80	3·56	exp $(\alpha, 6)$	$Q(v)$	f
	4·16	3·57	None	MCSCF	g
HKr	6·08	3·70	exp $(\alpha, 6)$	$Q(v)$	f
HXe	6·81	3·95	exp $(\alpha, 6)$	$Q(v)$	f

[a] MCSCF: multiconfiguration selfconsistent field calculation. $Q(v)$: total cross section.
[b] Gengenbach et al. (1973b).
[c] Gengenbach et al. (1973c).
[e] Das and Wahl (1971).
[f] Bickes et al. (1973a).
[g] Das et al. (1974).

favours the results of Table VII although they do not quite agree. The remaining discrepancies can be traced back to a deviation of the absolute values of the total cross section by 15% which lies within the error bound that can reasonably be associated with the evaluation of the data and with the calculations. Similar considerations hold for HHe. A careful analysis of all data available gives reasonable parameters for the repulsive wall and r_m, but cannot decide between ε-values varying between $\varepsilon = 0.39$ and 0.96 meV (Gengenbach et al., 1973b). An additional measurement of the total cross section down to 400 m/s (Gengenbach et al., 1973c) shows the Ramsauer–Townsend effect from which the value $\varepsilon = 0.46$ meV could be determined, which is in agreement with the theoretical calculation of Das and Wahl (1971). It is obvious that the analysis of the differential cross section would improve this situation. Recent measurements of orbiting resonances for HXe will also yield better information on the inner attractive wall of the potential (Welz et al., 1974). Notwithstanding the difficulties which prove to be an inherent part of the weak interaction the first promising steps to solve this problem have been done.

B. Diatomic Systems (more than one potential)

The interaction by way of more than one potential always occurs if the interacting atoms are not in S-states and if the total angular momentum J and m_J are not selected. This happens with atoms in the ground state or excited states. New results from molecular beam measurements of the elastic scattering for such species are available for $O(^3P)$ on Ar (Aquilanti *et al.*, 1973), for K on *I* (Fluendy *et al.*, 1970; Hack *et al.*, 1971; Kaufmann *et al.*, 1974), for Hg*($^3P_2$) on K, Na and Rb (Davidson *et al.*, 1973), for He*($^{2\,3}S$) on He (Morgenstern *et al.*, 1973), and for He*(2^1S) on He and Ar*($^3P_2$) on Ar (Haberland *et al.*, 1973b). In all cases oscillations were observed either in the total [O(^{3}P)–Ar] or the differential cross section (all other cases). Although the evaluation of the data is in a preliminary stage for most of these examples at the present time we shall give a short outline of the results for the alkali halides (V.B.1) and the metastable species (V.B.2).

1. Alkali Halides

The alkali halide molecules are well known examples for the crossing of an "ionic" potential showing a deep minimum with a "covalent" potential of the neutral species with a very shallow well. The scattering experiments have been performed in order to get insight in the problem whether the neutral particles follow the diabatic (covalent) curve (strong coupling) or the adiabatic (ionic) curve (weak coupling). In both cases the elastic channel, the neutral species, is observed after the collision. The measurements of Fluendy *et al.* (1970) for K($^2S_{1/2}$) on I($^2P_{3/2}$) show an oscillation pattern at very small angles which was attributed to an edge effect of the crossing of the steep ionic potential and the very flat covalent potential, thus indicating an adiabatic behaviour. The other two experiments show several peaks at larger angles which were attributed to the rainbow scattering of the diabatic potentials alone. No perturbation from the adiabatic potentials could be detected. Kaufmann *et al.* (1974) synthesized the measured cross section by an assumption of four different interacting potentials with different weights, thus accounting for several neutral potentials because of the coupling of the angular momentum due to Hund's case (c). The well depths of the potentials are $\varepsilon = 18\cdot5$, $16\cdot0$, $10\cdot5$ and $3\cdot2$ meV whereas r_m varies from $3\cdot85$ to $4\cdot82$ Å. Similar numbers are estimated by Hack *et al.* (1971) and Rosenkranz (1970) by qualitative arguments, yielding $\varepsilon = 28$ to $1\cdot9$ meV for the covalent potentials. This result is supported by molecular beam measurements looking at the ionic channel (Moutinho *et al.*, 1971) where only a small amount of ions is found for KI. It should be noted that any adiabatic behaviour would most likely occur at large impact parameters which are not probed by rainbow scattering. A recent measurement of the differential cross section of spin selected K on unpolarized I confirms the diabatic behaviour (Carter and Pritchard, 1974). However, the results strongly indicate a coupling scheme for the angular momentum which only leads to two potentials. The positions of the minima of these potentials were estimated from rapid oscillations to be near $5\cdot8$ Å. For a high resolution experiment on the differential cross section for Na$^+$ produced by NaI scattering see Delvigne and Los (1973).

2. Metastable Atoms

Differential cross section experiments have been performed with $Hg^*(^3P_2)$ beams scattered from Na, K, Rb (Davidson *et al.*, 1973). The metastables are produced by electron bombardment. The two other species which are excited decay ($^3P_1$) or are assumed to be populated with much less probability ($^2P_0$). Five different potentials occur from the combination of the two atoms: $^4\Pi_{5/2}$, $^4\Pi_{3/2}$, $^4\Pi_{1/2}$, $^4\Sigma_{3/2}$, $^4\Sigma_{1/2}$. Neither the deep lying $^2\Sigma_{1/2}$ nor the excited $^2\Sigma^*_{1/2}$ state will correlate with the $^3P_2$-state (see Section V.A). The apparent interference structure of the measured cross section over the whole angular range leads to the following conclusions. (1) Either a single potential or a group of potentials that are very similar are responsible for the structure. (2) At least two branches of the deflection function are present. Thus, quenching cannot affect the structure to a large amount. A detailed analysis will give more information on the potentials.

Several potentials are also involved when metastable He-atoms are scattered from He. This case is shown in Fig. 20, following Morgenstern *et al.*

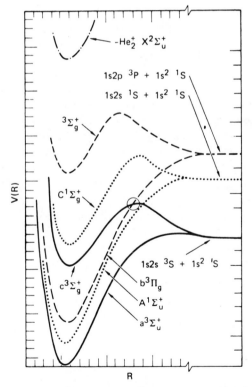

Fig. 20. Schematic potential curves for He_2 (Morgenstern *et al.*, 1973).

(1973). Due to the symmetry of the wavefunctions the potentials between the metastable and the ground state are split into gerade and ungerade curves for both the He*(2^3S) + He(1^1S) and He*(2^1S) + He(1^1S). The He$_2$(A $^1\Sigma_u^+$) curve is known from spectroscopy to have a well depth of 2·5 eV. Differential cross section experiments with metastable He-atoms can give information on the deep lying minima of the potentials involved only if the collision energy is high enough to overcome the barriers. Otherwise the interaction at large separations is sampled. Two different experiments have been reported. Haberland et al. (1973b) have measured the differential cross section for He*(2^1S) on He in the thermal energy range. Thus the long range parts of the $^1\Sigma_g^+$ and $^1\Sigma_u^+$ potentials are probed by the measurement. The metastable He beam was produced by electron excitation. The measured cross section shows a peak at 180° due to exchange and symmetry oscillations. Irregularly superimposed oscillations and a general fall-off of the cross section was attributed to orbiting contributions of the $^1\Sigma_u$-state. Thus the barrier height in the $^1\Sigma_u$ potential should be equal to or smaller than the collision energy of 63 meV.

Morgenstern et al. (1973) report measurements of the differential cross section of the metastable triplet state He*(2^3S) in the energy range between 5 and 10 eV. The metastable atoms are produced by charge exchange of He$^+$ ions with Cs-atoms. Therefore one should expect three different rainbow peaks, one originating from the well of the $^3\Sigma_u$ potential and two from the well and the barrier of the $^3\Sigma_g$-state. Fig. 21 displays the results. The reduced differential cross section $\rho = I(\vartheta)\vartheta \sin \vartheta$ is plotted versus τ. Only one of these peaks is clearly to be seen at 160 eV-deg which could be attributed to the minimum of the $^3\Sigma_u$-state ($\varepsilon = 2\cdot00$ eV). Although a lot of structure was observed between $\tau = 40$ and $\tau = 190$ eV-deg it could not be attributed to the rainbow peaks mentioned. A strong interaction between the $^3\Sigma_g^+$ and the $^3\Pi_g$-state originating from He(2^3P) and He(1^1S) proved to be the best explanation. The large peak growing at the higher energies at $\tau = 90$ eV-deg as well as the decrease in the magnitude of the main rainbow at $\tau = 160$ eV-deg was predicted by a corresponding three-state calculation. These examples show the great variety and the possibilities of atom–atom interactions which, in general, are not as easily interpreted as the examples of Section V.A.

C. Polyatomic Systems

1. Alkali–Molecule Systems

A considerable amount of data has been accumulated for alkali atom–molecule collisions. The velocity dependence of the total cross section has been measured by Kramer and LeBreton (1967), Helbing and Rothe (1968) and Rothe and Helbing (1969, 1970a, 1970b). For recent results along these lines see Düren et al. (1972) and Hermann et al. (1973). The quenching of the

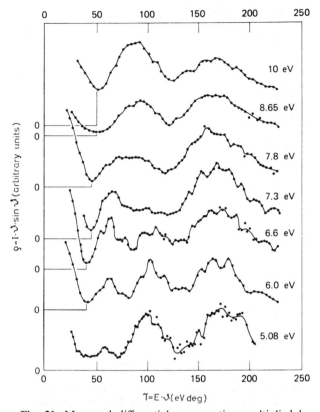

Fig. 21. Measured differential cross section multiplied by $\vartheta . \sin \vartheta$ for the scattering of He*(2^3S) by He(1^1S) in the centre-of-mass system (Morgenstern *et al.*, 1973) as a function of the reduced angle $\tau = E\vartheta$. The rainbow structure at $\tau = 160 \, eV \, deg$ is the only one not disturbed by a potential crossing.

glory undulations has been found for many systems. Specifically, systems which contain chemically reactive molecules and asymmetric fluorocarbons or hydrocarbons are severely affected whereas symmetrical tops and tri-atomic molecules have only been quenched partially. Diatomic systems and spherical tops exhibit no or weak quenching. Measurements of the differential cross sections have been reviewed by Ross and Greene (1970). For recent publications see Harris and Wilson (1971) and Sloan *et al.* (1972). The dif-ficulties in interpreting the data have been discussed in detail in Section III.C. For several cases anisotropy parameters and opacity functions could be extracted but, in general, only ε and r_m have been determined for an assumed spherically symmetric model potential since the main source for an exact determination of the potential, the oscillations, is lacking. An example where,

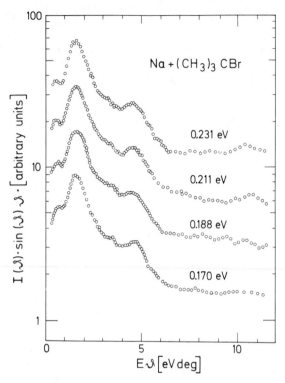

Fig. 22. Measured differential cross section multiplied
by $\vartheta \cdot \sin \vartheta$ for Na-$(CH_3)_3$CBr in the centre-of-mass
system (Buck *et al.*, 1975). Two well resolved rainbow
structures are displayed.

on the contrary, an additional structure appears in the cross section is
displayed in Fig. 22. A high resolution measurement of the elastic differential
cross section for sodium atoms on the symmetric top molecule $(CH_3)_3$CBr is
plotted versus the reduced angle $\tau = E \cdot \vartheta$ (Buck *et al.*, 1975). Apart from a
well resolved rainbow structure at $\tau = 1.7$ eV . deg, a second structure
appears with primary and secondary maximum. The smith plot (Smith *et al.*,
1966) also exhibits for the second structure the same energy dependence as is
valid for rainbow scattering. Similar results have been obtained for
$(CH_3)_3$CCl and $(CH_3)_3$CI, whereas the cross sections of Na on the spherical
tops $(CH_3)_4$C, CBr$_4$ and CCl$_4$ do not display any secondary structure. Thus
it is obvious to attribute these features to a double rainbow as predicted for
molecules scattering with large anisotropic potentials (Harris and Wilson,
1971). Unfortunately the general fall-off the cross section does not agree
with the results of their simple calculation valid in the first order HEA which
only account for a P_2-term (see Equation 28). Extensive calculations where

WKB-phase shifts are used for the isotropic part of the potential and phase shifts in the sudden limit for the anisotropic part (Cross, 1967) produced cross sections which are also in quantitative agreement with the experimental results (Buck et al., 1975). It proved necessary to introduce a large P_1-contribution to the potential in order to get this agreement for the scattering of symmetrical top molecules on atoms. Thus this type of measurements seems to provide a reliable method for the determination of the anisotropic part of the potential.

Only a few elastic scattering experiments of molecules which contain an alkali atom have been performed. Recently published results on the scattering of alkali dimers on molecules show that the rainbow structure in the differential cross section is only slightly less resolved than the structure for the monomers, thus indicating a weak anisotropy of the long range potential (Hardin and Grice, 1973). For a discussion of total cross section measurements of alkali halides on rare gas atoms with a Boltzmann rotational state distribution and state selected beams see David et al. (1973).

2. Rare Gas–Molecule Systems

The velocity dependence of the total cross section of rare gas atoms scattered from several molecules has been measured by Helbing et al. (1968) and Butz et al. (1971a) (Ne, Ar, Kr, Xe–H_2), by Butz et al. (1971b) (He–H_2, –N_2, –O_2, –NO, –CO, –CH_4 and CO_2), and by Aquilanti et al. (1973) (Ar–O_2). Furthermore the system HeH_2 has attracted most of the interest (Butz et al., 1971a; Gengenbach and Hahn, 1972; Lilenfeld et al., 1972). Apart from HeH_2 all the other systems display a well resolved glory structure. A quenching of the glory amplitudes has only been observed for HeCO$_2$. Differential cross sections have been reported for ArN$_2$ (Anlauf et al., 1971; Cavallini et al., 1971a; Kalos and Grosser, 1972), for Ar and Kr scattered from N_2 and O_2 (Tully and Lee, 1972) and, very recently, for the systems Ar on CO_2 and N_2O (Farrar et al., 1973c). Preliminary results have also been obtained for Ar and Ne scattered from H_2O (Bickes et al., 1973c). Most of these systems show a rainbow structure for which the degree of quenching increases from the diatomic molecules O_2 and N_2 to the triatomics N_2O and CO_2. The last systems have also been studied by velocity analysis where an appreciable amount of rotational excitation has been found. These observations together with the quenching of the rainbow amplitudes make it difficult to extract reliable potentials from such measurements. Anisotropic and inelastic effects do not play such an important role in interpreting the total cross sections. The glory undulations, however, do not determine the potential in a unique way. They are mainly sensitive to the product εr_m while the rainbow scattering should give reliable ε values. Hence, the potential parameters given in Table VIII should be regarded with these caveats. The system of HeH_2 has been examined in some details where theoretical

TABLE VIII
Potential parameters for rare gas–molecule systems[i]

System	ε [meV]	r_m [Å]	Potential form	Data	Reference
He–H$_2$	1·341	3·38	MSV	$Q(v)$	a
Ne–H$_2$	3·00	3.30	LJ 12–6	$Q(v)$	b, c
Ar–H$_2$	6.25	3·33	LJ 12–6	$Q(v)$	b, c
Kr–H$_2$	6·80	3.65	LJ 12–6	$Q(v)$	b, c
Xe–H$_2$	7.43	3·90	LJ 12–6	$Q(v)$	b, c
He–N$_2$	2·37	3.5	LJ 12–6	$Q(v)$	d
Ar–N$_2$	10·2–11·9	3.9	LJ 20–6	$I(\vartheta)$	e, f, g, h
Kr–N$_2$	13·4	3.9	LJ 20–6	$I(\vartheta)$	h
He–O$_2$	2·68	3·4	LJ 12–6	$Q(v)$	d
Ar–O$_2$	12·6	3·9	LJ 20–6	$I(\vartheta)$	h
Kr–O$_2$	14·2	4·05	LJ 20–6	$I(\vartheta)$	h
He–NO	2·37	3·70	LJ 12–6	$Q(v)$	d
He–CO	2·37	3·50	LJ 12–6	$Q(v)$	d
He–CH$_4$	2·68	3·35	LJ 12–6	$Q(v)$	d
He–CO$_2$	3·55	3·50	LJ 12–6	$Q(v)$	d

[a] Gengenbach and Hahn (1972).
[b] Helbing et al. (1968).
[c] Butz et al. (1971a).
[d] Butz et al. (1971b).
[e] Anlauf et al. (1971).
[f] Cavallini et al. (1971a).
[g] Kalos and Grosser (1972).
[h] Tully and Lee (1972).
[i] Recommended values; for an explanation of the data see Table V.

calculations of the short and long range part of the potential have been included in the analysis (Gengenbach and Hahn, 1972; Lilenfeld et al., 1972). Theoretical calculations yield for the spherically averaged potential $\varepsilon = 1\cdot4$ meV and $r_m = 3\cdot32$ Å (Tsapline and Kutzelnigg, 1973) which is in agreement with the potential of Gengenbach and Hahn (1972). This potential has also been used as the spherically averaged part for fitting measurements of spin lattice relaxation, rotational relaxation and Raman line shapes (Shafer and Gordon, 1973). Recent measurements of the differential cross sections of Ne, Ar and Kr on H_2 which show well resolved quantum oscillations will hopefully remove some of the uncertainties of these potentials (Bickes et al., 1974).

3. Hydrogen–Molecule Systems

Similar conclusions as for rare gas–molecule systems can be drawn when the scattering of hydrogen atoms and molecules from other molecules is considered. Hydrogen atom beams are difficult to produce. In addition the B-values are appreciably small so that measurements of the total cross section do not display any oscillation pattern: H on H_2 (Gengenbach et al., 1974a), and H on H_2, D_2, N_2, O_2, CO, NO, CO_2, C_2H_2 (Bickes et al., 1974). Differential cross sections for DH_2 which show well resolved quantum oscillations have been measured by Bassi et al. (1973). An analysis of the total cross section measurements for $H-H_2$ where also previous measurements (Stwalley et al., 1969) have been included has been presented by Gengenbach et al. (1974a). They fitted an ESV-potential to the data where the theoretical dispersion coefficients were used as additional information. The results are listed in Table IX. The evaluation of the other data is in progress.

TABLE IX
Potential parameters for hydrogen interactions

Systems	ε [meV]	r_m [Å]	Potential	Data	Reference
H_2-H_2	3·00	3·49	MSV	$I(\vartheta)$, p-H_2	Farrar and Lee (1973)
	3·12	3·30	LJ 12–6	$Q(v)$, n-D_2(HD)	Butz et al. (1971b)
	2·93	3·45	Compound	$I(\vartheta)$, n-H_2	Dondi et al. (1973)
	3·10	3·42	MSV	$Q(v)$, n-H_2	Gengenbach et al. (1974)
	3·05	3·50	exp–6	$Q(v)$, p-H_2	Bauer et al. (1975a)
$H-H_2$	2·32	3·60	MSV	$Q(v)$	Gengenbach et al. (1975)

Much more data have been accumulated for hydrogen molecule–molecule systems. Again the potential capacity B is low (30 to 120), but now the measurements of the total cross section of D_2(HD) on N_2, O_2, NO, CO, CO_2, CH_4 (Butz et al., 1971b) and of D_2 on N_2, CH_4, C_2H_6, C_3H_4, C_2H_2 (Aquilanti et al., 1971) display, in general, the first glory oscillation. Differential cross sections have been reported for D_2-N_2 (Winicur et al., 1970), for $H_2(D_2)-O_2$, CO, CH_4, SF_6, NH_3 (Kuppermann et al., 1973) and for H_2-NH_3, H_2O (Bickes et al., 1973b). Several quantum oscillations have been resolved for all systems. A special attraction has been attributed to the system H_2-H_2. Measurements of the velocity dependence of the total cross section (Butz et al., 1971b; Gengenbach et al., 1974) and of the differential cross section (Dondi et al., 1972; Farrar and Lee, 1973) are available. Fig. 23 illustrates the results of the last two groups. The upper curve measured for normal H_2-H_2 by the Genoa group shows clearly resolved quantum oscillations. The lower curve obtained for para-H_2-H_2 by the Chicago group displays, in addition, symmetry oscillations due to the identity of the particles. The potential parameters which have been obtained from the different measurements are listed in Table IX. Recent measurements of the velocity dependence of the total cross section for p-H_2-H_2 where identical

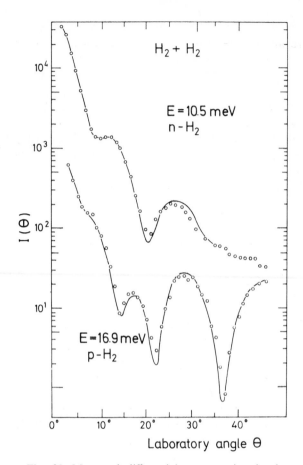

Fig. 23. Measured differential cross section in the laboratory system for n-H_2–H_2 (Dondi *et al.*, 1972) and *p*-H_2–H_2 (Farrar and Lee, 1973). Both systems display well resolved quantum oscillations. The *p*-H_2–H_2 system shows additional oscillations due to the indistinguishability of the particles. The solid lines are calculated from best fit potentials.

particle oscillations show up (Bauer *et al.*, 1975) provide additional information on the attractive well and support the potential of Gengenbach *et al.* (1974).

The results which have been obtained for the other systems may be summarized as follows:

(1) No quenching has been observed both for the differential and the total cross section. This indicates that anisotropy effects should be very small for these systems. Fig. 24 displays the results for the total

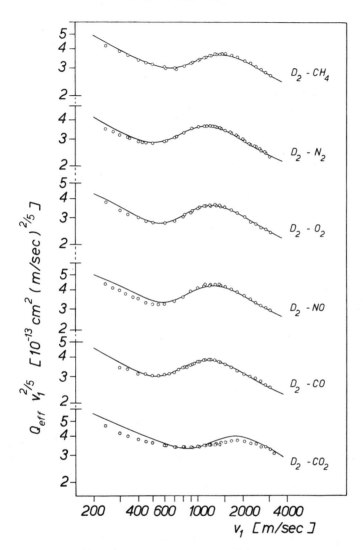

Fig. 24. Total cross section multiplied by $v_1^{0.4}$ as a function of the primary beam velocity v_1 for the scattering of D_2 by various molecules. The solid line is calculated from a LJ 12–6 potential. Deviations only occur for the system D_2–CO_2.

cross section of D_2 on various molecules (Butz *et al.*, 1971b). Only the system D_2–CO_2 is marked by a deviation from the cross section calculated by means of a simple L.J. 12–6 potential.

(2) The results for H_2 and the isotopes D_2 and HD are the same to within the error limits.

(3) The information content of the present measurements is restricted. Simple three parameter potentials are sufficient to fit the data. It should be noted that the total cross section is mainly sensitive to the product εr_m as is the differential cross section to r_m. Therefore we have preferred to list such results in Table X which account for this fact in evaluating

TABLE X
Potential parameters for hydrogen molecule–molecule systems[e]

System	[meV]	r_m [Å]	Data	Reference
H_2–N_2	5·45	3·68	$Q(v), I(\vartheta)$	a, b, c
H_2–O_2	5·60	3·73	$Q(v), I(\vartheta)$	b, d
H_2–CO	5·20	3·77	$Q(v), I(\vartheta)$	b, d
H_2–CH_4	6·20	3·88	$Q(v), I(\vartheta)$	a, b, d
H_2–NO	5·91	3·64	$Q(v)$	b
H_2–CO_2	8·71	3·34	$Q(v)$	b
H_2–NH_3	9·8	3·78	$I(\vartheta)$	d
H_2–SF_6	10·4	4·54	$I(\vartheta)$	d

[a] Aquilanti et al. (1971).
[b] Butz et al. (1971b).
[c] Winicur et al. (1970).
[d] Kuppermann et al. (1973).
[e] The potential form used for all values is LJ 12–6.

the data (Kuppermann et al., 1973). Obviously this is only possible for systems where measurements of the differential and the total cross section are available. Very recent measurements of the differential cross section, where an additional quantum oscillation could be resolved could not be fitted by a simple three parameter potential (Bickes et al., 1973b). This indicates that the present analysis of these systems should be regarded only as a first step to the true potential. Obviously the next steps will follow in the near future.

Acknowledgements

I am greatly indebted to many colleagues for letting me have their results prior to publication. I am grateful to H. Pauly for several helpful suggestions. Finally, I would like to thank W. L. Dimpfl, P. McGuire and D. Pust for reading parts of the manuscript.

References

A. A. Abrahamson (1963). Phys. Rev., 130, 693.
N. Abuaf, J. B. Anderson, R. P. Andres, J. B. Fenn and D. R. Miller (1967). Advan. App. Mech. Supp., 5, 1317.
Z. S. Agranovich and V. A. Marchenko (1963). The Inverse Problem of Scattering Theory, Gordon and Breach, New York.
A. C. Allison and A. Dalgarno (1967). Proc. Phys. Soc. (London), 90, 609.

I. Amdur, J. E. Jordan and S. O. Colgate (1961), *J. Chem. Phys.*, **34**, 1525.

J. B. Anderson, R. P. Andres and J. B. Fenn (1966). *Advan. Chem. Phys.*, **10**, 275.

K. G. Anlauf, R. W. Bickes, Jr. and R. B. Bernstein (1971). *J. Chem. Phys.*, **54**, 3647.

V. Aquilanti, G. Liuti, F. Veechio-Cattivi and G. G. Volpi (1971). *Mol. Phys.*, **21**, 1149.

V. Aquilanti, G. Liuti, E. Luzzati, F. Veechio-Cativi and G. G. Volpi (1972a). *Z. Phys. Chem.*, **79**, 200.

V. Aquilanti, G. Liuti, F. Veechio-Cattivi and G. G. Volpi (1972b). *Chem. Phys. Lett.*, **15**, 305.

V. Aquilanti, G. Liuti, F. Veechio-Cattivi and G. G. Volpi (1973). *Disc. Far. Soc.*, **55**, 187.

R. A. Arndt and M. H. MacGregor (1966). *Meth. Comput. Phys.*, **6**, 253.

D. J. Auerbach (1974). *J. Chem. Phys.*, **60**, 4116.

D. Auerbach, C. Detz, K. Reed and L. Wharton (1971). In L. M. Branscomb *et al.* (Ed.), *Abstracts of Papers of the VIIth ICPEAC*, Amsterdam, p. 541.

A. Barcellona, M. G. Dondi, V. Langomarsino, F. Tommasini and U. Valbusa (1973). *Proc. IV Int. Symp. on Mol. Beams*, Cannes, p. 443.

J. A. Barker, M. L. Klein and M. V. Bobetic (1970). *Phys. Rev.*, **B2**, 4176.

J. A. Barker, R. A. Fisher and R. O. Watts (1971). *Mol. Phys.*, **21**, 657.

J. A. Barker, M. V. Bobetic and M. L. Klein (1972). *Phys. Rev.*, **B5**, 3185.

J. A. Barker, R. O. Watts, J. K. Lee, T. P. Schafer and Y. T. Lee (1974). *J. Chem. Phys.*, **61**, 308.

P. Barwig, U. Buck and H. Pauly (1975). *Bericht 104/1975 MPI Strömungsforschung*, Göttingen, Germany.

D. Bassi, M. G. Dondi, R. Penco, F. Tommasini, F. Torello and U. Valbusa (1974). *Fourth International Conference on Atomic Physics, Abstract of Papers*, p. 524.

W. Bauer, R. W. Bickes, B. Lantzsch, J. P. Toennies and K. Walaschewski (1975a). *Chem. Phys. Lett.*, **31**, 12.

W. Bauer, B. Lantzsch, K. Walaschewski and J. P. Toennies (1975b). *Chem. Phys. Lett.* (to be published).

D. E. Beck, (1968). *Mol. Phys.*, **14**, 311.

D. Beck (1970). 'Enrico Fermi', In Ch. Schlier (Ed.), *Proceedings of the International School of Physics Course XLIV*, Academic Press, New York, p. 1.

D. Beck and H. D. Loesch (1966). *Z. Physik*, **195**, 444.

J. Benn and G. Scharf (1967). *Helv. Phys. Acta*, **40**, 271.

H. G. Bennewitz, H. Busse, H. D. Dohmann, D. E. Oates and W. Schrader (1972a). *Z. Physik*, **253**, 435.

H. G. Bennewitz, H. Busse, H. D. Dohmann, D. E. Oates and W. Schrader (1972b). *Phys. Rev. Lett.*, **29**, 533.

H. G. Bennewitz, W. aufm Kampe and W. Schrader (1973). Private communication.

R. B. Bernstein (1963). *J. Chem. Phys.*, **38**, 2599.

R. B. Bernstein (1966). *Advan. Chem. Phys.*, **10**, 75.

R. B. Bernstein and J. T. Muckermann (1967). *Advan. Chem. Phys.*, **12**, 389.

R. B. Bernstein and R. A. LaBudde (1973). *J. Chem. Phys.*, **58**, 1109.

M. V. Berry (1966). *Proc. Phys. Soc. (London)*, **89**, 479.

M. V. Berry (1969). *J. Phys.*, **B2**, 381.

M. V. Berry and K. E. Mount (1972). *Repts. Progr. Phys.*, **35**, 315.

P. J. Bertoncini and A. C. Wahl (1970). *Phys. Rev. Lett.*, **25**, 991.

P. J. Bertoncini and A. C. Wahl (1973). *J. Chem. Phys.*, **58**, 1259.

R. W. Bickes, and R. B. Bernstein (1970). *Rev. Sci. Instr.*, **41**, 759.

R. W. Bickes, B. Lantzsch, J. P. Toennies and K. Walaschewski (1973a). *Disc. Far. Soc.*, **55**, 167.

R. W. Bickes, G. Scoles and K. M. Smith (1973b). *Proc. IV. Int. Symp. on Mol. Beams*, Cannes.

R. W. Bickes, G. O. Este, G. Scoles and K. M. Smith (1973c). *Chemical Physics Research Report CP15*, University of Waterloo, Canada.

R. W. Bickes, Jr., G. Scoles and K. M. Smith (1974). *Chemical Physics Research Report*, University of Waterloo, CP20.

M. V. Bobetic and J. A. Barker (1970). *Phys. Rev. B*, **2**, 4169.

J. F. Boyle (1971). *Mol. Phys.*, **22**, 993.

J. W. Bredewout (1973). *Doctoral thesis*, University of Leiden, Netherlands.

U. Buck (1971). *J. Chem. Phys.*, **54**, 1923.

U. Buck (1974). *Rev. Mod. Phys.*, **46**, 369.

U. Buck and H. Pauly (1968). *Z. Physik*, **208**, 390.

U. Buck and H. Pauly (1971). *J. Chem. Phys.*, **51**, 1929.

U. Buck, K. A. Köhler and H. Pauly (1971). *Z. Physik*, **244**, 180.

U. Buck, M. Kick and H. Pauly (1972). *J. Chem. Phys.*, **56**, 3391.

U. Buck, M. G. Dondi, U. Valbusa, M. L. Klein and G. Scoles (1973a). *Phys. Rev.*, **A8**, 2409.

U. Buck, M. Düker, H. Pauly and D. Pust (1973c). *Proc. IV Int. Symp. on Mol. Beams*, Cannes, p. 70.

U. Buck, H. O. Hoppe, F. Huisken and H. Pauly (1974). *J. Chem. Phys.*, **60**, 4925.

U. Buck, F. Gestermann and H. Pauly (1975). *Chem. Phys. Lett.*, in press.

D. L. Bunker and E. A. Goring-Simpson (1973). *Disc. Far. Doc.*, **55**, 93.

H. P. Butz, R. Feltgen, H. Pauly, H. Vehmeyer and R. M. Yealland (1971a). *Z. Physik*, 247, 60.

H. P. Butz, R. Feltgen, H. Pauly and H. Vehmeyer (1971b). *Z. Physik*, **247**, 70.

P. Cantini, M. G. Dondi, G. Scoles and F. Torello (1972). *J. Chem. Phys.*, **56**, 1946.

G. M. Carter and D. Pritchard (1974). *Fourth International Conference on Atomic Physics*, Abstract of Papers, p. 279.

M. Cavallini, L. Meneghetti, G. Scoles and M. Yealland (1970a). *Phys. Rev. Lett.*, **24**, 1469.

M. Cavallini, G. Gallinaro, L. Meneghetti, G. Scoles and U. Valbusa (1970b). *Chem. Phys. Lett.*, **7**, 303.

M. Cavallini, M. G. Dondi, G. Scoles and U. Valbusa (1971a). *Chem. Phys. Lett.*, **10**, 22.

M. Cavallini, M. G. Dondi, G. Scoles and U. Valbusa (1971b). *Entropie*, **42**, 236.

M. Cavallini, L. Meneghetti, G. Scoles and M. Yealland (1971c). *Rev. Sci. Instr.*, **42**, 1759.

C. H. Chen, P. E. Siska and Y. T. Lee (1973). *J. Chem. Phys.*, **59**, 601.

C. Chester, B. Friedmann and F. Ursell (1957). *Proc. Camb. Phil. Soc.*, **53**, 599.

D. Coffey, D. C. Lorents and F. T. Smith (1969). *Phys. Rev.*, **187**, 201.

J. N. L. Connor and M. S. Child (1970). *Mol. Phys.*, **18**, 653.

R. P. Creaser, R. P. English and J. L. Kinsey (1973). *J. Chem. Phys.*, **58**, 1321.

R. J. Cross (1967). *J. Chem. Phys.*, **47**, 3724.

R. J. Cross (1968). *J. Chem. Phys.*, **49**, 1976.

R. J. Cross (1970). *J. Chem. Phys.*, **52**, 5703.

G. Das and A. C. Wahl (1971). *Phys. Rev.*, **A4**, 825.

G. Das, A. F. Wagner and A. C. Wahl (1974). *J. Chem. Phys.*, **60**, 1885.

R. David, W. Spoden and J. P. Toennies (1973). *J. Phys. B*, **6**, 897.

T. A. Davidson, M. A. D. Fluendy and K. P. Lawley (1973). *Disc. Far. Soc.*, **55**, 158.

V. DeAlfaro and T. Regge (1965). *Potential Scattering*, Wiley, New York.

G. A. L. Delvigne and J. Los (1973). *Physica*, **67**, 166.

C. Detz and L. Wharton (1970). Unpublished, see C. Detz, *doctoral thesis*, University of Chicago, U.S.A.

K. K. Docken and T. P. Schafer (1973). *J. Mol. Spectr.*, **46**, 454.

M. G. Dondi, U. Valbusa and G. Scoles (1972). *Chem. Phys. Lett.*, **17**, 137.

R. Düren, R. Feltgen, W. Gaide, R. Helbing and H. Pauly (1965). *Phys. Lett.*, **18**, 282.

R. Düren, G. P. Raabe and Ch. Schlier (1968). *Z. Physik*, **214**, 410

R. Düren, A. Frick and Ch. Schlier (1972). *J. Phys. B*, **5**, 1744.

J. H. Dymond and E. B. Smith (1969). *The Virial Coefficient of Gases*, Clarendon, Oxford.

W. Eastes and D. Secrest (1972). *J. Chem. Phys.*, **56**, 640.

J. M. Farrar and Y. T. Lee (1972). *J. Chem. Phys.*, **56**, 5801.

J. M. Farrar and Y. T. Lee (1973). *J. Chem. Phys.*, **57**, 5492.

J. M. Farrar, T. P. Schafer and Y. T. Lee (1973a). *AIP Conf. Proc. (USA)*, **11**, 279.

J. M. Farrar, Y. T. Lee, V. V. Goldman and M. L. Klein (1973b). *Chem. Phys. Lett.*, **19**, 359.

J. M. Farrar, J. M. Parson and Y. T. Lee (1973c). *Proc. IV Int. Symp. on Mol. Beams*, Cannes, p. 214.

R. Feltgen, H. Pauly, F. Torello and H. Vehmeyer (1973). *Phys. Rev. Lett.*, **30**, 820.

R. Feltgen, H. Kirst, K. A. Köhler, H. Pauly, F. Torello and H. Vehmeyer (1974). Unpublished.

O. B. Firsov (1953). *Zh. Eksp. Teor. Fiz.*, **24**, 279.

M. A. D. Fluendy, D. S. Horne, K. P. Lawley and A. W. Morris (1970). *Mol. Phys.*, **19**, 659.

K. W. Ford and J. A. Wheeler (1959). *Ann. Phys. (N.Y.)*, **7**, 259.

R. J. Gallagher and J. B. Fenn (1972). *J. Chem. Phys.*, **60**, 3492.

I. M. Gelfand and B. M. Levitan (1951). *Isvest. Akad. Nauk SSSR Ser. Math.*, **15**, 309 (*Ann. Math. Soc. Transl.*, **1**, 253).

R. Gengenbach and Ch. Hahn (1972). *Chem. Phys. Lett.* **15**, 604.

R. Gengenbach, Ch. Hahn and W. Welz (1973a). Unpublished.

R. Gengenbach, Ch. Hahn and J. P. Toennies (1973b). *Phys. Rev.*, **A7**, 98.

R. Gegenbach, J. P. Toennies, W. Welz and G. Wolf (1973c). *Disc. Far. Soc.*, **55**, 186.

R. Gengenbach, Ch. Hahn and J. P. Toennies (1974a). *J. Chem. Phys.* (in press).

R. Gengenbach, Ch. Hahn and J. P. Toennies (1974b). *Theor. Chim. Acta*, **34**, 199.

R. B. Gerber (1973). *J. Phys. A.*, **6**, 770.

R. B. Gerber and M. Karplus (1970). *Phys. Rev.*, **D1**, 998.

T. L. Gilbert and A. C. Wahl (1967). *J. Chem. Phys.*, **47**, 3425.

W. Gläser and F. Gompf (1969). *Nucleonik*, **12**, 153–159.

R. G. Gordon and Y. S. Kim (1972). *J. Chem. Phys.*, **56**, 3122.

D. W. Gough, E. B. Smith and G. C. Maitland (1974). *Mol. Phys.*, **27**, 867.

E. F. Greene and E. A. Mason (1972). *J. Chem. Phys.*, **57**, 2965.

E. F. Greene and E. A. Mason (1973). *J. Chem. Phys.*, **59**, 2651.

H. G. L. Guerin and B. C. Eu (1974). *Mol. Phys.*, **27**, 401.

H. Haberland, F. P. Tully and Y. T. Lee (1973a). *8th Int. Symposium on Rarefied Gas Dynamics*, Abstr. p. 533.

H. Haberland, C. H. Chen and Y. T. Lee (1973b). In S. J. Smith and K. Walters (Eds.), *Atomic Physics 3*, Plenum Press, New York, p. 339.

W. Hack, F. Rosenkranz and H. G. Wagner (1971). *Z. Naturf.*, **26a**, 1128.

D. R. Hardin and R. Grice (1973). *Mol. Phys.*, **26**, 1321.

R. M. Harris and J. F. Wilson (1971), *J. Chem. Phys.*, **54**, 2088.

R. M. Harris and D. R. Herschbach (1973). *Disc. Far. Soc.*, **55**, 121.

E. F. Hayes, Ch. A. Wells and D. J. Kouri (1971). *Phys. Rev. A.*, **4**, 1017.

R. K. B. Helbing (1968). *J. Chem. Phys.*, **48**, 472.

R. K. B. Helbing (1969a). *J. Chem. Phys.*, **50**, 493.

R. K. B. Helbing (1969b). *J. Chem. Phys.*, **50**, 4123.

R. K. B. Helbing (1969c). *J. Chem. Phys.*, **51**, 3628.

R. Helbing and H. Pauly (1964). *Z. Physik*, **179**, 16.

R. K. B. Helbing and E. W. Rothe (1968). *J. Chem. Phys.*, **48**, 3945.

R. Helbing, W. Gaide and H. Pauly (1968). *Z. Physik*, **208**, 215.

J. Hermann, E. Hierholzer, R. Maier, J. Mania, Ch. Schlier and A. Schultz (1973). *Abstracts of Papers, VIII ICPEAC* (Eds., Beograd, B. Cobic and M. Kurepa), p. 45.

I. V. Hertel and W. Stoll (1974). *J. Phys. B.*, **7**, 570.

G. Herzberg (1950). *Molecular Spectra and Molecular Structure. I. Diatomic Molecules*, Van Nostrand Reinhold Comp., New York, p. 358.

W. F. Heukels and J. van de Ree (1972). *J. Chem. Phys.*, **57**, 1393.

E. Hylleraas (1963). *Ann. Phys. (N.Y.)*, **25**, 309.

J. E. Jordan and I. Amdur (1967). *J. Chem. Phys.*, **46**, 165.

F. Kalos and A. E. Grosser (1972). *Can. J. Chem.*, **50**, 892.

H. Kanes, H. Pauly and E. Vietzke (1971). *Z. Naturf.*, **26a**, 689.

K. J. Kaufmann, J. Lawter and J. L. Kinsey (1974). *J. Chem. Phys.*, **60**, 4016.

M. Kennedy and F. J. Smith (1969). *Mol. Phys.*, **16**, 131.

M. L. Klein, T. R. Koehler and R. L. Gray (1973). *Phys. Rev.*, **B7**, 1571.

R. Klingbeil (1972). *J. Chem. Phys.*, **56**, 132.

R. Klingbeil (1973). *J. Chem. Phys.*, **59**, 797.

H. L. Kramer and P. R. LeBreton (1967). *J. Chem. Phys.*, **47**, 3367.

A. Kuppermann, R. J. Gordon and M. J. Coggiola (1973). *Disc. Farad. Soc.*, **55**, 145.

V. Lagomarsino, D. Bassi, E. Bertok, M. DePaz and F. Tommasini (1973). *Proc. IV Int. Symp. on Mol. Beams*, Cannes, p. 438.

P. R. LeBreton and H. L. Kramer (1969). *J. Chem. Phys.*, **51**, 3627.

R. D. Levine (1972). *J. Chem. Phys.*, **57**, 1015.

Y. T. Lee, J. D. McDonald, P. R. LeBreton and D. R. Herschbach (1970). *Rev. Sci. Inst.*, **40**, 1402.

H. V. Lilenfeld, J. L. Kinsey, N. C. Lang and E. K. Parks (1972). *J. Chem. Phys.*, **57**, 4593.

B. Liu and A. D. McLean (1973). *J. Chem. Phys.*, **59**, 4557; note that the ε-value given in the paper was corrected by the authors to $\varepsilon = 0.908$ meV.

J. J. Loeffel (1968). *Ann. Inst. Henri Poincaré*, **4**, 339.

P. McGuire (1974). *Chem. Phys.*, **4**, 483.

P. McGuire and D. A. Micha (1972). *Int. J. Quant. Chem.*, **6**, 111.

P. McGuire and D. Kouri (1974). *J. Chem. Phys.*, **60**, 2488.

G. C. Maitland (1973). *Mol. Phys.*, **26**, 513.

G. C. Maitland and E. B. Smith (1971). *Mol. Phys.*, **22**, 861.

G. C. Maitland and E. B. Smith (1972). *J. Chem. Eng. Data*, **17**, 150.

G. C. Maitland and E. B. Smith (1973a). *Chem. Soc. Rev.*, **2**, 181.

G. C. Maitland and E. B. Smith (1973b). *Chem. Phys. Lett.*, **22**, 443.

G. Marenco, A. Schutte, G. Scoles and V. Tommasini (1972). *J. Vac. Sci. Tech.*, **9**, 824.

R. Mariott and D. A. Micha (1969). *Phys. Rev.*, **180**, 120.

A. Martin (1969). *Nuovo Cimento*, **59A**, 131.

E. A. Mason, J. T. Vanderslice and C. J. G. Raw (1964). *J. Chem. Phys.*, **40**, 2153.

W. H. Miller (1968). *J. Chem. Phys.*, **48**, 464.

W. H. Miller (1969a). *J. Chem. Phys.*, **50**, 3124.

W. H. Miller (1969b). *J. Chem. Phys.*, **51**, 3631.

W. H. Miller (1971a). *J. Chem. Phys.*, **54**, 4174.

D. R. Miller and D. F. Patch (1969), *Rev. Sci. Instr.*, **40**, 1566.

T. M. Miller (1971b). *Entropie*, **42**, 69.

M. H. Mittleman and H. Tai (1973). *Phys. Rev.*, **A8**, 1880.

H. U. Mittmann, H. P. Weise, A. Ding and A. Henglein (1971). *Z. Naturf.*, **26a**, 1112.

R. Morgenstern, D. C. Lorents, J. R. Peterson and R. E. Olson (1973). *Phys. Rev.*, **A8**, 2373.

K. E. Mount (1973). *J. Phys.*, **B6**, 1397.

A. M. C. Moutinho, J. A. Aten and J. Los (1971). *Physica*, **53**, 471.

J. M. Mullen and B. S. Thomas (1973). *J. Chem. Phys.*, **58**, 5216.

R. G. Newton (1962). *J. Math. Phys.*, **3**, 75.

R. G. Newton (1966). *Scattering of Eaves and Particles*, McGraw-Hill Book Co., New York, Chap. 20.

R. G. Newton (1968). *J. Math. Phys.*, **9**, 2050.

R. G. Newton (1972). 'Mathematics of Profile Inversion'. In L. Colin (Ed.), *NASA Technical Memorandum X-62*, p. 52.

C. Y. Ng, Y. T. Lee and J. A. Barker (1974). *J. Chem. Phys.*, **61**, 1996.

T. J. P. O'Brien and R. B. Bernstein (1969). *J. Chem. Phys.*, **51**, 5112.

R. E. Olson and R. B. Bernstein (1968). *J. Chem. Phys.*, **49**, 162.

R. E. Olson and R. B. Bernstein (1969). *J. Chem. Phys.*, **50**, 246.

R. E. Olson and F. T. Smith (1971). *Phys. Rev.*, **A3**, 1607.

J. M. Parson, T. P. Schafer, F. P. Tully, P. E. Siska, Y. C. Wong and Y. T. Lee (1970a). *J. Chem. Phys.*, **53**, 2123.

J. M. Parson, T. P. Schafer, P. E. Siska, F. P. Tully, Y. C. Wong and Y. T. Lee (1970b). *J. Chem. Phys.*, **23**, 3755.

J. M. Parson, P. E. Siska and Y. T. Lee (1972). *J. Chem. Phys.*, **56**, 1511.

H. Pauly (1974). *Physical Chemistry, an Advanced Treatise*, Vol. VIB, Academic Press, New York, Chap. 8, p. 553.

H. Pauly and J. P. Toennies (1965). *Advan. At. Mol. Phys.*, **1**, 195.

H. Pauly and J. P. Toennies (1968). *Methods of Experimental Physics*, 7A, Academic Press, New York, p. 227.

J. Penta, C. R. Mueller, W. Williams, R. E. Olson and P. Chakraborti (1967). *Phys. Lett.*, **A25**, 658.

P. E. Phillipson (1962). *Phys. Rev.*, **125**, 1981.

D. E. Pritchard (1972). *J. Chem. Phys.*, **56**, 4206.

D. E. Pritchard, G. M. Carter, F. Y. Chu and D. Kleppner (1970). *Phys. Rev.*, **A2**, 1933.

R. T. Prosser (1969). *J. Math. Phys.*, **10**, 1819.

H. Rabitz (1972). *J. Chem. Phys.*, **57**, 1718.

E. A. Remler (1971). *Phys. Rev.*, **A3**, 1949.

W. G. Rich, S. M. Bobbio, R. L. Champion and L. Doverspike (1971). *Phys. Rev.*, **A4**, 2253.

R. E. Roberts and J. Ross (1970). *J. Chem. Phys.*, **53**, 2126.

M. Rosen and D. R. Yennie (1964). *J. Math. Phys.*, **5**, 1505.

F. Rosenkranz (1970). *Doctoral thesis*, University of Göttingen, Germany.

J. Ross and E. F. Greene (1970). 'Enrico Fermi'. In Ch. Schlier (Ed.), *Proceedings of the International School of Physics, Course XLIV*, Academic Press, New York, p. 86.

E. W. Rothe and L. H. Veneklasen (1967). *J. Chem. Phys.*, **46**, 1209.

E. W. Rothe and R. K. B. Helbing (1969). *J. Chem. Phys.*, **50**, 3531.

E. W. Rothe and R. K. B. Helbing (1970a). *J. Chem. Phys.*, **53**, 1555.

E. W. Rothe and R. K. B. Helbing (1970b). *J. Chem. Phys.*, **53**, 2501.

P. C. Sabatier (1965). *Nuovo Cimento*, **37**, 1180.

P. C. Sabatier (1966). *J. Math. Phys.*, **7**, 1515.

P. C. Sabatier (1972a). Mathematics of Profile Inversion, L. Colin (ed.), *NASA Technical Memorandum X-62*, pp. 5–14.

P. C. Sabatier (1972b). *J. Math. Phys.*, **13**, 675.

P. C. Sabatier and F. Qugen Van Puy (1971). *Phys. Rev.*, **D4**, 127.

H. F. Schaefer III, D. R. McLaughlin, F. E. Harris and B. J. Alder (1970). *Phys. Rev. Lett.*, **25**, 988.

T. P. Schafer, P. E. Siska and Y. T. Lee (1971). *Abstracts of Papers of the VIIth ICPEAC*, North-Holland, Amsterdam, p. 546.

R. Schieder, H. Walther and L. Wöste (1972). *Opt. Commun.*, **5**, 337.

Ch. Schlier (1969). *Ann. Rev. Phys. Chem.*, **20**, 191.

A. Schutte, D. Bassi, F. Tommasini and G. Scoles (1972). *Phys. Rev. Lett.*, **29**, 979.

A. Schutte, D. Bassi, F. Tommasini and G. Scoles (1974). *J. Chem. Phys.*, **62**, 600.

R. Shafer and R. G. Gordon (1973). *J. Chem. Phys.*, **58**, 5422.

P. E. Siska, J. M. Parson, T. P. Schafer, F. P. Tully, Y. C. Wong and Y. T. Lee (1970). *Phys. Rev. Lett.*, **25**, 271.

P. E. Siska, J. M. Parson, T. P. Schafer and Y. T. Lee (1971). *J. Chem. Phys.*, **55**, 5762.

K. Sköld (1968). *Nucl. Inst. and Meth.*, **63**, 114.

T. M. Sloane, S. Y. Tang and J. Ross (1972). *J. Chem. Phys.*, **57**, 2745.

F. T. Smith (1969). In S. Geltman *et al.* (Ed.) Lectures in Theoretical Physics: Atomic Collision Processes, Vol. XIC, Gordon and Breach, New York, p. 95.

F. T. Smith, R. P. Marchi and K. G. Dedrick (1966). *Phys. Rev.*, **150**, 79.

F. T. Smith, R. P. Marchi, W. Aberth, D. C. Lorents and D. Heinz (1967). *Phys. Rev.*, **161**, 31.

F. T. Smith, H. H. Fleischmann and R. A. Young (1970). *Phys. Rev.*, **A2**, 379.

G. Spindler, H. Ebinghaus and E. Steffens (1973). *Proc. IV Int. Symp. on Mol. Beams*, Cannes, p. 391.

G. Starkschall and R. G. Gordon (1972). *J. Chem. Phys.*, **56**, 2801.

W. J. Stevens, A. C. Wahl, M. A. Gardener and A. M. Karo (1974). *J. Chem. Phys.*, **60**, 2195.

E. C. G. Stueckelberg (1932). *Helv. Phys. Acta*, **5**, 369.

W. C. Stwalley (1971). *J. Chem. Phys.*, **55**, 170.

W. C. Stwalley (1972). Unpublished.

W. C. Stwalley and K. L. Kramer (1968). *J. Chem. Phys.*, **48**, 5555.

W. C. Stwalley, A. Niehaus and D. R. Herschbach (1969). *J. Chem. Phys.*, **51**, 2287.

Y. Tanaka and K. Yoshino (1970). *J. Chem. Phys.*, **53**, 2012.

J. P. Toennies (1974). *Physical Chemistry, an Advanced Treatise*, Vol. VIA, Academic Press, New York, Chap. 5, p. 227.

B. Tsapline and W. Kutzelnigg (1973). *Chem. Phys. Letts.*, **23**, 173.

F. P. Tully and Y. T. Lee (1972). *J. Chem. Phys.*, **57**, 866.

G. B. Ury and L. Wharton (1972). *J. Chem. Phys.*, **56**, 5832.

G. Vollmer (1969). *Z. Physik*, **226**, 423.

G. Vollmer (1971). *Z. Physik*, **243**, 92.

T. P. Tsien, G. A. Parker and R. T. Pack (1973). *J. Chem. Phys.*, **59**, 5373.

G. Vollmer and H. Krüger (1968). *Phys. Letts.*) **28A**, 165.

W. Welz, G. Wolf and J. P. Toennies (1974). *J. Chem. Phys.*, **61**, 2461.

H. Wilsch (1972). *J. Chem. Phys.*, **56**, 1412.

D. H. Winicur, A. L. Moursund, W. R. Devereaux, L. R. Martin and A. Kuppermann (1970). *J. Chem. Phys.*, **52**, 3299.

G. Wolken, W. H. Miller and M. Karplus (1972). *J. Chem. Phys.*, **56**, 4930.

SCATTERING FROM ORIENTED
MOLECULES

J. REUSS

Katholieke Universiteit NIJMEGEN, The Netherlands

CONTENTS

I. Introduction 389
II. The Angular Dependent Potential 390
III. State Selection of Molecular Beams 395
IV. Measurements of the Anisotropy of the Total Collision Cross Section . . . 397
V. Theory Needed for Interpretation 401
VI. Results 406
   A. The Non-Glory Contribution 406
   B. The Glory Contribution 408
Acknowledgements 414
References 414

I. INTRODUCTION

Scattering from oriented molecules as discussed in this review involves molecular beam experiments where at least one of the collision partners has cylindrical symmetry and where the symmetry axis is preferentially oriented in space before the collision.

There are two ways to orient molecules in an external field. In *high electric fields* it is possible to virtually align polar molecules (in low rotational states). Theoretically, the Stark-effect was thoroughly analysed by Brouwer (1930) who investigated how the solution of the Schrödinger equation for a polar diatomic in an electric field approaches the two-dimensional oscillator wavefunction where the molecule simply oscillates about the direction of the external field. Alignment along this pattern has never been used in molecular beam scattering experiments; fields of suitable strength can easily be obtained for diatomics with relatively large electric dipole moments, though.

In comparatively *small electric fields* molecules selected with respect to their rotational quantum numbers j and m_j rotate in space but have preferential orientations of their molecular axis. The preferential orientations depend on the chosen set of quantum numbers j and m_j and are defined with respect to the direction of the external field. The method of state selection is well known to molecular beam spectroscopists and is not restricted to polar diatomics, Ramsey (1956). In Section 3 we shall review some recent developments concerning state selection of molecular beams.

If (preferentially) oriented molecules are used in scattering experiments they will provide a source of information for the angular dependent part of the intermolecular potential (AIP). In Section 2 we shall discuss how little is known of the AIP.

Bennewitz et al. (1964) were the first to get this information on the AIP from total collision cross section measurements. In the last three sections we shall confine ourselves to this type of measurements and discuss a more recent experimental set up (Section IV) and the results obtained so far (Section VI). In Section V we shall discuss the theory which connects the AIP with the orientation dependent part of the total collision cross section. Our main technique will be the first order distorted wave approximation (d.w.a.).

The velocity dependence of the total collision cross section is a restricted source of information regarding the intermolecular potential (IP), Bernstein (1973). Much more information is contained in differential cross section measurements which therefore form a great challenge for future scattering experiments with oriented molecules.

There has been a predecessor to the Bennewitz-experiment; Berkling et al. (1962) have measured the total cross section for collisions between oriented $^2P_{3/2}$ Ga atoms with noble gas molecules. This experiment was never taken up again although there is a whole group of atoms with anisotropic electronic structure which can be likewise oriented. The information which can be obtained from scattering experiments with oriented atoms is essentially the same as with oriented molecules.

Recently experiments of chemical reactions involving oriented molecules were published, Brooks et al. (1966, 1969 and 1973) and Beuhler et al. (1966 and 1969). Perhaps this is the most interesting field for the future of scattering experiments from oriented beams. Information is obtained about preferential orientations for which chemical reactions occur.

II. THE ANGULAR DEPENDENT PART OF THE INTERMOLECULAR POTENTIAL

Even for very simple systems little is known about the AIP; this is true both theoretically and experimentally. We shall indicate the difficulties and the possibilities, discussing the system He–H_2 which forms the best studied system having an angular dependent IP.

Measurements on this system were done by Gengenbach et al. (1972) who determined the velocity dependence and the absolute value of the total collision cross section. From these measurements the isotropic part of the intermolecular potential (IIP) was obtained, see Fig. 1. Recently Riehl et al.

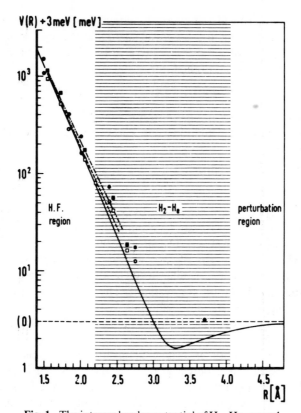

Fig. 1. The intermolecular potential of H_2–He, according to Gengenbach (1972). In the HF region (Hartree–Fock) $-----$ corresponds to the isotropic potential fit of Kraus (1965), $-\cdot-\cdot-$ to the fit of Gordon (1970); ■ and □ are points with $\theta = 0°$ and $90°$ as calculated by Krauss (1965); ● and ○ are corresponding points as calculated by Gordon (1970). The angle θ is formed by the molecular axis of H_2 and the line connecting the He-atom with the centre of the H_2-molecule. In the HF region, the angular dependence of the intermolecular potential has thus been calculated from first principles. In the perturbation region, the angular dependence is determined semi-empirically, for instance by Langhoff (1971). The most interesting intermediate region, however, is not yet accessible to reliable calculations.

(1973) have published experimental results on spin lattice relaxation with an interpretation in terms of an AIP, for He–H_2 (and Ne–H_2). The discussion of Riehl (1973) is based on an IIP which differs significantly from the one obtained by Gengenbach (1972). The AIP proposed by Riehl (1973) is therefore not to be considered as an ultimate one as the authors themselves remark. Nevertheless, the first step was taken and the hunt is on.

Theoretically, at close distances Krauss et al. (1965) and Gordon et al. (1970) have proposed a potential of the form

$$V_c = A \exp(-\alpha R) \cdot [1 + q_r P_2(\cos \theta)] \tag{1}$$

with $q_r(\text{Krauss}) = 0.1896$ and $q_r(\text{Gordon}) = 0.2657$.

From perturbation calculations one knows that at large distances the potential has the form

$$V_L = -C_6[1 + p_{2.6}P_2(\cos \theta)]R^{-6} - C_8[1 + p_{2.8}P_2(\cos \theta) + p_{4.8}P_4(\cos \theta)]R^{-8} \tag{2}$$

Precise values for C_6 were calculated by Victor et al. (1970) and Langhoff et al. (1971). The value of C_8 can be estimated from calculations on similar systems, see Gengenbach (1972). The most recent value of $p_{2.6}$ was calculated by Langhoff (1971) for He–H$_2$, $p_{2.6} = 0.101$. The coefficients $p_{2.8}$ and $p_{4.8}$ can be calculated in a similar way; to our knowledge, however, no results of such a calculation are available. The only other way to estimate crudely the magnitude of the coefficients is to use the oscillator model, which gives for the system He–H$_2$ the values $p_{2.8} = 0.13$ and $p_{4.8} = 0.009$, Reuss (1967).

The procedure of both experimental groups, Gengenbach (1972) and Riehl (1973), was to fit a potential to their results which is as far as the experiments permit in agreement with the theoretical predictions at large and short distances and which is determined numerically in the form of a spline function in the region around the potential minimum. This procedure has the consequence that in the neighbourhood of the minimum of the IIP no theoretical links exist anymore leading from the IIP to at least a formal expression for the AIP. In Fig. 1 we have distinguished three regions and indicate that in the middle one, theory is of no help (See, however, the calculation by Tsapline et al. (1973) on H$_2$–He.) Nevertheless it is just this middle region which is of great importance for thermal collisions. To find the potential curve one has to use purely empirical data to spline the AIP from the short range region with its theoretically estimated angular dependence to the long range region with its angular dependence accurately calculated from perturbation theory. Riehl et al. (1973) use as input their experimental spin lattice relaxation results and the rotational relaxation measurements of Jonkman et al. (1968). For the AIP the following parameters are obtained; $q_r = 0.32, p_{2.6} = 0.1, p_{2.8} = 0.095$. The value of $p_{4.8}$ is of no influence on the experiments of Riehl (1973). If one defines

$$V = V_{\text{IIP}} + V_{\text{AIP}}P_2(\cos \theta) \tag{3}$$

one find that the minima in the R-dependent functions V_{IIP} and V_{AIP} occur at different positions, R_m and $R_{m,a}$. Riehl (1973) obtains $R_{m,a} = 4.25$ Å and $R_m = 3.71$ Å.

Besides the disagreement with Gengenbach (1972) which concerns the IIP, the AIP does not entirely satisfactorily reproduce the rotational relaxation measurements of Jonkman et al. (1968); furthermore, the short range anisotropy parameter q_r is not in agreement with theoretical prediction. The same is true for the experimental factor α at short distances, see Equation (1).

Very recently Shafer et al. (1973) have extended the analysis of Riehl (1973) using essentially the IIP of Gengenbach (1972) and, as additional experimental input, the Raman line shapes of May (1961) and Cooper (1970). It was possible to explain all experimental material very well by a potential function for the AIP which is in satisfying agreement with theoretical requirements at long and short range.

Much less is known about the AIP (and IIP) for all other anisotropic systems. In order to extract information from measurements of the orientation dependent part of the total collision cross section under these unfavourable circumstances one proceeds as follows: for the IIP one uses an empirical two parameter potential, the Lennard-Jones 12–6 potential, with the parameters determined from independent scattering experiments. These potentials are implemented by an angle dependent part so that one has, for instance for the case of atom–non-polar diatom collisions (3) with

$$V_{IIP} = \varepsilon\left[\left(\frac{R_m}{R}\right)^{12} - 2\left(\frac{R_m}{R}\right)^6\right] \tag{4}$$

and

$$V_{AIP} = \varepsilon\left[q_{2\cdot12}\left(\frac{R_m}{R}\right)^{12} - 2q_{2\cdot6}\left(\frac{R_m}{R}\right)^6\right]$$

$$= \varepsilon_a\left[\left(\frac{R_{m,a}}{R}\right)^{12} - 2\left(\frac{R_{m,a}}{R}\right)^6\right] \tag{5}$$

Here $q_{2\cdot12}$ and $q_{2\cdot6}$ stand for adjustable parameters, the so-called anisotropy parameters. Equations (4) and (5) must not be used at very short and very large distances. Actually the meaning of the anisotropy parameters is not much more than a description of the position and depth of the anisotropic minimum, i.e. the minimum of the function V_{AIP}; one finds for the depth $\varepsilon_a = \varepsilon q_{2\cdot6}^2/q_{2\cdot12}$ and for the position $R_{m,a} = R_m(q_{2\cdot12}/q_{2\cdot6})^{1/6}$.

From the discussion of Section V it will become clear that the glory undulation of the orientation dependent part of the total collision cross section mainly determines $q_{2\cdot12}/q_{2\cdot6}$ which in turn allows us to calculate the position of the anisotropic minimum $R_{m,a}$ from a known R_m value. If in addition it is possible to get information on $q_{2\cdot6}$, the depth of the anisotropic minimum ε_a can be calculated.

The parameter $q_{2\cdot6}$ in (5) does not necessarily assume the long range value $p_{2\cdot6}$ which can be calculated in good approximation by perturbation theory, Buckingham (1967), Victor (1970), Langhoff (1971). The experimental values of Stolte (1973) differ by as much as a factor of 1·5 from the long range values of $p_{2\cdot6}$. One has to keep in mind that the $q_{2\cdot6}$ values of Stolte (1973), too, are not determined at $R \approx R_{m,a}$ but roughly at $R \approx 2R_{m,a}$, as shall be discussed in detail in Section V.

The $p_{2\cdot6}$ values found from perturbation theory possess the property that they are approximately independent on the secondary non-state selected collision partner. In Section VI.A we shall discuss whether the experimentally determined $q_{2\cdot6}$ values share this property with the long range parameter $p_{2\cdot6}$.

The value of $q_{2\cdot12}$ is entirely open to speculation. There are no theoretical predictions, either for its absolute value or for the trend if collisions of the same selected molecule are considered with different collision partners.

To justify the use of the simple angular dependence of (4) and (5), some arguments can be put forward which are discussed here in connection with progressively more complicated colliding systems. If collisions between *non-polar diatoms* and atoms of spherical symmetry are considered electro-static multipole terms and induction terms are absent or entirely negligible. The angular dependence can be expressed as a series of terms each of which is proportional to a Legendre polynomial of even order. Thus, the lowest order neglected in (3) would be proportional to $P_4(\cos\theta)$. However, this term and higher ones contribute neither to the quantity measured by Bennewitz (1964), Stolte (1973) and Moerkerken (1973) nor to the spin relaxation time as determined by Riehl (1973), at least not in lowest order. The influence of a '$P_4(\cos\theta)$ term' on the orientation dependent part of the total collision cross section can in principle be measured if rotational states with $j > 3/2$ are selected, see also Reuss (1967). Because in this review we restrict the discussion to first order d.w.a. (see Section V) and almost always to selected states with $j \leq 3/2$ the angular dependence of the intermolecular potential may be reduced to the form of (3).

If collisions are considered between a *polar diatom* and an atom of radial symmetry, angular dependent long range induction terms are present. Their influence on the total collision cross section is discussed by Stolte (1972) and found either to be negligible or to change only the effective value of $p_{2\cdot6}$.

The fact that the centre of rotation of the polar diatomic does not coincide with the centre of charge of the electron cloud gives rise to extra terms in the AIP, Ree (1971); Legendre polynomials of odd order appear. Their contribution to the orientation dependent part of the total cross section has been shown to vanish in first order, Stolte (1972). Again, therefore, the angular dependence of (3) appears to be satisfactory for the interpretation of the total collision cross section measurements.

For *polar molecules* colliding with *polar molecules* quite a number of angular dependent interaction terms are already present in the long range part of the IP, Buckingham (1967). For total collision cross section measurements Stolte (1972) has shown that these extra terms are of little influence on the orientation dependent part. Their effect through higher order perturbations can be judged from the calculations of Kuijpers (1973) to be small. An exception is formed by the dipole–dipole interaction of two symmetric top-like molecules. Experimentally, too, a large orientation dependent part of the total collision cross section was found for the system $NO–CF_3H$ by Stolte (1973). For the time being it seems wise to turn to the study of simpler systems where qualitative questions still remain to be answered.

III. STATE SELECTION OF MOLECULAR BEAMS

The Stark effect and the Zeeman effect of molecules in an inhomogeneous external field serve to select molecules due to their different rotational quantum states and are thus used to produce molecules with well defined preferential orientations in a beam; molecules with different orientation are differently deflected in these fields. With regard to state selectors one can distinguish between simple deflection devices like Stern–Gerlach magnets (and their electrical analogues) and multipole fields where molecules with a well-defined Stark or Zeeman-effect are focused into the detector. In both cases the state selector works as a filter enhancing the relative number of molecules in a certain quantum state.

Historically, magnetic deflection formed the main device in the early investigations of spectroscopists, Ramsey (1956). The strong fields necessary to measurably deflect molecules in general cause a nearly complete decoupling between the nuclear angular momenta and the rotational angular momentum. Therefore, molecules with different sets of quantum numbers j and m_j may be equally deflected; j and m_j determine the rotational angular momentum and its projection onto the direction of the external field. In order to produce oriented molecules in a particular rotational state one needs some additional means besides the deflecting magnet.

Moerkerken *et al.* (1970) applied an RF field to H_2-molecules of a molecular beam passing through a fairly conventional arrangement of A-, B- and C-fields. Molecules in a well-defined rotational state undergo a transition into a state with different Zeeman-effect when they pass through the C-field where the RF field is applied (see Fig. 2). This combination of deflecting fields and spectroscopic techniques permits the production of a beam of preferentially oriented non-polar molecules. The scattering chamber is also shown in Fig. 2 where the beam of selected molecules can be attenuated for determination of the total collision cross section.

Where possible it seems preferable to use multipole focusers because here the absolute number of selected molecules can also be enhanced

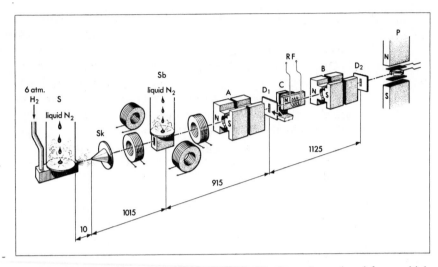

Fig. 2. The H_2-machine of Moerkerken (1970). The beam is produced from a high pressure source (S); a skimmer (Sk) is placed 10 mm downstream to obtain a nearly monochromatic beam of high intensity; the scatter-box (Sb) is surrounded by two pairs of coils to produce a B-field of variable orientation; the beam passes through A-, C- and B-fields, which together with the slit D_1 function as the state selector; in the C-field transitions can be induced by a r.f.-coil (RF); the detector slit (D_2) is followed by a uhv-Penning detector (P). The numbers below indicate distances in mm.

considerably. An intensity gain of a factor of 100 and more is reported by Bennewitz (1964) and Stolte (1972). For molecules with quadratic Stark effect, fourpole fields were used; for molecules with linear Stark effect, sixpole fields.

There are some promising recent developments concerning the technique of state selection in molecular beams. Everdij *et al.* (1973) were able to improve the focusing of polar diatomics using a combination of fourpole and sixpole fields. The general idea is that thereby the lens error of the four-pole due to an imperfect quadratic Stark effect can be partially corrected. To underline the possible importance of such a combination we remind the reader that really a sharp image of the source opening must be formed by the selected molecules in the detector plane in order to achieve a well defined and possibly high angular resolution. The requirements in this respect are much higher than is common in molecular beam spectroscopy.

For a long time there have been speculations about the feasibility of applying alternating gradient methods to molecular beam focusing, Auerbach *et al.* (1966); recently experimental results were published by Kakati *et al.* (1971) and Gunther (1972) where polar molecules with negative Stark-effect were focused, too, though still with rather low intensity. The state selector consisted of a number of twopole fields in series with geometries turned by 90° around the beam axis with respect to each other. An interesting application of this technique would be that a beam of polar diatomics in the $^1\Sigma$-ground state with the rotational quantum number $j = m_j = 0$ can be obtained which except for the possibility of excitations has the properties of a beam of molecules of spherical symmetry. Consequently, the AIP can be switched off except for inelastic events. Comparison with scattering cross sections obtained from molecular beams with thermal occupation of the rotational levels would provide information about the AIP.

Selection of a single quantum state of the symmetric top-like molecule NO for a scattering experiment was reported by Stolte et al. (1972a); until then only a mixture of states with nearly equal $m_j K / j(j + 1)$ was selected, Kramer et al. (1965), Brooks et al. (1969) and Brooks (1973). Here K stands for the projection of the rotational angular momentum j on the molecular symmetry axis.

Other focusing devices employed by molecular beam spectroscopist, Laine et al. (1971), Basov et al. (1961) and Becker (1963) cannot be utilized in molecular beam machines for scattering experiments due to their inferior lens properties.

IV. MEASUREMENTS OF THE ANISOTROPY OF THE TOTAL COLLISION CROSS SECTION

In Fig. 3 the apparatus is shown with which Stolte (1972) has performed measurements of this type; NO molecules were selected with the help of electrostatic sixpole fields, in accordance with their linear Stark effect in strong fields. The source slit of 0·05 mm width has an image formed by the selected NO molecules in the plane of the detector slit which has an experimental width of 1·4 mm f.w.h.m.; this width includes all disturbing effects like the magnification factor (about 1·8), the imperfect linear Stark effect of the NO molecules in the selected state, the finite width of the transmission of the velocity selector ($\Delta v/v = 7\%$ f.w.h.m.) in combination with the chromatic lens errors and the directional dependence of the maximum transmitted velocity of the velocity selector. The $j = m_j = \Omega = 3/2$ state was selected where Ω is the projection of the electronic angular momentum on the molecular axis. The hyperfine structure of NO influences the situation only slightly.

The heart of the apparatus is the scattering region where a secondary beam crosses the path of the NO molecules in a magnetic field of variable direction. The collision partners possess a well-defined direction of average relative velocity so that the magnetic field can be made parallel and perpendicular to this direction. What is observed is the difference in attenuation of the primary beam for the two orientations of the magnetic field or (translated into cross sections) the anisotropy defined by

$$A = (\sigma_\| - \sigma_\perp)/\sigma^{(0)} \tag{6}$$

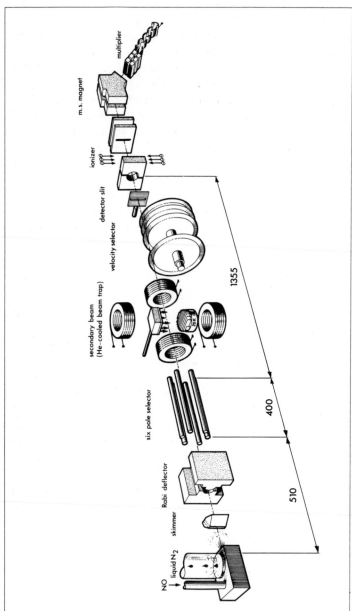

Fig. 3. The NO-beam is formed in the liquid N_2 cooled source at left and passes through a slit-skimmer into the state selector, which consists of a deflection magnet and a sixpole. Molecules in the selected state with a velocity distribution determined by the velocity selector form a sharp image of the source opening in the plane of the detector slit. The secondary beam enters the scattering region from above through a multichannel effuser and is cryopumped. At the right the ionizer, m.s. magnet and multiplier of the detector are shown. Distances are indicated in mm.

Here $\sigma_{\parallel}(\sigma_{\perp})$ is the total cross section measured with the magnetic field parallel (perpendicular) to the direction of the average relative velocity; $\sigma^{(0)}$ stands for the orientationally and glory-averaged total cross section $8\cdot083\,(2\varepsilon R_m^6/\hbar v)^{2/5}$. As shall be discussed in Section V the anisotropy A can be split into a glory and a non-glory contribution, in first order d.w.a.

$$A = A_{ng} + A_g \tag{7}$$

The physical implications of both orientations are shown in Fig. 4, where the meaning of the preferential orientation of the NO molecule is portrayed (in an exaggerated way) by its precession cone.

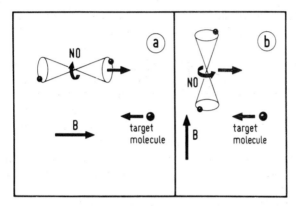

Fig. 4. Collision of state selected NO molecules. The $j = \Omega = m_j = 3/2$ state is shown, with m_j defined along the B-field; a) corresponds to a measurement of $\sigma_{\parallel}$, b) to a measurement of $\sigma_{\perp}$. Due to the dominant influence of the dispersion forces one has $\sigma_{\parallel} < \sigma_{\perp}$.

Because a very small effect is observed in the total collision cross section one has to correct most carefully for a number of effects. We follow Stolte (1972) in the discussion of the various corrections. First one normally deals with molecules which possess a hyperfine structure such that the rotational quantum numbers j and m_j describe the rotational state of the molecules properly only in the high field limit. Therefore, in the actual fields of the scattering region a mixing of pure high-field states has to be taken into account resulting in a correction of about 10% in the case of NO, with respect to the anisotropy A. An inhomogeneity of the orienting field in the scattering region results in an extra correction for A, which amounts to about 1% in the practical case of NO.

Next one must consider the motion of the target gas. For a secondary beam and for a scattering chamber the velocity distribution of the target molecules produces a distribution of relative velocities with respect to magnitude and with respect to direction. The anisotropy measurements are sensitive to both; the influence of the distribution of absolute values comes from the dependence of the total cross section on the relative velocity and the influence of the distribution of directions comes from the fact that we investigate as anisotropy A the effect of an external field in the scattering region whose direction is chosen $0°$ or $90°$ with respect to the relative velocity. If the relative velocity is only defined within a certain angular range this clearly will cause corrections to the measured anisotropy. Even if the glory undulation of the total collision cross section is neglected, the calculation of this combined correction is rather complicated. For the case of a secondary beam effusing from a multichannel array Ragas (1972) has shown that for NO molecules of a velocity of 500 m/s scattered from Ar and Xe the correction amounts to about 5 %. Here the secondary beam source was assumed to be cooled as far as possible. For a secondary beam twice as fast as the primary NO beam, the correction becomes as large as 30 %. It turns out to be very advantageous to adjust the orientation field in the scattering region such that its direction is parallel or perpendicular to the most probable direction of the relative velocity. In this way one measures the smallest and the largest value of the total collision cross section and the corrections are minimal due to the extremum properties of the measured quantities.

No systematic study of this correction is known to us which includes the glory undulation of the anisotropy, only a rather crude method is discussed by Schwartz (1973).

The next important correction to the measured anisotropies arises from the pure long range origin of the non-glory part of the anisotropy A_{ng}, Stolte (1972). At small impact parameters scattering takes place anyhow irrespective of the orientation of the primary beam molecule. However, at large impact parameters and small deflections it can happen that with one orientation the event is registered as a scattering but not with the other orientation. Stolte (1972) derives for slit geometry

$$C_{\gamma,\Delta} = \frac{1 - 2\sqrt{2/3}\,0.06646k_1[\sigma^{(0)}(v_1)]^{1/2}\gamma}{1 - 0.06646k_1[\sigma^{(0)}(v_1)]^{1/2}\gamma} \qquad (8)$$

Here, $C_{\gamma,\Delta}$ is defined as the ratio of the corrected anisotropy A_{ng} to the uncorrected one; γ stands for the lab-angle under which in the plane of the detector slit the full width at half intensity of the primary beam is seen from the scattering centre; the wave number k_1 is calculated in the lab-system, for the primary beam molecules; $\hbar k_1 = m_1 v_1$. For NO this angular resolution correction amounts to about 10 %, Stolte (1973).

In Section VI we shall apply this correction to the results of Bennewitz (1969) when slit geometry was used and the angle γ was quoted. Stolte (1972) has shown that some conclusions were significantly altered after application of this correction.

We should like to mention one interesting point of the work of Stolte (1972). Instead of working with a central beam stop in order to get rid of non-focused molecules in other than the selected state a magnetic two-pole field is used with on–off modulation. Its purpose is to deflect the molecules in the selected state sufficiently so as not to pass through the detector slit if the deflection field is on. At the same time the signal due to the molecules in other states is not altered because these have a rather broad and flat distribution in the plane of the detector slit. This technique is applied to the paramagnetic $^2\Pi_{3/2}$-state of NO. An extension to diamagnetic molecules is discussed by Stolte (1972); here the source slit should be displaced cyclically from its centre position whereby the same effect is obtained as with the magnetic deflection of the NO molecules.

The measurements discussed so far are not restricted to a particular selected state. Bennewitz et al. (1969) report measurements on CsF with the $j = 1$, $m_j = 0$ or the $j = 2$, $m_j = 0$ state being selected. Theoretically those states with the highest preferential orientation should be selected. More practical are considerations of thermal occupation numbers and easy focusability. It is more difficult to produce a pure beam of molecules in higher rotational states because the Stark effect of neighbouring states differs less and consequently their discrimination involves more effort.

The use of a nozzle beam may help to increase the occupation of the low lying rotational states. In the case of NO, supersonic expansion by means of a nozzle had a negative effect, Stolte (1972), because here measurements were done on the excited state of the $^2\Pi$-doublet; the occupation of the selected state has been observed to decrease.

V. THEORY NEEDED FOR INTERPRETATION

In this section we shall deal with the theoretical work connecting the IP to the quantities which are determined in the total collision cross section measurements. Although the path leading from the IP to experimental quantities involves nothing but the solution of the Schrödinger equation there are no really satisfactory programs at hand for this problem. The angular dependence of the IP opens a great number of channels through which a collision may proceed. Instead of being confronted with the task of solving a (large number) of independent radial Schrödinger equations one has (in principle infinite) set(s) of such equations which are connected within each set by the AIP, Arthurs (1960). These sets contain energetically forbidden channels the influence of which may be judged to be small, and it has become the practice to truncate the sets to finite size. However, there is an infinite number of these energetically forbidden channels and their influence is not yet clear.

Even the allowed channels (open channels) present a horrible task if taken into account properly. This is partially an economic problem of how much computer time one wants to spend on the solution of a problem, but partially also a problem of numerical stability of the oscillating solution of a large set of coupled equations. Other ways of circumventing the solutions of the Schrödinger equation and working instead with sets of integral equations derived from the Lippman–Schwinger equation suffer from similar problems, Sams *et al.* (1969).

Concerning the total collision cross section measurements one may be hopeful that the full apparatus of many coupled equations to be solved is not needed, but that perturbation theory suffices to cope with the influence of the AIP; this perturbation theory must treat the IIP rigorously. The vehicle for such an approach is the d.w.a. Concerning the orientation dependent part of the total collision cross section, the AIP shows up in first order, i.e. by terms proportional to $q_{2.6}$ and $q_{2.12}$, in contrast to the cross section for rotational excitation and to the orientationally averaged total cross section where only terms in $q_{2.6}^2$ and $q_{2.12}^2$ and higher orders occur. This fact may be taken as the first indication that d.w.a. is a suitable means for the calculation of the orientation dependent part of the total collision cross section.

The common problem with perturbation theory is that it is only practical in first order and that it is rather hard to really estimate its accuracy as a first order approximation. One way to investigate its validity is to perform exact close coupling calculations on special systems where the number of participating channels is not exceedingly large. Reuss *et al.* (1969) have found remarkably good agreement with the results of d.w.a. for the system H_2–Ar below the threshold for rotational excitations (with all energetically forbidden channels being neglected). Similar findings were reported for instance by Riehl (1968) and Erlewein *et al.* (1968). Another approach to test the validity of d.w.a. employs the high energy approximation (h.e.a.) Glauber (1958), which was already used by Bennewitz (1964) for the interpretation of the anisotropy measurements. This approximation incorporates higher order terms in an approximate way (straight path trajectories, vanishingly small excitation energies) and it can therefore be used to estimate the influence of higher order terms. Extensive non-glory calculations of the total collision cross section for different types of interactions were recently published by Kuijpers *et al.* (1973) demonstrating that normally the first order distorted wave approximation suffices.

Within the framework of the linear d.w.a. one finds for the potential of (4) and (5) that the orientation dependent part $\Delta\sigma_{\parallel}$ of the total collision cross section can split into two contributions

$$\Delta\sigma_{\parallel} = \Delta\sigma_{\mathrm{g}} + \Delta\sigma_{\mathrm{ng}} \tag{9}$$

This quantity $\Delta\sigma_{\parallel}$ normally vanishes if orientationally averaged collisions are discussed. It often forms only 1 % of the total cross section σ, see the practical case shown in Fig. 5.

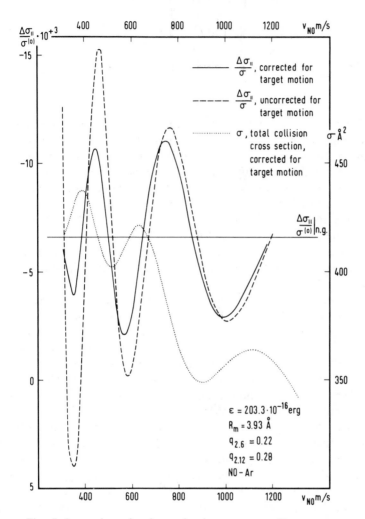

Fig. 5. Isotropic and anisotropic glory structure. For a system resembling NO–Ar in the experimental set up of Stolte (1972), the total collision cross section σ and its orientation dependent part $\Delta\sigma_{\parallel}/\sigma$ are calculated in d.w.a. as functions of the primary beam velocity v_{NO}. Extremes of σ coincide with zeros of $\Delta\sigma_{\parallel}/\sigma$. The orientation dependent part oscillates around the average value $(\Delta\sigma_{\parallel}/\sigma^{(0)})|_{ng}$.

The selected state is characterized by $j = \Omega = m_j = 3/2$.

Franssen (1973) finds for the glory contribution

$$\Delta\sigma_g = \frac{8\pi\mu}{\hbar^2 k^3} \frac{3m_j^2 - j(j+1)}{(2j+3)(2j-1)} \frac{j(j+1) - 3\Omega^2}{j(j+1)} \frac{l_0}{2}\left(\frac{\pi}{\eta_0''}\right)^{1/2} \frac{\varepsilon R_m}{2^{1/6}} \sin\left(2\eta_0 - \frac{\pi}{4}\right)$$

$$\times \{q_{2\cdot 12}[S(12, 0, E, b_0) - 3S(12, 2, E, b_0)]$$

$$- q_{2\cdot 6}[S(6, 0, E, b_0) - 3S(6, 2, E, b_0)]\} \tag{10}$$

In this expression μ stands for the reduced mass, l_0 for the orbital angular momentum for which the phase shift η assumes its maximum value η_0; the absolute value of the second derivative of the phase shift with respect to l at $l = l_0$ is denoted by η_0''; (10) is linear in the anisotropy parameters $q_{2\cdot 6}$ and $q_{2\cdot 12}$ of (5); the quantities $S(n, l - l', E, b)$ are integrals which are calculated and tabulated by Franssen (1973) for a range of values of the reduced energy E and the reduced collision parameter b with $b_0 = 2^{1/6}(l_0 + 1/2)/(R_m k)$. The parameter n takes the values 12 and 6 for the Lennard-Jones potential actually used; for $l - l'$ one has to insert the values 0 and 2, the difference of the orbital angular momentum of channels which are coupled by the interaction of (5).

The projection m_j of the rotational angular momentum j has to be taken with respect to the direction of an orientational field in the scattering region which is assumed to point parallel to the direction of the relative velocity. The factor $[3m_j^2 - j(j+1)][j(j+1) - 3\Omega^2][j(j+1)(2j+3)(2j-1)]^{-1}$ contains the dependence of $\Delta\sigma_g$ on the rotational state being selected, i.e. the quantum numbers of the rotational angular momentum; Ω stands for the projection of the electronic angular momentum on the axis of the considered molecule. For diatomics in a $^1\Sigma$-ground state Ω is zero.

As (10) shows $\Delta\sigma_g$ is an undulating function with regard to its dependence on the relative velocity of the collision partners; its phase is shifted by $\pi/2$ with respect to the isotropic glory contribution of the total collision cross section, see Fig. 4. The amplitude of $\Delta\sigma_g$ depends on the V_{AIP}, its magnitude being governed by the anisotropy parameters. The repulsive anisotropy proportional to $q_{2\cdot 12}$ and the attractive anisotropy proportional to $q_{2\cdot 6}$ counteract each other.

For the non-glory contribution of (9) one finds

$$\Delta\sigma_{ng} = \sigma^{(0)} \frac{3m_j^2 - j(j+1)}{(2j+3)(2j-1)} \frac{j(j+1) - 3\Omega^2}{j(j+1)} \frac{1}{10} q_{2\cdot 6} \tag{11}$$

In (11) the factor $1/10$ remains the same if the inverse power of the attractive terms in (4) and (5) are changed from 6 to 5, it becomes $2/21$ if one takes 7 instead of 6, Miller (1969). The repulsive part is of negligible influence on $\Delta\sigma_{ng}$.

The combined effect of $\Delta\sigma_g$ and $\Delta\sigma_{ng}$ is shown in Fig. 5. More useful for the interpretation of the measured orientational dependency of the total collision cross section is an expression where the difference is formed between $\sigma_\parallel$ and $\sigma_\perp$; according to Stolte (1972) and Kuijpers (1973)

$$\sigma_\parallel - \sigma_\perp = \tfrac{3}{2}\Delta\sigma_g + \tfrac{3}{2}\Delta\sigma_{ng} \tag{12}$$

$$= (A_g + A_{ng})\sigma^{(0)} \tag{13}$$

That the anisotropy A is a source of only restricted information can be seen from (10) and (11). For $\Delta\sigma_g$ only the solution of the radial Schrödinger equation enters with $l \approx l_0$. The corresponding collision parameter is $b_0 \approx 1\cdot1R_m$, Olson et al. (1968). Because the influence of the IP is largest at the moment of closest approach, we may conclude that $\Delta\sigma_g$ is sensitive to the AIP mostly at $R \approx 1\cdot1R_m$.

In (11) the main influence comes from $l \approx l_1$ with corresponding collision parameters, Reuss (1964) and (1965) and Stolte (1972),

$$b_1 = (6\pi\mu\varepsilon R_m^6/8\hbar^2 k)^{1/5} = 0\cdot4(\sigma^{(0)})^{1/2} \tag{14}$$

For most systems $1\cdot8R_m \lesssim b_1 \lesssim 2\cdot6R_m$, in the thermal energy region. The experimental values $\Delta\sigma_{ng}/\sigma^{(0)}$ thus determine the ratio V_{AIP}/V_{IIP} at $R \approx b_1$.

According to (10) and (11) both $\Delta\sigma_{ng}$ and $\Delta\sigma_g$ vanish for $j = 0$ and $j = 1/2$. For $j > 3/2$ other terms occur if a $P_4(\cos\theta)$-dependence is included in (3); this opens, in principle, the means to investigate the influence of different parts of the IP successively by selecting states with increasing rotational quantum numbers j.

Our (10) and (11) are of limited value for systems like H_2–Ar where the de Broglie wavelength is rather large at thermal energies and where only a few partial waves contribute to the cross section. Here exact quantum mechanical calculations must be performed, Reuss et al. (1969).

In linear d.w.a. the total collision cross section equals the elastic cross section, i.e. all transition with Δj and/or $\Delta m_j \neq 0$ are neglected. These transitions influence $\Delta\sigma_\parallel$ through terms at least quadratic in $q_{2\cdot6}$ and $q_{2\cdot12}$. As already mentioned, therefore, only the glory contribution can be affected by transitions to a measurable degree. Preliminary results of sudden approximation calculations of $\Delta\sigma_g$ show, however, that for the example of NO–Ar at relative velocities between 500 and 650 m/s the glory contribution. $\Delta\sigma_g$ is changed by less than 2% through quadratic contributions, Kuijpers (1974).

Bernstein et al. (1966) have brought the sudden approximation (s.a.) into a very suitable form for the investigation of molecular collisions. For straight path trajectories the high energy approximation and s.a. become identical. For curved trajectories s.a. gives simpler expressions. The trajectories are calculated from classical mechanics. Consequently, for the glory impact parameters b_0 a more realistic trajectory can be used in which the combined attraction and repulsion result in no final deflection at all but where the nearest distance of approach may be considerably smaller than b_0. As far as first order contributions in $q_{2.12}$ and $q_{2.6}$ to the anisotropy A are concerned, s.a. gives results identical with the d.w.a. results of Franssen (1973). This type of calculations enables one to really judge whether and when the linear approach of (10) is justified. Moreover, in the case of $NO-CO_2$ it may replace the linear approach and answer the question of how much of the glory is quenched by inelastic processes, Kuijpers (1974).

VI. RESULTS

A. The Non-Glory Contribution

The results of the Bonn group, Bennewitz (1969) and (1969a), are corrected as far as possible for the effects discussed in Section IV. The angular resolution correction appeared to be especially important because this correction is velocity dependent and if applied compensates a spurious velocity dependence of the anisotropy A for the system CsF-Ar, Stolte (1972). The values of A_{ng} and $q_{2.6}$ of Table I were obtained from the results of the Bonn group using the correction factor $C_{\gamma,\Delta}$ (8). This angular resolution correction is judged to be negligible for the old results of the Bonn group, Bennewitz et al. (1964), where circular diaphragms were used.

With regard to the Bonn results a problem was created by the fact that ratios of the velocity of the primary beam molecules to the velocity of the secondary beam molecules were used which were often much smaller than 1. This yields very unfavourable kinematics because the secondary beam is far from unidirectional although effusing from a multichannel array. At the moment we see no way to take large kinematics effects into account and have therefore disregarded the data of TlF-He, TlF-Ne and TlF-CH$_4$. The magnitude of correction is much enhanced due to the fact that the two field orientations in the scattering region at which the Bonn measurements were performed were fixed and could not be chosen, as nearly as possible, parallel and perpendicular to the direction of the relative velocity. Consequently, the corrections have to be applied not to a maximum and a minimum value of the total collision cross section but to intermediate values where a first order correction is already present.

In Table I the values for A_{ng} are not directly comparable because different states were selected in different experiments. Using (11) we can extract the anisotropy parameter $q_{2.6}$, see column 4. For those systems where the glory effect has been observed the average anisotropy has been obtained by a suitable averaging procedure, Schwartz (1973). The H$_2$-systems investigated

TABLE I

The non-glory results. When glory undulations were observed the average value over a full glory oscillation was calculated. The anisotropy parameter $q_{2.6}$ is connected to A_{ng} by (11), yielding $A_{ng} = -3q_{2.6}/50$ for $j = 1, m_j = 0$; $-3q_{2.6}/100$ for $j = m_j = \Omega = 3/2$; $-3q_{2.6}/70$ for $j = 2, m_j = 0$. The Unsöld-approximation gives for the long range anisotropy parameter $p_{2.6} = (\alpha_{\parallel} - \alpha_{\perp})/3\bar{\alpha}$ (last column). The value of A_{ng} for NO-CH$_3$F is put in parenthesis because it is uncorrected and its correction is as yet unknown. The values for b_1/R_m are calculated with the help of (14).

System	Selected state	$A_{ng} \cdot 10^3$ corrected anisotropy	$q_{2.6}$ anisotropy parameter	b_1/R_m	$p_{2.6}$	Reference
NO–CO$_2$	$j = m_j = \Omega = 3/2$	-8.6 ± 0.5	0.29 ± 0.02		0.16	Stolte (1973)
NO–CS$_2$	$j = m_j = \Omega = 3/2$	-6.1 ± 0.9	0.20 ± 0.03		0.16	Stolte (1973)
NO–N$_2$O	$j = m_j = \Omega = 3/2$	-7.6 ± 1.3	0.25 ± 0.05		0.16	Stolte (1973)
NO–N$_2$	$j = m_j = \Omega = 3/2$	-6.4 ± 0.2			0.16	Kessener (1975)
NO–Ar	$j = m_j = \Omega = 3/2$	-6.5 ± 0.5	0.22 ± 0.02	2.1	0.16	Schwartz (1973)
No–Kr	$j = m_j = \Omega = 3/2$	-7.1 ± 0.2	0.24 ± 0.01	2.2	0.16	Schwartz (1973)
NO–Xe	$j = m_j = \Omega = 3/2$	-6.6 ± 0.2	0.22 ± 0.01	2.4	0.16	Schwartz (1973)
NO–CCl$_4$	$j = m_j = \Omega = 3/2$	-4.2 ± 0.4	0.14 ± 0.01		0.16	Stolte (1973)
NO–SF$_6$	$j = m_j = \Omega = 3/2$	-5.3 ± 0.2	0.18 ± 0.01		0.16	Kessener (1975)
NO–CF$_3$H	$j = m_j = \Omega = 3/2$	(-16.7 ± 0.2)			0.16	Stolte (1973)
TlF–Ar	$j = 1; m_j = 0$	-14.0 ± 0.7	0.23 ± 0.01	2.6		Bennewitz (1969)
TlF–Kr	$j = 1; m_j = 0$	-14.0 ± 1.5	0.23 ± 0.03	2.5		Bennewitz (1969)
TlF–Xe	$j = 1; m_j = 0$	-15.8 ± 0.3	0.26 ± 0.05			Bennewitz (1969)
CsF–Ar	$j = 2; m_i = 0$	-15.2 ± 0.6	0.355 ± 0.01			Bennewitz (1969)

by Moerkerken *et al.* (1973) did not yield A_{ng}-values because here the experimental velocity range has been too small compared to a full glory undulation. The CO_2 molecule is itself very anisotropic and known to possess a very large inelastic cross section. Consequently, one expects that the interference effect causing the glory undulation is nearly absent so that one measures directly A_{ng}. Moerkerken (1973) has not detected any velocity dependence in this case and neither has Stolte (1973) found a velocity dependence of the anisotropy for CO_2 colliding with NO. However, recent extension by Zandee (1975) of the velocity range reveals that the anisotropy A rises for higher velocities and assumes positive values. Therefore, the H_2–CO_2 results shown in Fig. 7 cannot be interpreted with $A_g = 0$.

A somewhat similar situation existed with respect to the results for H_2–N_2; here, too, no velocity dependence of the anisotropy was detected by Moerkerken (1974). However, the measured anisotropy was positive and could very well be explained as a broad glory maximum. This case is discussed in Section VI.B.

With the help of Table I we can now answer the question of how far $q_{2.6}$ depends on the nature of the scattering partner. The long range parameter $p_{2.6}$ was found to be nearly independent, Victor (1970), Langhoff (1971) and Buckingham (1967). For TlF one finds $q_{2.6} = 0.23$ for all three heavy noble gases, within the experimental uncertainty. Incidentally, the same value is found for NO colliding with the three heavy noble gases; if the other gases are included in the comparison one finds that the average value of $q_{2.6} = 0.25$ lies within 2.5 standard deviations from the experimental results, with the exception of NO–CCl_4. We conclude, therefore, that the structure of the secondary beam molecules influences somewhat the measured anisotropy but that within 20% one can expect a constant value for $q_{2.6}$. The exception NO–CCl_4 we find hard to explain. Perhaps the ratio b_1/R_m (see 14) is already rather large in this case so that the smaller $p_{2.6}$ value is more approached than in the other cases. The fact becomes apparent that with the help of the scattering experiments discussed here, the parameter $q_{2.6}$ of (5) is determined at the distance $R \approx b_1$, which depends on the particular system and the experimental velocity range. To really determine the relevant ratio b_1/R_m one must know the IIP parameters ε and R_m. We have included only some estimated values of b_1/R_m in Table I because the values of ε and R_m are uncertain for the other systems.

B. The Glory Contribution

The first to accomplish a determination of the angular dependent repulsive part of the IP was again the Bonn group, Bennewitz *et al.* (1969), who published a $q_{2.12}$ value for CsF–He. As discussed in Section VI.A we refrain from an analysis of this system because of a rather uncertain and large kinematics correction.

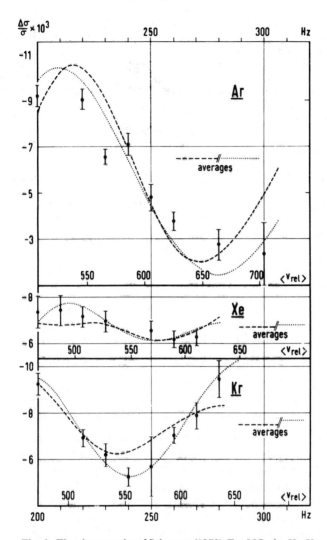

Fig. 6. The glory results of Schwartz (1973). For NO–Ar,,Kr, Xe the anisotropy $A_g + A_{ng} = \Delta\sigma/\sigma$ is plotted as function of the relative velocity. Of the two curves, $-----$ belongs to the upper set, $\cdots\cdots$ to the lower set of parameters, in table 2. Corresponding average values are shown as $----\|\cdots$ and are equal to $-3q_{2.6}/100$. The selected state of NO is characterized by $j = \Omega = m_j = 3/2$. (Reproduced from Schwartz (1973), with permission of North Holland Publishing Company).

In Fig. 6 glory measurements of the anisotropy by Schwartz *et al.* (1973) are shown together with a fit for a Lennard-Jones 12–6 potential using values of the anisotropy parameters as shown in Table II.

TABLE II

The glory results. For the NO-results two choices of εR_m lead to two different fits; at this moment it is uncertain which choice should be preferred. For the H_2-results the εR_m-values of Helbing (1968) and Butz (1971) are given in parentheses to be compared with the εR_m-values used for the $q_{2 \cdot 12}/q_{2 \cdot 6}$ fit. The H_2–He results are obtained from the IIP and AIP of Riehl (1973), and from the IIP of Gengenbach (1972) and the AIP of Shafer (1973). The $R_{m,a}/R_m$ values were calculated from (23). The R_m values for NO were derived from Tully (1973) and from the values of Hirschfelder (1965) using combination rules; for H_2 from Helbing (1968) and Butz (1971).

System	$\varepsilon R_m \cdot 10^{14}$ erg . Å	$q_{2 \cdot 12}/q_{2 \cdot 6}$	$R_{m,a}/R_m$	R_m Å	Reference
NO–Xe	$11.4 \pm 0 \cdot 15$	$1 \cdot 63 \pm 0 \cdot 05$	$1 \cdot 08$	$4 \cdot 3$	Schwartz (1973)
	$9.76 \pm 0 \cdot 15$	$1 \cdot 56 \pm 0 \cdot 05$	$1 \cdot 075$	$4 \cdot 2$	Schwartz (1973)
NO–Kr	$9.35 \pm 0 \cdot 05$	$1 \cdot 44 \pm 0 \cdot 05$	$1 \cdot 06$	$4 \cdot 05$	Schwartz (1973)
	$8 \cdot 25 \pm 0 \cdot 05$	$1 \cdot 77 \pm 0 \cdot 05$	$1 \cdot 10$	$3 \cdot 96$	Schwartz (1973)
NO–Ar	$8 \cdot 15 \pm 0 \cdot 1$	$1 \cdot 09 \pm 0 \cdot 05$	$1 \cdot 03$	$3 \cdot 93$	Schwartz (1973)
	$6 \cdot 73 \pm 0 \cdot 1$	$2 \cdot 00 \pm 0 \cdot 05$	$1 \cdot 12$	$3 \cdot 85$	Schwartz (1973)
H_2–Xe	$5 \cdot 3 \pm 0 \cdot 1$ (4.64 ± 0.12)	$1 \cdot 1 \pm 0 \cdot 2$	$1 \cdot 02$	$3 \cdot 90$	Moerkerken (1973, 1974)
H_2–Kr	$4 \cdot 5 \pm 0 \cdot 1$ $(3.97 \pm 0 \cdot 1)$	$1 \cdot 1 \pm 0 \cdot 2$	$1 \cdot 02$	$3 \cdot 65$	Moerkerken (1973, 1974)
H_2–Ar	$3 \cdot 6 \pm 0 \cdot 1$ $(3 \cdot 37 \pm 0 \cdot 14)$	$1 \cdot 2 \pm 0 \cdot 2$	$1 \cdot 03$	$3 \cdot 34$	Moerkerken (1973, 1974)
H_2–N_2	$3 \cdot 15 \pm 0 \cdot 1$ $(3 \cdot 15 \pm 0.1)$	$1 \cdot 2 \pm 0 \cdot 3$	$1 \cdot 03$	$3 \cdot 42$	Moerkerken (1973, 1974)
H_2–He			$1 \cdot 15$	$3 \cdot 71$	Riehl (1973)
			$1 \cdot 14$	$3 \cdot 38$	Shafer (1973)

The H_2–noble gas systems were investigated by Moerkerken *et al.* (1970 and 1973). Their choice was dictated by the consideration that in the velocity range of the experiment (1000–2000 m/s), inelastic processes could be ruled out and strict quantum mechanical calculations could be compared with the measurements, e.g. Reuss *et al.* (1969). Moreover, a large body of experimental results exists on the total collision cross section, Helbing (1968), and on the differential cross section, Kuppermann (1973). A relatively easy and secure intepretation was therefore hoped for. In Fig. 7 we present the latest results of the Nijmegen group.

From the amplitude of A_g mainly the ratio $q_{2 \cdot 12}/q_{2 \cdot 6}$ can be determined. The reason is that in (10) the two contributions proportional to $q_{2 \cdot 12}$ and $q_{2 \cdot 6}$ nearly cancel each other for all systems investigated so far.

In Table II the results obtained by Schwartz (1973), Moerkerken (1973) and Moerkerken (1974) are collected, together with the products εR_m which were assumed in the determination of $q_{2 \cdot 12}/q_{2 \cdot 6}$. Concerning the NO results two choices were made for εR_m, the upper one based on the differential cross section measurements by Tully (1972) on the systems N_2–Ar, N_2–Kr,

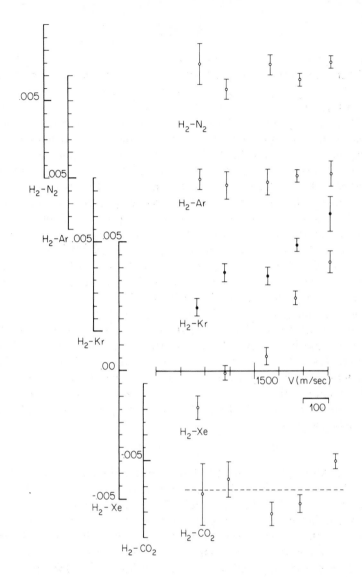

Fig. 7. The H_2 results of Moerkerken (1974). The anisotropy $A_g + A_{ng}$ is plotted as a function of the relative velocity v for the selected state $j = m_j = 1$. A sign error has been corrected in the original work.

O_2–Ar and O_2–Kr; the other one based on combination rules and the potential parameters found in Hirschfelder (1965). Although the first choice is the only one based upon experiments on related systems one has to keep in mind that differential cross section measurements explore a different region of the IP than do the total collision cross section measurements. In other cases (Butz, 1971 and Kupperman, 1973) differences of about 30% were found for the product εR_m, the value from the differential cross section being higher. Therefore, the decision which set of values ultimately has to be taken cannot yet be made; independent total cross section measurements are needed for this purpose.

Regarding the H_2 values, independent total cross section measurements by Helbing (1968) and Butz (1971) resulted in a rather precise determination of the parameter εR_m (see Table II). Moerkerken (1974) has used R_m values about 15% smaller in order to find a good fit for the $q_{2.12}/q_{2.6}$ ratios. Moerkerken (1973) suggests that the origin of the discrepancy may be attributed to the fact that energetically forbidden channels are neglected in the calculation and that these channels may play an important role especially in the case of H_2–Xe where the available energy in the centre of mass is largest. Recently, evidence is accumulating mainly from an analysis of spectra of H_2–noble gas-complexes (Leroy, 1974, Gordon, 1973, and from elastic scattering Kupperman (1973) that the Helbing and Butz values for εR_m may be 15% too small. The analysis of Moerkerken has therefore been repeated; preliminary considerations indicate that a fit with the new information about the IIP leads to the smaller $q_{2.12}/q_{2.6}$ values of Table II. The new fit has been obtained after a sign correction in the work of Moerkerken (1974). The results are in fair agreement with Leroy (1974). The $q_{2.6}$-term and the $q_{2.12}$-term still nearly compensate each other with slight dominance of the $q_{2.6}$-term. In the fits of Fig. 7 a slight dominance of the $q_{2.12}$-term is present. In Table II the entry $R_{m,a}/R_m$ is calculated with the help of (see Section II)

$$R_{m,a}/R_m = (q_{2.n}/q_{2.6})^{1/(n-6)} \qquad (23)$$

with $n = 12$. Although this quantity has a direct meaning with respect to the curve $V_{AIP}(R)$, it is hard to form an intuitive picture about its value for different systems. At $R = R_{m,a}$ the radial force between the two colliding molecules is uninfluenced by the orientation of the selected molecule.

At the end of this review it seems proper to ask what has been learned from anisotropy measurements of the total collision cross section with respect to the AIP. In Fig. 8 the V_{IIP} and the V_{AIP} are sketched as functions of R. We assume that the V_{IIP} is known from other experiments. The anisotropy measurements produce two pieces of information, i.e. A_{ng} and A_g. With this information, the whole curve V_{AIP} cannot be determined, but some features can be unambiguously established. From (23) we see that measurements of A_g determine the position of the minimum of V_{AIP}. Experience has shown that the value of $q_{2.12}/q_{2.6}$ is rather independent of the individual values of ε and R_m employed in the analysis, as long as the product εR_m remains constant (Moerkerken, 1974). Therefore, one needs an independent determination of εR_m of the V_{IIP} and a measurement of A_g to establish the value of $R_{m,a}/R_m$. Naturally, the absolute value of $R_{m,a}$ can only be found, then,

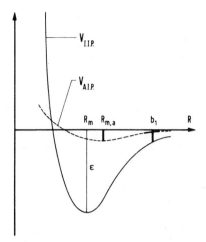

Fig. 8. The relevant pieces of information. The ratio $R_{m,a}/R_m$ determines the relative position of the minimum of V_{AIP}. The anisotropy parameter $q_{2.6}$ determine the value of V_{AIP} at $R = b_1$.

from an independent determination of R_m. The reader is reminded that although by a measurement of A_g a very limited range of the AIP is explored around the minimum position R_m, a two-parameter potential like the one of (4) may not be sufficient.

On the other hand a measurement of A_{ng} determines the ratio of V_{AIP}/V_{IIP} at $R = b_1 \approx 2R_m$. Here the intermolecular distance and the relative value of V_{AIP} at this distance are determined. Use is made of an at least approximate R^{-6} dependence of V_{IIP} and V_{AIP} at $R = b_1$. Stolte (1972) has shown that a small deviation from the inverse sixth power does indeed change the relation between A_{ng} and $q_{2.6}$ very little (see 11).

This essentially is all one gets from the anisotropy measurements discussed here. If one assumes a certain functional form to extrapolate from the value of V_{AIP} at $R = b_1$ to the position of the minimum at R_m one may be able to extend the description of V_{AIP} somewhat. At the moment, however, such an extension is purely guesswork.

Acknowledgements

What is worthwhile reading in the foregoing has emerged from discussions mainly with (in alphabetical order) H. A. Kuijpers, H. C. H. A. Moerkerken, M. Prior, H. L. Schwartz, S. Stolte and A. P. L. M. Zandee.

References

A. M. Arthurs and A. Dalgarno (1960). *Proc. Roy. Soc. A*, **256**, 540.

D. Auerbach, E. E. A. Bromberg and L. Wharton (1966). *J. Chem. Phys.*, **45**, 2160.

N. G. Basov and V. S. Zuev (1961). *Prob. Techn. Eks.*, **1**, 120.

G. Becker (1963). *Z. angew. Phys.*, **15**, 281.

H. G. Bennewitz, K. H. Kramer, W. Paul and J. P. Toennies (1964). *Z. Phys.*, **177**, 84.

H. G. Bennewitz, R. Gengenbach, R. Haerten and G. Müller (1969). *Z. Phys.*, **226**, 279.

H. G. Bennewitz and R. Haerten (1969a). *Z. Phys.*, **227**, 399.

K. Berkling, Ch. Schlier and P. Toschek (1962). *Z. Phys.*, **168**, 81.

R. B. Bernstein and K. H. Kramer (1966). *J. Chem. Phys.*, **44**, 4473.

R. B. Bernstein and R. A. La Budde (1973). *J. Chem. Phys.*, **58**, 110a.

R. J. Beuhler, R. B. Bernstein and K. H. Kramer (1966). *J. Am. Chem. Soc.*, **88**, 5331.

R. J. Beuhler and R. B. Bernstein (1969). *J. Chem. Phys.*, **51**, 5305.

P. R. Brooks and E. M. Jones (1966). *J. Chem. Phys.*, **45**, 3449.

P. R. Brooks (1969). *J. Chem. Phys.*, **50**, 5031.

P. R. Brooks (1973). *Faraday Disc.*, **55**.

F. Brouwer (1930). *Thesis*, Utrecht.

A. D. Buckingham (1967). In J. O. Hirschfelder (Ed.), *Intermolecular Forces*, Interscience Publishers, New York.

H. P. Butz, E. Feltgen, H. Pauly and H. Vehmeyer (1971). *Z. Phys.*, **247**, 70.

V. G. Cooper, A. D. May and B. K. Gupta (1970). *Can. J. Phys.*, **48**, 725.

W. Erlewein, M. von Seggern and J. P. Toennies (1968). *Z. Phys.*, **211**, 35.

J. J. Everdij, A. Huijser and N. F. Verster (1973). *Rev. Sci. Instr.*, **44**, 721.

W. Franssen and J. Reuss (1973). *Physica*, **63**, 313; *Physica*, **77**, 203 (1974).

R. Gengenbach and Ch. Hahn (1972). *Chem. Phys. Lett.*, **15**, 604.

R. J. Glauber (1958). *Lectures in Theoretical Physics*, Vol. I, Interscience, New York, p. 315.

M. D. Gordon and D. Secrest (1970). *J. Chem. Phys.*, **52**, 120.

R. G. Gordon (1973). Private communication.

F. Günther and K. Schügerl (1972). *Z. Phys. Chem.*, **80**, 155.

E. Helbing, W. Gaide and H. Pauly (1968). *Z. Phys.*, **208**, 215.

J. Hirschfelder, C. Curtiss and R. Bird (1965). *Molecular Theory of Gases and Liquids*, Wiley, New York.

R. M. Jonkman, G. J. Prangsma, T. Ertas, H. F. P. Knaap and J. J. M. Beenhakker (1968). *Physica*, **38**, 441.

D. Kakati and D. C. Lainé (1971). *J. Phys.*, **E4**, 269.

H. P. M. Kessener and J. Reuss (1975). *Chem. Phys. Lett.*, **31**, 212.

K. H. Kramer and R. B. Bernstein (1965). *J. Chem. Phys.*, **42**, 767.

M. Krauss and F. Mies (1965). *J. Chem. Phys.*, **42**, 2703.

A. Kuppermann, R. J. Gordon and M. J. Coggiola (1973). *Farad. Disc.*, **55**, 145.

J. W. Kuipers and J. Reuss (1973). *Chem. Phys.*, **1**, 64.

J. W. Kuipers and J. Reuss. *J. Chem. Phys.*, **4**, 277.

D. C. Lainé and R. C. Sweeting (1971). *Entropie*, **42**, 165.

P. W. Langhoff, R. G. Gordon and M. Karplus (1971). *J. Chem. Phys.*, **55**, 2126.

R. J. Le Roy and J. van Kranendonk (1974). *J. Chem. Phys.*, **61**, 4750.

A. D. May, V. Degen, J. C. Stroyland and H. L. Welsh (1961). *Can. J. Phys.*, **39** 1769.

W. H. Miller (1969). *J. Chem. Phys.*, **50**, 341.

H. C. H. A. Moerkerken, M. Prior and J. Reuss (1970). *Physica*, **50**, 499.

H. C. H. A. Moerkerken, A. P. L. M. Zandee and J. Reuss. *Chem. Phys. Lett.*, **23**, 320.

H. C. H. A. Moerkerken (1974). *Thesis*, Nijmegen.

R. E. Olson and R. B. Bernstein (1968). *J. Chem. Phys.*, **49**, 162.

J. A. A. Ragas (1972). *Internal Report*, Nijmegen.

Ramsey (1956). *Molecular Beams*, Oxford University Press, London.

J. van de Ree and R. G. Okel (1971). *J. Chem. Phys.*, **54**, 589.

J. Reuss (1964). *Physica*, **30**, 1465.

J. Reuss (1965). *Physica*, **31**, 597.

J. Reuss (1967). *Physica*, **34**, 413.

J. Reuss and S. Stolte (1969). *Physica*, **42**, 111.

J. W. Riehl, J. L. Kinsey, J. S. Waugh and J. H. Rugheimer (1968). *J. Chem. Phys.*, **49**, 5276.

J. W. Riehl, C. J. Fisher, J. D. Baloga and J. L. Kinsey (1973). *J. Chem. Phys.*, **58**, 4571.

W. N. Sams and D. J. Kouri (1969). *J. Chem. Phys.*, **51**, 4809.

P. B. Scott (1973). *J. Chem. Phys.*, **58**, 644.

R. Shafer and R. G. Gordon (1973). *J. Chem. Phys.*, **58**, 5422.

H. L. Schwartz, S. Stolte and J. Reuss (1973). *Chem. Phys.*, **2**, 1.

S. Stolte (1972). *Thesis*, Nijmegen.

A. Stolte, J. Reuss and H. L. Schwartz (1972). *Physica*, **57**, 254.

S. Stolte, J. Reuss and H. L. Schwartz (1973). *Physica*, **66**, 211.

B. Tsapline and W. Kutzelnigg (1973). *Chem. Phys. Lett.*, **23**, 173.

F. P. Tully and Y. T. Lee (1972). *J. Chem. Phys.*, **57**, 866.

A. Victor and A. Dalgarno (1970). *J. Chem. Phys.*, **53**, 1316.

A. P. L. M. Zandee (1975). *Private communication.*

ELECTRONIC EXCITATION IN COLLISIONS BETWEEN NEUTRALS

V. KEMPTER

Fakultät für Physik der Universität Freiburg/Br., Germany

CONTENTS

I. Introduction.	417
II. Non-Adiabatic Collisions	423
III. Experimental	424
A. General Remarks	424
B. Fast Beam Techniques	426
1. Beam Production by Neutralizing Ion Beams	428
2. Sputtered Beams	429
3. Seeded Supersonic Beams	430
IV. Experimental Results and their Interpretation	431
A. Total Cross Sections	431
1. Collisional Excitation: Processes (1) to (3)	432
a. Alkali–Rare Gases	432
b. Alkali–Alkali and Alkali–Mercury	439
c. Alkali–Molecules	442
d. Cadmium–Zinc	445
e. Rare Gas–Rare Gas	446
2. Chemiluminescence and Collisional Dissociation with Excitation: Processes (4), (5) and (6)	454
B. Differential Cross Sections	455
1. Amdur Type Experiments	455
2. Differential Cross Section Measurements	456
Acknowledgements	458
References	459

I. INTRODUCTION

The study of excitation in collisions between neutral atoms in the ground states is by no means a new field: Thomas's book[1] which covers this subject, lists some 30 reactions studied in beam experiments, mostly carried out after 1960. However, until very recently, only a few experiments of this type have been performed at low collision energies (<1 keV). The reason was that neutral beam intensities in this range were too low to allow studies in the threshold region of excitation in collisions between neutrals. Better detection techniques and the improvement of fast beam sources for the range between 1·0 and 100 eV gave the impetus for a number of studies at low energies, starting from about 1968.

TABLE I

Transvibronic reactions. Underline denotes translational energy, asterisk
electronic excitation and dagger vibrational excitation

$$
\underline{A} + B \rightarrow AB \rightarrow (AB)^*
\begin{cases}
\rightarrow A\ + B^* \\
\rightarrow A^* + B
\end{cases} (1) \\
\rightarrow A^* + B^* \quad (2) \\
\rightarrow A^+ + B\ + e \quad (3)
$$

$$
\left.
\begin{array}{l}
\rightarrow A\ + B^* \\
\rightarrow A^* + B
\end{array}
\right\} (1) \\
\left.
\begin{array}{l}
(1) \\
(2) \\
(3)
\end{array}
\right\}
\begin{array}{l}
\text{Collisional} \\
\text{Excitation}
\end{array}
$$

$\underline{A} + BC \rightarrow AB + C^*$	(4)	Chemiluminescence
$\underline{A} + BC \rightarrow (AB)^* + C$	(5)	
$\underline{A} + BC \rightarrow A + B + C^*$	(6)	Collisional dissociation with excitation
$A^* + BC \ \rightarrow \underline{A} + (BC)^\dagger$	(7)	Quenching
$A^* + BC \ \rightarrow A + (BC)^*$	(8)	Excitation transfer
$A^* + BC \ \rightarrow A + B + C^*$	(9)	
$A\ + (BC)^\dagger \rightarrow A^* + BC$	(10)	Collisional excitation by vibration

Table I shows the possible processes (1) to (6) which can result from conversion of translational into electronic energy. The excited states A* or B* can either be metastable or can decay by electron or photon emission. A or B can also denote molecules. Common to processes (1) to (3) is the transition from the ground state to some excited states of the quasimolecule AB formed during the collision. When the particles separate again, there are several possibilities:

The excited atoms A* or B* can decay when the particles are well separated, giving rise to the emission of photons or electrons whose energy is solely determined by atomic properties. When the excited atoms decay during the time the two particles are still close together, molecular emission will be observed. Of course, molecular effects do only occur, if the lifetime of the excited states is of the same order as the collision time. In (3) the excitation energy is high enough to ionize one of the colliding particles, very similar to Penning ionization.

Collisional excitation can be taken as part of the larger class of the 'transvibronic' reactions,[2] all involving interconversion of translational, vibrational and electronic energy; the more important processes are also listed in Table I. Several of these are related to each other by the principle of microreversibility;[19,20] take, for instance, (1), excitation in a collision between two atoms: if the cross section for excitation can be measured as a function of the collision energy, the cross section for collisional quenching, process (7), can be calculated from the former one.

Several excellent reviews on most of the processes (1) to (10) exist, covering the literature up to 1970. The reader is referred to[4,5] for collisional quenching, to [1,2,5,6] for collisional excitation, to [3,4,7,8] for excitation transfer and to[9] for chemiluminescence.

This paper tries to review studies of the processes (1) to (6) at collision energies below approximately 1 keV. The excitation mechanism will be discussed in relation to the other transvibronic reactions, in as far as the experimental results are conclusive.

A catalogue of the available measurements on the processes (1) to (6) is presented in Table II. It covers the data which appeared before July 1973.

TABLE II
Catalogue of available data

Projectile	Target	Excited particle	State	Collision energy (eV)	Data	Reference
Ar	Ar	Ar	$4s, 4s'$	11·5–16·3	E_{th}, Q_i	90
Ar	Ar	Ar	$4s, 4s'$	15–400	Q_{ij}	89
Ar	Ar	Ar	$3s^2 3p^5 nl$	20–450	E_{th}, Q_{ij}	92, 117
Ar	Ar	Ar	$3s^2 3p^4 nl, n'l'$; n, l, n', l' unspecified	200–3000	Q_i (rel)	88
Ar	Ar	Ar	$3s 3p^6 4s$	40–250	E_{th}, Q_i (rel)	83, 88
Ar	Kr	Ar, Kr	Autoionizing states of Ar and Kr	80–600	Q_i (rel)	83
Ar	Xe	Xe	Autoionizing Xe-states	80–600	Q_i (rel)	83
Ar	H_2	H	$n = 3, 4$	48–1400	Q_{ij} (rel)	86
		Ar	Unspecified	48–1400	E_{th}, Q_{ij} (rel)	86
Cd	Zn	Cd	$5^3 P_1$	15–370	E_{th}, Q_{ij} (rel)	32
He	He	He	He($1snl$); n, l unspecified	100–175	$\sigma_i(\vartheta)$	121
He	He	He	Simultaneous exc. of both atoms to He($1snl$); n, l unspecified	100–175	$\sigma_i(\vartheta)$	121
He	He	He	Sim. exc. of both atoms to He($1snl$); nl unspecified	1000–2000	$\sigma_i(\vartheta)$	74
He	He	He	He($1snl$)	20–600	E_{th}, Q_{ij}	92
He	H_2	He	Unspecified	230–11 500	Q_{ij} (rel)	85, 86
		H	$n = 3, 4$	230–11 500	Q_{ij} (rel)	85, 86
He	H_2	H	$n = 3, 4, 5$	367	Q_{ij} (rel)	37

TABLE II cont.

Projec-tile	Target	Excited particle	State	Collision energy (eV)	Data	Reference
He	O_2	O	Autoionizing states of O	250	Q_i (rel)	82
He	Ar	Ar	$5d, 5d', 6d$	20–700	E_{th}, Q_{ij}	92
He	Ar	Ar	$3s3p^64s$, $3s^23p^4nln'l'$	200–2000	Q_{ij} (rel)	83, 87
He	Ne	Ne	$3p', 3p$	920	Q_{ij} (rel)	37
He	Ne	Ne	$3p'$	50–600	E_{th}, Q_{ij}	92
He	Ne	Ne	Autoionizing Ne^--states	200–2000	Q_i (rel)	87
He	Kr	Kr	Singly + doubly excited auto-ionizing states	100–2000	Q_i (rel)	83, 87
He	Xe	Xe	Singly + doubly excited auto-ionizing states	100–2000	Q_i (rel)	83, 87
K	He	K	4^2P	30–80	E_{th}, Q_{ij} (rel)	63
K	He	K	4^2P	100–300	Q_{ij} (rel)	65
		K	5^2P	100–300	Q_{ij} (rel)	65
K	Ne	K	4^2P	100–1000	Q_{ij} (rel)	65
			5^2P	100–1000	Q_i (rel)	65
K	Ne	K	4^2P	100–260	E_{th}, Q_{ij} (rel)	63
K	Ar	K	4^2P	75–400	E_{th}, Q_{ij} (rel)	63
	Ar	K	$4^2P_{1/2}, 4^2P_{3/2}$	15–300	E_{th}, Q_{ij} (rel)	64, 66
	Ar	K	4^2P	150–1500	Q_{ij} (rel)	65
	Ar	K	5^2P	400–1500	Q_{ij} (rel)	65
K	Ar	Ar	$4s$	40–150	E_{th}, Q_{ij} (rel)	64, 114
	Ar	K	$4^2P_{3/2}$	60–200	P_{ij}	116
K	Kr	K	4^2P	100–560	E_{th}, Q_{ij} (rel)	63
K	Kr	K	4^2P	200–2000	Q_{ij} (rel)	65
			5^2P	200–2000	Q_{ij} (rel)	65
K	Kr	K	$4^2P_{1/2}, 4^2P_{3/2}$	20–300	E_{th}, Q_{ij} (rel)	64, 66
K	Kr	K	$4^2P_{3/2}$	100–300	P_{ij}	116
K	Xe	K	4^2P	100–800	E_{th}, Q_{ij} (rel)	63
K	Xe	K	$4^2P_{1/2}, 4^2P_{3/2}$	20–450	E_{th}, Q_{ij} (rel)	64, 66
		Xe	$7p$	50–300	E_{th}, Q_{ij} (rel)	64, 114
		Xe	$6s$	50–300	E_{th}, Q_{ij} (rel)	64, 114
	Xe	K	$4^2P_{3/2}$	100–250	P_{ij}	116

TABLE II cont.

Projectile	Target	Excited particle	State	Collision energy (eV)	Data	Reference
K	K	K	$4^2P_{1/2}, 4^2P_{3/2}$	1–50	E_{th}, Q_{ij}	69, 72
K	K	K	4^2P	100	$\sigma_i(\vartheta)$	42
K	Hg	K	$4^2P_{1/2}, 4^2P_{3/2}$	1–50	E_{th}, Q_{ij}	69, 73
K	H_2	K	4^2P	7–35	Q_{ij} (rel)	63, 2
K	HCl	K	4^2P	30–100	Q_{ij} (rel)	2
K	Cl_2	K	4^2P	20–300	E_{th}, Q_{ij} (rel)	63
K	N_2	K	4^2P	1·5–250	E_{th}, Q_{ij} (rel)	63, 30
	N_2	K	4^2P	1–15	E_{th}, Q_{ij}	75, 76
	N_2	K	4^2P	1–3	E_{th}, Q_{ij}	2, 62
	N_2	K	$4^2P_{1/2}, 4^2P_{3/2}$	2–50	Q_{ij} (rel)	78
	N_2	K	$4^2P_{1/2,3/2}$	5, 14	P_{ij}	79
	N_2	K	$6^2S, 5^2P$	10–50	Q_{ij} (rel)	77
	N_2	K	Unspecified	40–80	$\sigma_i(\vartheta)$	42
K	O_2	K	4^2P	2–250	Q_{ij} (rel)	63, 30
	O_2	K	4^2P	1–15	E_{th}, Q_{ij}	75, 76
	O_2	K	$4^2P_{1/2}, 4^2P_{3/2}$	12–31	Q_{ij} (rel)	78
	O_2	K	$6^2S, 5^2P$	10–50	Q_{ij} (rel)	77
K	NO	K	4^2P	2–20	Q_{ij} (rel)	30
	NO	K	4^2P	1–15	E_{th}, Q_{ij}	75, 76
	NO	K	$6^2S, 5^2P$	10–50	Q_{ij} (rel)	77
K	CO	K	Unspecified	40–80	$\sigma_i(\vartheta)$	42
	CO	K	4^2P	3–20	Q_{ij} (rel)	30
	CO	K	4^2P	1–15	E_{th}, Q_{ij}	75, 76
	CO	K	$6^2S, 5^2P$	10–50	Q_{ij} (rel)	77
	CO	K	$4^2P_{1/2}, 4^2P_{3/2}$	10–29	Q_{ij} (rel)	78
	CO	K	4^2P	1–3	E_{th}, Q_{ij} (rel)	62, 2
K	CO_2	K	4^2P	1–15	E_{th}, Q_{ij}	76
	CO_2	K	$4^2P_{1/2}, 4^2P_{3/2}$	17–37	Q_{ij} (rel)	78
	CO_2	K	$6^2S, 5^2P$	10–50	Q_{ij} (rel)	77
K	C_2H_4	K	4^2P	1–15	E_{th}, Q_{ij}	76
	C_2H_4	K	$5^2P, 6^2S$	10–50	Q_{ij} (rel)	77
K	NO_2	K	4^2P	1–15	E_{th}, Q_{ij}	76
	NO_2	K	$4^2P_{1/2}, 4^2P_{3/2}$	25–100	Q_{ij} (rel)	78
	NO_2	NO_2	$^2B_2$ or $^2B_1$	1–20	Q_{ij} (rel)	81

TABLE II cont.

Projec-tile	Target	Excited particle	State	Collision energy (eV)	Data	Reference
K	SO$_2$	K	4^2P	1–15	E_{th}, Q_{ij}	76
	SO$_2$	K	4^2P	1–25	E_{th}, Q_{ij} (rel)	2
K	C$_6$H$_6$	K	$5^2P, 6^2S, 4^2P$	10–50	Q_{ij} (rel)	77
Kr	Kr	Kr	$5p', 6p$	20–450	E_{th}, Q_{ij} (rel)	117
Kr	Kr	Kr	Unspecified auto-ionizing states	3000	Q_i (rel)	88
Kr	Ar	Ar, Kr	Ar$(3s3p^64s)$, Kr$(4s4p^65s)$	400–600	Q_i (rel)	83
Na	N$_2$	Na	3^2P	1–20	E_{th}, Q_{ij}	76
Na	O$_2$	Na	3^2P	1–20	E_{th}, Q_{ij}	76
Na	NO	Na	3^2P	1–20	E_{th}, Q_{ij}	76
Na	CO	Na	3^2P	1–20	E_{th}, Q_{ij}	76
Na	CO$_2$	Na	3^2P	1–20	E_{th}, Q_{ij}	76
Na	CS$_2$	Na	3^2P	1–20	E_{th}, Q_{ij}	76
Na	C$_2$H$_4$	Na	3^2P	1–20	E_{th}, Q_{ij}	76
Na	SO$_2$	Na	3^2P	1–20	E_{th}, Q_{ij}	76, 80
		SO$_2$	$^1A_1, ^3A_1$	1–20	E_{th}, Q_{ij} (rel)	80
Na	NO$_2$	Na	3^2P	1–20	E_{th}, Q_{ij}	76, 80
		NO$_2$	$^2B_2, ^2B_1$	1–20	E_{th}, Q_{ij} (rel)	80, 81
Na	K	K	$4^2P_{1/2}, 4^2P_{3/2}$	1–50	E_{th}, Q_{ij}	69, 72
Ne	Ne	Ne	$3p'$	30–1000	E_{th}, Q_{ij} (rel)	91
Ne	Ne	Ne	$4p, 3p'$	20–450	E_{th}, Q_{ij} (rel)	117
Ne	Ne	Ne	$3p'$	30–500	E_{th}, Q_{ij} (rel)	92
Ne	H$_2$	Ne	Unspecified	130–3600	Q_{ij} (rel)	86
Ne	H$_2$	H	$n = 3, 4$	130–3600	Q_{ij} (rel)	86
NO	O$_3$	NO$_2$	Unspecified	0·1–0·5	E_{th}, Q_{ij} (rel)	84

Q_i	level excitation cross section.
Q_{ij}	line emission cross section.
Q_i(rel) and Q_{ij}(rel)	relative values of level excitation and line emission cross sections.
$\sigma_i(\vartheta)$	differential level excitation cross section.
E_{th}	threshold for level excitation.
P_{ij}	polarization of emission.

II. NON-ADIABATIC COLLISIONS

Since the velocity of the atoms in the range under study is small compared to the velocity of the electrons, the particle wave functions change adiabatically except near crossings or pseudo-crossings of the molecular states formed by the colliding atoms. Transitions between molecular states, which lead to electronic excitation, are only possible near crossings or pseudo-crossings.

The problem is to calculate all molecular states leading to a given excited atomic state, and find out whether one or more of these are crossing the ground state. If all potential curves involved and the molecular wave-functions in the region of interaction are known, the transition probability between the states can be calculated by various models. A number of excellent reviews exist on this subject,[10,11,12,18,22] and so it will not be discussed here. For a short comprehensive discussion the reader is referred to the article on collisional ionization in this volume.[118]

In the case of a collision between two atoms one can often decide from the symmetry of the molecular states involved, whether a given atomic state can be excited or not. The selection rules for the most important cases of radial and rotational coupling are given in Herzberg's book.[14] Let Ω, Λ, Σ be the quantum numbers of the components of the total, the orbital and the spin angular momentum of the electrons along the internuclear axis. If the states belong to Hund's coupling case (a) or (b)

$$\Delta\Lambda = 0, 1 \tag{11}$$

If both states belong to case (a), to a good approximation as long as the interaction between the spin and the orbital angular momentum is not too great,

$$\Delta\Sigma = 0 \tag{12}$$

If the interaction is at such large internuclear distances that both states belong to Hund's case (c), the following rule for Ω holds:

$$\Delta\Omega = 0, 1 \tag{13}$$

Naturally in all symmetrical cases the parity is conserved:

$$u \rightarrow u, \qquad g \rightarrow g, \qquad u \nrightarrow g \tag{14}$$

One important rule for excitation in atom–molecule collisions comes from the Franck–Condon principle. When only small amounts of kinetic energy are converted into vibrational energy it seems reasonable that the non-adiabatic transition is governed by the Franck–Condon factor in the transition region. However, due to the interaction of the heavy particles these factors can be quite different from those for the undisturbed molecule. Bauer et al.[15] show that the interaction matrix element which determines the transition probability can be written as a product of the electronic part of the matrix element and the square root of the Franck–Condon factor.

Until now only a few excited molecular states have been calculated: He_2,[105] Ne_2,[106] $LiNa$,[70] Na_2,[71] some of the alkali–rare gas pairs,[107,108,109] and the heavy alkali atom pairs.[110,111] Some of them have only been calculated for large distances and are not useful in order to understand the mechanism for collisional excitation.

In almost all cases where crossings are at distances of the order of 1 Å or less, potentials are not known. Qualitative predictions about the transition mechanism can still be made by using the molecular orbital (MO) model. The MO energies depend on the internuclear distance which varies during the collision. The MO energies as a function of the internuclear distance are found from a correlation diagram connecting the united and separated atom limits. The rules from which these diagrams can be constructed are given in Ref. 16 for the homonuclear, and in Ref. 17 for the heteronuclear case. Transitions between MO's are possible at crossings between the diabatic molecular states made up from antisymmetrized products of the one electron MO functions. During the collision one or more electrons may be promoted from the orbitals they occupy into other, empty orbitals. When they remain in their new orbitals after separation of the colliding atoms, this leads to excitation of one or even both atoms. The selection rules for transitions between MO's caused by various types of interactions are given in Ref. 16. The probability for transitions between MO's can be calculated by means of similar models as discussed before for transitions between molecular states.[16,119,120] For radial coupling the Landau–Zener model has been applied. Lichten[16] and Barat and Lichten[17] point out that the concept of the diabatic molecular states may lead to unreliable predictions for outer shell excitation when the interactions between the MO's will be of the same order as their energy. Nevertheless the model seems to be very useful for qualitative considerations as will be shown below.

III. EXPERIMENTAL

A. General Remarks

The ideal machine for excitation studies should allow for the determination of the full differential cross section $\sigma(\theta, E, \Delta E)$ as a function of the scattering angle θ, the collision energy E and the energy loss ΔE. $\sigma(\theta, E, \Delta E)$ can be determined in several ways:

(a) By measuring the energy loss spectrum of the projectile beam as a function of the scattering angle.

(b) By measuring simultaneously the scattering angles of projectile and target atom. This method is very difficult to apply when both scattered particles are neutral atoms.

(c_1) Spectroscopic observation of the radiation emitted from the excited particle in conjunction with the scattering angle of the projectile.

(c_2) By measuring the energy spectrum of the electrons emitted in the decay of the excited particle in conjunction with the scattering angle of the projectile.

Most experiments concerned with excitation in collisions between neutrals do not determine full differential cross sections. Usually the total cross section, defined as

$$Q(E, \Delta E) = 2\pi \int \sigma(\theta, E, \Delta E) \sin \theta \, d\theta \qquad (15)$$

is measured. Measurements determining $Q(E, \Delta E)$, either by observing electron or photon emission, or measurements where ratios of Q's are determined, are described in Section IV.A.

Only a few studies of differential cross sections $\sigma(\theta, E, \Delta E)$ have been performed so far, and only method (a) has been applied. Measurements of σ are described in Section IV.B.

Table III gives an impression of what information on the excitation process can be gained from the various types of experiments. The references lead the reader to a number of papers which give the general theoretical background to the understanding of the experiments performed.

TABLE III

Experiments yielding information on excitation processes

Determination of the total cross section $Q(E, \Delta E)$	
Spectrum of emitted radiation. Energy distribution of the ejected electrons	Spectrum of excited states and the relative probabilities for their population[16,17]
Angular distributions of the emitted electrons	Excitation mechanism[119] Identification of the excited state
Polarization of the emission[1]	Relative cross sections for population of the magnetic sublevels[21,126]
Energy dependence of the excitation cross section	Properties of the potentials which are involved in the excitation process in the region of interaction[10,11,12,118,119]
a. absolute size	crossing distance, R_c
b. shape	coupling matrix element
c. threshold energy	potential energy at R_c, $V(R_c)$
Structure in the total excitation cross section	Long-range interaction between the excited states[67,68,123]

TABLE III cont.

Determination of differential cross sections $\sigma(\vartheta, E, \Delta E)$

Energy loss spectrum for a fixed scattering angle ϑ	Spectrum of the excited states and the relative probabilities for their population[16,17]
Angular distribution of the scattered particles for a fixed energy loss ΔE	Form parameters of the potentials involved Details of the excitation mechanism[10,11,12,118,96]
a. threshold value of $\tau = \vartheta E$	R_c, $V(R_c)$
b. shape of distribution	Coupling matrix element, form parameters
c. distortions in the angular distributions for elastic scattering	R_c, $V(R_c)$, coupling matrix element
d. oscillations in the inelastic cross section (Stueckelberg oscillations)	coupling matrix element, form parameters

Except for the beam production and its detection, the setup for the measurement of total cross sections is by no means different from what is commonly used in inelastic ion–neutral collision experiments and those between neutrals at high collision energies. Techniques for spectral analysis and the detection of light emission in the various wavelength regions are discussed in full detail in Ref. 1, and no detailed description will be given here. For a very detailed study on what requirements will have to be fulfilled in order to measure a true excitation cross section, the reader can also be referred to Ref. 1. A short description of an experiment for the study of autoionizing excited states is given in Section IV.A.

The measurement of differential elastic and inelastic cross sections is no different from what has been used in the study of pure elastic scattering,[23,24] vibrational excitation[24] and chemical reactions[25] in low energy collisions between neutrals. The author feels however that this may be the place to give a short summary of the techniques of producing and detecting fast neutral beams in the range from 1·0 to 1000 eV energy.

B. Fast Beam Techniques

Only three methods proved to yield beams of an intensity high enough to perform excitation experiments: beam production by ion neutralization, sputtered beams and seeded supersonic beams. The merging beam technique[26] does not seem to be suited for the measurement of optical excitation cross sections because of the very long beam interaction path.

Figure 1 shows a comparison of the beam intensities which can be reached by means of the various methods in their specific energy range. The graph is

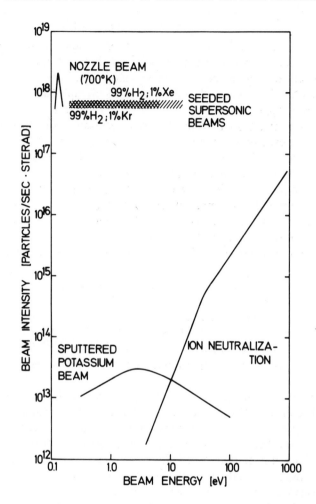

Fig. 1. Comparison of fast beam sources. Beam intensities were estimated from published data by the present author.

not much different from Fig. 16 in Ref. 24 which gave the state of the art in 1968. However the number of applications of the various techniques has been going up very fast.

As one can see there is no method which covers the whole range from 1·0 to 1000 eV; the sputtered alkali beams have an energy distribution which covers the range from below 1 eV to over 100 eV, but the intensities and the energy resolution which can be reached at energies above 50 eV are not satisfactory.

1. Beam Production by Neutralizing Ion Beams

The neutralization of a fast ion beam is the oldest and by far the simplest method to achieve high-energy neutral beams. It is still the only method which is practical to use at energies higher than 50 eV. Very detailed reviews are available on this topic,[27,28] and so only the principles and a short summary of newer applications will be given here. Ions of a given species are produced by electron bombardment, surface ionization or other methods. The ions are accelerated to the desired beam energy and, after collimation, directed into a charge exchange cell. Some fraction of the ions is neutralized by charge exchange reactions; the remaining ions are swept out of the beam by an electric field. In order to achieve high intensities one is using near resonant or, if possible, resonant charge exchange.

In some cases a mass spectrometer is needed between ion source and charge transfer cell in order to select the desired ion species.

This method has been applied to beam experiments for several groups of substances:

metals with low melting point: alkalis,[29,30,112] mercury,[31] cadmium and zinc[32]

permanent gases: N_2,[33] rare gases[34,35,36]

reactive atoms: H, O, N and others[28]

metastable atoms and long lived molecules: rare gases[37,38,40] $N_2(B^3\Pi_g)$.[41]

This list is by no means complete, but tries to give one or two examples for each group.

The method of producing fast beams by charge transfer has several nice features. The energy of the beam can be varied fast by changing the lens potentials of the optical system between ion source and charge transfer cell. Furthermore the beam can be pulsed,[41] the pulse width being as short as 50 nsec.[42] This implies that the life-time of excited products or the energy loss spectrum of the scattered particles can be measured.

The energy resolution of the beam is reasonably good; usually $\Delta E = $ 2–5 eV over the whole energy range and no velocity selector will be needed. The cross sections for non-resonant processes are lower than those for resonant charge exchange and, furthermore, particles formed in non-resonant processes will mainly be scattered into large angles. Provided the acceptance angle in the forward direction is sufficiently small, the neutral beam will be clean, even in cases where no mass spectrometer is used, and this means that some 99 % of the atoms will be in the same electronic state.[35,36,74]

The current of slow ions produced in neutralizing the fast ion beam is a measure for the fast neutral beam intensity, so that no special detector system is necessary to determine the fast beam intensity versus collision energy.

Towards lower energies the beam intensity decreases sharply because of space charge effects; in most experiments this limits the useful range to energies higher than 40 eV. The most intense beams at low energies have been produced for the alkali metals. Here a fairly high intensity ion beam of small energy spread can be produced by surface ionization. No mass spectrometer is needed between the ion source and the charge transfer region, and the distance between the ionizer filament and the charge transfer cell can be made as short as 1 mm. Lacmann and Herschbach[30] report a potassium beam intensity of $2 . 10^{13}$ atoms sec^{-1} sterad^{-1} at 10 eV beam energy. The energy spread was estimated to be $\Delta E = 0.5$ eV over the whole energy range from 1 to 100 eV. In this particular arrangement space charge effects seem to be minimized, and the beam intensity seems to be the highest one which can be reached by the charge transfer method.

Sometimes the hot cathode filament of the ion source may disturb measurements where a low light background is needed; Dworetsky et al.[39] have built an electron bombardment ion source which still gives a high ion current even though the cathode temperature is only near 1000°C.

2. Sputtered Beams

A fast neutral beam can be produced by bombarding solid targets with an intense ion beam of 3 to 50 keV energy. About 95 % of the sputtered particles are leaving the target as neutrals.[43] The sputtered neutral atoms have a very broad energy distribution which for most metals seems to be peaked at a few eV energy;[43] therefore a velocity selector is needed in order to get a monoenergetic beam.

This method can be applied to practically all metals and certain non-metallic compounds too. Beams of the following metals have been produced and studied in some detail: alkalis (Li, Na and K),[44,45,46,47,52] Ba, Al, Ti and Ag,[51] and indium.[47] It has also been shown that a halogen atom beam can be produced by sputtering silver halides[49] or alkali halides.[50,124]

Common to all sputtering machines is a high power ion source either of the Penning type or of the Plasmatron type, capable of delivering ion currents of the order of 5 mA at approximately 15 keV draw-out voltage. By means of a lens system, the ions are collected to a beam and focused in a rectangular spot on the target. The current density at the target at 15 keV ion energy is typically 30 mA/cm^2; the spot size is 1×15 mm^2. A beam of sputtered atoms is formed by a slit system and enters the velocity selector.

Two methods are available for the velocity selection of the beam. By pulsing the bombarding ion beam, a pulsed neutral beam will be formed, and a time of flight method may be applied. In many applications the beam or the reaction products can be analysed directly by a multichannel analyser.[113]

In applications where a monoenergetic beam is needed, mechanical chopping by means of one slotted wheel is sufficient.[45,52] In all time-of-flight methods the velocity resolution $\Delta v/v$ is proportional to the beam velocity; typical values for a potassium beam are $\Delta v/v = 0.05 . (E)^{1/2}$, where E is in eV. A better energy resolution at higher energies can be achieved by using a mechanical Fizeau type selector familiar from molecular beam experiments at thermal energies.[24] This method has been used up to 40 eV for potassium with an energy independent velocity resolution of $\Delta v/v = 10\%$.[46] It has the disadvantage that there is no possibility for a time-of-flight analysis of the reaction products nor for lifetime measurements on excited products.

A serious disadvantage of the sputtered beams is the low intensity caused by the need of a velocity selector.

Furthermore it has been found that sputtered alkali beams contain a certain amount of dimers, usually more than 10% at energies below 1 eV.[48] At energies $\gtrsim 10$ eV a background of fast metastable Ar-atoms seems to be present when alkali metals are sputtered by Ar-ions.[53,54] The intensity of excited Ar-atoms is probably of the order of 10% of the alkali beam intensity at 20 eV.[53]

Except for the alkali beams, where surface ionization can be used, the weak beam is difficult to detect. Cohen et al.[51] succeeded in detecting Ba and Al beams with a specially designed electron bombardment detector. In spite of these problems sputtering is for many substances, like light metal atoms and any other material of high melting point, the only way to produce a fast neutral beam.

3. Seeded Supersonic Beams

A high energy beam of heavy atoms or molecules can be obtained through supersonic expansion of a gas containing a small fraction of the desired species mixed with light molecules at high temperatures.[55,56]

Typically the gas is expanded through a 0·1 mm diameter nozzle which can be heated up to about 1000°C. The nozzle pressure has to be of the order of 2000 torr. The core of the beam passes through a skimmer with 0·5 mm diameter opening whose position in respect to the nozzle can be varied. The nozzle–skimmer distance is usually of the order of 5 mm. The beam can be further collimated by a slit.

One pump of 6000 l/sec pumping speed for the nozzle–skimmer region and another one of 2000 l/sec pumping speed for the skimmer–collimator region have been found to be sufficient for satisfactory operation at the given dimensions and pressure values.

For sufficiently high pressures in the expansion region, both heavy and light molecules will reach the same final velocity v which is given by[57]

$$v = [2\gamma kT/(\gamma - 1)m]^{1/2} \tag{16}$$

γ is the specific heat ratio $\gamma = c_p/c_v$, T is the nozzle temperature, and m is the average mass $m = \alpha m_{light} + \beta m_{heavy}$, α and β being the concentrations of heavy and light component respectively. For H_2 as carrier gas, $\gamma = 5/3$, and a mixture of 1% Xe and 99% H_2, one has $m = 0·01 m_{Xe} + 0·99 m_{H_2} = 3·29$, and the Xe energy obtained, $5/2(m_{Xe}/m)kT$, is a factor m_{Xe}/m higher than that obtained by expanding pure Xe.

Usually the velocity equilibrium is not complete, and the ratio of velocities, v_{Xe}/v_{H_2}, called slippage ratio, is of the order of 0·9.[59] The highest Xe energies which have been reached so far are around 15 eV.[59,60]

The velocity distribution has been measured in two ways: Parks and Wexler[59] have used a time-of-flight method combined with a specially designed electron bombardment detector, while Haberland et al.[60] made a retarding field analysis of the ionized fragments produced by electron bombardment of the primary beam. For Kr and Xe the velocity resolution has been found to be 7% f.w.h.m. or better over the whole velocity range from 0·2 to 5 eV for Kr and up to 10 eV for Xe.

Because of the very high beam intensity, typically 10^{17} to 10^{18} atoms/ sterad sec, there is no difficulty in detecting seeded beams, and simple electron bombardment detectors may be used.

One disadvantage of the seeded beams is that the beam energy cannot be varied very fast, which may be important for very accurate measurements of a total cross section. For changing the energy of the beam one either changes the mixture ratio[59] or the nozzle temperature.[58]

So far molecular beam experiments have been performed with seeded beams of the following species: Kr,[59] Xe,[58,59] NO,[84] N_2, CO,[62] and CH_3I.[61]

IV. EXPERIMENTAL RESULTS AND THEIR INTERPRETATION

A. Total Cross Sections

If not stated otherwise, all measurements described in this section have been performed with a scattering chamber and a combination of an interference filter and a cooled photomultiplier for spectroscopic analysis. The observation region viewed by the photomultiplier is usually of the order of several *cm*, and only states with a lifetime of less than 10^{-7} sec have been studied. This ensures that the reported emission and excitation cross sections are independent of the size of the observation region. Only those measurements which have been performed under single collision conditions are discussed.

In most studies of this section the influence of the polarization of the emission has been neglected. If polarization is present, the anisotropy in the emission must be taken into account when calculating an emission cross section.[1] However, it has been shown[21] that due to the hyperfine structure of the alkali atoms the polarization of the sodium and potassium resonance lines is less than 20%. In these cases the influence of the polarization on the angular distribution of the emitted photons can be neglected.

1. Collisional Excitation: Processes (1) to (3)

a. Alkali–rare gases. The cross sections for collisional quenching of alkali atoms in the lowest excited states $^2P_{1/2,3/2}$ in thermal collisions with rare gases are known to be negligibly small.[4,5] It follows from the principle of microreversibility,[19,20] that the threshold for excitation of the resonance states will be well above the endoergicity of the process which would be 1·6 eV for potassium and 2·1 eV for sodium.

Anderson et al.[63] were the first to measure cross sections for excitation of potassium to the resonance states $4^2P_{1/2,3/2}$ for beam energies from 25 to 600 eV. They used a fast potassium beam produced by resonant charge exchange. The spectral resolution of the optical detection system was about 70 Å. The threshold energies they extrapolated are 30, 110, 65, 75 and 50 eV for K + He, Ne, Ar, Kr and Xe. The measured cross sections were compared with those calculated with the Landau–Zener formula; reasonable agreement could be found when assuming appropriate values for the parameters R_c, the crossing radius, and $v_0 = 2\pi|H_{12}|^2/\hbar\,\Delta S_{12}$ ($\Delta S_{12} = \partial/\partial R|H_{11} - H_{22}|_{R_c}$).[118]

Kempter et al.[64] showed that the threshold energies are in fact much lower; Fig. 2 gives the cross section for emission of the fine structure component

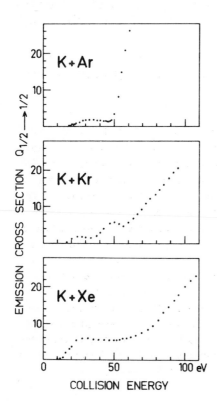

Fig. 2. Cross sections for emission of $K(4^2P_{1/2} \to 4^2S_{1/2})$ in collisions of K with Ar, Kr and Xe.

$K(4^2P_{1/2} \to 4^2S_{1/2})$, $Q_{1/2 \to 1/2}$, for K + Ar, Kr and Xe in the threshold region. Similar results were obtained for the other fine structure component. The interference filters employed were 20 Å f.w.h.m. wide. This ensures that the measured radiation is entirely due to emission from the states $4^2P_{1/2}$ or $4^2P_{3/2}$. The measurements show that the apparent thresholds are below 20 eV in all cases. The sharp rise coincides with the threshold energies given in Ref. 63 and indicates that at higher energies a different coupling mechanism may be responsible for the population of the resonance states.

Kempter et al.[66] determined relative cross sections for emission of the two fine structure components $4^2P_{3/2} \to 4^2S_{1/2}$ and $4^2P_{1/2} \to 4^2S_{1/2}$ for K + Ar, Kr and Xe. The fast potassium beam was produced by resonant charge exchange. Range of beam energies was from about 10 to 600 eV. Interference filters of 20 Å f.w.h.m. were employed to separate the two doublet components. The measurements are shown in Fig. 3a–c; all cross section ratios $Q_{3/2 \to 1/2}/Q_{1/2 \to 1/2}$ are corrected for the transmission of the other doublet line. The relative cross sections show a sharp rise, approximately at the same energy where the cross sections themselves rise sharply (see Fig. 2). Towards higher energies the relative cross sections decrease slowly to values near 2·0. In all cases the cross section ratio is oscillating, the oscillations being regular on a $1/v$-scale.

A simple model has been proposed[66] in order to explain the occurrence of the oscillations, and the gross structure of the ratio which is markedly different for the three systems. Fig. 4a shows a correlation diagram for the alkali–rare gas systems for large internuclear distances. The atomic states have been connected with the molecular ones by realizing that for slow collisions the projection of the total angular momentum along the internuclear axis is conserved. m_j is therefore a good quantum number, and $|m_j|$ corresponds to Ω at smaller distances. As long as the collision can be described by the adiabatic potential curves of Fig. 4a, $^2P_{1/2}$ is only populated by way of $A^2\Pi_{1/2}$, while $^2P_{3/2}$ will be populated from $A^2\Pi_{3/2}$ and $B^2\Sigma$. No selection rule for Ω is governing the interaction between X and A since Hund's coupling case (b) applies for X and case (a) for A. $^2P_{1/2}$ and $^2P_{3/2}$ will therefore be populated equally as long as the collision energy is insufficient for the population of $B^2\Sigma$. When B is populated too, the cross section ratio will rise, since B correlates with $^2P_{3/2}$ only. In Fig. 4b it is assumed that the interaction between X and B is much stronger than the interaction between X and A which is caused by rotational coupling. At sufficiently high collision energies the adiabatic approximation breaks down, and the molecular states will mix completely according to the appropriate Clebsch–Gordan coefficients connecting molecular and separated atom states.[115] In the sudden approximation the cross section ratio turns out to be 2·0, the value which one would get from the statistical weight of the fine structure levels. This could explain why the ratios show maxima, and decrease towards 2·0 at high energies.

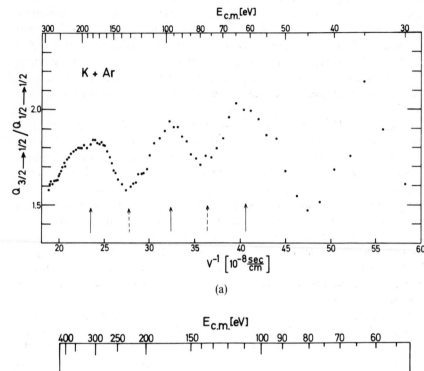

(a)

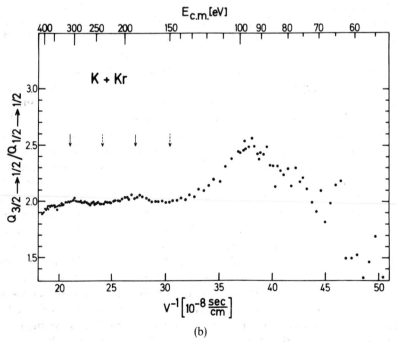

(b)

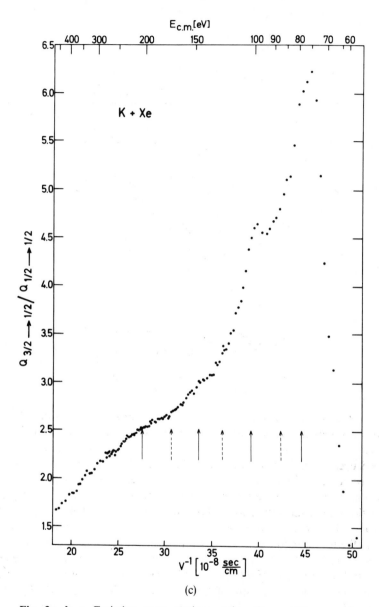

(c)

Fig. 3a, b, c. Emission cross sections ratios, $Q_{3/2 \rightarrow 1/2}/Q_{1/2 \rightarrow 1/2}$, for K + Ar, Kr and Xe.

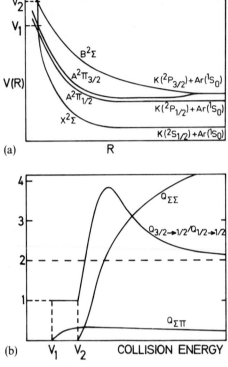

Fig. 4a, b. Model to explain excitation in collisions of K with rare gases.

The interaction between all three excited states $A^2\Pi_{1/2,3/2}$ and $B^2\Sigma$ at large distances is responsible for the excitation transfer between the fine structure levels of the alkali atoms in collisions with rare gas atoms.[3,115] This is because of the rearrangement of the angular momenta in going from the molecular to the atomic states.[115] Since the excited states are populated coherently in the inner interaction region, phase interference effects arise from the additional interaction at large distances (Rosenthal effect).[67,68,123] This model has been used before for the interpretation of interference structure in the total excitation cross section in ion–atom collisions. By adopting these results,[67,68,123] the area between the interacting potentials between the inner and outer region of interaction, $\overline{(\Delta E\,\Delta R)}$, can be determined from the distance between the extrema of the oscillations. One gets $\overline{(\Delta E\,\Delta R)}$ = 4·9, 6·7 and 7·5 eV . Å for K + Ar, Kr and Xe, respectively. From the size of $\overline{(\Delta E\,\Delta R)}$ it can be concluded that the interaction which causes the oscillation is between $A^2\Pi$ and $B^2\Sigma$. Because the population of the fine structure states is non-statistical, one could ask whether the substates $|3/2 \pm 1/2\rangle$ and $|3/2 \pm 3/2\rangle$ of $^2P_{3/2}$ which correlate with $B^2\Sigma$ and $A^2\Pi_{3/2}$ respectively, are

populated equally. This can be checked by measuring the degree of polarization of the transition $^2P_{3/2} \to {}^2S_{1/2}$. Preliminary results give 1·5, 2·5 and
5·0% polarization for K + Ar, Kr and Xe, respectively in the energy range
between 120 and 200 eV.[116] From the polarization measurements and the
ratio measurements of[66] it is possible to determine the population of all
magnetic substates of the fine structure components separately.

Cross sections for emission at 4050 Å, corresponding to the transition
K($5^2P \to 4^2S$), have been measured for K + He, Ne, Ar and Kr for beam
energies between 400 and 3000 eV.[65] The cross section for emission at
4050 Å is much smaller than that for emission of the resonance doublet at
7680 Å. The cross section ratio at 1000 eV lab energy is $Q(4^2P \to 4^2S)/$
$Q(5^2P \to 4^2S)$ = 88·4, 44·4, 29·8 and 10·7 for K + He, Ne, Ar and Kr.
However, no additional studies were made to ensure that no emission from
excited target states is contributing at this wavelength. Emission from excited
target states was studied in Refs. 64, 114. The u.v. emission in collisions of
potassium with Ar and Xe was detected by converting the ultraviolet photons
into visible light by a sodium salicylate crystal. The visible light was detected
by a photomultiplier through a 70 Å wide interference filter. In both cases
the apparent threshold for u.v. emission coincides with the sharp rise in the
cross section for emission of the resonance doublet. When the cross section
for u.v. emission is normalized to that for emission of the resonance lines, the
cross section ratio rises linearly from zero at the apparent threshold to about
0·1 at 200 eV.

Emission from Xe($7p$) at 4680 Å was studied from about 70 to 200 eV.[64]
The cross section Q_{4680} normalized to that for emission of the resonance
lines Q_{7680} is shown in Fig. 5. Again the apparent threshold coincides with

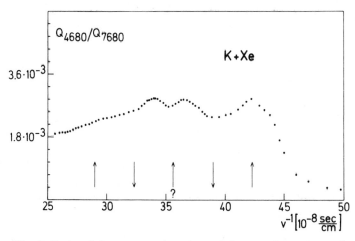

Fig. 5. Ratio of the cross sections for emission at 4680 and 7680 Å,
Q_{4680}/Q_{7680}, for K + Xe.

the sharp rise in the cross section for emission of the resonance lines. Since the cross section for emission at 7680 Å is smooth, the observed structure, which appears to be regular on a $1/v$-scale, is due to the emission at 4680 Å.

Some suggestions concerning the excitation mechanism can be made in terms of crossings between diabatic molecular orbitals. The correlation diagram for K + Ar is shown in Fig. 6. It has been constructed by using the

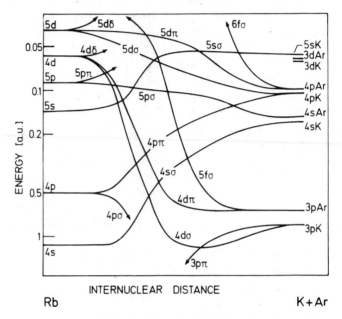

Fig. 6. Diabatic molecular orbitals for K + Ar (schematic).

rules given in Ref. 17, and is assumed to be similar to the symmetrical case Ar + Ar, for which the correlation diagram is given in Ref. 16. The excitation of the lowest excited states can solely be explained by transitions of a $5f\sigma$-electron to higher vacant orbitals:

The excitation of Ar(4s) is due to the single transition $5f\sigma \to 5p\sigma$.

Excitation of the potassium resonance states is a two step process. The first step is the transition $5f\sigma \to 4p\pi$ or $5f\sigma \to 5d\sigma$, leading to the formation of an ionic intermediate $K^-(3p^6, 4s, 4p)Ar^+(3p^5)$. In the second step the $4s\sigma$-electron jumps back into the vacancy in the $5f\sigma$-orbital leading to the formation of $K(4^2P) + Ar(3p^6)$. This can be written as follows

$$K(4^2S) + Ar(^1S) \to \ldots (5f\sigma)^2 4s\sigma^2\Sigma \to \ldots (5f\sigma)(4s\sigma)4p\pi^2\Pi$$
$$\to \ldots (5f\sigma)^2 4p\pi^2\Pi \to K(4^2P_{1/2,3/2}) + Ar(^1S) \tag{17}$$

$$K(4^2S) + Ar(^1S) \rightarrow \ldots (5f\sigma)^2 4s\sigma^2\Sigma \rightarrow \ldots (5f\sigma)(4s\sigma)5d\sigma^2\Sigma$$
$$\rightarrow \ldots (5f\sigma)^2 5d\sigma^2\Sigma \rightarrow K(4^2P_{3/2}) + Ar(^1S) \tag{18}$$

Another mechanism which would lead to formation of $K(4^2P)$ by way of population of the molecular state $B^2\Sigma$ would be a transition $5f\sigma \rightarrow 5p\sigma$, leading to $K(4^2S) + Ar(3p^54s)$, followed by a double transition $5p\sigma \rightarrow 5d\sigma$, $4s\sigma \rightarrow 5f\sigma$:

$$K(4^2S) + Ar(^1S) \rightarrow \ldots (5f\sigma)^2 4s\sigma^2\Sigma \rightarrow \ldots (5f\sigma)(4s\sigma)5p\sigma^2\Sigma$$
$$\rightarrow \ldots (5f\sigma)^2 5ds^2\Sigma \rightarrow K(4^2P_{3/2}) + Ar(^1S) \tag{19}$$

The correlation which is found between the threshold for formation of $Ar(4s)$ and the sharp rise in the cross section for emission of the potassium resonance lines, indicates that the second mechanism may be the most important one at higher energies.

b. Alkali–alkali and alkali–mercury. In contrast to the potassium–rare gas pairs, low thresholds for excitation of the potassium resonance states have been found for collisions of potassium and sodium with potassium and potassium with mercury.[69] The bandwidth of the interference filters was about 70 Å; in some measurements 20 Å wide filters have been used to make sure that the emission is due to the resonance line emission. All measurements below 10 eV collision energy were made with a sputtered alkali beam, and those at higher energies with a beam produced by resonant charge exchange.

For $K + K$ the threshold for emission of the potassium resonance lines $4^2P_{3/2,1/2} \rightarrow 4^2S_{1/2}$ is $1 \cdot 6$ ($\pm 0 \cdot 4$) eV, for $Na + K$ $2 \cdot 0$ ($\pm 0 \cdot 5$) eV. From measurements with the charge exchange source, it is found that the excitation cross section becomes constant above 10 eV collision energy. No measurements with the charge exchange source have been performed for $Na + K$. For both systems the absolute value of the cross section is about $3 \, Å^2$ at 10 eV. No emission was found in either case at 4050 and 6930 Å which should occur if the higher excited potassium states 6^2S and 5^2P were populated. Cross sections for emission from these states are at least two orders of magnitude lower than emission from 4^2P.

Also no emission was found at 5890 Å in $Na + K$ collisions. An upper limit of the cross section for excitation of $Na(3^2P)$ is 1 % of that for excitation of $K(4^2P)$.

The results indicate that a 'crossing' of potential curves will be near $R_c = 2$ Å in both cases, and that the potential energy at the crossing, $V(R_c)$, is near $1 \cdot 6$ eV. The existence of a crossing between $a^3\Sigma_u^+$ and the lowest $^3\Pi_u$ state was recently confirmed in 'ab initio' calculations for Na_2,[71] and was already found before for $Na + Li$[70] for the analogous states $a^3\Sigma^+$ and $^3\Pi$. In both cases the crossing is not much above the asymptotic limit of the potential curves ($\leq 0 \cdot 6$ eV), and is found around $R_c = 1 \cdot 75$ Å.

Additional information about the excitation mechanism comes from measurements of the relative population of the two fine structure states $^2P_{3/2}$ and $^2P_{1/2}$.[72] 20 Å wide interference filters were employed to separate the two doublet components. In both cases one does not find the statistical value 2·0 for the ratio of the emission cross sections, $Q_{3/2 \to 1/2}/Q_{1/2 \to 1/2}$, in the studied range from 2·5 to 50 eV. For K + K the ratio is 1·25, independent of energy, while for Na + K the ratio decreases from 2·5 at 2·0 eV to 1·6 at 40 eV.

These measurements can also be explained by assuming a single crossing between the $a^3\Sigma$ and $^3\Pi$ states. Fig. 7a gives the correlation diagram of the u-states emerging from $K(4^2P) + K(4^2S)$. The relative order of the states at large distances is taken from Ref. 110, at small distances from Ref. 71. At

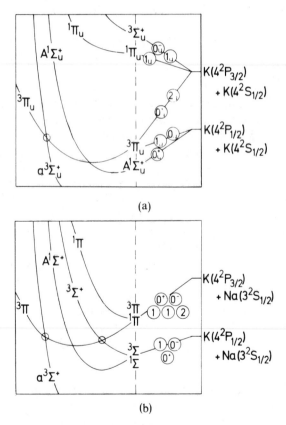

(a)

(b)

Fig. 7a, b. Simplified correlation diagrams for some of the lowest excited molecular states of K + K and Na + K. The horizontal axis indicates the inter-nuclear distance, the vertical axis the energy of the molecular states.

low collision energies the adiabatic states should be a good approximation. Since Hund's coupling case (b) applies for a $^3\Sigma_u$, and case (a) for $^3\Pi_u$, there is no selection rule for Ω; all substates $^3\Pi_\Omega$ will be populated equally. If no interaction between $^3\Pi_\Omega$ and other states leading to $^2P_{3/2}$ or $^2P_{1/2}$ is effective, the relative population of the two fine structure states should be 1·0.

If another interaction, besides that between $a^3\Sigma_u$ and $^3\Pi_u$, leads to population of ^2P_j, the population ratio would be altered considerably. Obviously there is only a very small interaction between $^3\Pi_u$ and $A^1\Sigma_u$ because of the spin change, even though there is a crossing between the two states. Probably interaction between the substates of $^3\Pi_u$ at large distances is responsible for the deviation of the measured value from 1·0.[110] The energy independence of the ratio is further evidence for just one region of interaction.

It is not completely clear why the ratio is energy dependent for Na + K. The correlation diagram for the states emerging from $K(4^2P) + Na(3^2S)$ is given in Fig. 7b. The relative order of the states at small distances is expected to be similar to K + K; the only low-lying crossing is between $a^3\Sigma^+$ and the lowest $^3\Pi$ state. At large distances the relative order is taken from Ref. 111. The essential difference to K + K seems to be that there is an interaction at large distances between $^3\Pi$ and the lowest $^3\Sigma^+$ leading to $K(4^2P) + Na(3^2S)$ asymptotically. This $\Sigma\Pi$-interaction becomes the stronger, the higher the collision energy. At low energies $^3\Pi$ correlates with $^2P_{3/2}$, while at high energies there is a high probability for a transition to $^2P_{1/2}$ by rotational coupling. This could explain why the population ratio decreases in going to higher energies.

The apparent threshold for emission of the potassium D-lines in K + Hg collisions is 4·0 ($\pm$0·4) eV; the absolute size of the emission cross section at 10 eV has been estimated[69] to 3 Å^2. Again no emission from excited potassium states other than 4^2P has been observed up to 100 eV collision energy; upper limits for emission at 4050 and 6930 Å are 1 % of that at 7680 Å. No emission from excited mercury states at 5440 and 5780 Å was observed either.

The threshold near 4·0 eV indicates that excited mercury states may be involved in the excitation to $K(4^2P)$. The similar system H + Hg has bound excited states.[125] Probably one or several of the molecular states arising from $K(4^2S) + Hg(^3P_{0,1,2})$ are interacting with both the initial and final channels $K(4^2S) + Hg(^1S)$ and $K(4^2P_{1/2,3/2}) + Hg(^1S)$, but see Ref. 78.

The relative population of the fine structure states $K(4^2P_{1/2})$ and $K(4^2P_{3/2})$ was determined between 6 eV and 40 eV;[78] the ratio is strongly energy dependent, and rises from values near 1·0 at 6 eV to those higher than 2·5 at 40 eV. The reasons for this behaviour are not well understood yet; the correlation diagram for the states emerging from $K(4^2P) + Hg(^1S_0)$ is similar to that for the alkali–rare gas pairs shown in Fig. 4a. At low energies the states $^2\Pi_{1/2,3/2}$, correlating with $|1/2 \pm 1/2\rangle$ of $^2P_{1/2}$ and $|3/2 \pm 3/2\rangle$ of $^2P_{3/2}$, seem to be populated predominantly, yielding a relative population of 1·0.

At higher energies $B^2\Sigma$, correlating with $^2P_{3/2}$ only, seems to become increasingly important for the excitation mechanism. This means that the cross section ratio will rise, similar to the case of the potassium–rare gas pairs.

c. Alkali–molecules. The large cross sections for collisional quenching of the alkali resonance states by most molecules[4,5] lead us to expect that the reverse process of collisional excitation will have threshold energies at or near the endoergicity of the process. Extensive studies of excitation of potassium and sodium in collisions with several small molecules have been performed in the last few years.[2,30,62,63,75,76,77] Except for Ref. 62, where a secondary beam was employed, all measurements were performed with a scattering chamber; the spectral resolution was in most cases about 70 Å. The following general picture has been found (see Fig. 8):

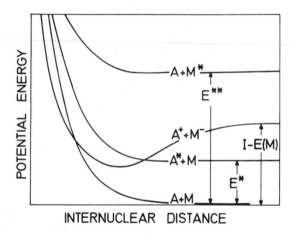

Fig. 8. Schematic potential energy curves for interaction of alkali atoms with molecules.

Almost certainly the configurations $A + M$ and $A^* + M$, where A denotes an alkali atom and M a molecule, do not interact directly; the coupling between them is made by the ion-pair configuration $A^+ + M^-$. The excitation functions can be divided in two groups according to whether a rapid onset of the excitation cross section near threshold and a maximum at low energies is found or not.[2]

The first class contains all those molecules for which the excitation energy of the alkali atom, E^*, is lower than the difference between the ionization energy I and the electron affinity $E(M)$ of the molecule $[E^* < I - E(M)]$, and where in addition the change in bond length in going from M to M^- is very small (<0.1 Å). Examples are N_2, O_2, NO, CO, SO_2, CO_2, the unsaturated hydrocarbons, and the olefinic and aromatic hydrocarbons. In all

cases large cross sections, of the order of 5 Å^2 at 10 eV collision energy, and
low thresholds, 2 eV or even less above the endoergicity E^*, are found, indicating that the interaction with the ion-pair configuration is at such large
distances that the nuclear repulsion is not yet effective. Lacmann and
Herschbach[30] were able to show that the maxima in the excitation cross
section at low energies for K + O$_2$, NO, SO$_2$ and others correlate with the
thresholds for collisional ionization in these cases. Fig. 9 shows results for

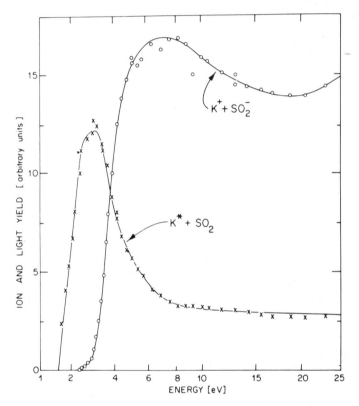

Fig. 9. Cross sections for collisional ionization, and for emission
of K($4^2P \to 4^2S$) for K + SO$_2$ (from Ref. 2).

K + SO$_2$. A sharp maximum, followed by a marked decrease towards
higher energies occurs in the emission cross section, when the channel for
collisional ionization opens. When $I - E(M) < I$, but still $I - E(M) > E^*$,
high yields for collisional ionization are found,[30,46] while in cases where
$I - E(M) \gtrsim I$, they are very low; this indicates that when the ion-pair
configuration 'cuts across' all Rydberg states, inelastic collisions mostly
lead to excitation and not to ionization.

Further evidence for the importance of the ion-pair configuration comes from $A + NO_2$, where $I - E \lesssim E^*$ for K, but $I - E(M) > E^*$ for Na. Consequently the cross section for excitation of potassium is much lower than that for sodium.[76]

Kempter et al.[77] studied the emission from the higher excited states $K(6^2S)$ at 6930 Å, and $K(5^2P)$ at 4050 Å relative to that from $K(4^2P)$ for several molecules in the energy range from 10 to 50 eV. In all cases the cross section for emission from higher excited states is only a few percent of that from $K(4^2P)$.

The second class comprises molecules for which either $I - E(M) < E^*$ or the bond length of M is considerably different from that of M^-.[2,30] Examples are Cl_2, HCl, the saturated hydrocarbons, and also probably H_2.

When $I - E(M) < E^*$, the ion-pair configuration does not couple the initial and final channel, and the threshold will be at much higher energies. When the bond lengths are too different, the Franck–Condon factors are unfavourable for a transition $A^+ + M^- \rightarrow A^* + M$ in the interaction region.[2,6,15] Consequently the excitation cross section will be small.

The relative populations of the potassium resonance states $4^2P_{1/2}$ and $4^2P_{3/2}$ have been determined[78] for collisions with N_2, CO, NO, O_2, CO and NO_2 in the energy range between 10 and 50 eV. Within the error limits the population ratio is $2.0 (\pm 0.2)$ in all cases, and is independent of energy.

The polarization of the transition $4^2P \rightarrow 4^2S$ has been measured for $K + N_2$ at 5 and 14 eV collision energy.[79] The doublet components were not separated. The emission was found to be unpolarized within the experimental error which was estimated to about 2%.

Even though collisional excitation of alkali atoms in collisions with diatomic molecules has been studied to a large extent experimentally, and also the excitation mechanism seems to be understood, no quantitative calculations have been performed yet. In principle the same models and techniques as are used for collisional quenching[15] and ionization[118] could be applied.

Excitation of the target molecule has been found[80] for $Na + SO_2$ and NO_2, and has been studied later on in more detail for $Na + NO_2$.[81] The identity of the excited molecular species could be demonstrated by measuring the lifetime of the transitions under question. A pulsed beam of sputtered sodium, the pulse width being 10 μsec, was employed and the time between the excitation of the target gas and the observation with the photomultiplier was varied. The variation of the emission intensity with the delay time gives the lifetime of the excited state. The spectral analysis was made with interference filters of 70 Å bandwidth; some measurements have been made with sharp-cut filters.

In all cases studied radiation from long-lived states has been found. The lifetime depends on the wavelength of the emitted radiation, and is of the order of 40 μsec.[80,81] It was shown that the emission is due to excited NO_2

and SO_2 states, most likely $NO_2(^2B_2$ and/or $^2B_1)$ and $SO_2(^1A_1$ and/or $^3A_1)$. Threshold energies for excitation of these states are about 3 eV for Na + SO_2[80] and 3.3 ($\pm$0.3) eV for Na + NO_2.[81]

In a later study[53] it was also shown that in collisions of K and Na with several diatomic molecules, molecular emission can be detected when the collision energy is >8 eV. For N_2 the radiation could be identified as emission from the first positive group $N_2(^3\Pi_g \rightarrow {}^3\Sigma_u^+)$. All these measurements were made with Na and K beams produced by sputtering with 25 keV Ar^+ ions. When the sputtering source was replaced by a charge exchange source, no similar emission was observed.[77] The authors suggest that the molecular excitation in collision of diatomic molecules with alkali beams produced by sputtering with Ar^+-ions may be caused by Ar-metastables produced in the process of sputtering. This is supported by the fact that the mass spectrum of the ions produced by electron bombardment of a Ar^+-sputtered Li-beam shows an Ar^+-peak when the mass analyser is not very well aligned with the beam axis. It is suggested that the Ar^+-ions are produced by collisions of fast metastable Ar-atoms with metal parts of the detection system.[54] This also implies that the measurements for NO_2 and SO_2 should be repeated with an alkali beam produced by charge exchange in order to establish the identity of the exciting atoms. For this reason no discussion of the mechanism for excitation of NO_2 and SO_2 will be given here.

For all diatomic molecules E^{**} is of the order of 8 eV for emission in the visible range. Since $I - E(M) < E^{**}$ the ion-pair configuration $A^+ + M^-$ will not couple the configurations $A + M$ and $A + M^*$ at low energies. Therefore in this picture no excitation of diatomic molecules by alkali atoms is expected at low collision energies. For NO_2 and SO_2 $I - E(M) > E^{**}$, so that molecular excitation could be possible even at low collision energies.

d. Cadium–zinc. Excitation of Cd and Zn has been reported[32] in collisions of Cd with Zn. Cd^+-ions were produced by electron bombardment and after acceleration, neutralized by resonant charge transfer. The useful range of beam energies was from about 10 to 1000 eV. Light emission was observed under 90° to the beam axis by a monochromator; no values for the spectral resolution were given in the Conference Abstract.

Emission from several excited states of Cd as well as from Zn was observed; however only the cross section for the transition $Cd(5^3P_1 \rightarrow 5^1S_0)$ was reported in some detail. The excitation threshold is about 15 eV, and the cross section shows a sharp maximum at about 45 eV collision energy. No values for the absolute value of the emission cross section are given. The authors do not discuss whether cascade from higher excited states and the very long lifetime of $Cd(5^3P_1)$ could have any influence on the shape of the emission cross section. It is interesting to speculate on the excitation mechanism, even though not enough is known about this system. Since the ground states of both atoms are ^1S-states, but the excited states studied were ^3P-states, one would expect to find a very small cross section for excitation of these states because of Wigner's spin conservation rule.[6] A possible direct crossing at small internuclear distances between $X^1\Sigma$ and any triplet states leading to separated atom states $^3P_J + {}^1S_0$ should therefore not be very effective in populating excited triplet states.

However an excited singlet state $^1\Sigma$ resulting from $^1P + {}^1S_0$ could be populated effectively provided there is an avoided crossing with the $X^1\Sigma$. The interaction of the excited $^1\Sigma$ with some of the triplet states could be at such large distances that Hund's coupling (c) applies. Transitions will then be possible from $0^+(^1\Sigma)$ to $0^+(^3\Pi)$, leading to $Cd(5^3P_1) + Zn(^1S_0)$. Probably rotational coupling between $0^+(^1\Sigma)$ and $1(^3\Sigma)$ is also effective in populating $Cd(^3P_1)$. Since the excited $^1\Sigma$ state would also cross all states leading to $Cd(^1S_0) + Zn(4^3P_1)$, this model could explain why emission from excited Zn-states is also observed.

e. Rare gas–rare gas.   Excitation in collisions between rare gas atoms has been studied by two different methods depending on whether the excited states are stable or not against autoionization. In the first case conventional optical spectroscopy can be applied while in the second case one has to measure the energy distribution of the ejected electrons in order to identify the excited states. If not mentioned otherwise, the fast rare-gas beams of this Section were produced by resonant charge exchange.

The spectrum emitted from excited rare gas atoms is rather complex: studies in the visible range require a resolution of about 10 Å in order to isolate a single line. Until recently small beam intensity and low detection sensitivity limited optical studies to high collision energies.

Most studies reported below have been made with insufficient resolution in order to measure a true emission cross section. The influence of polarization of the emission on the cross section has been neglected entirely; on the other hand it has been shown that the emission from collisionally excited rare gas atoms can be highly polarized.[1]

Ultraviolet emission has been studied in collisions between two ground state Ar-atoms.[89] The primary beam was produced by charge exchange between Ar^+-ions and hydrogen molecules. The collision chamber contained Ar-gas of approximately 10^{-4} torr, and an open multiplier for detecting the ultraviolet photons. Care was taken not to detect positive or negative charged particles produced by collisions; the multiplier was also protected against fast scattered neutrals. Since no spectral analysis of the emitted photons could be made, photons of all wavelengths less than 1500 Å were detected. The energy dependence of the emission cross section was measured between 15 and 400 eV collision energy. A lower limit of 10^{-17} cm^2 at 100 eV, and $2 \cdot 10^{-20}$ cm^2 at 15 eV was obtained from geometrical considerations. The emission cross section has a similar shape to that for negative charge production.[90] Structure has been found between 60 and 100 eV, also similar to that in the cross section for negative charge production.[90]

Formation of long-lived Ar-atoms was studied with the same experimental setup between threshold and about 16 eV collision energy.[90] The long-lived atoms were detected by Penning ionization of acetylene. The cross section for production of metastables is practically constant over the range studied from 11·5 to 16·3 eV. The absolute magnitude is $3·6 (\pm 2·4) . 10^{-20}$ cm^2.

Emission from $Ne(2p^5 3p')$ at 6266 Å was studied in collisions between ground state Ne-atoms from the apparent threshold up to about 2 keV.[91] The spectral analysis was made by interference filters; no values of the spectral resolution were given. At 60 eV the emission cross section rises sharply; a weak tail is observed below 60 eV, the apparent threshold is estimated to be about 40 eV by the present author. The cross section has a similar shape as found in[89] for the u.v.-emission in Ar + Ar. No estimation of the absolute cross section has been made.

Emission from excited He, Ne and Ar-atoms has been studied for several combinations of rare gases.[92] He + He, He + Ne, He + Ar, Ne + Ne, Ne + Ar and Ar + Ar were studied from the apparent threshold up to about 500 eV. The spectroscopic analysis was made with interference filters of 70 Å bandwidth (for more details about the studied wavelengths see Table IV(a) and IV(b). Except for He + Ar the apparent threshold energies are between 25 and 60 eV. The smallest cross sections which could be detected were of the order of 10^{-21} cm^2. The shape of the emission cross sections is similar to what was found in Refs. 89 and 91.

With the same experimental arrangement, emission from the decay of $Ne(2p^5 4p)$ at 3472 and 3454 Å, of $Ar(3p^5 5p')$ at 4050 Å and of $Kr(4p^5 6p)$ at 4290 Å has been studied for Ne + Ne, Ar + Ar and Kr + Kr, and has been compared with the emission from the decay of $Ne(3p')$, $Ar(4p')$ and $Kr(5p)$.[117] The relative cross sections show a complicated structure, part of which seems to be regular on a $1/v$-scale. The structure has been interpreted tentatively as due to long-range interaction of the states given above with the almost degenerate ones arising from the $Ne(3d')$, $Ar(4d)$ and $Kr(6d)$ configurations.[117] This mechanism has been used before for the interpretations of structure in ion–atom collisions[67,68,123] and has been applied in this article to the interpretation of the structure found in alkali–rare gas collisions.

Excitation in collisions of ground state helium atoms with He, Ne, H_2 and N_2 at 1100 eV beam energy has been studied.[37] Part of the radiation emitted along the beam path was focused onto the entrance slit of a spectrograph equipped with an image intensifier tube. The spectral resolution was about 20 Å. No light emission from He was found in He + He nor in He + Ne collisions. Only excitation of the heavier rare gas atom has been found for He + Ne. Most lines centered around 6000 Å, those arising from the $Ne(2p^53p)$ and $Ne(2p^53p')$ configurations being about equally strongly excited. The results for He + H_2 are discussed in the section on chemiluminescence.

The importance of the above measurements[37] lies mainly in the fact that the emisssion intensity factor, which represents a relative emission cross section at 1100 eV beam energy, can be converted into an absolute cross section by comparing the data for He^+ + Ne and H_2 with those of other groups for the same systems at the same energy. The absolute cross section for emission of the 5945 Å line in collisions of He^+ with Ne at about 100 eV beam energy is $3\cdot8 \cdot 10^{-19}$ cm^2.[93] The intensity factor $3\cdot0$[37] corresponds to $3\cdot8 \cdot 10^{-19}$ cm^2. An additional check is provided by the relative cross sections for H_α emission in collisions between He and H_2.[85,86] This data can be normalized by using other data[94] for the excitation of the 3s, 3p and 3d-levels of hydrogen under the assumption that cascades are negligible in populating these levels.[85,86] The data for He + H_2[37] can now be normalized by using the calibrated ones.[85,86] This calibration leads to about the same correspondence as given above.

Autoionizing states have been studied by analysing the energy spectra of the ejected electrons.[34,87,88,83] A representative experiment is described in Ref. 87. The fast neutral beam was crossed at right angles by a thermal secondary beam formed by effusion from a multichannel array. The electrons formed in the scattering centre were energy analysed by a combined retarding field and 127° electrostatic cylindrical analyser. The detector unit was rotatable around the secondary beam in a plane containing the primary beam. The energy resolution was typically 100 meV. Energy calibration of the spectrometer could be made by measuring photoelectrons of known energy together with those formed in the collision process. Excitation could be studied in the range between 100 and 5000 eV collision energy.

In collisions of He with Ar, Kr and Xe only autoionizing states of the heavier rare gas atom are excited.[87] Singly and doubly excited states, as for example $Ar(3s3p^64s)$ and $Ar(3s^23p^4nl, n'l')$, have been found with comparable probability. For He + Ne, no autoionizing states of the neutral atoms have been observed. For none of the studied systems the excitation cross section was measured as function of the collision energy, nor were estimates made of the absolute size of the excitation cross sections.

With the same apparatus the symmetrical systems He_2, Ne_2, Ar_2 and Kr_2 were studied.[88] The prominent process in collisions between He-atoms at low energies, besides formation of negative excited He^--ions, is the simultaneous formation of two excited $He(1snl)$-atoms. When the atoms are still close together, molecular ionization very similar to a Penning ionization process occurs and accounts for the emission of electrons:[88]

$$He(1s2s) + He(1s2s) \rightarrow He^+(1s) + He(1s^2) + e^- \qquad (20)$$

The peaks of the electron spectra are broad and reflect the shape of the potential curve between the two excited atoms. For Ne_2 only the formation of excited $Ne^-(2p^5 3s^2)$-ions has been found.

For Ar_2 the simultaneous formation of two excited $Ar(3p^5 4s)$-atoms again was found; as for He_2 molecular ionization occurs and accounts for the emission of electrons. Both singly and doubly excited states of Ar are populated, like $Ar(3s3p^6 4s)$ and $Ar(3p^4 nln'l')$. The prominent process at low energies is autoionization of $Ar(3s3p^6 4s)$ into the ion ground state. The cross section for formation of $Ar(3s3p^6 4s)$ was measured as a function of the collision energy from the apparent threshold at 42 eV up to 250 eV. For the interpretation of the results it is assumed that the excitation process can be explained by a single crossing at R_c, and that the potential energy at R_c is equal to the threshold energy, $V(R_c) = 42$ eV. By using the theoretical curves[13] for the ground state of Ar_2, R_c was determined to $R_c = 1\cdot15$ Å. By fitting the measured cross section with the Landau–Zener formula, the absolute value of the excitation cross section is estimated to be $1\cdot14$ Å^2 at 150 eV. On the other hand the total cross section for negative charge production at this energy has been found[90] to be only 10^{-16} cm^2. This would imply that practically all ionization in Ar_2 is due to autoionization of $Ar(3s3p^6 4s)$. Hammond et al.,[34] also by measuring electron spectra, come to the conclusion that only 15 % of the formed Ar^+-ions come from the decay of autoionizing states. From the analysis of their ion spectra they find two crossings between the ground state of Ar_2 and states leading to $Ar^+(^2P_{1/2,3/2})$ + $Ar(^1S_0)$ at about $27\cdot8$ and $37\cdot8$ eV. For Kr_2 autoionizing states have been found, but no identification of the states involved has been made yet.[88]

The same systems were studied[83] between 80 and 600 eV beam energy. The fast rare gas beams were not monoenergetic, but contained all energies up to the energy of the primary ions. The energy resolution of the electron spectrometer was 30 to 40 meV.

The results agree in general with those of Gerber et al.;[87,88] differences are only found in the relative peak intensities. Due to the better resolution employed[83] more certain peak assignments could be made. It was confirmed that in collisions of He with Ar, Kr and Xe both singly and doubly excited states of the heavy atom occur. For Ar + Ar excitation of $Ar(3s3p^6 4a)$ was dominant; no autoionizing states were found for Kr + Kr. Recently, collisions of Ar with Kr and Xe were studied. For Ar + Kr both $Kr(4s4p^6 5s)$ and $Ar(3s3p^6 4s)$ were observed. For Ar + Xe one of the most dominant peaks is due to the decay of $Xe(5s5p^6 6s)$. No cross sections were determined, nor was the energy dependence of the processes studied systematically.

Table IV summarizes all information which is available on the threshold energies and the size of the emission cross sections for the various rare gas combinations. Also the excited states are shown which could be contributing at the wavelength studied. The lowest states listed are expected to be responsible for the main contribution to the emission cross section.

TABLE IV(a)

Apparent excitation thresholds and absolute emission cross sections for several combinations of rare gases

System	Studied wave-length	Contributing excited states	Apparent threshold [eV]	Emission cross sections		Reference
				Estimated absolute size [cm^2]	Collision energy [eV]	
Ar–Ar	<1500	4s, 4s'	15	$2 . 10^{-17}$	150	89
Ar–Ar		4s, 4s'	11·5	$3·6 . 10^{-20}$	11·5–16·3	90
Ar–Ar	4050	5p'	27	$5·2 . 10^{-18}$	150	92
Ar–Ar	6280	5d, 5d', 6d	29	$8·8 . 10^{-19}$	150	92
Ar–Ar	6930	4p', 4d'	32	$3·0 . 10^{-18}$	150	92
Ar–Ar	7020	4p', 6s	31	$8·1 . 10^{-18}$	150	92
Ar–Ar		Ar(3s3p^{6}4s)	42	$1·14 . 10^{-16}$(*)	150	88
Ne–Ne	3430	4p	32	$2·7 . 10^{-17}$	150	92
Ne–Ne	6280	3p'	35	$1·5 . 10^{-16}$	150	92
Ne–Ne	6266	3p'	40(***)	—	—	91
Ne–Ne	7020	3p	33	$3·9 . 10^{-17}$	—	92
Ne–Ar	4050	Ar(5p')	40	$1·5 . 10^{-19}$	150	92(****)
	6280	Ne(3p'), Ar(5d) Ar(5d'), Ar(6d)	34	$1·5 . 10^{-18}$	150	92
He–Ar	6280	Ar(5d), Ar(5d') Ar(6d)	127	$1·7 . 10^{-19}$	150	92
He–Ne	5876 5882	He(3^{3}D), Ne(3p')	—	$1·5 . 10^{-18}$	920	37
	5852	Ne(3p')	—	$6·7 . 10^{-19}$	920	37(**)
	6096	Ne(3p')	—	$5·0 . 10^{-19}$	920	37
	6143	Ne(3p)	—	$1·1 . 10^{-18}$	920	37
	6280	Ne(3p')	50	$2·3 . 10^{-18}$	150	92
	6334	Ne(3p)	—	$3.8 · 10^{-19}$	920	37
	6402	Ne(3p')	—	$1·4 . 10^{-18}$	920	37
	6507	Ne(3p)	—	$1·1 . 10^{-18}$	920	37

(*) No direct measurement; estimated by using the Landau–Zener formula.
(**) Data of[37] were normalized by using data of.[93]
(***) Estimated from[91] by the present author.
(****) Emission cross sections in[92] are calculated from equations (2–5) in[1]; the sensitivity of the photon detection system at the studied wavelength has been used for the estimate of the absolute values. A lifetime correction for emission from the excited projectile atom has been applied.

TABLE IV(b)

Apparent excitation thresholds and absolute excitation cross sections for He–He, and Ne–Ne

System	Studied wave-length	Contributing excited states	Apparent threshold [eV]	Excitation cross sections(*)		Reference
				Estimated absolute size [cm^2]	Collision energy [eV]	
He–He	584	2^1P	21	$3.9 \cdot 10^{-17}$	150	92
	7065	3^1S	60	$9 \cdot 3 \cdot 10^{-19}$	150	92
	3888	3^3P	55	$8 \cdot 5 \cdot 10^{-19}$	150	92
	6678	3^1D	60	$4 \cdot 4 \cdot 10^{-19}$	150	92
	5876	3^3D	58	$1 \cdot 0 \cdot 10^{-18}$	150	92
	5047	4^1S	50	$2 \cdot 2 \cdot 10^{-19}$	150	92
	4713	4^3S	60	$1 \cdot 3 \cdot 10^{-19}$	150	92
Ne–Ne	6929	$2p^5(^2P_{3/2})3p[3/2]$	32	$8 \cdot 7 \cdot 10^{-17}$	150	92
	7020	$2p^5(^2P_{3/2})3p[1/2]$	33	$3 \cdot 3 \cdot 10^{-17}$	150	92

(*) Values are excitation cross sections in the sense of equations (2–20) in[1]; cascade contributions are neglected.

The mechanisms leading to excitation in collisions between He-atoms have been discussed in terms of crossings between diabatic molecular orbitals.[74,88,121] Mechanisms for excitation in collisions between the heavier rare gases have not yet been proposed. However, some suggestions have been made for the ionic systems Ne_2^+ and Ar_2^+[16,73] which were studied at low energies experimentally.[73] Within the framework of the MO model, conclusions about transitions between MO's in the ionic systems should also remain valid for the neutral systems. Although correlation diagrams for the mixed systems can be constructed by means of rules,[17] we will limit ourselves to the symmetrical systems He_2, Ne_2 and Ar_2.

A correlation diagram for the lowest diabatic MO's of He_2 is given in Ref. 16, and the correlation diagram for some of the diabatic molecular states in Ref. 74. The correlation diagram of Fig. 10 shows in addition some of the higher lying MO's; it is constructed by using the rules given in Ref. 16. The following experimental facts must be explained: excitation of either one or both of the colliding atoms to He(2^1P) is a very likely process. This follows from the results[92] given in Table IV(b) and the measured differential cross sections[121] which will be described in some detail in Section IV.B. Excitation of states with the main quantum number $n > 2$ is much less likely. Since singlet and triplet states are about equally strong excited, the conservation of the total spin requires that mainly simultaneous excitation of both atoms must occur. The authors of Ref. 121 suggest that excitation

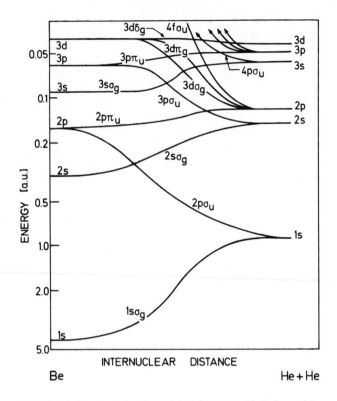

Fig. 10. Diabatic molecular orbitals for He + He (schematic).

of He($1s2p$) may be due to rotational coupling between the $2p\sigma_u$ and $2p\pi_u$ orbitals:[121]

$$He(1s^2) + He(1s^2) \rightarrow (1s\sigma_g)^2(2p\sigma_u)^2\ {}^1\Sigma_g^+ \rightarrow (1s\sigma g)^2 2p\sigma_u 2p\pi_u{}^2\Pi_g$$
$$\rightarrow He(1s^2) + He(1s2p) \qquad (21)$$

Another mechanism which would also allow for population of higher excited states could be the double transition $(2p\sigma_u)^2 \rightarrow (2s\sigma_g)^2$, followed by the double transition $2(s\sigma_g)^2 \rightarrow (nl\sigma_u)(2p\sigma_u)$, where $l = 1, 3, 5, \ldots$ and $n = 3, 4, \ldots$

$$He(1s^2) + He(1s^2) \rightarrow (1s\sigma_g)^2(2p\sigma_u)^2\ {}^1\Sigma_g^+ \rightarrow (1s\sigma_g)^2(2s\sigma_g)^2\ {}^1\Sigma_g$$
$$\rightarrow (1s\sigma_g)^2(2p\sigma_u)(nl\sigma_u)\ {}^1\Sigma_g \rightarrow He(1s^2) + He(1s2p \text{ or higher levels}) \qquad (22)$$

Simultaneous excitation of both He-atoms to He($1s2s$) can also be explained by the double transition $(2p\sigma_u)^2 \rightarrow (2s\sigma_g)^2$:[88]

$$He(1s^2) + He(1s^2) \rightarrow (1s\sigma_g)^2(2p\sigma_u)^2\ {}^1\Sigma_g^+ \rightarrow (1s\sigma_g)^2(2s\sigma_g)^2\ {}^1\Sigma_g^+$$
$$\rightarrow He(1s2s) + He(1s2s) \qquad (23)$$

Rotational coupling between $2p\sigma_u$ and $2p\pi_u$ gives rise to simultaneous excitation to He(1s2p):[88]

$$\text{He}(1s^2) + \text{He}(1s^2) \rightarrow (1s\sigma_g)^2(2p\sigma_u)^2\ {}^1\Sigma_g^+ \rightarrow (1s\sigma_g)^2(2p\pi_u)^2\ {}^1\Pi_g,\ {}^1\Delta_g$$

$$\rightarrow \text{He}(1s2p) + \text{He}(1s2p) \tag{24}$$

The MO model offers no mechanism with a large cross section for simultaneous excitation of both atoms, one of them to He($n^{1,3}$L) with $n > 2$. This is due to the fact that $2p\sigma_u$ does not interact strongly with orbitals other than $2p\pi_u$ and probably $2s\sigma_g$.

Correlation diagrams for Ne$_2$ and Ar$_2$ have been published.[16] Excitation mechanisms for Ne$_2$ and Ar$_2$ do have to explain the very low threshold energies (30 eV for Ne(3p'),[92] <15 eV for Ar(4s),[89,90] and 27·5 eV for Ar(4p')[92]). From the low thresholds we may conclude that, at least at low collisions energies, excitation of Ne(3p) and Ar(4s) is due to single electron excitation in one of the colliding atoms.

Excitation of Ne(3p) and Ar(4p) could, as in Ne$_2^+$ and Ar$_2^+$,[73] be explained by transitions $4f\sigma_u \rightarrow 3p\pi_u$ and $5f\sigma_u \rightarrow 4p\pi_u$, respectively. The transition $4d\pi_g \rightarrow 4s\sigma_g$ probably accounts, again by analogy to Ar$_2^+$,[73] for excitation of Ar(4s). By analogy to He$_2$, simultaneous excitation of both atoms to Ar($3s^23p^54s$) should be due to the transition $(5f\sigma_u)^2 \rightarrow (4s\sigma_g)^2$.

A method has been described[11] to get information on the crossing distance R_c from the position of the maximum in the excitation cross section. It is assumed that the excitation process can be described within the framework of the Landau–Zener model, and that the shape of the two potentials involved is known in the vicinity of R_c.

When R_c is small, both assumptions are usually not fulfilled. Let us ask instead if one can get information on R_c from the measured threshold behaviour of the cross section. Sharp thresholds only occur if $V(R_c)$ is smaller or equal to the endoergicity of the process, determined through the conservation laws. No sharp thresholds will be observed when the potential energy at the crossing R_c, $V(R_c)$, is larger than the endoergicity. This is because of the tunnelling between the potentials involved;[10] it follows that in general no information can be extracted from the measured apparent threshold energies since they are determined by the detection sensitivity of the apparatus.

However, often a maximum of the excitation cross section is found at low energies and a sharp, almost linear fall-off towards lower energies. It seems safe to extrapolate the sharp decrease linearly to zero, assuming that the weak tail at still lower energies is due to tunnelling. This extrapolated threshold energy is set equal to $V(R_c)$; since for most rare gas combinations $V(R)$ is well known for the molecular ground state X$^1\Sigma_g^+$,[122,13] R_c can be determined. Values received by this procedure should represent lower limits for R_c. For the crossing between $4f\sigma_u$ and $3p\pi_u$ in Ne_2 one finds 1·75 a.u. from results already published,[91] 2·7 a.u. for that between $4d\pi_g$ and $4s\sigma_g$ in Ar_2,[89] and 2·6 a.u. for that between $5f\sigma_u$ and $4p\pi_u$.[92] For the crossing which is responsible for excitation of $Ar(3s3p^64s)$ a value of 2·3 a.u. was derived[88] using a similar method.

2. Chemiluminescence and collisional dissociation with excitation: processes (4), (5) and (6)

Most chemiluminescent reactions studied in molecular beam experiments so far are exothermic and proceed without activation energy.[9] The only beam study of a chemiluminescent reaction for which translational energy seems to be necessary in order to initiate the reaction has been performed for $NO + O_3 \rightarrow (NO_2)^* + O_2$.[84] A NO-beam of hyperthermal energy was produced by the seeded beam method; the beam energy was varied by changing the mixing ratio H_2/NO. The total cross section for light emission was determined by normalizing the total emitted light intensity as detected by the photomultiplier to the primary beam intensity. It was not checked whether the shape of the emitted spectrum changed over the studied energy range. An apparent threshold of 0·13 ($\pm$0·02) eV was found, and the energy dependence of the emission cross section was measured from threshold up to 0·5 eV collision energy. No estimate for the absolute value of the emission cross section was given.

In collisions of He, Ne and Ar with H_2, emission of the H_α and H_β lines of atomic hydrogen has been studied.[85,86] The cross sections are compared with those for emission of some lines of the corresponding rare gases and cross sections for production of H^+ and H_2^+. It was not specified which lines of the rare gases were studied. Ions were produced in an arc source, and accelerated to energies between 0·5 and 30 keV. Neutrals were formed by charge transfer, and were directed into the collision chamber. A lens system was used to focus the emission onto the entrance slit of a monochromator, and the light was detected by a photomultiplier. No values for the spectral resolution were given. No attempt was made to assess the influence of metastables in the neutral beam on the emission cross sections.

Emission cross sections were measured between 0·048 and 1·4 keV for $Ar + H_2$, 0·13 and 3·6 keV for $Ne + H_2$, and 0·23 and 11·5 keV for $He + H_2$. At low energies the cross sections for H_α and H_β emission become larger than those for emission from the rare gas atoms, and seem to have lower thresholds too. This has explicitly been shown for $Ar + H_2$ where the cross section for H_α emission is still $> 10^{-18}$ cm^2 at the apparent threshold of the cross section for emission of Ar I lines (50 eV). No other thresholds could be measured. There seems to be no correlation between the rare gas emission and emission from excited hydrogen atoms, since the shape of the cross sections is quite different. The cross sections for H_α emission relative to those for emission from the rare gases increase in going from He to Ar. No estimation of the absolute cross section is given in the final paper;[86] the values as published[85] are between 10^{-19} and 6×10^{-18} cm^2 for all three systems.

Light emission in collisions of He with H$_2$ was studied[37] for a He-beam energy of 1100 eV. The apparatus has been described in the foregoing section. Emission from excited states of the hydrogen atom was found to be much stronger than emission from excited He-states which could not be detected at all. No emission from excited states of the hydrogen molecule could be found.[86,37]

Excitation of autoionizing states of the oxygen atom in collisions of fast helium atoms in the ground state with oxygen molecules has been reported.[82] The apparatus has been characterized briefly in Section IV.A (see Ref. 83). The reported measurements have been performed at about 250 eV collision energy. The identification of the excited states was made by measuring the spectrum of the electrons emitted in the decay of the autoionizing states. The spectrum was compared with that obtained from collisions of thermal metastable helium atoms with oxygen. No measurements of the energy dependence of the excitation cross sections were reported, nor were values given of the absolute size of the cross sections.

B. Differential Cross Sections

1. Amdur Type Experiments

In an Amdur type experiment one determines the classical deflection function rather than the differential scattering cross section; one measures the incomplete total scattering cross section, $Q(E, \vartheta)$, as a function of the collision energy.[27,28] $Q(E, \vartheta)$ is a function of the angular resolution of the detector. The resolution is given by ϑ, the angle through which a particle must be scattered in order to miss the detector. If ϑ is not too small, $Q(E, \vartheta)$ can be expressed through the classical deflection function $b(\vartheta, E)$:[27]

$$Q = \pi b^2(\vartheta, E) \tag{25}$$

In order to sample the potential in the range between 1 and 10 eV, in general collision energies of 500 to 5000 eV are needed. Since ϑ is small, the deflection function $b(\vartheta, E)$ in this energy range is to a good approximation given by

$$E\vartheta = CV[b(\vartheta)] \tag{26}$$

C has values between 3 and 5 for reasonable potentials.

It has been shown[28] that the total cross section will be a smooth function of the energy in the range given above as long as scattering is caused by only one potential. In a semilogarithmic plot the energy dependence of the $Q(E, \vartheta)$ will give an almost straight line whose slope is connected with the steepness of the potential in the studied range. Breaks in the $Q(E, \vartheta)$-curves are always caused by abrupt changes in the potential function, and could therefore yield information on curve crossings. However at the time no completely satisfactory interpretation of the results from scattering of O-atoms on several molecules, like NO and CO, where those breaks have been observed could be given.[95,28]

2. Differential Cross Section Measurements

More information on excitation processes can be gained from measurements of the differential cross section of the elastically or inelastically scattered particles. The phenomena expected in the differential cross sections will be the same as found in the corresponding ion–atom scattering experiments. They have been demonstrated most beautifully in the standard case $He^+ + Ne$ which has been studied thoroughly both experimentally and theoretically.[97,99,100,101,102] A fully developed theory exists for the interpretation of differential cross sections in cases where excited states play an important role.[10,96,97,98]

Similar type experiments for the study of excitation in collision between neutrals have been hampered by the difficulty of detecting the weak beam fluxes which are expected in a differential scattering experiment.

The differential elastic cross section for scattering of potassium on several halogen compounds has been measured between 1 and 100 eV[103,104] in order to get information on the curve crossing between the ionic and covalent configuration in these systems. This experimental approach could also be used in cases where electronic excitation is important.

More information will be gained from the measurement of the energy loss spectra of the scattered particles as a function of the scattering angle and the collision energy. This technique has been used to study electronic excitation in collisions of Na and K-atoms with rare gas atoms and several molecules.[42] A fast, monoenergetic alkali beam was produced by resonant ion neutralization. In order to perform a time-of-flight analysis of the scattered particles, the alkali beam was pulsed by sweeping the ion beam over the entrance slit of the charge exchange chamber. With a pulse width of 50 nsec and a flight time of 37 μsec, the energy resolution which could be achieved was $\Delta E/E = 0.75\%$. Time-of-flight spectra were taken at energies between 100 and 200 eV and angles between 0 and about 3° for (K, Na) + (N_2, CO), K + K and K + rare gases.

For the rare gases no inelastic scattered particles at all could be found in the studied range of energies and angles, confirming the optical results which give very small excitation cross sections below about 50 eV collision energy.[63,64]

For K_2 a large inelastic peak was found at 100 eV collision energy shifted by 1·6 eV off the elastic peak. This peak is interpreted as due to excitation of

$K(4^2S)$ to the lowest excited states $K(4^2P_{1/2,3/2})$. At angles larger than $2°$ this peak becomes higher than the elastic peak. No excitation of higher excited potassium states could be detected. The results are in qualitative agreement with those from optical studies.[69]

For $K + N_2$ and CO the time-of-flight spectra are dominated by the elastic peak at all angles. Only energy losses smaller than 5 eV were found, indicating that excitation of the K atom is the only inelastic process occurring with high probability besides rotational and vibrational excitation of the molecule. This again confirms the optical results where no electronic excitation of the molecules has been found.[77] For $K + N_2$ an interpretation was tried under the assumption that vibrational excitation of the N_2-molecule can be neglected. The authors come to the conclusion that at low energies excitation of $K(4^2P)$ is the dominant process. At energies higher than 50 eV excitation of states with an energy of about 3·5 eV becomes dominant. On the other hand the optical results show that at these energies the total cross section for emission from $K(4^2P)$ is still one order of magnitude stronger than emission from any other excited K-state.[77] The authors[77] also came to the conclusion that the cross section for direct excitation of $K(4^2P)$ is much larger than the cross section for population of any other higher excited states. The elastic differential cross section, and differential cross section for excitation of $K(4^2P)$, $K(5^2S)$ and $K(3^2D)$, where the latter two cannot be separated, were derived for $40 < E\vartheta < 220$ eV deg from the energy loss spectra. All the cross sections decrease towards larger $E\vartheta$-values. For $K + CO$ no interpretation of the results in terms of electronic excitation of only the K-atom could be given.

For $Na + N_2$ and CO the inelastic processes occur with much smaller probability in the studied range of angles and energies; no results are reported in Ref. 42. This is a little surprising since the optical studies on these systems seem to indicate that the total excitation cross sections of the states $Na(3^2P_{1/2,3/2})$ are of the same order or larger than those for excitation of $K(4^2P_{1/2,3/2})$: the corresponding emission cross sections are of the same order of magnitude.[76]

The energy loss spectra of the ions produced in collisions between He-atoms have been studied[74] as a function of the scattering angle and the collision energy. The fast He-beam was produced by resonant charge exchange. The reaction products were produced in a collision chamber. Scattered particles regardless of their energy and charge state were detected by secondary electron emission and the ejected electrons counted with an electron multiplier. It was assumed that all types of particles are detected with the same efficiency. In addition ions were counted, after being energy analysed by a 127° electrostatic selector, by a second electron multiplier. The angular resolution was 0·1°, the energy resolution 2/1000 typically. With this apparatus the sum of the differential cross sections for all processes, and the differential cross sections for specific ionization processes could be measured for energies around 1 keV.

From the energy loss spectra it was concluded that the ionization processes are due to autoionization of excited molecular states during the collision according to (20), and to the decay of those molecular states which dissociate into a He^- and a He^+ ion. Differential cross sections for several inelastic channels were measured between 1 and 3 keV in the range $1 < E\vartheta(=\tau) < 25$ keV deg. The results are discussed in terms of crossings between quasi-molecular states. The angular distributions for $\tau > 3$ keV deg can be explained by taking into account rotational coupling between the ground state $(1s\sigma_g)^2(2p\sigma_u)^2 X^1\Sigma_g^+$ and the excited states $(1s\sigma_g)^2(2p\sigma_u)(2p\pi_u)^1\Pi_g$ and $(1s\sigma_g)^2(2p\pi_u)^{21}\Delta_g$ besides radial coupling between $X^1\Sigma_g^+$ and $(1s\sigma_g)^2$. $(2s\sigma_g)^2\,^1\Sigma_g^+$ and $(1s\sigma_g)^2(2s\sigma_g)(3s\sigma_g)^1\Sigma_g^+$. The excitation mechanisms in $He +$ He collisions are discussed in some more detail in Section IV.A.

The energy loss spectra of the scattered He-atoms in collisions between He-atoms have been measured between 100 and 175 eV collision energy.[121] Elastic cross sections have been determined between about $500 < \tau < 7800$ eV deg, inelastic cross sections between $2700 < \tau < 7800$ eV deg. A pulsed He-atom beak is produced by resonant charge exchange. The energy loss spectrum of the scattered neutral atoms is measured by a time-of-flight method. The scattered atoms are detected with an open multiplier by secondary electron emission. It is assumed that all scattered particles are detected with the same efficiency. Absolute differential cross sections are obtained by fitting the measured elastic differential cross section to a theoretical one at small angles where pure elastic scattering can be assumed.

Only single excitation of either one or both of the colliding atoms has been found. No specific states could be resolved. No doubly excited He-atoms were observed. At $\tau = 6200$ eV deg the cross sections for elastic scattering, for single and double excitation have about the same size. Excitation mechanisms, based on the MO model, were proposed in order to explain the observed processes. They have been discussed before in Section IV.A in connection with the total cross section measurements for the same system.

Acknowledgements

Support of the work in Freiburg came from the Deutsche Forschungs-gemeinschaft. The author wishes to thank his coworkers in the various stages of the experiments, in particular G. Schuller, W. Mecklenbrauck, M. Menzinger, P. Le Breton, J. Lorek, W. Koch, C. Schmidt, B. Kübler and L. Zehnle.

The author is indebted to Prof. Ch. Schlier and Dipl.-Phys. W. Mecklen-brauck for many stimulating discussions and critical reading of the manuscript. He also thanks Miss G. Dienst for her skilful preparation of the drawings.

References

1. E. W. Thomas, *Excitation in Heavy Particle Collisions*, Wiley-Interscience.
2. D. R. Herschbach, In *Chemiluminescence and Bioluminescence* (Eds J. Lee, D. M. Hercules and M. J. Cornier), Plenum Press, N.Y., 1973.
3. L. Krause, *VII ICPEAC, Invited Talks and Progress Reports* (Eds T. R. Govers and F. J. de Heer), North-Holland, Amsterdam, 1972, p. 65.
4. R. Seiwert, *Springer Tracts in Modern Physics*, **47**, 143 (1968).
5. P. L. Lijnse, Rijksuniversiteit Utrecht, *Report i*, 398 (1972).
6. F. R. Gilmore, E. Bauer and J. W. McGowan, *J. Quant. Spectr. Rad. Transfer*, **9**, 157 (1969).
7. R. S. Berry, in *Molecular Beams and Reaction Kinetics* (Ed. Ch. Schlier), Academic Press, 1970, p. 228.
8. D. H. Steadman and D. W. Setser, *Progress in Reaction Kinetics*, Vol. 6, 193 (1973).
9. T. Carrington and J. C. Polanyi, *Physical Chemistry, Series One*, 9, 135 (1973).
10. E. E. Nikitin and M. Ya. Ovchinnikowa, *Sov. Phys. Uspekhi*, **14**, 394 (1972).
11. R. E. Olson, F. T. Smith and E. Bauer, *Appl. Optics*, **10**, 1848 (1971).
12. E. E. Nikitin, in *Chemische Elementarprozesse* (Ed. H. Hartmann), Berlin, 1968, p. 43.
13. R. G. Gordon and Y. S. Kim, *J. Chem. Phys.*, **56**, 3122 (1972).
14. G. Herzberg, *Spectra of Diatomic Molecules*, Van Nostrand Co., 1950, p. 240.
15. E. Bauer, E. R. Fisher and F. R. Gilmore, *J. Chem. Phys.*, **51**, 4173 (1969).
16. W. Lichten, *P.R.*, **164**, 131 (1967).
17. M. Barat and W. Lichten, *Phys. Rev.*, **A6**, 211 (1972).
18. Th. F. O'Malley, *Adv. At. Mol. Phys.*, **7** (1971).
19. L. D. Landau and E. M. Lifshitz, *Quantum Mechanics*, Pergamon Press, London, 1959, Chap. XV, § 116.
20. A. Dawydow, *Quantum Mechanics*, Pergamon Press, London, 1965, § 108.
21. I. C. Percival and M. J. Seaton, *Phil. Trans. Roy. Soc. London, Ser. A*, **251**, 113 (1958).
22. J. B. Delos and W. R. Thorson, *Phys. Rev.*, **A6**, 728 (1972).
23. U. Buck, this book.
24. H. Pauly and J. P. Toennies, in *Meth. Exp. Phys.* (Ed. L. Marton), Vol. 7, part A, Academic Press, 1968, p. 227.
25. Y. T. Lee, *VII ICPEAC, Invited Talks and Progress Reports* (Eds T. R. Govers and F. J. de Heer), North-Holland, 1972, p. 341.
26. R. H. Neynaber, in *Meth. Exp. Phys.* (Ed. L. Marton), Vol. 7, Part A, Academic Press, 1968, p. 476.
27. I. Amdur, in *Meth. Exp. Phys.*, Vol. 7, Part A, Academic Press, p. 341.
28. V. B. Leonas, *Sov. Phys. Uspekhi*, **15**, 266 (1972).
29. M. Hollstein and H. Pauly, *Z. Phys.*, **196**, 353 (1966).
30. K. Lacmann and D. R. Herschbach, *Chem. Phys. Lett.*, **6**, 106 (1970).
31. T. R. Powers and R. J. Cross, *J. Chem. Phys.*, **56**, 3181 (1972).
32. A. N. Zavilopulo, I. P. Zapesochnyi, O. B. Shpenik and I. P. Kirlik, *VII ICPEAC Abstr. Papers* (Eds L. M. Branscomb *et al.*) North-Holland, Amsterdam, 1971, p. 597.
33. N. G. Utterback and G. Miller, *Rev. Sci. Instr.*, **32**, 1101 (1961).
34. R. H. Hammond, J. M. S. Henis, E. F. Greene and J. Ross, *J. Chem. Phys.*, **55**, 3506 (1971).

35. L. G. Piper, L. Hellemans, J. Sloan and J. Ross, *J. Chem. Phys.*, **57**, 4742 (1972).
36. R. C. Amme and P. O. Haugsjaa, *Phys. Rev.*, **177**, 230 (1969).
37. M. Hollstein, A. Salop, J. R. Peterson and D. C. Lorents, *Phys. Lett.*, **32A**, 327, (1970).
38. M. L. Coleman, R. Hammond and J. W. Dubrin, *Chem. Phys. Lett.*, **19**, 271 (1973).
39. S. Dworetsky, R. Novick, W. W. Smith and N. Tolk, *Rev. Sci. Instr.*, **39**, 1721 (1968).
40. C. Rebick and J. Dubrin, *J. Chem. Phys.*, **55**, 5825 (1971).
41. M. Hollstein, D. C. Lorents, J. R. Peterson and J. R. Sheridan, *Can. J. Chem.* **47**, 1858 (1969).
42. E. Gersing, H. Pauly, E. Schädlich and M. Vonderschen, *Disc. Far. Soc.*, **55**, 211 (1973).
43. R. Behrisch, *Springer Tracts in Modern Physics*, **35**, 301 (1964).
44. J. Politiek, P. K. Rol, J. Los and P. K. Ikelaar, *Rev. Sci. Instr.*, **39**, 1147 (1968).
45. E. Hulpke and Ch. Schlier, *Z. Physik*, **207**, 294 (1967).
46. A. P. M. Baede and J. Los, **52**, 422 (1971).
47. E. Hulpke and V. Kempter, *Z. Physik*, **197**, 41 (1966).
48. A. P. M. Baede, W. F. Jungmann and J. Los, *Physica*, **54**, 459 (1971).
49. F. Schmidt-Bleek, G. Ostrom and S. Datz, *Rev. Sci. Instr.*, **40**, 1351 (1969).
50. H. M. Windawi and C. B. Cooper, *Phys. Lett.*, **43A**, 491 (1973).
51. R. B. Cohen, C. E. Young and S. Wexler, *Chem. Phys. Lett.*, **19**, 99 (1973).
52. V. Kempter, Th. Kneser and Ch. Schlier, *Z. Physik*, **248**, 264 (1971).
53. V. Kempter and P. R. LeBreton, to be published.
54. S. Wexler, private communication.
55. J. B. Anderson, *Entropie*, **18**, 33 (1967).
56. N. Abuaf, J. B. Anderson, R. P. Andres, J. B. Fenn and D. G. H. Marsden, *Science*, **155**, 997 (1967).
57. L. D. Landau and E. M. Lifshitz, *Fluid Mechanics*, Pergamon Press, London, 1959, p. 315.
58. F. P. Tully, Y. T. Lee and R. S. Berry, *Chem. Phys. Lett.*, **9**, 80 (1971).
59. E. K. Parks and S. Wexler, *Chem. Phys. Lett.*, **10**, 245 (1971).
60. H. Haberland, F. P. Tully and Y. T. Lee, *Rarefied Gas Dynamics, Eighth Symp.*, 1972, p. 533.
61. M. E. Gersh and R. B. Bernstein, *J. Chem. Phys.*, **56**, 6131 (1972).
62. H. J. Loesch, R. A. Larsen, J. R. Krenos and D. R. Herschbach, *J. Chem. Phys.*, to be published.
63. R. W. Anderson, V. Aquilanti and D. R. Herschbach, *Chem. Phys. Lett.*, **4**, 5 (1969).
64. V. Kempter, B. Kübler, W. Mecklenbrauck, *J. Phys. B*, **7**, 2375 (1974).
65. P. Andresen, *Bericht 121/72*, Max Planck-Institut für Strömungsforschung Göttingen, 1972.
66. V. Kempter, B. Kübler and W. Mecklenbrauck, *J. Phys. B*, **7**, 149 (1974).
67. H. Rosenthal and H. M. Foley, *Phys. Rev. Lett.*, **23**, 1480 (1969).
68. S. V. Bobashev, *VII ICPEAC, Invited Talks and Progress Reports* (Eds T. R. Govers and F. J. de Heer), North-Holland, 1972, p. 38.
69. V. Kempter, W. Koch, B. Kübler, W. Mecklenbrauck and C. Schmidt, *Chem. Phys. Lett.*, **24**, 117 (1974).
70. P. J. Bertoncini, G. Das and A. C. Wahl, *J. Chem. Phys.*, **52**, 5112 (1970).
71. P. J. Bertoncini, *J. Chem. Phys.*, to be published.
72. V. Kempter, B. Kübler, W. Mecklenbrauck, W. Koch and C. Schmidt, *Chem. Phys. Lett.*, **24**, 597 (1974).

73. M. Barat, J. Baudon, M. Abignoli and J. C. Houver, *J. Phys.*, **B3**, 230 (1970).
74. M. Barat, D. Dhuicq, R. François, C. Lesech and R. McCarroll, *J. Phys.*, **B6**, 1206 (1973).
75. V. Kempter, W. Mecklenbrauck, M. Menzinger, G. Schuller, D. R. Herschbach and Ch. Schlier, *Chem. Phys. Lett.*, **6**, 97 (1970).
76. V. Kempter, W. Mecklenbrauck, M. Menzinger and Ch. Schlier, *Chem. Phys. Lett.*, **11**, 353 (1971).
77. V. Kempter, B. Kübler, J. Lorek, P. R. LeBreton and W. Mecklenbrauck, *Chem. Phys. Lett.*, **21**, 164 (1973).
78. V. Kempter, W. Koch and C. Schmidt, *J. Phys. B*, **7**, 1306 (1974).
79. W. Koch, *Diplomathesis*, Universität Freiburg, 1973.
80. P. R. LeBreton, W. Mecklenbrauck, A. Schultz and Ch. Schlier, *J. Chem. Phys.*, **54**, 1752 (1971).
81. V. Kempter, J. Lorek and T. K. Dastidar, *Chem. Phys. Lett.*, **16**, 310 (1972).
82. V. Čermák, J. Srámek, *VIII ICPEAC, Abstr. Papers* (Eds B. C. Čobić and M. V. Kurepa), Institute of Physics, Beograd, Yug., 1973, p. 534.
83. V. Čermák, M. Smutek and J. Srámek, *J. Electron. Spectrosc.*, **2**, 1 (1973).
84. A. E. Redpath and M. Menzinger, *Can. J. Chem.*, **49**, 3063 (1971).
85. V. A. Gusev, G. N. Polyakova, V. F. Erko, Y. M. Fogel and A. V. Zats, *VI ICPEAC, Abstr. Papers* (Ed. I. Amdur), MIT Press, Cambridge, Mass., 1969, p. 809.
86. G. N. Polyakova, V. A. Gusev, V. F. Erko, Y. M. Fogel and A. V. Zats, *Sov. Phys. JETP*, **31**, 637 (1970).
87. G. Gerber, R. Morgenstern and A. Niehaus, *J. Phys.*, **B5**, 1396 (1972).
88. G. Gerber, R. Morgenstern and A. Niehaus, *J. Phys.*, **B6**, 493 (1973).
89. P. O. Haugsjaa and R. C. Amme, *Phys. Rev. Lett.*, **23**, 633 (1969).
90. P. O. Haugsjaa and R. C. Amme, *J. Chem. Phys.*, **52**, 4874 (1970).
91. N. H. Tolk, C. W. White, S. H. Dworetsky and D. L. Simms, *VII ICPEAC, Abstr. Papers* (Eds L. M. Branscomb et al.), North-Holland, Amsterdam, 1971, p. 584.
92. L. Zehnle, *Diplomathesis*, Universität Freiburg, 1974.
93. D. Jaecks, F. J. de Heer and A. Salop, *Physica*, **36**, 606 (1967).
94. V. A. Ankudinov, S. V. Bobashev and E. P. Andreev, *Sov. Phys., JETP*, **25**, 236 (1967).
95. Y. N. Belyaev, N. V. Kamyskow and V. B. Leonas, *Sov. Phys., DOKLADY*, **13**, 551 (1968).
96. L. P. Kotova and M. Ya. Ovchinnikova, *Sov. Phys. JETP*, **33**, 1092 (1971).
97. R. E. Olson and F. T. Smith, *Phys. Rev.*, **A3**, 1607 (1971).
98. B. C. Eu and T. P. Tsieu, *Phys. Rev.*, **A7**, 648 (1973).
99. F. T. Smith, R. P. Marchi, W. Aberth, D. C. Lorents and O. Heinz, *Phys. Rev.*, **161**, 31 (1967).
100. D. Coffey, D. C. Lorents and F. T. Smith, *Phys. Rev.*, **187**, 201 (1969).
101. S. M. Bobbio, L. D. Doverspike and R. L. Champion, *Phys. Rev.*, **A7**, 526 (1973).
102. J. Baudon, M. Barat and M. Abignoli, *J. Phys.*, **B3**, 207 (1970).
103. V. Kempter, Th. Kneser and Ch. Schlier, *J. Chem. Phys.*, **52**, 5871 (1970).
104. B. S. Duchart, M. A. D. Fluendy and K. P. Lawley, *Chem. Phys. Lett.*, **14**, 129 (1972).
105. J. C. Browne and F. A. Matsen, *Advances Chemical Physics*, Vol. XXIII, 1973, p. 161.
106. B. I. Schneider and J. S. Cohen, *VIII ICPEAC, Abstr. Papers* (Eds B. C. Čobić and M. V. Kurepa), Institute of Physics, Beograd, Yug., 1973, p. 49.
107. J. Pascale and J. Vandeplanque, *J. Chem. Phys.*, **60**, 2278 (1974).

108. M. Krauss, P. Maldonado and A. Wahl, *J. Chem. Phys.*, **54**, 4944 (1971).
109. W. E. Baylis, *J. Phys. Chem.*, **51**, 2665 (1969).
110. E. I. Dashevskaya, A. I. Voronin and E. E. Nikitin, *Can. J. Phys.*, **47**, 1237 (1969).
111. E. I. Dashevskaya, E. E. Nikitin, A. I. Voronin and A. A. Zembekov, *Can. J. Phys.*, **48**, 981 (1970).
112. R. K. B. Helbing and E. W. Rothe, *J. Chem. Phys.*, **51**, 1607 (1969).
113. A. Beiser, *Diplomathesis*, Universität Freiburg, 1973.
114. V. Kempter, B. Kübler and W. Mecklenbrauck, to be published.
115. E. E. Nikitin, *J. Chem. Phys.*, **43**, 744 (1965).
116. H. Alber, V. Kempter and W. Mecklenbrauck, *J. Phys. B.* 8, to be published.
117. V. Kempter and L. Zehnle, to be published.
118. A. P. M. Baede, this book, Chapter 10.
119. M. E. Rudd and J. H. Macek, *Case Studies in Atomic Physics*, Vol. 3, 1972, p. 47.
120. Q. C. Kessel and B. Fastrup, *Case Studies in Atomic Physics*, Vol. 3, 1973, p. 137.
121. R. Morgenstern, M. Barat and D. C. Lorents, *J. Phys. B.*, **6**, L330 (1974).
122. F. T. Smith, *VII ICPEAC, Invited Talks and Progress Reports* (Eds T. R. Govers and F. J. de Heer), North-Holland, 1971, p. 1.
123. H. Rosenthal, *Phys. Rev.*, **A4**, 1030 (1971).
124. G. P. Können, J. Grosser, A. Haring, A. E. de Vries and J. Kistemaker, *Rad. Eff.*, **21**, 171 (1974).
125. T. L. Porter, *J. Opt. Soc. Amer.*, **52**, 1201 (1962).
126. U. Fano and J. H. Macek, *Rev. Mod. Phys.*, **45**, 553 (1973).

CHARGE TRANSFER BETWEEN NEUTRALS AT HYPERTHERMAL ENERGIES

A. P. M. BAEDE*

F.O.M.-Instituut voor Atoom- en Molecuulfysica, Kruislaan 407, Amsterdam/Wgm., The Netherlands

CONTENTS

Introduction. 464
I. Theoretical Considerations 465
  A. Two-Particle Collisions 465
    1. The Born–Oppenheimer Approximation; Diabatic and Adiabatic States . 465
    2. The LZS-Model; Solution of the Coupled Equations 469
    3. Total Inelastic Cross Sections 477
    4. Criticism of the LZS-Theory; Alternative Theories 479
    5. The Non-Diagonal Coupling Term 482
  B. Extension of the Theory to More Dimensions 485
    1. Conical Crossings 485
    2. Symmetry Considerations 487
    3. Trajectory Calculations. 489
    4. The Franck–Condon Factor Model 490
    5. Two-Particle Approximation 491
    6. Threshold Behaviour 492
II. Review of Experimental Work 493
  A. Introduction 493
    1. Introductory Remarks 493
    2. Instrumental 494
    3. Experimental 494
  B. Two-Particle Collisions 495
    1. Mutual Neutralization 495
    2. Total Charge Transfer Cross Sections 495
    3. Differential Cross Sections 499
  C. Atom–Molecule Collisions 506
    1. Total Cross Sections; Alkali–Halogens 507
    2. Differential Cross Sections; Alkali–Halogens 517
    3. Electron Transfer between Alkali Atoms and Non-Halogen Molecules . 521
    4. Non-Alkali Charge Transfer Processes 527
    5. Thresholds and Electron Affinities 529
Acknowledgements 530
References 531

* Present address: K.N.M.I. (Royal Dutch Meteorological Institute), De Bilt, The Netherlands.

I. INTRODUCTION

The development of hyperthermal neutral beam sources, some eight years ago, has disclosed a new field of beam research on charge transfer processes between neutral particles in their electronic ground state. In particular, charge transfer with low endoergicity of the order of 1 eV turned out to be very efficient and therefore has been studied extensively between its threshold and, say, 50 eV. The special interest of this field lies in its close relationship with chemical reaction kinetics and, from a theoretical point of view, its suitability to tell us more about diabatic behaviour at the crossing of potential energy surfaces.

At first only collisions between alkali atoms and halogen molecules were studied. This work could be considered as a natural extension of the beam research on reactive collisions between these particles at thermal energies. Once it was shown that these charge transfer cross sections are very large (between 10 and 100 Å^2), the field rapidly expanded. Other electronegative molecules were investigated and the improvement of detection techniques made it possible to study non-alkali collisions. This research revealed chemi-ionization phenomena such as associative and reactive charge transfer. Because the relative velocities of the nuclei involved are much smaller than the electron velocities, it is clear that charge transfer takes place effectively only at so-called pseudo-crossings between potential energy surfaces. This stimulated the study of atom–atom charge transfer, the more so as the theory of such collisions dates from 1932 without having been tested quantitatively in a satisfactory way. The study of collisions between alkali atoms and halogen atoms proved to be an excellent opportunity to carry out this test.

Table I summarizes the different charge transfer processes which have been observed between ground state particles. In almost all processes which have been studied, the energy required for these endoergic processes results primarily from the relative translational energy, but in a few cases the contribution of the vibrational energy of the target molecule XY has been investigated.

TABLE I

Observed charge transfer processes between atoms M and atoms or molecules XY

$M + XY \rightarrow M^+ + XY^-$	(1)	(Non-dissociative) charge transfer (ionization)
$\rightarrow M^- + XY^+$	(2)	
$\rightarrow M^+ + X^- + Y$	(3)	Dissociative charge transfer (ionization)
$\rightarrow M^+ + (X + Y) + e^-$	(4)	Ionization
$\rightarrow MX^+ + Y + e^-$	(5)	
$\rightarrow MX^+ + Y^-$	(6)	Reactive chemi-ionization
$\rightarrow MX + Y^+ + e^-$	(7)	
$\rightarrow MXY^+ + e^-$	(8)	Associative chemi-ionization
$\rightarrow M + X^+ + Y^-$	(9)	Collision induced polar dissociation

In this article we will review the theoretical as well as the experimental work devoted to these processes. In the first Section a résumé will be given of the theory of diabatic transitions at crossings between molecular states. The interaction between two particles is treated comprehensively and is followed by a brief discussion of crossings of multidimensional potential surfaces. Section II reviews the experimental work. Because the instrumental techniques are not essentially different from those applied to beam research on electronic excitation we will refer to the article on this subject in this volume.

This article is not at all meant as a review of a more or less closed field of research. On the contrary the field is rapidly developing and much theoretical as well as experimental progress may be expected in the near future. It is hoped that this article will contribute a little to a better understanding of the physical background and will stimulate further research.

I. THEORETICAL CONSIDERATIONS

A. Two-Particle Collisions

Here we will discuss the theory of transitions between molecular states at intermolecular distances at which these states approach each other closely or cross each other. At first we will confine ourselves to two-particle collisions. Later on we will discuss how the theory can be extended to three or more particle systems.

1. The Born–Oppenheimer Approximation; Diabatic and Adiabatic States

The interaction between two particles is described quantum mechanically by the Schrödinger equation

$$H\Psi = E\Psi \tag{1}$$

in which the Hamiltonian H can be written as

$$H = H_{el} + T_n \tag{2}$$

$$H_{el} = T_e + U \tag{3}$$

Here T_n is the kinetic energy operator describing the motion of the nuclei and T_e is the corresponding electronic kinetic energy operator; U represents the mutual coulombic repulsive and attractive interaction between the two nuclei and all electrons. In the Born–Oppenheimer (B.O.) approximation it is assumed that the electronic and nuclear motion can be separated because of the large mass difference and the corresponding large difference in velocity. In general this can be accomplished by expanding the wave function Ψ in some complete basis set $\{\phi_i\}$, which leads in a well-known way to a set of coupled equations for the nuclear wave functions χ_i (O'Malley, 1971).

Neglect of the terms which result from the operation of T_n on the wave functions ϕ_i leads to the B.O. set of coupled equations:

$$[T_n + H_{ii}(R) - E]\chi_i(R) = -\sum_{i \neq j} H_{ij}\chi_j(R) \tag{4}$$

with

$$H_{ij} = \langle \phi_i | H_{el} | \phi_j \rangle \tag{5}$$

in which R represents the internuclear distances in contrast with r, standing for all electronic coordinates. This set can be transformed by partial wave expansion into a set of coupled equations for the radial wave functions for every angular momentum l. One way to construct the complete basis set $\{\phi_i\}$, well known to molecular spectroscopists and theoretical chemists, is solving the equation

$$H_{el}\phi_i(r; R) = E_i(R) \cdot \phi_i(r; R) \tag{6}$$

for each internuclear configuration R. Using this set of functions the B.O. set will clearly be decoupled. The eigenvalues $E_i(R)$ are normally considered as the interatomic potentials, determining the nuclear motion. The states defined by (6) are called adiabatic states. A well-known theorem (Landau and Lifshitz, 1967) asserts that adiabatic states of the same symmetry cannot cross each other. This leads to a so-called avoided crossing, like the one shown in Fig. 1.

It is clear that the functions $\phi_i(r; R)$ are not eigenfunctions of H. This means that H is non-diagonal in the basis $\{\phi_i\}$ and that the states ϕ_i are coupled by the terms involving the nuclear kinetic energy operator which we neglected above. This violation of the B.O. approximation can cause transitions between adiabatic molecular states. The coupling between these states is especially important at internuclear distances where $\phi_i(r; R)$ is strongly dependent on R, because the kinetic energy is represented by derivatives with respect to R. This is the case around avoided crossings. Physically this corresponds to the fact that if two adiabatic states of equal symmetry come together, the non-crossing rule forces the corresponding wave functions to change their character considerably. If however two particles approach each other along one of these states with high velocity, the time spent in the avoided crossing region might be much too short to give the electrons the opportunity to adjust their positions. This means that the system violates the non-crossing rule and moves along a so-called diabatic potential curve (Lichten, 1967). Thus although the concept of adiabatic states is very useful in molecular physics it is not always the most appropriate starting point in dynamical collision problems. However it turns out to be much more difficult to put the concept of diabatic states on the same rigorous mathematical base as the adiabatic concept (Smith, 1969b), but from the point of view of

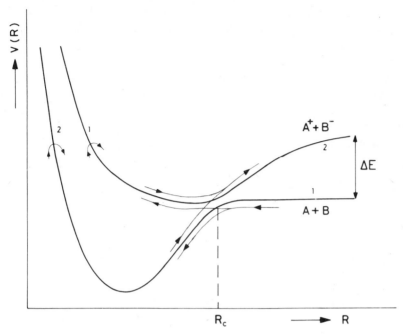

Fig. 1. Adiabatic states of the same symmetry with an avoided crossing at
$R = R_{\mathrm{c}}$.

physical intuition this concept of diabatic states is often much more appealing. Considering again Fig. 1, we see that curve (2), which at large distance
represents the Coulombic interaction between the ions $A^+ + B^-$, and
curve (1) representing the van der Waals interaction between the neutral
particles $A + B$, are almost smooth curves, interrupted only over a short
distance. This strongly suggests the choice of these states (1) and (2) as the
diabatic states, which requires the construction of crossing connections
around R_{c}. Besides their failure to obey the non-crossing rule, a general
property of the electronic wave functions ϕ_i corresponding to such states is
their weak dependence on the internuclear distance R. (The consequences of
this dependence, especially if it is not negligibly small, have been investigated by Oppenheimer, 1972.) This permits us now without restriction to
apply the B.O.-approximation so that (4) describes the behaviour of the
nuclear wave function. Consequently, transitions at an avoided crossing
between adiabatic states are violations of the B.O.-approximation, but from
the point of view of diabatic states this approximation remains completely
valid, transitions arising from the fact that the diabatic wavefunctions are
not eigenfunctions of the electronic Hamiltonian.

A few attempts have been made to put the concept of diabatic states upon
the same firm mathematical base as the adiabatic concept (Lichten, 1967);

Smith, 1969b; O'Malley, 1971), but their choice and construction (Lewis and Hougen, 1968) remains mainly a matter of physical intuition. F. T. Smith showed that the collision problem is defined by only three matrices: the potential matrix $[H_{ij}]$ defined above, a radial momentum matrix and a matrix containing only angular momentum operators. Whereas Smith defined the adiabatic states in a common way by the requirement that $[H_{ij}]$ is diagonal for all R, coupling being caused by the non-diagonal elements of the momentum operators, he proposed to define different diabatic representations by diagonalizing the different momentum operators. The application of these ideas in a molecular reference frame seems to raise problems, however (G. Nienhuis and F. T. Smith, private communication).

Corresponding with these radial and angular momentum representations, two important types of coupling between diabatic states can be distinguished: radial and angular coupling. The more important one for our discussion is the radical coupling caused by the non-diagonal elements of $[H_{ij}]$ which are different from zero if both states have the same symmetry. In Hund's case (a) notation, this means that the parity g, u and Λ, the component of the orbital angular momentum along the internuclear axis, are conserved. Neglect of magnetic interactions, such as spin–spin and spin–orbit, moreover restricts the transitions to states with equal spin component Σ. Coupling of states of different symmetry ($\Delta\Lambda = \pm 1$) however, which is important in collisional excitation (Kempter, 1974) but play only a minor role in collisional ionization (Delvigne and Los, 1973), can be caused by angular coupling and in particular by the velocity dependent Coriolis coupling, arising from the angular part of the Hamiltonian. A summary of the selection rules can be found in Section II of the article on electronic excitation in this volume.

Before now discussing the solution of the coupled equations (4) we introduce another important approximation viz. the two-state approximation. Anticipating that in the inelastic processes which we will discuss here only two diabatic states ϕ_1 and ϕ_2 play a role, we limit the expansion of the complete wave function to these two wave functions. This approximation seems justified if transitions are confined to the crossing region and if all other states remain far from these two states for all R. Now one can show that these wave functions can be chosen real if magnetic interactions are neglected. In that case the relation $H_{ij} = H_{ji}$ holds. The adiabatic states are easily found by diagonalizing the electronic Hamiltonian H_{el}, so one obtains

$$E_{1,2} = \tfrac{1}{2}\{H_{11} + H_{22} \pm [(H_{22} - H_{11})^2 + 4H_{12}^2]^{1/2}\} \tag{7}$$

It is clear from this formula that diabatic and adiabatic states only differ substantially if the quotient $H_{12}/(H_{22} - H_{11})$ is not too small, i.e. only in the vicinity of the crossing point R_c, which is defined by the relation $H_{11}(R_c) = H_{22}(R_c)$.

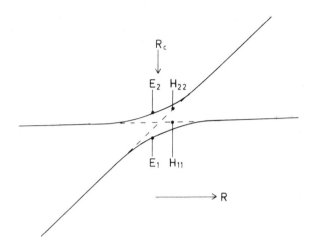

Fig. 2. Diabatic and adiabatic states around an avoided crossing.

In Fig. 2 the behaviour of diabatic and adiabatic states is depicted. One observes that the energy gap between the two adiabatic states at the crossing point R_c is equal to $2H_{12}$. From (7) it is clear that the crossing becomes real [i.e. $E_1(R_c) = E_2(R_c)$] if H_{12} is identical to zero. If H_{el} is electrostatic it can be shown (Landau and Lifshitz, 1967) that $H_{12} = 0$ if and only if both wave functions ϕ_1 and ϕ_2 have different symmetry and multiplicity, as we have seen already.

In this context it is worthwhile to stress an important difference between diabatic coupling and adiabatic coupling. As we have seen adiabatic states are coupled by the nuclear momentum operator and therefore the coupling term in general shows a singular peaked behaviour around the crossing point R_c (see Berry, 1957; Grosser, 1973, and also Oppenheimer, 1972). The diabatic states in contrast are coupled by H_{12} which does not show any particular behaviour at R_c but is in general a decreasing function of R. This of course does not imply that transitions are possible far from R_c, because as we have seen, it is the parameter $H_{12}/H_{11} - H_{22}$ which determines the width of the region, where transitions are possible.

From now on we will assume that we are dealing with a set of two crossing diabatic states $\{\phi_i\}$ for all R and an interaction matrix element $H_{12}(R) = \langle \phi_1 | H_{el} | \phi_2 \rangle$, at least at $R = R_c$. In the next section we will discuss the well-known Landau–Zener–Stueckelberg (henceforth abbreviated as LZS-) model in order to solve the coupled equations (4).

2. The LZS-Model; Solution of the Coupled Equations

Already in 1932, independently of each other, Landau, Zener and Stueckelberg proposed an expression for the inelastic transition probability at the crossing of two potential curves. Since then, and especially during the last ten years, many papers have been devoted to their work, criticizing their starting points, extending their model or simply affirming or denying the validity of their results. But the significance of their work has never been reduced to that of only an historically interesting piece of pre-war quantum-mechanics. On the contrary, as we will see, it becomes more and more clear that the LZS-theory also presents a quantitatively valid description of many experimental results. For that reason and because of its relative mathematical and physical simplicity, we will discuss the inelastic collisions in which we are interested on the basis of this theory.

Let us outline first their simplifying assumptions, together called the LZS-model:

1. The radial velocity v_l is supposed to be constant around R_c

$$v_l = \text{constant} \tag{8}$$

2. The diabatic states are supposed to be linearly dependent on R and therefore on t around R_c

$$H_{22} - H_{11} = \alpha t \tag{9}$$

3. Moreover it is assumed that

$$H_{12}(R) = H_{12}(R_c) \tag{10}$$

Equations (9) and (10) correspond with the assumption that transitions are restricted to a small region around the crossing point.

Now Zener (1932) starts with the time dependent Schrödinger equation and, assuming that the colliding particles follow a classical trajectory, he finds a set of coupled equations from which he derives the probability p_l of staying at the diabatic potential energy curve during the passage of R_c (the diabatic transition probability). This probability is dependent on the angular momentum l via the radial velocity v_l at R_c:

$$p_l = \exp\left(\frac{-2\pi H_{12}^2(R_c)}{\hbar v_l \, d/dR \, |H_{11} - H_{22}|_{R_c}}\right) \equiv \exp\left(-2\pi\delta_l\right) \tag{11}$$

This is the well known Landau–Zener formula. Landau (1932) derived the same expression, starting from the stationary Schrödinger equation and using a semi-classical approximation. Russek (1971) succeeded in casting the effect of the angular coupling in a formula analogous to this equation. He found that the diabatic transition probability, caused by angular coupling, is given by

$$p_l = \exp\left(-\frac{2\pi\omega^2|L_{12}|^2}{\hbar v_l \, d/dR \, |H_{11} - H_{22}|_{R_c}}\right) \tag{12}$$

in which ω is the angular velocity at R_c and $L_{12} = \langle\phi_1|L_\perp|\phi_2\rangle$ with $L_\perp$ being the component of the electronic angular momentum operator perpendicular to the plane of the collision trajectory. Realizing that during a collision the point R_c is passed twice and that there are two possible ways to go from the initial to the final state, we find for the transition probability from state (1) to state (2) or conversely (see Fig. 1):

$$P(v_l) = 2p_l(1 - p_l) \tag{13}$$

In this formula the quantum mechanical behaviour of the colliding particles is not taken into consideration. This means that interference between particles following different collision paths, but arriving at the same scattering angle, is neglected. Stueckelberg (1932), however, using a WKB type of treatment, arrived at an expression which does allow for interference. In the following we shall indicate the model which allows for nuclear interference with 'LZS', the other one with 'LZ'. It is outside the scope of this article to treat his derivation in any detail. A complete and critical survey can be found in Thorson et al. (1971) while a simple but surveyable introduction is presented by Raabe (1971). Because recent charge transfer experiments have shown that the inelastic cross sections reveal a strongly oscillatory structure, indicative of interference (phenomena which have been observed much earlier in elastic and inelastic ion–atom collisions, for example Coffey et al., 1969) we will discuss here the results of Stueckelberg's treatment, adapted somewhat to the present state of our knowledge. This treatment is far from suited to deal with all details of the problem. For example, if the distance of closest approach is comparable with R_c, the LZS-model becomes invalid, because both crossings cannot be separated and tunneling effects start playing a role. This problem has been discussed thoroughly by the Nikitin group (Bykhovskii et al., 1965, Nikitin, 1968) and by Delos and Thorson (1972) and will be omitted here.

The coupled equations (4) and (5), truncated by the two-state approximation, describe the relative motion of two particles at two crossing diabatic potential energy curves, their coupling being responsible for inelastic collisions. From initial state (1), (see Fig. 1), the inelastic channel (2) can be reached along the two paths depicted in the diagram. The phase shift difference along the two paths will cause interference and will result in oscillations in the cross sections. The whole process is described once the S-matrix element for each angular momentum l is specified, so the aim of the theory will be to derive expressions for the elements of this matrix from the solutions of the coupled equations. Using a WKB-type of treatment, together with the LZS-model assumptions, Stueckelberg (1932) derived essentially the following expressions, although not in the same form (see also Raabe, 1973):

$$S^l_{12} = \{p_l(1 - p_l)\}^{1/2}[e^{i(\eta_{1l} + \eta_{3l} + \chi_l)} - e^{i(\eta_{2l} + \eta_{4l} - \chi_l)}] \tag{14}$$

$$S^l_{11} = p_l e^{2i\eta_{1l}} + (1 - p_l) e^{2i(\eta_{2l} - \chi_l)} \tag{15}$$

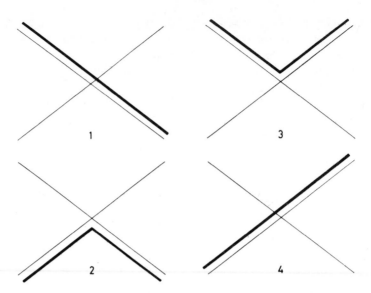

Fig. 3. The four integration paths i for the calculation of the semi-classical phase shifts η_{il}.

in which p_l is given by (11) and η_{il} is the semi-classical phase shift along path i in Fig. 3. The residual phase shift χ_l (Smith, 1969a) was assumed to be zero by Stueckelberg. Thorson and Boorstein (1965) however, found its asymptotic behaviour for $p_l \to 0$:

$$\chi_l \to \frac{\pi}{4} \quad \text{if } p_l \to 0 \tag{16}$$

Later an expression for χ_l was derived by Kotova (1969) and independently by Child (1969):

$$\chi_l = \delta_l \ln \delta_l - \delta_l - \arg \Gamma(i\delta_l) - \frac{\pi}{4} \tag{17}$$

in which δ_l is defined by (11).

The validity of this formula is restricted by the requirement that the impact parameter is much smaller than R_c, or more precisely (Nikitin and Reznikov, 1972): the phase shift difference $\eta_{1l} + \eta_{3l} - \eta_{2l} - \eta_{4l}$ should be much greater than 1 which is violated if $l \simeq kR_c$. For an unrestricted treatment we again refer the reader to the paper of Delos and Thorson (1972). Although this residual phase might be important in actual calculations, it will be neglected here because it has no important physical significance.

From the knowledge of the S matrix all information concerning elastic and inelastic cross sections can be obtained, via expressions for the scattering amplitudes, by applying the following formulae (Mott and Massey, 1965).

For the elastic and inelastic scattering amplitudes

$$f_{el}(\theta) = \frac{1}{2ik_1} \sum_l (2l + 1)(S_{11}^l - 1)P_l(\cos \theta) \tag{18}$$

$$f_{in}(\theta) = \frac{1}{2ik_2} \sum_l (2l + 1)S_{12}^l P_l(\cos \theta) \tag{19}$$

from which the differential cross sections are calculated:

$$\sigma_{el}(\theta) = |f_{el}(\theta)|^2 \tag{20}$$

$$\sigma_{in}(\theta) = |f_{in}(\theta)|^2 \tag{21}$$

while the total cross sections are found as follows:

$$Q_{el} = \frac{\pi}{k_1^2} \sum_l (2l + 1)|1 - S_{11}^l|^2 \tag{22}$$

$$Q_{in} = \frac{\pi}{k_1^2} \sum_l (2l + 1)|S_{12}^l|^2 \tag{23}$$

In these formulae k_1 and k_2 denote the asymptotic wavenumbers in the elastic and inelastic channel respectively.

Although it is clear that elastic scattering can be treated adequately by this formalism as well, we will limit ourselves here to a discussion of the inelastic cross sections.

Let us abbreviate the exponents in (14) as:

$$\begin{aligned} 2\eta_a^{(l)} &= \eta_{1l} + \eta_{3l} + \chi_l \\ 2\eta_b^{(l)} &= \eta_{2l} + \eta_{4l} - \chi_l \end{aligned} \tag{24}$$

The suffixes a and b refer to the two possible inelastic paths. Now by replacing summation by integration and approximating the Legendre polynomials $P(\cos \theta)$ by their Laplace expansion (valid for $l \cdot \theta \gg 1$), the following expression is obtained for the inelastic scattering amplitude (19)

$$f_{in}(\theta) = -\frac{1}{k_2(2\pi \sin \theta)^{1/2}} \int_{l=0}^{l_m} (l + \tfrac{1}{2})^{1/2} \{p_l(1 - p_l)\}^{1/2}$$
$$\times \{e^{i\phi_a^+(l)} - e^{i\phi_a^-(l)} - e^{i\phi_b^+(l)} + e^{i\phi_b^-(l)}\} \, dl \tag{25}$$

in which l_m is the maximum l for which the system can reach R_c, i.e. for which inelastic collisions are possible at all. The phase angles are given by

$$\begin{aligned} \phi_a^\pm(l) &= 2\eta_a(l) \pm \left\{(l + \tfrac{1}{2})\theta + \frac{\pi}{4}\right\} \\ \phi_b^\pm(l) &= 2\eta_b(l) \pm \left\{(l + \tfrac{1}{2})\theta + \frac{\pi}{4}\right\} \end{aligned} \tag{26}$$

From (25) we see that all angular momenta l contribute to scattering at an angle θ. Classically however only one or at most a few collision parameters will contribute. So our classical intuition suggests that the integrals in (25) might possibly be approximated by a sum over a few selected regions of angular momentum. Mathematically we observe that the integrand is a rapidly oscillating function of l, except in those l-regions in which the phase shifts are stationary, i.e. around the angular momentum l_i, defined by

$$\left(\frac{d}{dl}\phi_{a,b}^{\pm}\right)_{l_i} = 0 \qquad (27)$$

It was Ford and Wheeler (1959) who realized that this observation is nothing but the mathematical expression of our classical intuition. They skilfully applied this idea to elastic scattering on a simple potential curve, but Matsuzawa (1968) remarked that it applies as well to scattering on two crossing curves. In the lowest order stationary phase approximation the phase shifts $\phi_{a,b}^{\pm}$ are developed around the points of stationary phase at l_i and approximated by a parabola. The inelastic scattering amplitude can be written as a sum over separate contributions:

$$f_{in}(\theta) = \sum_i f_i(\theta) \qquad (28)$$

each contribution apart from the sign being given by

$$f_i(\theta) = \pm\frac{1}{k_2(2\pi\sin\theta)^{1/2}}\int_0^{l_m} (l+\tfrac{1}{2})^{1/2}\{p_l(1-p_l)\}^{1/2}$$
$$\times \exp i\left[2\eta_i \pm \left\{(l+\tfrac{1}{2})\theta + \frac{\pi}{4}\right\} + \eta_i''(l-l_i)^2\right]dl \qquad (29)$$

in which η_i is either $\eta_a(l_i)$ or $\eta_b(l_i)$ and η_i'' its second derivative with respect to l at l_i. Taking $l+\tfrac{1}{2}$ and p_l constant and equal to their value at l_i in the small l-range over which the integrand contributes significantly, we find after integration:

$$f_{in}(\theta) = \sum_i \left[2p_{l_i}(1-p_{l_i})\frac{l_i+\tfrac{1}{2}}{2k_2 b^2\sin\theta\eta''}\right]^{1/2} e^{i\beta_i} = \sum_i (\sigma_i)^{1/2} e^{i\beta_i} \qquad (30)$$

with

$$\beta_i = 2\eta_i - 2\eta_i'(l_i+\tfrac{1}{2}) - \left(2-\frac{\eta_i''}{|\eta_i''|}-\frac{\eta_i'}{|\eta_i'|}\right)\frac{\pi}{4} \qquad (31)$$

σ_i is the classical inelastic scattering cross section. From (27) we easily derive the following important so called semi-classical relation:

$$\left(\frac{d}{dl}\eta_{a,b}\right)_{l_i} = \pm\tfrac{1}{2}\theta(l_i) \qquad (32)$$

relating the scattering angle θ with the stationary l-value via the semi-classical phase shift, the sign being positive or negative in case of repulsive or attractive deflections respectively. Oppositely this relation can be used to calculate the phase shifts from a known deflection function.

Using (21), (30) and (31) we see that the inelastic differential cross section will contain oscillating terms of the type

$$2(\sigma_i \sigma_j)^{1/2} \cos (\beta_i - \beta_j) \tag{33}$$

so for a further discussion of the inelastic cross section we need information about the behaviour of the phases. Let us therefore examine a hypothetical inelastic process:

$$A + B^* \to A^* + B$$

In Fig. 4 the deflection function with the two branches a and b, corresponding with the two inelastic paths, is shown together with the phase shifts $\eta_{a,b}$ and $\phi_{a,b}^{\pm}$. From these last pictures we see that up to six different stationary

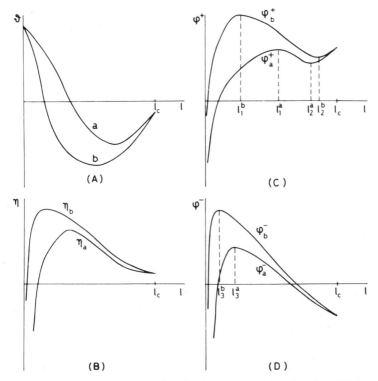

Fig. 4. Qualitative pictures of (A) the classical deflection function, (B) the phase shifts $\eta_{a,b}$ and (C, D) the phases $\phi_{a,b}^{\pm}$ of the hypothetical processes
$$A + B^* \to A^* + B.$$

points l_i can contribute to the scattering amplitude. The reader who is familiar with the theory of elastic scattering (to those who are not, the paper of Bernstein, 1966, is strongly recommended), will recognize that two independent rainbow systems with supernumeries and rapid oscillations are possible from interference of the stationary contributions of l_1^a, l_2^a and l_3^a on one hand and l_1^b, l_2^b and l_3^b on the other hand. (It should be remarked here that the lowest order stationary phase approximation breaks down near the rainbow angles because there the stationary points l_1 and l_2 coalesce so that a parabolic approximation of ϕ^+ becomes meaningless. This breakdown can be observed from (30), for the rainbow angle corresponds to $\eta'' = 0$. Here a cubic expansion of ϕ^+, leading to Airy functions, or a method devised by Berry (1966) should be used. Besides the two rainbow systems, interference is possible between stationary contributions at l_i^a and l_j^b. These are the so-called Stueckelberg oscillations, which correspond with interference between trajectories along two different potentials. The angular position of the different interference systems can be inferred from the classical deflection function (Fig. 4A). We see that with increasing scattering angle we will meet the two complete rainbow systems together with rapid oscillations and superimposed on it several systems of Stueckelberg oscillations. Beyond the second rainbow the pattern simplifies considerably because the stationary $l_{1,2}^{a,b}$ have disappeared leaving only the Stueckelberg oscillations due to l_3^a and l_3^b.

Substituting from (31) in (33) the angular spacing between successive maxima of an interference system due to the stationary points l_i and l_j is found to be given to a first approximation by

$$\Delta\theta = \frac{2\pi}{l_i \pm l_j} \tag{34}$$

the $+$ sign being valid if one of the interfering trajectories is a repulsive one and the other an attractive one while the $-$ sign holds if both trajectories are of the same type.

The applicability of the theory, outlined above, on charge transfer processes remains to be investigated. It is well known that Coulombic phase shifts diverge as $\ln 2kR$ for large R. Delvigne and Los (1973), applying this theory on the process $Na + I \rightarrow Na^+ + I^-$, assumed that this factor could be neglected because only phase shift differences are relevant. Their work will be discussed in more detail in Section II. Suffice it here to refer to Fig. 14 showing the classical deflection function from which we observe that here only one rainbow system will be present. The complete calculated differential cross section is shown in Fig. 15. The very complicated interference pattern is to be compared with the experimental results depicted in Fig. 13. It is clear that the present technique is not yet able to resolve all details.

The final goal is in fact the reverse of the method outlined above: the intention is to obtain information about potential parameters from the experimental results. No systematic treatment of this problem is known yet. If sufficient experimental information is available, the diabatic potentials might be constructed in principle via inversion procedures such as those developed by Buck and Pauly (1968) or Düren *et al.* (1968). For the time being however, as long as our experimental knowledge is far from complete, we have to resort to trial and error methods such as a systematic variation of unknown potential parameters until an optimal fit is obtained of the amplitude wavelength and position of the interference pattern.

3. *Total Inelastic Cross Sections*

From (14) together with (23) we easily obtain, after a few trivial approximations, the following expression for the total inelastic cross section:

$$Q_{in} = 8\pi \int_0^{b_c} p_b(1 - p_b) \sin^2 \tau(b) b \, db \tag{35}$$

in which $\tau(b) = \eta_a(b) - \eta_b(b)$ and b_c is the impact parameter corresponding to a distance of closest approach, equal to R_c. Replacing the rapidly oscillating $\sin^2$ term by its averaged value we obtain the well known expression:

$$Q_{in} = 4\pi \int_0^{b_c} p_b(1 - p_b) b \, db \tag{36}$$

which can be rewritten:

$$Q_{in} = 4\pi R_c^2 \left[1 - \frac{H_{11}(R_c)}{E} \right] G(\lambda) \tag{37}$$

Here $G(\lambda)$ is the integral

$$G(\lambda) = \int_1^\infty e^{-x/\lambda}(1 - e^{-x/\lambda}) x^{-3} \, dx \tag{38}$$

with

$$\lambda = \frac{v}{K} \left[1 - \frac{H_{11}(R_c)}{E} \right]^{-1/2}$$

and

$$K = \frac{2\pi H_{12}^2}{\hbar \, d/dR \, |H_{11} - H_{22}|_{R_c}}$$

The integral has a maximum value of 0·113 at $\lambda = 2·358$ and is shown in Fig. 5.

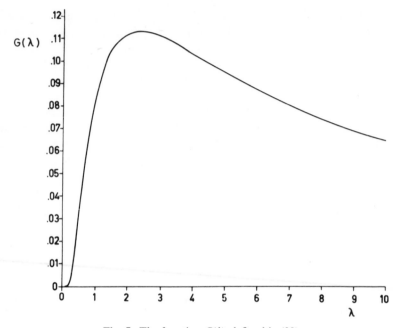

Fig. 5. The function $G(\lambda)$, defined in (38).

A comparison of experimental total cross sections with (37) yields information on the quantity $H_{12}(R_c)$. This has been the most important source of information on this quantity, especially for numerous ion–atom collisions. Even in the case that the experimental energy dependence of the total section does not agree with the theoretical one, the position of the maximum has been used for this purpose. We will come back to this point in Section I.A.5 in connection with a more general discussion on the non-diagonal matrix element.

If $\tau(b)$ exhibits a stationary point at $b = b_0$ given by

$$\left(\frac{d\tau(b)}{db}\right)_{b=b_0} = 0 \tag{39}$$

then the random phase approximation, used to derive (36) is not valid. This case has been treated by Olson (1970). He showed that at high velocities (39) is fulfilled if the difference potential $H_{11} - H_{22}$ possesses and extremum somewhere in the range within R_c. In that case the stationary phase approximation may be employed to evaluate the integral (35), yielding the following expression:

$$Q_{in} = \bar{Q}_{in} - \frac{4p_{b_0}(1 - p_{b_0})\pi^{3/2}b_0}{|d^2\tau(b)/db^2|_{b=b_0}^{1/2}} \cos\left[2\tau(b_0) \pm \tfrac{1}{4}\pi\right] \tag{40}$$

The $+$ sign is valid if the extremum in the difference potential is a maximum, otherwise the $-$ sign should be used. The second term in (40) produces oscillations on the smooth $\bar{Q}_{in}$, which is given by (37).

It can be shown that, neglecting the residual phase $\chi(l)$, the stationary phase shift difference $\tau(b_0)$ is inverse proportional to the velocity in the limit of large v:

$$\tau(b_0) = \text{const} \cdot v^{-1} \tag{41}$$

which implies that the distance between the extrema is proportional to v^{-1} in this limit. From (40) with the $+$ sign, we see that the extrema can be labeled with an index N, such that

$$\tau(b_0) = -(N - \tfrac{3}{8})\pi \tag{42}$$

where $N = 1, 2, 3, \ldots$ are the indices for the maxima, while $N = 1\cdot5, 2\cdot5, 3\cdot5, \ldots$ refer to the minima. So once the indices are assigned we have absolute values of $\tau(b_0)$ as a function of v. If one of the diabatic potentials is known and the other one parametrized, this knowledge enables us to fix one of the parameters. No systematic treatment of this problem however is known to the author. Other information, concerning b_0 itself and $\tau''(b_0)$, which is crudely related to the curvature of the difference potential at its extremum, can be obtained from the accurate measurement of the amplitude of the oscillations. The reader's attention is drawn to the formal analogy of this theory with the theory of the glory oscillations in elastic total cross section and to the analogy with the oscillations in total resonance exchange cross sections (Smith, 1966). Combining (41) and (42), it follows that a plot of the index N as a function of v^{-1} should have an intercept at $N = 3/8$ for $v^{-1} = 0$. This has an interesting consequence for, as we remember, we have neglected the residual phase $\chi(l)$. Reintroducing this phase, formula (41) reads

$$\tau(b_0) = \text{const} \cdot v^{-1} + \chi(b_0)$$

So the experimental intercept provides us with a check on the asymptotic value of χ_l given by (16), for the intercept is now given by

$$N = \frac{3}{8} - \frac{1}{\pi} \lim_{v \to \infty} \chi(b_0)$$

Olson (1970) reports a disagreement with the theoretical value $\pi/4$ of this asymptotic value in the case of the asymmetric alkali ion–alkali atom experiments of Perel and Daley (1971).

In this and previous sections we limited the discussion exclusively to the LZS theory. Let us now discuss a few alternative approaches.

4. Criticism of the LZS-Theory; Alternative Theories

Many papers have been written which criticize the LZS-theory. The strong limitations imposed by the LZS-model and their consequences have especially been the subject of numerous discussions. We shall not discuss them all in detail. For a recent discussion of the region of applicability of the LZS-model we refer to Nikitin (1970). A review of different attempts to modify the LZS-theory by changing one or more of its assumptions and to widen its applicability is given by Delos and Thorson (1972). In Section I.A.2 we met already one of the fundamental shortcomings of the LZS-theory, viz. its invalidity if the distance of closest approach is near R_c. A second error worthy of mention is the high velocity limit of the transition probability (13) which behaves like v^{-1} while Bates (1966) has shown that it should behave like v^{-2}. Thorson et al. (1971) made a detailed study of Stueckelberg's derivation. He showed that this derivation is subjected to rather severe limitations. Eu (1971), (1972a), (1972b) solved the two-channel and multi-channel coupled equations by a uniform WKB-method. The transition probability found in this way reduces to the LZS-expression in the limit of weak coupling, but remains asymptotically larger than zero if the coupling element increases. A comparison of this theory with LZS and with a numerical two-state solution (Eu and Tsien, 1972) proved that the uniform WKB method is superior. On the other hand different authors, choosing alternative ways to solve the coupled equations, found reasonable agreement between their results and the LZS-theory except of course at impact parameters near R_c. For example Olson and Smith (1971) and Nikitin and Reznikov (1972) compared the LZS theory with a distorted wave calculation and with an exact numerical solution of the coupled equations for the process $He^+ + Ne \rightarrow He^+ + Ne^*$ at 709 eV. Good agreement between the different methods was found. Below we will meet other examples from which support can be derived for the validity of the LZS-theory. Therefore in general it seems that the LZS-theory has a much wider applicability than one should expect in view of the restrictions of its starting points.

In discussing the alternative theoretical approaches let us limit ourselves to those which have been applied directly to processes in which we are interested in this article, but first of all let us stress once more the importance of the work of Delos and Thorson (1972). They formulated a unified treatment of the two-state atomic potential curve crossing problem, reducing the two second-order coupled equations to a set of three first-order equations. Their formalism is valid in the diabatic as well as the adiabatic representation and also at distances of closest approach near R_c. Moreover the problem of the residual phase $\chi(l)$ is solved implicitly. They were able to show that a solution of the three first-order 'classical trajectory' equations is not sensitive to all details of the potentials and the coupling term, but to only one function which therefore can be used readily for modelling assumptions. The resulting equations should be solved numerically. Their method has been applied now to the problem of the elastic scattering of $He^+ + Ne$ (Bobbio et al., 1973) but unfortunately not yet to any ionization problem.

Let us now review briefly other curve crossing theories, having been applied to ionizing neutral–neutral collisions, without however claiming any completeness.

Important contributions to the understanding of the role of curve crossing in alkali atom–halogen atom scattering have been published by Child. At first Child (1969) was interested in the influence of the curve crossing on the elastic scattering at thermal energies. Using the LZS-model he derived essentially the same expressions, given in this paper. Later (Child, 1971) he combined an impact parameter method of Dubrovskii (1964), arriving at an S-matrix expression without LZS restrictions. In the case of a covalent ionic model, with a Coulombic form of $H_{12}(R)$, he was able to show that his S-matrix reduced to the LZS-result, at least in the energy range below, let us say, 10 eV. An extension of his theory for systems with many crossing points is given by Woolley (1971).

Bandrauk (1969) used a distorted wave Born approximation to calculate the inelastic cross sections of atomic alkali halogen collisions. He found a forwardly peaked differential cross section without oscillations. His total cross section decreases with $E^{-1/2}$ which is the high energy asymptotic behaviour of the LZ cross section (37). The magnitude of the cross section is much larger than the LZ-result, however. His results were not essentially different if instead of a Coulombic $H_{12}(R)$ he used a constant or screened Coulombic interaction.

Recently Bandrauk (1972) has shown that the time dependent equations coupling the two-state amplitudes have an analytic solution for a constant potential crossing a Coulombic one, together with a Coulombic $H_{12}(R)$. For systems with a large reduced mass (such as K + Br) he finds agreement again with a distorted wave approximation and with the LZS theory.

The availability of electronic computers has opened the possibility of solving numerically the coupled equations (4) and (5). Different numerical methods have been developed for example by Johnson and Levine (1972) for adiabatic potentials and by Gordon (1969) for diabatic potentials. The latter has been applied by De Vries and Kuppermann (1971) in a study of inelastic s-wave scattering of $K + I \rightarrow K^+ + I^-$. Fig. 6 shows the calculated, strongly oscillating, total cross section for this process close above the threshold of 1·27 eV. They used two simplified diabatic potentials and a constant coupling of finite width ΔR. Varying ΔR the charge transfer cross section was found to be oscillatory dependent on this parameter but becoming constant for $\Delta R > 10$ Å. For a width of this magnitude the transition probability was found to be in agreement with the LZS result. It is interesting to note however, that at higher energies the energy dependence deviates from the LZS-behaviour and becomes in better agreement with Bates' (1966) prediction. In a later section we will come back to this work of De Vries and Kuppermann.

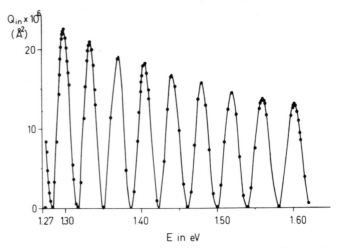

Fig. 6. Charge transfer cross section for s-wave scattering of the process $K + I \to K^+ + I^-$ as a function of the initial relative kinetic energy. Notice the sharp onset at threshold. (De Vries and Kuppermann, 1971).

Reddington *et al.* (1973) performed a numerical solution of the time dependent coupled equations for $K + I$ in the energy range 20–100 eV. He reproduced all features which we discussed in the previous section. Risking some monotony, we cite here their conclusion that the simple LZ-formula accounts well for the numerically computed transition probabilities.

Numerical solutions of the coupled equations are not necessarily restricted to the two-state approximation. A three-state calculation on atomic alkali-oxygen collisions has been performed by Van den Bos (1971). The cross sections are found to be from 30 to 60 % larger than the LZ results, in fairly good agreement with the experiments of Woodward (1970).

5. The Non-Diagonal Coupling Term

A problem that remains almost undiscussed so far is the determination of the coupling matrix element $H_{12}(R_c)$. If sufficiently accurate calculations of the adiabatic states are available this quantity is obtained directly since, as can be seen from (7), it is half of the energy gap between the adiabatic states at the pseudo-crossing point. See for example a recent *ab initio* calculation on LiF by Mulder *et al.* (1974) and Kahn *et al.* (1974). If one only possesses (estimated) diabatic states other methods have to be invoked. With (5) the quantity $H_{12}(R)$ can be calculated if one has sufficiently reliable wave functions. In the case of long range crossings, the ones in which we are interested in this paper, the active electron is to a good degree of approximation localized at one of the colliding atoms for internuclear distances around R_c. It can be described therefore by an atomic orbital wave function and, in particular, by its asymptotic form. This method was applied with a good deal of success on alkali halides by Magee (1940), (1952) and Berry (1957). The rise of the electronic computer made it possible to improve these calculations by using more realistic and therefore more complicated wave functions or asymptotic approximations (Van den Bos, 1970); Grice and Herschbach, 1974; Mulder *et al.*, 1974). A discussion of this subject and recent calculations can be found in Grice and Herschbach (1974). Their results for alkali halides and for alkali–halogen molecule interactions in collinear orientation are shown in Fig. 7.

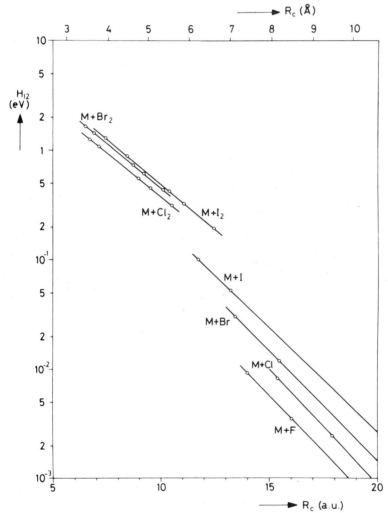

Fig. 7. Calculated coupling matrix element $H_{12}(R_c)$ in eV as a function of the crossing distance R_c (Å). Points indicate values for various alkali atoms (leftmost point for Li, etc.). The M–X$_2$ values are valid for collinear orientation. (Grice and Herschbach, 1974.)

Oppositely, as we have indicated in Section I.A.3, the value of $H_{12}(R_c)$ can be obtained experimentally by analysing total cross section measurements. Let us consider a collision of the type $A + B \rightarrow A^+ + B^-$. Approximating the covalent potential

$$H_{11}(R) = 0$$

and the ionic potential

$$H_{22}(R) = -\frac{e^2}{4\pi\varepsilon_0 R} + \Delta E$$

in which ΔE is the endoergicity of the process, we obtain from (37) and (38) the following expression, relating $H_{12}(R_c)$ with the velocity v_{max} at which the maximum of the total cross section is found.

$$H_{12}(R_c) = \left[\frac{\hbar e^2}{2 \cdot 36 \times 8\pi^2 \varepsilon_0 R_c^2} \right]^{1/2} v_{max}^{1/2} \qquad (43)$$

For ion–atom collisions an analogous expression can be derived. Hasted and Chong (1962) established an empirical relationship between H_{12} and R_c on the basis of data available at that time. They found that all values, theoretical and experimental, of $H_{12}(R_c)$ cluster around a curve which shows a decreasing and more or less exponential behaviour with increasing R_c. However, a large number of data obtained since then showed that this relationship was too simple. Recently Olson et al. (1971) succeeded in finding a new relationship, based upon earlier work by Smirnov (1965a, 1967) and Rapp and Francis (1962), which represents 83 % of the present data within a factor of three. A physical basis can be given to this relationship by the assumption that only the asymptotic behaviour of the hydrogenic wave function is important. This new relationship can be written in the following reduced form:

$$H_{12}^* = 1 \cdot 0 R_c^* \exp(-0 \cdot 86 R_c^*) \qquad (44)$$

with

$$H_{12}^* = \frac{H_{12}}{I_1^{1/2} \cdot I_2^{1/2}} \quad \text{and} \quad R_c^* = \frac{\alpha + \gamma}{2} R_c$$

in which $\alpha^2/2$ is equal to the ionization potential I_1 of the transferred electron in the reactant and $\gamma^2/2$ is the analogous ionization potential I_2 of the product. All quantities are in atomic units. In Fig. 8 (44) is shown as a full line, together with some recently determined values, which will be discussed in Section II.

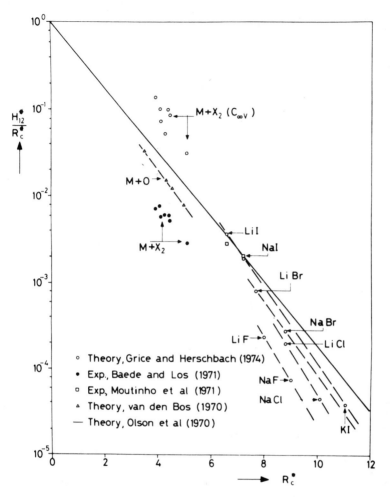

Fig. 8. The ratio of the reduced coupling matrix element H_{12}^* to R_c^* vs. the reduced crossing distance R_c^*. Full line shows the relation of Olson et al. (1971). Recent theoretical and experimental values are included.

B. Extension of the Theory to More Dimensions

1. Conical Crossings

The discussion up till now has been concerned with two colliding atomic particles, the motion of which can be described on one-dimensional potential energy curves. If however, one or both of the colliding particles are molecules, the collision process takes place on multidimensional potential surfaces. If $N > 2$ atoms are involved, we have $s = 3N - 6$ independent relative coordinates and an s-dimensional surface. The crossing of such multi-dimensional surfaces has been treated first by Von Neumann and Wigner (1929) and by Teller (1937), while recently Razi Naqvi and Byers Brown (1972) and Razi Naqvi (1972) have contributed much to our understanding of the problem. We again consider (7) and realize that we can formulate the one-dimensional avoided crossing problem in the following way: in order to have a real crossing ($E_1 = E_2$) at the point R_c, two independent conditions have to be fulfilled:

$$H_{11}(R_c) = H_{22}(R_c), \qquad H_{12}(R_c) = 0 \qquad (45)$$

However, we have only one parameter R_c available and therefore a real crossing is impossible unless $H_{12} \equiv 0$ from symmetry considerations.

Now Teller (1937) argued that in multidimensional situations we still have these two conditions to be fulfilled while we have s parameters available. Therefore two s-dimensional surfaces should cross following an intersection with dimension $s - 2$. If they have different symmetry the intersection will have a dimension $s - 1$, while if spin orbit coupling must be taken into account, it can be shown that H_{12} is a complex quantity. In that case formula (45) represents not two but three conditions and the dimension of the intersection would be one lower. Let us consider now the case $s = 2$ and let us suppose that spin–orbit coupling is negligible. From the previous considerations it follows then that two crossing surfaces of the same species will have a point intersection. What is the shape of the surfaces near that intersection? If we assume that all H_{ij} are linearly dependent on the generalized coordinates R, we see that formula (7) represents a double cone with its vertex in the point R_c determined by conditions (45). Such intersections are called conical intersections and we will see below that these are also present in the alkali atom–diatomic molecule interactions which we have studied experimentally.

Recently Razi Naqvi (1972) was able to show that Teller's argument, just like the related argument of Von Neumann and Wigner (1929) is based on a circular reasoning. He concludes that the argument that the two conditions will not be satisfied simultaneously does not follow unless it is shown that both adiabatic states cross, which is to be proved. Instead Razi Naqvi proved in a logically consistent way that in fact in the general case $s + 1$ conditions must be fulfilled for crossing, from which the general validity of the non-crossing rule should follow for multidimensional potential energy surfaces, at least if only symmetry conserving displacements should be possible on these surfaces. The simple fact however that also non-totally symmetric displacements are possible suffices to lower the number of independent symmetry conditions to be fulfilled, which reestablishes the concept of conical crossings. For a full account of this interesting work we refer to the original papers. We will see below that conical crossings play an important role in the atom–diatomic molecule interactions which have been studied experimentally.

2. Symmetry Considerations

If we want to know the character of the crossing it is necessary to know the symmetry and the species (i.e. the combination of symmetry properties including the multiplicity) of the quasi-molecular states involved. The symmetry is determined by the geometrical configuration during the interaction and the species depends on the species of the component atoms and molecules. Let us first consider the atom–diatomic molecule interactions which we are discussing here. The geometrical configuration (see Fig. 9)

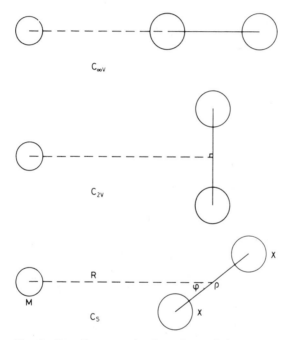

Fig. 9. Coordinates and orientations of the system M–X₂ with corresponding point groups.

can be represented by the angle ϕ between the axis of the diatomic molecule and the line between the atom and the centre of mass of the diatomic molecule. (For the meaning of the group theoretical symbols used below, we refer to Herzberg, 1966.) The symmetry point group for an arbitrary angle ϕ is C_s. If the quasi-molecule is linear then the symmetry point group is $C_{\infty v}$ and if $\phi = 90°$ then, for the collisions with homonuclear molecules M–X₂, the symmetry point group is C_{2v}. In case the diatomic molecule consists of two different isotopes (for example ^{79}Br–^{81}Br), the symmetry is strictly speaking not C_{2v}. But because the electronic wavefunction is hardly affected by this isotope effect and because the nuclear spin does not play any role in the collision, we can safely assume that the following considerations are valid. This however is not the case for collisions M–XY in which XY is CO or, even worse, IBr.

The species can be found from the species of the component interaction partners. For the alkali–halogen molecule interactions, for example, these are given by

$$M(^2S_g) + XY(^1\Sigma_g^+) \rightarrow M^+(^1S_g) + XY^-(^2\Sigma_u^+) \qquad (46)$$

For N_2 and CO in contrast we have

$$M(^2S_g) + XY(^1\Sigma_g^+) \rightarrow M^+(^1S_g) + XY^-(^2\Pi_g) \qquad (47)$$

The suffix g or u in brackets is defined only for homonuclear molecules.

TABLE II

The species of the different quasi-molecular states in form (46) and (47). The point group C_{2v} and the g and u in brackets are only defined for homonuclear molecules. The underlined species in the lowest row refer to one surface; the not-underlined species to the other. The $^2\Pi$ species refers to both.

	C_{2v}	$C_{\infty v}$	C_s
$M(^2S_g) + XY(^1\Sigma_g^+)$	$^2A_1$	$^2\Sigma^+$	$^2A'$
$M^+(^1S_g) + XY^-(^2\Sigma_u^+)$	$^2B_2$	$^2\Sigma^+$	$^2A'$
$M^+(^1S_g) + XY^-(^2\Pi_g)$	$^2A_2, ^2B_2$	$^2\Pi$	$^2A', ^2A''$

In Table II we give the species for the different possible situations. From this table we learn for example that an alkali–halogen collision in C_{2v} symmetry never leads to ionization because the species of both states is different. If however, we change the angle ϕ a little, C_s symmetry results and transitions become possible. This means that the potential energy surfaces as a function of R and ϕ exhibit a conical crossing with vertex at $\phi = 90°$ and $R = R_c$. This two-dimensional description of the crossing is of course not complete: the position R_c of the crossing is a function of the third independent co-ordinate, the internuclear distance ρ of the diatomic molecule. This dependence can be a very strong one, for example in the case of alkali–halogen molecule interaction, as we will see in Section II.

In the case of collisions with triatomic molecules, such as NO_2, in general the only symmetry operation is the identity operation I. This means that the only requirement of two interacting surfaces is that they have the same multiplicity.

From the previous discussion it became clear that the description of inelastic atom–molecule collisions on crossing potential surfaces is a very difficult problem. Moreover trajectories on such surfaces are critically dependent on details of the surfaces and, in contrast with two particle collisions, the crossing region may be passed several times. In the next few paragraphs we will give a brief discussion of several methods which have been proposed to simplify this task.

3. Trajectory Calculations

In this electronic computer era, a powerful technique is offered by the classical trajectory calculations, initiated by Wall et al. (1958) and developed by Blais and Bunker (1963) (see Bunker, 1970). The aim of this method is to infer the characteristics of a colliding system by following the behaviour of the classical trajectories of the system on its potential energy surface. The dependence of a collision property on one of the initial parameters can be studied by systematically scanning this parameter, but statistical properties such as cross sections are determined by choosing the initial conditions in a random (Monte Carlo) way, so that the corresponding trajectories form a representative sample of the collisions. A few years ago such calculations have been reported on reactive collisions between alkali atoms and halogen molecules (Kuntz et al., 1969 ; Blais, 1969 ; Godfrey and Karplus, 1968) on the lowest adiabatic surfaces for which empirical expressions are used.

For a complete description of the processes in which we are interested however, two or even more crossing surfaces are needed, together with the knowledge of their coupling as a function of the different coordinates. This information can be obtained either *ab initio* or parametrized in a physically intuitive way. The crossing region may be passed several times by one trajectory. At every pass the transition probability has to be calculated. This can be performed either by solving the coupled semiclassical equations (Tully and Preston, 1971 ; Miller and George, 1972) or by using a multidimensional extension of the LZ formula (11), such as the one proposed by Nikitin (1968) or by Tully and Preston (1971). The latter found good agreement between both methods in their analysis of the $H^+ + H_2$ or $H^+ + D_2$ inelastic collisions. Their method of calculating the potential energy surfaces, known as the method of 'diatomics in molecules' (DIM), seems both powerful and relatively simple, while the non-adiabatic coupling is obtained in a straightforward way. As yet it has not been applied to more complicated molecule systems.

Of the systems treated in this paper, potential energy surfaces of only the simplest alkali–halogen molecule combination $Li + F_2$ have been calculated with a corrected *ab initio* method yielding a wealth of information which will be discussed later in connection with the experimental information concerning related systems (Balint–Kurti, 1973, and this volume,

Chapter 4). This system should be considered as a serious candidate for future trajectory calculations. Unfortunately experimental information is not yet available, but might be expected soon because there seem to be no serious experimental obstacles. Further non-empirical potential surfaces are not known except an estimate involving a number of approximations of the K + Cl_2 surface by Nyeland and Ross (1971). The relative lack of reliable information on the potential energy surfaces is probably the cause of the fact that so far, only a few alkali–halogen molecule trajectory calculations at higher energies have been reported. Düren (1973) calculated K/Br_2 trajectories on two much simplified crossing surfaces, not allowing for dissociation of the Br_2^--ion. Some of his results will be discussed later. McDonald et al. (1973) did K/I_2 trajectory calculations at hyperthermal energies on the lowest adiabatic surface. This of course prohibits ionization but other inelastic processes can be studied. Zembekov (1971) and Zembekov and Nikitin (1972) used only the ionic part of the potential surface. Their results will be discussed in Section II. Finally, for completeness, the trajectory study of Na*/N_2 quenching by Bjerre and Nikitin (1967), involving three crossing surfaces, should be mentioned.

A final point of interest, related with the subject of the previous paragraph is the dependence of the coupling element H_{12} on the relative orientation of the particles. The two lowest adiabatic surfaces of the alkali–halogen molecule system for example do cross in the perpendicular C_{2v} orientation whereas crossing is symmetry forbidden in the linear $C_{\infty v}$ configuration. This means also that H_{12} should be zero in the latter case although it has a finite value H_{12}^0 in the former. Going from one orientation to the other H_{12} will continuously vary between H_{12}^0 and zero. This is a clear example of conical crossing and this behaviour is found indeed in actual calculations. For large R_c it has been found that H_{12} is given by

$$H_{12} = H_{12}^0 \cos \phi \qquad (48)$$

in which ϕ is the angle between the internuclear halogen axis and the line between the atom and the centre of mass of the halogen (Fig. 9) (Anderson and Herschbach, 1974; Zembekov, 1973). Zembekov moreover derived an expression for H_{12}^0. In the case of alkali–nitrogen molecule collisions (47), H_{12} equals zero for $\phi = 0°$ and $\phi = 90°$. Expressions for H_{12} in this case are given by Andreev (1972) and Andreev and Voronin (1969). The importance of all this is that the transition probability is exponentially dependent on H_{12}^2 and is therefore strongly orientation dependent. Later we will see that certain experimental features can be explained by this fact.

4. The Franck–Condon Factor Model

Studying the Na–N_2 quenching problem, Bauer et al. (1969) reduced the multidimensional problem to the crossing of an array of one dimensional curves via a network of avoided crossings, each corresponding to a transition between two definite vibrational states of the colliding molecule XY and the corresponding molecular ion XY^-. Assuming that the vibrational wave functions of XY and XY^- are independent of the M–XY distance R, they showed that at each crossing (11) is applicable provided that H_{12} is replaced by

$$H_{12}(R_c) = H_{12}^{el}(R_c) \cdot q_{vv'}^{1/2} \qquad (49)$$

in which H_{12}^{el} is the electronic non-diagonal matrix element and $q_{vv'}$ is the corresponding F.C.-factor (see also Magee, 1952). Fisher and Smith (1971), applying this method to a wide variety of quenching reactions, claimed considerable success in explaining different aspects of these reactions. Equation (49) has been applied on the system K/Br_2 by Zembekov (1973) who calculated $q_{vv'}$ as well as the angular dependent $H_{12}^{el}(R_c)$. It has been suggested however by different authors that this method is only applicable if the crossings can be considered to be independent of each other. In that case a simple classical LZ treatment can be applied at each crossing and phase interference effects between successive crossings can be neglected at least if no internal reflections are possible (Woolley, 1971). In the case of alkali–halogen interactions however, the independence of the crossings is doubtful. Not only is the vibrational spacing of the same order of magnitude as the electronic matrix element $H_{12}^{el}(R_c)$, (Baede, 1972a), but also the traversing time from one crossing to the next is so short that it causes an energy uncertainty large enough to encompass most of the vibrational states of the negative halogen molecular ion (Kendall and Grice, 1972). Therefore Kendall and Grice argue that in fact an average electron matrix element $H_{12}^{el}(R_c)$ alone is responsible for the electronic transition while the probability of obtaining a molecular ion in the vibrational state v' is proportional to the Franck–Condon factor $q_{vv'}$:

$$P_{vv'} = P_{LZ}q_{vv'} \qquad (50)$$

in which P_{LZ} is the normal LZ-transition probability, using $H_{12}^{el}(R_c)$. This formula was also the starting point of Mountinho's analysis of his measurements on the temperature dependence of total ionization cross sections to be discussed later (Moutinho et al., 1971). Child (1973) has given a theoretical justification and a quantitative validity criterion of (50).

Finally we note that a statistical diffusion-type treatment for the case of a very dense network of crossings is given by Mittleman and Wilets (1967).

5. Two-Particle Approximation

A last method to be mentioned here of simplifying the description of polyatomic interactions is a simple modification of the two-particle curve crossing model. The two-particle LZ cross section (13) is based upon the assumption that both crossings to be passed are completely equivalent, i.e. they are passed under equal physical circumstances. In collisions of an atom with a diatomic molecule however, one can in general not expect that this holds, because the internal state of the diatomic particle can change between both crossings by vibrational excitation or stretching of the bond length. So an expression, analogous to (13) can be formulated, taking into account this effect by choosing at each crossing a new set of parameters, determining that crossing. This necessitates a few model assumptions, enabling us to fix the internal state of the molecule or molecular ion at different stages of the collision. For example it has been suggested (Baede *et al.*, 1973) that during dissociative ionization of an alkali–halogen molecule collision complete dissociation of the molecular ion takes place in between both crossings, if the electron jumps at the first crossing. This means that the second crossing can be treated as a simple diatomic one, the parameters of which are well known. Together with some supplementary assumptions a few remarkable experimental observations could be explained at least qualitatively with this model.

6. Threshold Behaviour

In connection with the endoergicity of the processes which we are studying here, one problem remains to be discussed, viz. the threshold-behaviour of the total inelastic cross sections. This problem will turn out to be important because from the experimental threshold, information about the electron affinity can be obtained.

Wigner (1948) showed via phase space arguments that endothermic reactions without activation energy of the type $A + B \rightarrow C^+ + D^-$ have a stepfunction threshold. A recent discussion of this problem can be found in Grosser (1973). The same result can be obtained from simple classical arguments (Parks *et al.*, 1973). This law implies that the threshold region has zero width and that the inelastic charge transfer cross section at threshold is equal to the LZ-cross section, given by (37). This threshold discontinuity can be observed for coulombic s-wave scattering in Fig. 6, which shows numerical calculations of de Vries and Kuppermann (1971). Experimental confirmation is offered by Moutinho *et al.* (1971), who showed that the total cross section for the processes $Na, Li + I \rightarrow Na^+, Li^+ + I^-$ fits very well with the LZ cross section down to very near threshold. In the case of the process $A + BC \rightarrow A^+ + BC^-$ one might argue that such a step function will be found for every vibrational transition becoming possible with increasing energy. The result should be that the real threshold behaviour is a staircase of small-spaced step functions, the height of each step of which is determined by the transition probabilities between both vibrational states, i.e. the Franck–Condon factor.

For processes in which one neutral and two oppositely charged particles are formed such as dissociative ionization $(A + BC \rightarrow A^+ + B^- + C)$ or collision induced polar dissociation $(A + BC \rightarrow A + B^- + C^+)$ Maier (1964) derived the following threshold law:

$$Q \sim \frac{(E - E_{thr})^n}{E_i^{1/2}} \tag{51}$$

in which E is the total (kinetic + internal) energy of the reactants, E_{thr} is the threshold energy and E_i is the initial relative kinetic energy, all values in the centre of mass system. This expression contains an adjustable parameter $1 \leq n \leq 2$, in contrast with the expression recently derived by Levine and Bernstein (1971) for the threshold behaviour of collision induced dissociation:

$$Q \sim \frac{(E - E_{thr})^{5/2}}{E_i} \tag{52}$$

Analysis of experimental results on the basis of these laws are not conclusive. Experimental results reported by Maier (1965) seem to confirm Levine's expression but on the other hand measurements by Parks et al. (1973) show that the exponent is generally smaller than Levine's value 5/2. For a comprehensive discussion of these problems we refer the reader to the paper of Parks and his colleagues.

Summarizing we might say that the step function threshold of reactions yielding two oppositely charged products is well established, but that no satisfying threshold law is known at this moment for dissociative ionizing collisions yielding three product particles.

II. REVIEW OF EXPERIMENTAL WORK

A. Introduction

1. Introductory Remarks

In this chapter a review will be given of the experimental work on the charge transfer processes, indicated in Table I, in which diabatic transitions at avoided crossings play an important role. Let us confine ourselves to processes investigated by beam techniques. We leave aside a systematic treatment of elastic scattering experiments in which the total and differential cross sections have been influenced by curve crossings. We will however refer to this work if necessary for the understanding of the inelastic processes. Moreover we exclude a discussion of charge transfer and chemi-ionization in collisions between neutral electronically excited particles. Suffice it to refer to Berry (1970).

Collisional dissociation (process (9) of Table I) differs essentially from the other three particle processes in the sense that the diabatic transitions, responsible for ionization, take place inside the molecule XY. Therefore this process is more related with two-particle collisions than with three-particle collisions. We will pay some attention to collisional dissociation within the framework of the two-particle collisions, but for a more extensive discussion we refer to Wexler (1973). Some attention will be given moreover to the process of mutual neutralization, in which two oppositely charged ions collide and form two neutral products.

2. Instrumental

The crucial step towards the feasibility of charge transfer experiments in the eV energy range was made by the development of sources producing sufficiently intense energy selected beams. These techniques have been discussed by Wexler (1973) and by Kempter (1974) elsewhere in this book. At first only alkali beams were used because they could be detected easily with a surface ionization detector. A quantitative measurement of the beam intensity was hampered by the fact that the detection efficiency was unknown at superthermal energies. This problem was solved by Politiek and Los (1969) and Cuderman (1971a, 1971b). Nowadays techniques for the detection of other than alkali beams have been developed. Negative surface ionization can be applied to detect halogen beams (Können et al., 1973). Moreover efficient electron impact detectors are available now. Apart from the beam production and detection, the experimental set up, necessary for the measurement of total and differential cross sections is conventional and does not deserve an extensive discussion. A few machines have been described extensively in the literature (Moutinho, 1971; Leffert et al., 1972; Wexler, 1973; Können et al., 1974b). The interested reader is referred to these papers and to Fluendy and Lawley (1973) for a review of techniques.

3. Experimental

The first ionization experiment of interest to us was performed by Bydin and Bukhteev (1960) and Bukhteev et al. (1961) at energies between 150 and 2200 eV. But only after the development of a beam source using cathode sputtering and the improvement of the charge exchange source, did it become possible to perform these measurements down to threshold and to understand better their physical meaning and their relation to chemistry (Baede et al., 1969; Helbing and Rothe, 1969). Since then this field has shown a fast development. At first experimental feasibility restricted the measurement of total ionization cross sections to alkali atoms on rather simple molecules, but nowadays differential as well as total cross sections are being measured and the range of collision partners extends from electronegative atoms on one side to complicated organic molecules on the other. Moreover interactions with non-alkali atoms are now under investigation by different groups.

From an historical point of view the interaction between alkali atoms and simple molecules, in particular the halogen molecules, have a right to be discussed first, but from a physical point of view it seems appropriate to start with atom–atom scattering experiments, which are closer connected with the theoretical discussion of the previous section. Therefore let us give up historical priorities and see what can be learned from these experiments.

B. Two-Particle Collisions

1. Mutual Neutralization

Only after the development of the difficult merging and inclined beam techniques did it become possible to obtain experimental information on mutual neutralization at low energies. Now much information is available, partly on molecular ion interactions (Peterson et al., 1971), but mostly on atomic ions. The last category is the most important from a theoretical point of view, because it is much more suited to direct comparison with the LZ-theory. The interpretation is hampered by the fact that in general more than one covalent state must be taken into account. For example in the thoroughly investigated reaction between H^+ and H^- ions (Rundel et al., 1969; Olson et al., 1970; Moseley et al., 1970; Gaily and Harrison, 1970; Janev and Tančić, 1972) two covalent states are important, but in the $He^+ + H^-$ reaction ten states have to be incorporated. A first step to unravel such confused situations has been taken by Weiner et al. (1970 and 1971), who observed the light emitted by the neutral atom in the exit channels of the $Na^+ + O^-$ reactions, in this way identifying the channel.

In general the LZ theory gives a good qualitative description of the measurements. Quantitative agreement could be obtained in the case of $H^+ + H^-$, below 10 eV by Olson et al. (1970) using a value of H_{12} calculated with the formula of Smirnov (1965b), but also at higher energies by Janev and Tančić (1972) using an expression of Janev and Salin (1972).

Unfortunately there seem to be no atom–atom combinations in which both the ionization and the mutual neutralization have been studied. Only the alkali–oxygen system has been studied by both techniques, but as we will see not much mutually supplementing information could be obtained in this case.

2. Total Charge Transfer Cross Sections

Not many results have been reported so far on atom–atom charge exchange collisions in the eV range (see Tables IV and VI), but those which have been obtained are extremely interesting. Total cross sections have been measured for Cs on O and Na, Li on I, whereas differential cross sections only have been reported for Na + I.

Woodward (1970), has measured the production of ions following the process:

$$Cs(^2S) + O(^3P) \rightarrow Cs^+(^1S) + O^-(^2P)$$

at energies between 200 and 2000 eV, thus covering the range in which in this case the maximum LZ transition probability can be expected. In this experiment the O beam was the fast beam, formed by laser photodetachment of O^- ions. Besides cross sections for ground state $O(^3P)$ atoms, cross sections could be obtained for excited $O(^1D)$ atoms as well, which however correspond to an extremely large $R_c \simeq 33$ Å. Finally, by comparing Cs^+ and O^- signals, the cross section for the production of free electrons:

$$Cs + O \rightarrow Cs^+ + O + e^-$$

could be obtained, be it with a large degree of uncertainty. The results are shown in Fig. 10 together with the LZ-calculation for the ground state process of van den Bos (1970). Because it was assumed that rotational coupling can be neglected, ionization only results from collisions between states of equal symmetry. This is taken into account by introducing a weighing

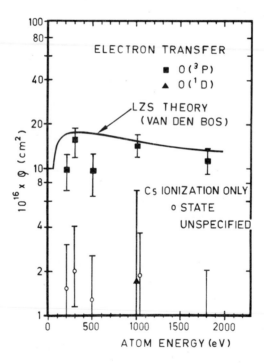

Fig. 10. Comparison of theoretical (full line, van den Bos, 1970) and experimental (squares, Woodward, 1970) charge transfer cross sections for the process $Cs + O(^3P) \rightarrow Cs^+ + O^-$. Experimental cross sections for the process $Cs + O(^1D) \rightarrow Cs^+ + O^-$ (triangle) and $Cs + O \rightarrow Cs^+ + O + e^-$ (open circles) are also included.

factor of 1/3. Besides this LZ-calculation, a numerical three-state calculation has been reported by van den Bos (1971), yielding from 30 to 60% higher cross sections. These calculations show that the contributions of the $p_{\pm 1}$ states of oxygen is still negligible in this energy range. Recently a calculation by Vora *et al.* (1973) has been reported, yielding cross sections likewise higher than the LZ results. In this calculation the transition is not necessarily restricted to the crossing point, but transitions outside the crossing do not appear to play an important role in this energy range. In contrast with the LZ-formula, this calculated cross section exhibits the right asymptotic high energy dependence. Unfortunately the experimental uncertainty prohibits a choice between these three theories.

Supplementary information on the alkali-atomic oxygen interaction was obtained by the study of the inverse process of mutual neutralization, mentioned earlier in Section II.B.1. Mosely *et al.* (1971) measured total neutralization cross sections for $Na^+ + O^-$, whereas Weiner *et al.* (1971) could distinguish between various possible excited Na states involved by observing the emitted light. The neutralization cross section is dominated by the process

$$Na^+ + O^- \rightarrow Na(3d) + O(^3P)$$

due to the near degeneracy of these two levels at infinity, whereas the observed $Na(3p - 3s)$ emission is ascribed to cascading from the $3d$ level. This means that this technique in this case provides no information on the interaction between states with small crossing distances such as the one corresponding with the Na and O ground states.

As stated by Ewing *et al.* (1971) the alkali halide molecules MX provide the archetype of the curve crossing problem. The covalent coupling between the ground state atoms $M(^2S_{1/2})$ and $X(^2P_{3/2})$ results in eight molecular states, only one of which is of the same species $^1\Sigma^+$ as the ionic states corresponding to $M^+(^1S_0) + X^-(^1S_0)$. This implies that a weighing factor of 1/8 should be introduced and moreover that elastic scattering, which can proceed along all available potential curves, is much more complicated than inelastic scattering. The relevant states of NaI, the most thoroughly studied interaction, is depicted in Fig. 11. It is clear that the crossing distance of the ionic state with the excited $Na(^2S_{1/2}) + I(^2P_{1/2})$ state is too large to play any significant role. This suggests that the two-state approximation holds in this case to a very high degree.

Total ionization cross sections for Na + I and Li + I have been measured by Moutinho *et al.* (1971), using a sputtered alkali beam and a secondary I beam, produced by thermal dissociation of I_2 at 1000°K. This temperature is high enough to exclude the influence of remaining I_2 molecules. The cross sections of the other alkali iodide systems are too small to be measurable in this energy range due to their very large crossing distances. By varying

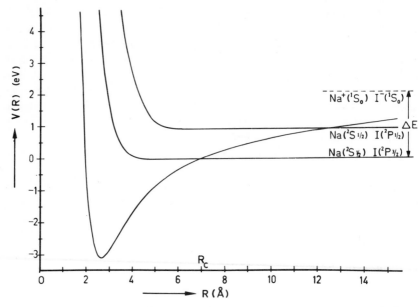

Fig. 11. Lowest NaI adiabatic potential curves, all of the same species $^1\Sigma^+$.

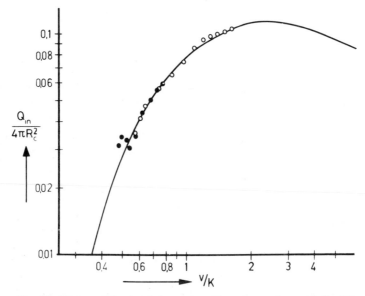

Fig. 12. Fitting of the total charge transfer cross section with the LZ cross section. From this fitting a value of K and, from this, of $H_{12}(R_c)$ is obtained (see equation 36). ○, Na + I, ●, Li + I (Moutinho *et al.*, 1971).

$H_{12}(R_c)$ the measured cross sections could very well be fitted with the LZ cross section (36). The results are shown in Fig. 12 whereas Table III gives the resulting values of the coupling energy, together with the theoretical values of Grice and Herschbach (1974) and those obtained from Olson's semi-empirical relation (44). The agreement is very good, indicating that the LZ-theory provides a quantitative description of these interactions, at least at the low energy side of the maximum. Extension of these measurements to higher energies would be desirable.

TABLE III

Crossing distance R_c (Å) and non-diagonal matrix element $H_{12}(R_c)$ (in eV) of LiI, NaI and KI

	LiI	NaI	KI	Reference
$H_{12}(R_c)$	8.5×10^{-2}	5.1×10^{-2}	—	Moutinho et al. (1971)
	9.5×10^{-2}	5.5×10^{-2}	2.7×10^{-3}	Olson et al. (1971)
	9.5×10^{-2}	5.0×10^{-2}	1.45×10^{-3}	Grice and Herschbach (1974)
R_c	6.2	6.9	11.3	

There are indications that the measured cross sections show undulations, as predicted by Olson (see equation 40). Unfortunately the resolution is too bad and the energy range too small to allow any conclusion to be drawn with respect to the potential energy curves.

3. Differential Cross Sections

The fact that the LZ-theory describes the total cross sections so well, apart from the undulations, indicates that the information which can be extracted from these experiments is rather limited and only concerns the crossing point. More information on the dynamics of the collisions can be expected from the differential cross section, as we have discussed in Section I.A.2.

Delvigne and Los (1973) measured the differential ionization cross section of Na + I, at energies between 10 and 60 eV. The alkali beam, produced by a charge exchange source, was crossed with a thermally dissociated iodine beam. The energy spread of the primary beam is about 0.5 eV, whereas the angular resolution of the detector varies between 0.3° at the smallest angles ($< 5°$) and 2.4° at the largest ($> 20°$). The polar differential cross section $I(\theta) \sin \theta$ for two different CM-energies E_i is shown in Fig. 13 as a function of the reduced scattering angle $\tau = E_i \theta$. From the theory of elastic scattering (Smith, 1969a) it is known that such a plot is energy invariant at least in lowest order. It can be seen that this law holds reasonably well, even for these inelastic interactions.

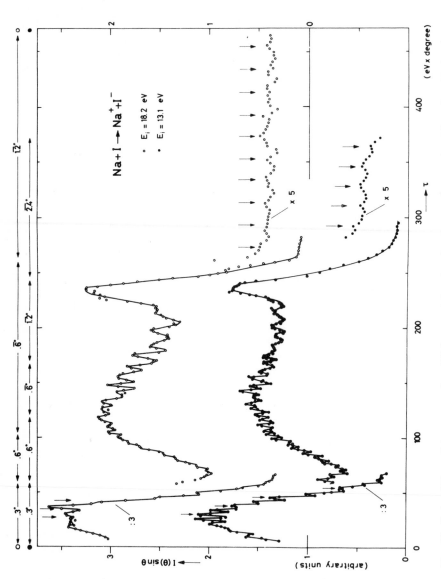

Fig. 13. Polar differential cross section (CM-system) for the charge transfer reaction Na + I → Na⁺ + I⁻. Notice the relative shift of the zero point of both curves and the multiplication factors of different parts of the

The ionic state is usually described by a Rittner potential.

$$U_{ion}(R) = -\frac{e^2}{R} - \frac{e^2(\alpha_{Na^+} + \alpha_{I^-})}{2R^4} - \frac{2e^2\alpha_{Na^+}\alpha_{I^-}}{R^7} - \frac{C_{ion}}{R^6}$$
$$+ A_{ion}\, e^{-R/\rho_{ion}} + \Delta E \tag{53}$$

whereas the covalent potential is assumed to be

$$U_{cov}(R) = -\frac{C_{cov}}{R^6} + A_{cov}\, e^{-R/\rho_{cov}} \tag{54}$$

All potential parameters are known (see Table I of Delvigne's paper), except those describing the repulsive part of the covalent potential. These have been obtained from the energy dependence of the wavelength of the oscillations beyond the rainbow at $\tau \simeq 240$, which are caused by Stueckelberg interference between particles scattered from the repulsive part of both potentials. Once the potentials were known the deflection function could be calculated. Fig. 14 shows the result for $E_i = 13\cdot1$ eV. Depending on the nature of the potential at distances smaller than R_c, Delvigne distinguishes ionic and covalent scattering. Corresponding with this distinction, two deflection branches, meeting at $b = R_c$ are present, namely the covalent branch and the ionic one. Although in contrast with the general case, discussed in Section I.A.2 (see Fig. 4), no covalent rainbow will be present, the flat part a will nevertheless cause a strong maximum. Fig. 15 shows the differential cross section calculated with the formalism of Section I.A.2, for $E_i = 13\cdot1$ eV and $H_{12}(R_c) = 6\cdot5 \, . \, 10^{-2}$ eV. The agreement is very satisfactory. Particularly the covalent maximum at $\tau \simeq 30$ and the ionic rainbow system $(100 < \tau < 300)$, separated by a deep minimum at $\tau \simeq 65$ corresponding with $b = R_c$, are reproduced but it is clear that the very complicated interference structure could not be resolved experimentally. Delvigne and Los therefore compared their results with a simplified calculation (Fig. 15b) in which the short wavelength contributions, due to interference between attractive and repulsive trajectories, are neglected. This approximation breaks down at very small angles and around the minimum. Outside these regions, however, a good agreement between observed and calculated positions of the oscillations is found. Two serious discrepancies remain however (see Fig. 16):

1) The experimental and theoretical smoothed differential cross sections (i.e. averaged over the oscillations) could only be brought into agreement for different collision energies if an energy dependent $H_{12}(R_c)$ was used.

2) The cross section at the minimum $(b = R_c, \tau = 65)$ is underestimated; this discrepancy increases with increasing energy.

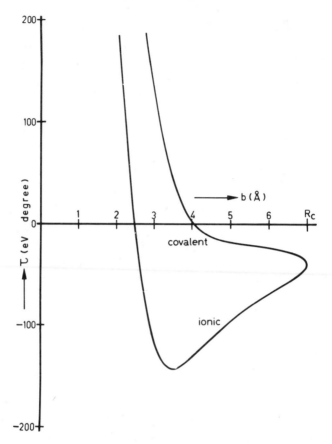

Fig. 14. The deflection function for the process Na + I →
Na$^+$ + I$^-$ at a relative energy of 13·1 eV. The covalent and
ionic branches are connected at $b = R_c$ (Delvigne and
Los, 1973).

These discrepancies could be explained by assuming that angular coupling
between the ionic $^1\Sigma^+$ and the $^1\Pi$ covalent state plays a role. By applying
(12) Delvigne and Los were able to remove the first discrepancy, using the
values $H_{12}(R_c) = 6\cdot5 \cdot 10^{-2}$ eV and $L_{12} = 4 \cdot 10^{-2}\hbar$. The second discrepancy
was reduced in this way but could not be removed completely, which might
be due to the breakdown of Russek's formula near $b = R_c$, whilst the LZS-
theory itself likewise loses its validity at this point. This also might partly
account for the only moderate agreement between $H_{12}(R_c)$, found in these
experiments and the values in Table II.

Further information concerning the role of the curve-crossing in alkali
halides was obtained from three other, widely different, sources.

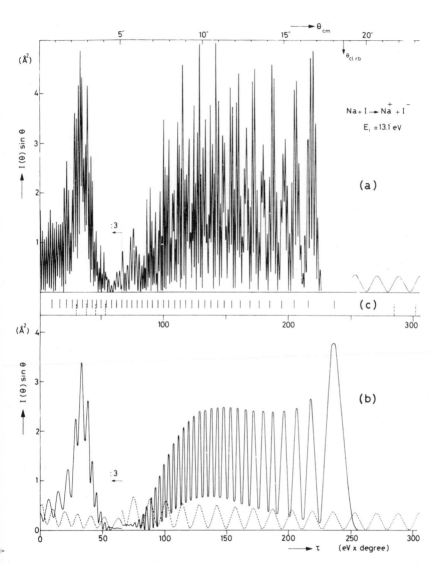

Fig. 15. Polar differential cross section calculated semi-classically for the charge transfer process $Na + I \rightarrow Na^+ + I^-$. (a) Calculation with the complete interference structure with omission of the primary rainbow. (b) Approximate semi-classical calculation taking into account only interferences from net repulsive and net attractive scattering. (c) The full bars indicate maxima observed experimentally for net attractive scattering, the dashed bars for net repulsive scattering. $H_{12}(R_c) = 0.065$ eV; angular coupling was neglected. (Delvigne and Los, 1973.)

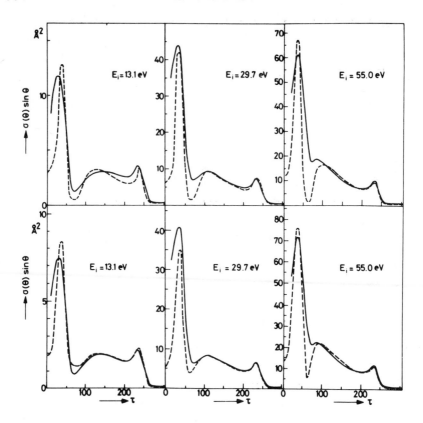

Fig. 16. Comparison of calculated (dashed curve) and measured (full curve) smoothed cross sections at three different energies, showing the improvement obtained by including rotational coupling. Upper pictures without rotational coupling $[H_{12}(R_c) = 0.05 \text{ eV}; L_{12} = 0]$ and lower pictures with rotational coupling $[H_{12}(R_c) = 0.065 \text{ eV}; L_{12} = 0.04\hbar]$. Notice the different vertical scales. (Delvigne and Los, 1973.)

Berry (1957) analysed the gas phase u.v. absorption spectra corresponding to transitions between the alkali-halide ground state and its first excited state. The KI spectrum shows a continuum, indicating a short lifetime of the vibrational states and consequently a large diabatic transition probability. NaI in contrast shows a bandspectrum, indicating a small transition probability. These, and similar experiments by Oppenheimer and Berry (1971) on alkali halides in inert matrices, permit an order of magnitude determination of the coupling matrix element $H_{12}(R_c)$. Recently chemiluminescence spectra of KI have been measured in a dilute flame experiment by Kaufmann *et al.*

(1974b). In combination with molecular beam elastic scattering data, to be discussed below, very accurate information is obtained on the KI excited state potential.

Information on the coupling matrix element could likewise be obtained from the observation of ions from dissociation of alkali-halide molecules, induced by collisions with inert gas atoms. Shock tube (Ewing *et al.*, 1971) as well as beam experiments (Tully *et al.*, 1971 and 1973; Piper *et al.*, 1972) have been reported. A recent review of this and related work has been given by Wexler (1973).

As we remarked above, the interpretation of elastic scattering between alkali and halogen atoms is severely hampered by the fact that all eight covalent states contribute. A comparison of three different experiments on the K/I system leads to the conclusion that indeed much remains to be clarified. Fluendy *et al.* (1970) measured the differential elastic K/I scattering at thermal energies with extremely high angular resolution. From a comparison with their calculations they conclude that the behaviour of this system is highly adiabatic. On the other hand Hack *et al.* (1971) measured differential as well as total differential cross sections for the same system. The energy dependence of the total cross section as well as the angular dependence of the differential cross section led them to the conclusion that the Coulomb potential has no significant influence which implies a diabatic behaviour. Kaufmann *et al.* (1974a), came to the same conclusion from their thermal K/I differential cross section measurements. From the $H_{12}(R_c)$ values of Grice and Herschbach (1974) adiabatic behaviour can indeed be excluded, except at the largest impact parameters. Both Hack and Kaufmann suggest that Fluendy's iodine beam was too cold to be sufficiently dissociated. The measurements of Hack and Kaufmann at least show some qualitative agreement. Both observe structure at small angles, although Hack fails to see structure at larger angles. Kaufmann, observing two rainbows (see Fig. 17), needs four potentials with appropriate weighing factors to fit his measurements. The fourth one with a very shallow minimum is necessary to reproduce the correct intensity ratio of wide and small angle scattering. Hack's structure might correspond with the first, small angle, rainbow. The quantitative analysis of both experiments reveals serious discrepancies, which make further experiments highly desirable, particularly in combination with other techniques. Moreover other combinations should be investigated. For example elastic scattering data on Na/I might reveal the influence of adiabatic scattering, on which accurate ionization data are available.

Concluding, the charge transfer technique has contributed much to our knowledge of the alkali halides. The consistency with other techniques is still rather poor, which may be a strong motivation for further research on this exciting field of beam physics.

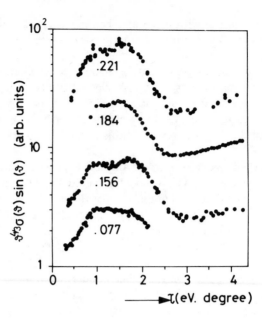

Fig. 17. Elastic polar differential cross section of K + I, multiplied by $\theta^{4/3}$, vs. $\tau = E\theta$ at four different energies (eV). Two distinct rainbows can be observed. (Kaufmann *et al.*, 1974a.)

C. Atom–Molecule Collisions

Although it became clear after the first experiments by Baede *et al.* (1969) and Helbing and Rothe (1969) that ionization could be observed in collisions of alkali atoms with many electronegative molecules, at present the most extensive information from an experimental as well as a theoretical point of view is available on alkali–halogen molecules collisions. Moreover the latter turn out to be far less complicated than most other alkali–molecule inter-actions. In the energy range between threshold and, say, 30 eV, halogen collisions have so far revealed elastic scattering and the production of halogen atomic and molecular ions. But NO_2 collisions, for example, exhibit not only elastic scattering and the production of O^- and NO_2^-, but also the formation of free electrons and photons from atomic as well as molecular excitation processes. So it does not seem surprising that the alkali–halogen interactions are understood now in greater detail than any of the other investigated systems, the more so as much more is known theoretically about their potential energy surfaces. Nevertheless, also the non-halogen molecule interactions allow some general conclusions to be drawn and some molecular properties to be derived.

Concluding, it seems appropriate first to discuss the total alkali–halogen ionization cross sections, next to extend the discussion to differential cross sections and finally to discuss other collision partners. A special section will be devoted to the determination and interpretation of the molecular electron affinities.

1. Total Cross Sections; Alkali–Halogens

During the last ten years the reactive and non-reactive interaction between alkali atoms and halogen molecules has been studied extensively at thermal energies. Reviews can be found in Herschbach (1966), Toennies (1968, 1974) and Kinsey (1972). Characteristic features of the reactive scattering are : the large size of the reactive cross section (100–200 Å^2), the forward product scattering and the large internal, particularly vibrational, excitation of the product. These experimental findings could satisfactorily be explained by the so-called harpoon model (Polanyi, 1932; Magee, 1940), which supposes an electron jump to occur from the alkali atom to the halogen molecule at an internuclear distance R_c, corresponding to the crossing of the lowest co-valent and ionic surfaces. Subsequent dissociation of the negative molecular ion and binding between both resulting atomic ions complete the reaction. The large reaction cross section can be understood simply from the rather large R_c, whereas the strong vibrational excitation of the products results from the highly attractive 'early downhill' character of the lowest adiabatic surface. Moreover it has been suggested that the sharp forward peaking can be understood from the plausible assumption that the reaction proceeds via a spectator stripping mechanism. Improvements of the simple harpoon model have been proposed by Herschbach (1966), Grice (1967) and Anderson (1968), whereas Edelstein and Davidovits (1971) could show that the magnitude of the reaction cross section is predicted quite accurately if a simple dynamical improvement is introduced.

Empirical potential energy surfaces based on these ideas, which have been used for trajectory studies indeed show an 'early downhill' character. The role of the spectator stripping mechanism is less clear from these studies, although it can be concluded that large product repulsion indeed favours backward scattering (Godfrey and Karplus, 1968). Kuntz et al. (1969) conclude that the possibility of charge migration within the negative molecular ion is essentially to obtain a sharp forward peaking. Blais (1968) however, who excluded charge migration, showed that agreement with experiment was only obtained by incorporating an induced dipole interaction between the departing X and the resulting charges M^+ and X^-.

Recent calculations of the potential energy surface for $Li + F_2$ by Balint-Kurti (1973) and for $K + Cl_2$ by Nyeland and Ross (1971) partially confirm and partially amend the earlier ideas. The calculated surfaces have indeed an early downhill character and show even some product attraction. On the

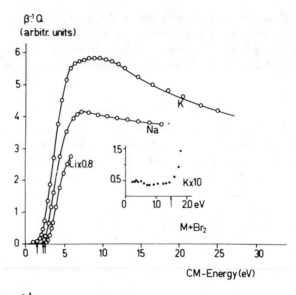

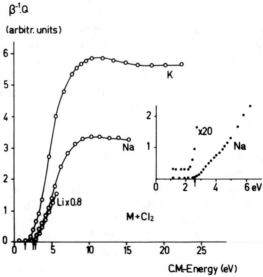

Fig. 18. Total cross sections Q for the production of negative ions in collisions of K, Na and Li on Br_2 and Cl_2, divided by the unknown energy dependent surface ionization detector efficiency β, vs. the CM-energy. The thresholds are indicated by arrows. The insets show enlargements of the thresholds. Notice the small structure at about 4 eV in the Na-Cl_2 cross section. The constant signal below threshold is due to the presence of dimers in the alkali beam (Baede and Los, 1971).

other hand all empirical surfaces assumed that the charge after the jump is situated at one of the halogen atoms, at best allowing for charge migration. Both calculations have indicated that this is definitely incorrect and that three-body interactions play a role on a large part of the potential surface.

Above 1 eV the reactive cross section probably decreases steeply (Van der Meulen *et al.*, 1973; McDonald *et al.*, 1973) to disappear between 2 and 5 eV. At about the same energy excitation and ionization become energetically possible, both of which are based upon the same electron jump. Anderson *et al.* (1969) showed however that excitation is not observed because the required second crossing is inaccessible below, say, 30 eV. Soon after the construction of alkali beam sources in the eV region it was realized that ionization measurements are very suited to test the electron jump mechanism. At first only total cross sections were measured (Baede *et al.*, 1969; Helbing and Rothe, 1969; Lacmann and Herschbach, 1970a; Baede and Los, 1971). Fig. 18 shows some results of Baede and Los (1971) on bromine and chlorine. All curves, including those on iodine not shown here, show the same general behaviour: a more or less sharp rise above threshold up to a broad maximum followed by a slow decrease. The magnitude of the maximum is estimated to be some tens of $Å^2$. Some curves exhibit a small structure close above threshold, which has been observed also by Leffert *et al.* (1973), but for which so far no satisfactory explanation has been given. These first experiments were soon followed by mass selection of the negative ions and the measurement of differential cross sections. A survey of all experiments which to the knowledge of the author have been published before the end of 1973 is given in Table IV.

TABLE IV

Catalogue of available data on alkali–halogen charge transfer collisions

Target	Projec-tile	CM-energy range (eV)	Data	EA (eV) (target)	References
Br_2	K	150–2200	$Q(E, m)$	—	Bukhteev *et al.* (1961)
	Rb	150–2200	$Q(E, m)$	—	
	Li	thr-7	$Q(E)$	2·55 ⎫	Baede *et al.* (1969)
	Na	thr-17	$Q(E)$	2·55 ⎬	Baede and Los (1971)
	K	thr-25	$Q(E)$	2·6 ⎭	Baede (1972b)
	Cs	thr-16·5	$Q(E)$	2·23 ± 0·1	Helbing and Rothe (1969)
	K	thr-30	$Q(E, T)$	—	Moutinho *et al.* (1971)
	Li	6·25	$\sigma(E)$	—	Delvigne and Los (1972)
	K	6·9, 10·35	σ	—	
	Li	thr-10	$Q(E, m)$	2·49	Baede *et al.* (1973)
	Na	thr-13	$Q(E, m)$	2·64	
	K	thr-25	$Q(E, m)$	2·49	
	Cs	thr-5	$Q(E)$	—	Leffert *et al.* (1973)
	Li	thr-14	$\sigma(E)$	2·6 ± 0·2	Young *et al.* (1974)

TABLE IV cont.

Target	Projectile	CM-energy range (eV)	Data	EA (eV) (target)	References
Cl_2	K	150–2200	$Q(E,m)$	—	Bukhteev et al. (1961)
	Rb	150–2200	$Q(E,m)$	—	
	Cs	150–2200	$Q(E,m)$	—	
	K	thr–20	$Q(E)$	$2\cdot6 \pm 0\cdot1$	Lacmann and Herschbach (1970a) EA: private communication
	Li	thr–5	$Q(E)$	$2\cdot45$ ⎫	Baede and Los (1971)
	Na	thr–15	$Q(E)$	$2\cdot5$ ⎬	Baede (1972b)
	K	thr–22	$Q(E)$	$2\cdot3$ ⎭	
	Li	thr–14	$\sigma(E)$	$3\cdot0 \pm 0\cdot3$	Young et al. (1974)
I	Li	thr–7	$Q(E)$	$3\cdot1 \pm 0\cdot1$	Moutinho et al. (1971), (1974)
	Na	thr–20	$Q(E)$	$3\cdot0 \pm 0\cdot1$	
	K	thr	—	$3\cdot2 \pm 0\cdot1$	
	Li_2	thr–3	$Q(E)$	—	Moutinho et al. (1971)
	Na_2	thr–3	$Q(E)$	—	
	K_2	thr–2	$Q(E)$	—	
	Na	13–85	$\sigma(E)$	—	Delvigne and Los (1973)
IBr	Li	thr–13	$Q(E,m)$	$2\cdot54$	Auerbach et al. (1973)
	Na	thr–13	$Q(E,m)$	$2\cdot59$	
	K	thr–15	$Q(E,m)$	$2\cdot54$	
ICl	Li	thr–13	$Q(E,m)$	—	Auerbach et al. (1973)
	Na	thr–16	$Q(E,m)$	$2\cdot45$	
	K	thr–20	$Q(E,m)$	$2\cdot39$	
I_2	Li	thr–7	$Q(E)$	$2\cdot3$ ⎫	Baede and Los (1971)
	Na	thr–20	$Q(E)$	$2\cdot55$ ⎬	Baede (1972b)
	K	thr–30	$Q(E)$	$2\cdot5$ ⎭	
	Li	thr–7	$Q(E,T)$	$2\cdot4 \pm 0\cdot1$	Moutinho et al. (1971)
	Na	thr–20	$Q(E,T)$	$2\cdot4 \pm 0\cdot1$	
	K	thr–35	$Q(E,T)$	$2\cdot6 \pm 0\cdot1$	
	Li_2	thr–3	$Q(E)$	—	
	Na_2	thr–4	$Q(E)$	—	
	K_2	thr–5	$Q(E)$	—	
	K	11·25	σ	—	Delvigne and Los (1972)
	Li	thr–13	$Q(E,m)$	$2\cdot44$	Baede et al. (1973)
	Na	thr–13	$Q(E,m)$	$2\cdot54$	
	K	thr–20	$Q(E,m)$	$2\cdot54$	
	K_2	∼0·5	Q (chemi-ionization)	—	Lin and Grice (1973)

Symbols: EA: electron affinity; thr: threshold energy; Q: relative total cross section; σ: relative differential cross section. Specification parameters: E: energy dependence; m: mass selection of negative ions; T: temperature dependence.

From the threshold of the total cross sections, the electron affinity (EA) can be calculated using the formula

$$E_{\text{thr}} = I(M) - A(XY) \simeq \frac{e^2}{4\pi\varepsilon_0 R_c} \tag{55}$$

in which $I(M)$ is the ionization potential of the alkali atom M and $A(XY)$ is the EA of XY. Although Fig. 19 and Table V show that the vertical EA, corresponding with a FC transition, differs strongly from the adiabatic one due to the rather large difference of the equilibrium distances, it can safely be assumed that the threshold corresponds to the latter.

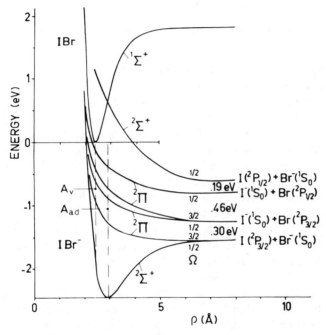

Fig. 19. Schematic representation of the potential energy curves of the ground state of IBr and the lowest states of IBr⁻. This picture is representative for all halogens. Notice however that in the case of a homonuclear halogen only two states remain asymptotically for large R. The very different adiabatic (A_{ad}) and vertical (A_v) electron affinities are indicated.

Different arguments can be put forward to support this assumption. In the first place experimental evidence was obtained by Moutinho et al. (1971), who showed that the threshold energy does not alter if the XY molecule is vibrationally excited. Secondly, the width of the lowest vibrational wave function of the XY molecule allows FC-transitions to vibrational states

TABLE V

Vertical electron affinities A_v(XY) and adiabatic electron
affinities A_{ad}(XY) of the halogens and well depth D_0(XY$^-$)
of their negative molecular ion ground states, all in eV.
The uncertainties of all values are in the order of ± 0.1 eV,
except A_v which has an uncertainty of ± 0.5 eV (Person,
1963; Baede, 1972; Auerbach et al., 1973)

XY	A_v	A_{ad}	D_0
Cl_2	1·3	2·45	1·31
Br_2	1·2	2·55	1·15
I_2	1·7	2·55	1·02
ICl	1·7	2·41	0·95
IBr	—	2·55	1·05

rather deep into the XY$^-$ well. This FC mechanism however cannot explain
the steep increase of the cross section above threshold (Kendall and Grice,
1972; Zembekov, 1971; Zembekov and Nikitin, 1972). It has been suggested
that non-vertical transitions might be important (Baede and Los, 1971), but
Zembekov has shown that the K$^+$ ion can act as an effective quencher of the
vibrationally excited ions, which makes the assumption of non-vertical
transitions superfluous. Before we discuss Zembekov's work in more detail,
let us consider the value of the quantity $H_{12}(R_c, \theta)$. An effective value of
$H_{12}(R_c)$ has been derived by Baede and Los (1971) from the position of the
maximum cross section by applying (43) and by using an R_c corresponding
to the vertical electron affinity (Person, 1963). Fig. 8 shows the values of
$H_{12}(R_c)$ so obtained together with the relation of Olson and calculated
values of Grice and Herschbach (1974) for collinear orientation. Clearly the
effective $H_{12}(R_c)$ is much smaller than either of the calculated values. In
fact Grice's values are so large that, taking into account the $\cos \theta$ dependence
for non-linear orientation, no diabatic transitions can be expected to occur at
the first crossing for almost all orientations except a small range near the
perpendicular one. Therefore, in spite of the fact that this perpendicular
orientation is favoured not only statistically but also dynamically (Düren,
1973; Balint-Kurti, 1973), Zembekov (1972) concluded that the collision
dynamics can be studied by calculating the trajectories on only the ionic
part of the lowest adiabatic surface. The magnitude of the cross sections
shows that almost every collision should lead to ionization. This implies
that the second crossing a diabatic transition is made with high probability,
which suggests a much larger distance R_c of the second crossing. Zembekov's
trajectory calculations show that close above threshold the highly vibrationally excited molecular ions are effectively quenched by the positive alkali ion.
Charge migration within the negative molecular ion plays an important

role in this process. Thus the second crossing is determined by an XY^- ion in a low vibrational state and therefore indeed by a large crossing distance due to the large corresponding electron affinity. Zembekov showed that this explains the sharp rise of the total cross section close above threshold. At energies far above threshold highly vibrating XY^- ions can also contribute which makes Zembekov's simple model invalid.

The adiabatic electron affinities $A_{ad}(XY)$ of the halogen molecules XY, measured by the Amsterdam group, are shown in Table V together with the vertical electron affinities of Person (1963). In this table the well depths $D_0(XY^-)$ of the negative molecular ions are also given, calculated from the relation

$$D_0(XY^-) = D_0(XY) + A_{ad}(XY) - A(X) \qquad (56)$$

All quantities at the right-hand side of this equation are known.

An interesting extension of the total cross section measurements was performed by Moutinho et al. (1971) by varying the temperature of the halogen beam and with that the internal energy of the halogen molecules. This has a strong effect near threshold as can be seen from Fig. 20. Although

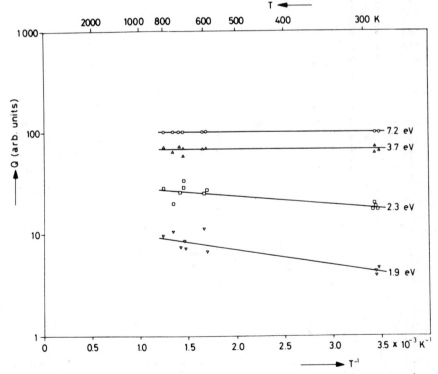

Fig. 20. Temperature dependence of the total $K + Br_2$ charge transfer cross sections at four different relative kinetic energies. (Moutinho et al., 1971.)

the threshold itself is not shifted to lower energies (indicating that the onset is determined by the adiabatic EA) the cross section close above threshold is enlarged considerably. Moutinho explained these results at least qualitatively by assuming that the electron jump is a FC-transition. Accepting for a while this FC-model, it is clear that at an energy E, near threshold, only transitions to those XY^- vibrational states v can lead to ionization which fulfil the relation.

$$E \geq I(M) - A_{vv'}(XY) \tag{57}$$

in which $A_{vv'}(XY)$ is the EA corresponding with a transition from vibrational state v' of XY to vibrational state v of XY^-.

By increasing the temperature from 300°K to 1000°K the number of vibrational states v' with a more than 5% population rises typically from 3 to 7. Now inspection of Fig. 19 shows that, in consequence of the steepness of the repulsive part of the XY^- potential energy curve, the transition probability to low XY^- states increases for high XY states, because as we saw in Section I.B.4 this probability is proportional to the FC-factor. At higher collision energies the influence of the vibrational state population is indirect only via the dependence of the LZ transition probability on the crossing distance.

Zembekov's results however are contradictory to the FC-model. Nevertheless it seems that the effect of strong quenching can at least partially account for Moutinho's results as well. In the FC-model the XY^- vibrational state distribution is an exact reflection of the original distribution of the XY vibrational states. If however the quenching of XY^- is very effective, the final XY^- vibrational distribution will be practically independent of the original one. The only information which is left in that case is the total (kinetic + internal) energy of the reactants, which suggests that the cross section might be determined completely by this total energy. This implies that a change of the internal energy has the same effect on the cross section as an equal change of the kinetic energy. As can be seen in Fig. 20 the cross section close above threshold is a very steep function of the kinetic energy. A small increase of the XY internal energy will therefore cause a considerable increase of the cross section. At larger kinetic energies this effect becomes negligible. Heating of the Br_2 gas from 300°K to 800°K increases the internal energy about 0·1 eV. The steepness of the cross section is so large that this should correspond with an increase of the cross section of about 40% at a kinetic energy of 1·9 eV. The observed increase is about twice as large. Thus although this model predicts an increase of the right order of magnitude, this implies that the quenching is not effective enough to eliminate the effect of the original Br_2 vibrational state distribution. Extension of Zembekov's calculations for different vibrational state distributions seems highly desirable.

Above it became clear that the observation of ionization near threshold implies that the value of the parameters governing the incoming and outgoing crossing are necessarily different. In Section 1.B.2 we pointed out that this fact can be taken into account by a simple extension of the two-particle LZ formula. Baede and Auerbach applied this method successfully on another observation (Baede et al., 1973; Auerbach et al., 1973). They performed mass selection of the negative halogen ions and found that the shape of the atomic ion cross section was characteristically dependent on the identity of the alkali collision partner: the potassium cross sections peak at energies close above threshold, whereas the sodium cross sections have either a smooth maximum at higher energies or no maximum at all (see the Br$^-$ cross section in Fig. 21). This could be explained by the assumption that those XY$^-$ ions which are formed in a dissociative state dissociate completely between both crossings. Therefore the second crossing is an 'atomic' one. In Section II.B.2 we saw that a very large difference exists between the transition probabilities in Na–X and K–X collisions in consequence of the much larger crossing distance of the latter system. This shifts the maximum cross section for X$^-$ production from K collisions to lower energies, as became clear from a simple modified LZ calculation.

An unexpected insight in the role of excited molecular ion states was obtained from experiments by Auerbach et al. (1973) on K collisions with mixed halogens ICl and IBr. Mass analysis of the negative ions revealed the production of I$^-$ ions, although the IX$^-$ ground state dissociates into I and X$^-$ (see Fig. 21). The authors interpreted their observation by assuming that not only transitions occur to the $^2\Sigma^+$ ground state, but also, at a much smaller crossing distance, to the first excited repulsive $^2\Pi_{1/2}$ state which indeed dissociates into I$^-$ and X. Although it is conceivable that the disturbing presence of K$^+$ results in I$^-$ from the IX$^-$ ground state, further experimental evidence was obtained from the threshold energy of the I$^-$ cross section. A lower bound of the negative atomic ion thresholds is given by the formula

$$E_{\text{thr}} \geq I(M) + D_0(XY) - A(X) \tag{58}$$

which $D_0(XY)$ is the dissociation energy of XY. All these values are accurately known. The measured thresholds are found to be in excellent agreement with this lower bound except the I$^-$ thresholds which are shifted upwards 0·1 eV (K + ICl) and 0·24 eV (K + IBr), respectively. This supports the assumption of a FC transition to the repulsive excited IX$^-(^2\Pi_{1/2})$ state.

In contrast with K collisions, Na produced almost no I$^-$ ions, and for Li essentially no negative iodine ions were observed. Auerbach et al. explained this by noticing that the diabatic transition probability at the first crossing is practically zero due to the small crossing distance. The ionization probability is therefore determined by the second crossing which, as explained above,

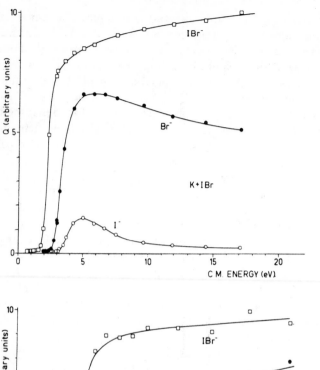

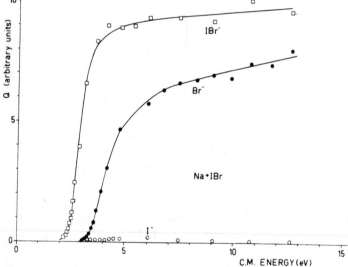

Fig. 21. Total cross sections for the production of negative ions in K + IBr and Na + IBr collisions. (Auerbach *et al.*, 1973.)

depends on the stretching of the molecular ion and thus on the collision time. The short collision time for Li results in a practically zero ionization probability whereas during a K collision the molecular ion has enough time to stretch. [These observations make it highly improbable that the small structure near threshold (see Fig. 18), also observed in collisions of Na with homonuclear halogens, may be attributed to the opening of the $X_2^-(^2\Pi_{g1/2})$ channel as has been suggested by Baede and Los, 1971.]

Charge transfer between alkali dimers and iodine atoms and molecules at energies between 1 and 5 eV has been observed by Moutinho et al. (1971). The cross section is comparable with those of the alkali atom reactions. At still lower energies chemi-ionization is observed by Lin and Grice (1973) in $K_2 + I_2$ collisions, the total cross section at 0·5 eV being about 1 Å^2. Energetically only two paths are accessible in this energy range:

$$K_2 + I_2 \rightarrow K^+ + KI + I^-$$

$$\rightarrow K_2I^+ + I^-$$

We will meet more examples of chemi-ionization in Sections II.C.3 and II.C.4.

2. Differential Cross Sections; Alkali–Halogens

As we have seen in Section II.B.2, measurements of the angular distributions of the ions produced, potentially provide us with more information on the dynamical aspects of the collision, i.e. about the potential energy surfaces involved. Because the total cross sections are large and the detection of slow ions is relatively simple, the measurements discussed in the previous section were soon followed by experiments aiming at the determination of the differential cross sections. Fig. 22 shows the results of Delvigne and Los (1972) of K on Br$_2$ at 6·9 and 10·35 eV. Similar results have been obtained for K/I$_2$ and Li/Br$_2$. The measurements do not show much oscillatory structure, which was imputed to the quenching effect of the internal degrees of freedom of the halogen molecules and the strong orientational dependence of the potential. Delvigne and Los compared their measurements with a simple classical two-particle calculation, assuming potentials of the same form (53 and 54) as for atom–atom scattering. The unknown parameters were derived from the measurements. Just as in the atom–atom case the contributions due to covalent and ionic scattering were found to be almost completely separated by a deep minimum at $\tau \simeq 150$. This is in contrast with the experiments which show only a very shallow minimum at this position. The three-body character of the interaction, neglected in the calculations, seems not entirely responsible for the discrepancy. Düren (1973) performed trajectory calculations on the two crossing surfaces using the so-called surface hopping technique (see Section I.B.3). He took into account the orientational dependence of the surfaces and their coupling, and replaced the halogen molecule potential by a harmonic potential, thus representing in a crude way the molecular internal degrees of freedom. These calculations show the same almost complete separation between ionic and covalent scattering. Fig. 23 shows an example of Düren's calculations. This

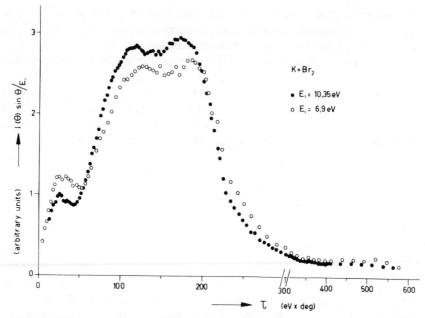

Fig. 22. Differential cross sections for K^+ production in $K + Br_2$ collisions at two different relative energies. For both energies equal units have been used on the ordinate. The classical rainbow is at $\tau \approx 275$. Notice the change of scale at $\tau = 300$. (Delvigne and Los, 1972.)

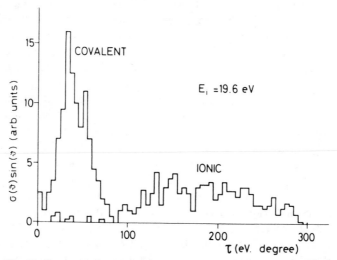

Fig. 23. Example of a differential cross section for K^+ production in $K + Br_2$ collisions, calculated by the trajectory method. In contrast with experiment, the covalent and ionic contributions are completely separated. (Düren, 1973.)

example was chosen because its potential parameters are in reasonable agreement with Delvigne's parameters. Los (1973) has suggested that this discrepancy might be caused by a distortion of the covalent surface by a crossing with a Coulombic surface corresponding to the first excited $X_2^-(^2\Pi_g)$ state.

Delvigne inferred that the ionic rainbow is to be sought at $\tau \simeq 300$ and certainly not at or near the angle of maximum intensity. At this point the three-body calculation of Düren is in good agreement with the experiments, in contrast with Delvigne's simple two-body calculation.

The observation of quantum mechanical structure was claimed by Fluendy et al. (1970) and Duchart et al. (1972) in their experiments on the elastic differential cross section of K on I_2 with extremely high angular resolution. They not only found considerable structure at $E_i = 100\,eV$ where the collision time is so short that the I_2 molecule will behave as a rigid particle, but even at thermal energies. The structure found in the latter case was attributed to edge diffraction either at the sharp transition of the adiabatic potential at R_c or at the edge of the black sphere representing the chemical reaction.

Differential cross sections for Li^+ production in collisions of Li on Cl_2 and Br_2, but also on HCl, HBr, O_2 and SF_6 have been measured by Wexler's group (Young et al., 1974), be it with considerably less angular resolution (Fig. 24). They find a correlation between the existence of a secondary maximum in the differential cross section and the energy difference ΔE at infinity between both diabatic states. As an example, Fig. 24 shows the differential cross sections for Li/Cl_2 (small $\Delta E \simeq 3.0\,eV$) and Li/HCl (large $\Delta E \simeq 5.4\,eV$). This effect is attributed to the dependence on ΔE of the shape of the classical deflection function near R_c, as shown in Fig. 25, and the corresponding rapid variation of the differential cross section around R_c. It seems that this interpretation can be brought into agreement with Delvigne's results. Fig. 25 suggests that a small ΔE favours the existence of two maxima around the minimum corresponding to $b = R_c$. With increasing ΔE the covalent maximum will diminish. A comparison of Delvigne's results on $Li + Br_2$ and $K + Br_2$ clearly shows this effect. Unfortunately Düren's (1973) calculations are not conclusive on this point because he separated the effects, in reality strongly correlated, of increasing ΔE and increasing $H_{12}(R_c)$.

From the previous and the present section it is clear that the study of ionization phenomena has contributed much to our understanding of alkali halogen interactions, especially on the point of the electron jump mechanism and the role of the excited states. However, our knowledge of the dynamics of the interaction is still rather unsatisfactory. From the experimental side this problem might be attacked by improving the angular measurements, particularly by incorporating energy loss and/or coincidence measurements;

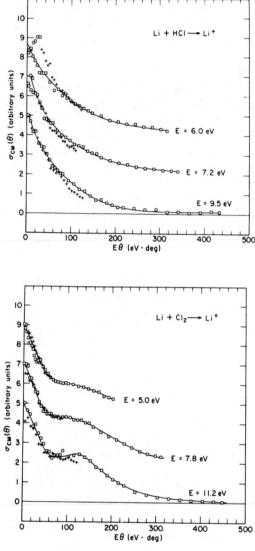

Fig. 24. Differential cross section (CM-system) for Li^+-production in $Li + HCl$ and $Li + Cl_2$ collisions, each at three different energies. The secondary maximum in the chlorine cross section is explained by the large Cl_2 electron affinity in contrast with the small electron affinity of HCl. (Young *et al.*, 1974.)

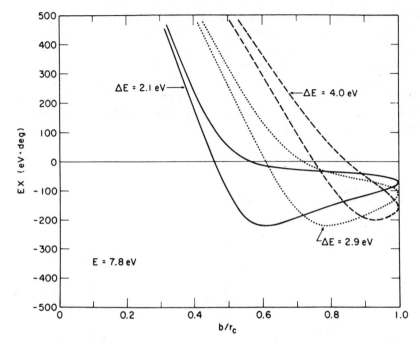

Fig. 25. Dependence of the classical deflection function for M + XY collisions on the asymptotic energy difference $\Delta E = I(M) - A(XY)$ between the covalent and ionic state. (Young *et al.*, 1974.)

theoretical contributions may be expected from improved trajectory calculations on realistic surfaces.

3. Electron Transfer between Alkali Atoms and Non-Halogen Molecules

Soon after the first experiments on halogen molecules, it became clear that alkali collisions with many other, electronegative and even electropositive, molecules lead to ionization. Many of these results have been published now, especially on relatively small inorganic molecules, whereas experiments on large organic molecules have been announced. Moreover the history of the chemical reaction beam research seems to repeat itself: after an 'alkali-age' of about five years the first non-alkali experiments have been reported, which will be discussed in the next section.

The results of the different groups are not always consistent. For example in the case of oxygen molecules a discrepancy of a factor of 100 has been reported in the magnitude of the cross section! Moreover, with increasing complexity of the systems our understanding decreases, which is demonstrated by the fact that in a few cases the interpretation of certain experimental results are in flat contradiction. Therefore in this immature state it seems inappropriate to discuss in detail all results which have been obtained, but rather to view some interesting aspects, which were not found in alkali–halogen collisions. A catalogue of the many studies which, to the knowledge of the author, have been published or were about to be published at the end of 1973, is given in Table VI. Conference abstracts and reports have not been included unless their results were not published elsewhere. High energy data ($E > 2.5$ keV) have been omitted. A discussion of the electron affinities is postponed until the next section.

TABLE VI
Catalogue of available data on alkali–non halogen charge transfer collisions

Target	Projectile	CM-energy range (eV)	Data	Observed particles	EA (eV) target	References
$C_4H_2O_3$	Cs	thr-20	$Q(E, m)$	$C_4H_2O_3^-$ $C_2H_2CO_2^-$ CO_2^-*	1.4 ± 0.2	Compton et al. (1974) Cooper and Compton (1972)
$C_4H_4O_3$	Cs	thr-20	$Q(E, m)$	Mass 28 $C_4H_2O_3^-$ $C_2H_4CO_2^-$ $C_2H_4CO^-$ CO_2^-	—	Compton et al. (1974) Cooper and Compton (1972)
CF_3I	K	thr-?	$Q(E, m)$	CF_3I^-, I^-, F^-	2.2 ± 0.2	McNamee et al. (1973)
CH_3Br	K	thr-30	$Q(E, T)$	—	-0.46 ± 0.1	Moutinho et al. (1974)
CH_3I	K	thr-?	$Q(E, m)$	I^-	0.3 ± 0.2	McNamee et al. (1974)
	Li	thr-7	$Q(E, T)$	—	0.29 ± 0.2	Moutinho et al. (1974)
	Na	thr-20	$Q(E, T)$	—	0.39 ± 0.1	
	K	thr-30	$Q(E, T)$	—	0.24 ± 0.1	
CO_2	K	thr-20	$Q(E)$	—	~ -2	Lacmann and Herschbach (1970)
	K	thr-15	$Q(a, E)$	—	~ -2	Baede and Los (1971)
	K	100–1000	$Q(a, E)$	—	—	Cuderman (1972)
CO_2	Cs	thr-90	$Q(E, m)$	CO_2^-*, O^-	-0.3 ± 0.2	Compton (1979)
Cs	$O(^3P, {}^1D)$	200–1800	$Q(a, E)$	ions, e^-	—	Woodward (1970)
D_2	Na	150–2200	$Q(a, E)$	—	—	Bydin and Bukhteev (1960)
	K	150–2200	$Q(a, E)$	—	—	
	Rb	150–2200	$Q(a, E)$	—	—	
	Cs	150–2200	$Q(a, E)$	—	—	
HBr	Li	thr-14	$\sigma(E)$	Li^+	—	Young et al. (1974)
HCl	K	thr-20	$Q(E)$	—	-0.8 ± 0.2	Lacmann and Herschbach (1970a)
	Li	thr-14	$\sigma(E)$	Li^+	—	Young et al. (1974)
H_2	Na	150–2200	$Q(a, E)$	—	—	Bydin and Bukhteev (1960)
	K	150–2200	$Q(a, E)$	—	—	
	Rb	150–2200	$Q(a, E)$	—	—	
	Cs	150–2200	$Q(a, E)$	—	—	
	K	150–1000	$Q(a, E)$	—	—	Cuderman (1972)
N_2	Na	150–2200	$Q(a, E)$	—	—	Bydin and Bukhteev (1960)
	K	150–2200	$Q(a, E)$	—	—	
	Rb	150–2200	$Q(a, E)$	—	—	
	Cs	150–2200	$Q(a, E)$	—	—	

TABLE VI cont.

Target	Projectile	CM-energy range (eV)	Data	Observed particles	EA (eV) target	References
	K	thr-20	$Q(E)$	—	~ -2	Lacmann and Herschbach (1970a)
	K	thr-15	$Q(a, E)$	e^-	~ -2	Baede and Los (1971)
	K	50–1000	$Q(a, E)$	—	—	Cuderman (1972)
	Na	thermal; $N_2^{\dagger}$	Q	—	—	Haug et al. (1973)
	K	thermal; $N_2^{\dagger}$	Q	—	—	
N_2O	Cs	thr-7	$Q(E, m)$	N_2O^-, O^-	-0.15 ± 0.1	Nalley et al. (1973)
NO	K	thr-20	$Q(E)$	—	0.0 ± 0.2	Lacmann and Herschbach (1970a)
	Cs	thr-7	$Q(E, m)$	NO^-, O^-	0.1 ± 0.1	Nalley et al. (1973)
NO_2	K	thr-100	$Q(E, m)$	—	—	Lacmann and Herschbach (1970b)
	Li	thr-16	$Q(a, E)$	—	— ⎫	Baede and Los (1971)
	Na	thr-14	$Q(a, E)$	—	2·45 ⎬	Baede (1972b)
	K	thr-20	$Q(a, E)$	—	2·55 ⎭	
	Cs	thr-2·5	$Q(E)$	—	2.50 ± 0.05	Leffert et al. (1973)
	Cs	thr-7	$Q(E, m)$	NO_2^-, O^-	2.5 ± 0.1	Nalley et al. (1973)
O_2	Na	150–2200	$Q(a, E)$	—	—	Bydin and Bukhteev (1960)
	K	150–2200	$Q(a, E)$	—	—	
	Rb	150–2200	$Q(a, E)$	—	—	
	Cs	150–2200	$Q(a, E)$	—	—	
	K	150–2200	$Q(a, E, m)$	O_2^-, O^-	—	Bukhteev et al. (1961)
	Rb	150–2200	$Q(a, E, m)$	O_2^-, O^-	—	
	Cs	150–2200	$Q(a, E, m)$	O_2^-, O^-	—	
	Na	thr-25	$Q(E, m)$	O_2^-, O^-, e^- NaO^+, Na^+	—	Neynaber et al. (1969)
	K	thr-20	$Q(E)$	—	0.5 ± 0.2	Lacmann and Herschbach (1970a)
	K	thr-100	$Q(E, m)$	?	?	Lacmann and Herschbach (1970b)
	Li	thr-5·5	$Q(a, E)$	ions, e^-	— ⎫	Moutinho et al. (1970)
	Na	thr-12	$Q(a, E)$	ions, e^-	0·4 ⎬	Baede (1972b)
	K	thr-15	$Q(a, E)$	ions, e^-	0·6 ⎭	
	Cs	thr-17	$Q(E, m)$	O_2^-, O^-	0.46 ± 0.05	Nalley and Compton (1971)
	K	20–1000	$Q(a, E)$	—	—	Cuderman (1972)
	Li	thr-14	$\sigma(E)$	Li^+	—	Young et al. (1974)
SF_6	Cs	thr-12	$Q(E, m)$	$SF_6^-, SF_5^-,$ F^-	$0.54 \begin{smallmatrix}+ 0.1 \\ - 0.17\end{smallmatrix}$	Compton and Cooper (1973)
	Li	thr-14	$\sigma(E)$	Li^+	—	Young et al. (1974)
	Li	thr-8	$Q(E, T, m)$ ⎫	SF_6^-, SF_5^-	0.49 ± 0.15	Hubers and Los (1974)
	Na	thr-20	$Q(E, T, m)$ ⎬	F_2^-, F^-	0.49 ± 0.1	
	K	thr-35	$Q(E, T, m)$ ⎭		0.48 ± 0.1	
SO_2	K	thr-100	$Q(E, m)$	?	?	Lacmann and Herschbach (1970b)
	K	thr-25	$Q(E)$	?	1·3	Herschbach (1973)
TeF_6	Cs	thr-12	$Q(E, m)$	TeF_6^-, TeF_5^- TeF_4^-, F^-	$3.34 \begin{smallmatrix}+ 0.1 \\ - 0.17\end{smallmatrix}$	Compton and Cooper (1973)
	Li	thr-14	$\sigma(E)$	Li^+	—	Young et al. (1974)

Symbols: EA: electron affinity; thr: threshold energy; Q: total cross sections; σ: differential cross section. Specification parameters: a: absolute value of cross section; E: energy dependence; m: mass selection of negative ions; T: temperature dependence.

The alkali–oxygen molecule system is the most thoroughly investigated non-halogen system. Nevertheless large experimental discrepancies still exist and the interpretation is often ambiguous. Moutinho *et al.* (1970) report total ionization cross sections (Fig. 26) of about 5 Å² at 10 eV, but

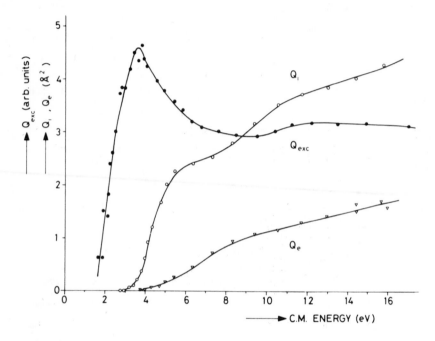

Fig. 26. Interaction of K atoms with O_2 molecules. Total cross section (Å²) for the production of negative ions (Q_i) and electrons (Q_e). (Moutinho *et al.*, 1971.) Total cross section (arbitrary units) for K resonance excitation (Q_{exc}). (Lacmann and Herschbach, 1970.)

Neynaber *et al.* (1969), using the difficult merging beam technique, and Cudermann (1972b) measured cross sections as being a factor of 100 smaller. Although the LZ theory favours Moutinho's values, the measurements of Bukhteev *et al.* (1961) agree better with the other ones. However, the order of magnitude as well as the energy dependence of the K/O_2 cross section measured by Lacmann and Herschbach (1970a) are in good agreement with Moutinho's measurements.

The oxygen measurements show a few interesting features, observed in other non-halogen collisions as well, which we will discuss briefly below:

1). Besides negative ions, free electrons are also produced,

2). the energy dependence of the cross section exhibits a clear structure,

3). the alkali excitation cross section shows a peak near the ionization threshold.

Each of these points is now amplified in turn.

By switching a magnetic field in their collision chamber on and off, the Amsterdam group observed the production of free electrons with O_2 and NO_2, in contrast with halogen collisions (Baede and Los, 1971), Moutinho et al., 1971). This is shown in Fig. 26. In the case of O_2 this can easily be understood from the relative position of the $O_2(^3\Sigma_g^-)$ ground state and the lowest $O_2^-(^2\Pi_g)$ state. Because the equilibrium distance of both states is almost equal and the EA is small (0·43 eV) vibrationally excited O_2^- molecules with more than 0·43 eV vibrational energy will be unstable against auto-detachment by overlap with the O_2 vibrational wave functions. For NO_2 the same mechanism might be active, although the EA is larger (2·5 eV). From Fig. 19 it is clear that no electron production via this mechanism can be expected from halogen collisions.

In collisions with the electropositive molecules N_2 and CO only electrons have been observed. It is well known that the ground states of N_2^- and CO^- auto-detach within 10^{-14} sec (Bardsley and Mandl, 1968). The likewise electropositive HCl (Lacmann and Herschbach, 1970a) however, produces Cl^- ions (electron production has not yet been investigated in this case). The ion production can be understood from the fact that the HCl^- ground state is repulsive and dissociates in a short time, making autodetachment improbable.

The total ionization cross sections of O_2 (Fig. 26), NO_2 (Baede and Los, 1971; Nalley et al., 1973) and NO (Lacmann and Herschbach, 1970a; Nalley et al., 1973) all show a distinct structure, the origin of which is not well known. Moutinho et al. (1970) suggested that the O_2 structure might be due to the opening of the dissociative $O_2^-(^2\Pi_u)$ channel. However the very small O^- cross section, found by Nalley and Compton (1971) makes this suggestion unlikely.

Baede and Los (1970) concluded from their K, Na and Li/NO_2 measurements that the observed structure cannot be attributed to the opening of the $NO_2^-(^3B_1)$ channel. In contrast with this conclusion, Nalley et al. (1973) did attribute the considerable structure of their deconvoluted (see Section II.C.5) Cs/NO_2 cross section to the opening of excited state channels. It is clear that further research is desirable.

In O_2, NO and NO_2 collisions a peak in the alkali excitation cross section was observed at the ionization threshold (Lacmann and Herschbach, 1970a), see Fig. 26. This was explained by an interesting mechanism called 'internal reflection'. This means that just below the ionization threshold many trajectories try to escape along the ionic channel. Because this attempt is unsuccessful, these trajectories will reflect from the Coulombic wall and finally find their way out along one of the energetically open channels, for example the excitation channel. As soon as the energy is above the ionization threshold, these trajectories will escape along this channel at the expense of the competing ones. In the case of the halogens this effect is not observed because, as remarked above, the excitation channel is inaccessible in this energy range.

The triatomic molecules NO_2 and N_2O show the influence of the difference between the bond angles of the parent molecule and its negative ion. NO_2 (Nalley et al., 1973) produces almost exclusively NO_2^- ions with only a small fraction of O^- ions. N_2O (same reference) however, shows the opposite. It is true that, in going from the NO_2 to the NO_2^- ground state, the bond angle has to decrease from $134.1°$ to $115 \cdot a°$, but during the collision the NO_2 molecule has time enough to adjust its position. Moreover, this bending needs only 0.18 eV. The N_2O bond angle however decreases from linear to $\sim 134°$ which needs 1.39 eV. This effect has consequences for the measured EA. Whereas NO_2 readjusts easily, yielding the adiabatic EA, N_2O will probably exhibit a threshold corresponding to the vertical one.

A few results on polyatomic halogen compounds have been reported. The chemical reaction of alkali atoms with methyl iodide

$$M + CH_3I \rightarrow MI \rightarrow CH_3$$

offers a typical and widely studied example of a class of reactions characterized by backward scattering of the products and cross sections much smaller than for harpoon reactions. This implies that small impact parameters are involved, suggesting a small crossing distance and correspondingly a small electron affinity of the molecule. Experiments of Brooks and Jones (1966) and Beuhler and Bernstein (1969) with oriented molecules, moreover, have shown that a reaction only occurs if the alkali atom hits the iodine side of the CH_3I molecule. No such orientational dependence is evident for CF_3I (Brooks, 1969). In agreement with all these findings McNamee et al. (1973) observed that CH_3I, colliding with alkali atoms, yields I^- only, whereas the threshold corresponds to an almost zero vertical EA at $300°K$ (Moutinho et al., 1974) due to the fact that the repulsive CH_3I^- potential crosses the molecular potential near its minimum. Moutinho argues that this vertical EA at room temperature corresponds to a higher vibrational state ($v \simeq 3$) of the molecule. This situation favours a strong dependence on temperature of the cross section near threshold. This indeed was found by Moutinho et al. (1973). Methyl bromide behaves essentially the same, be it that a small shift of the relative positions of the potential curves yields a slightly negative EA at $300°K$. Trifluoro methyliodide yields I^- ions as well, and also a small fraction of CF_3I^- and, above, 10 eV, F^-.

Two hexafluoride compounds have been investigated by Hubers and Los (1974) and Cooper and Compton (1973). SF_6 yields predominantly SF_5^-, with only a small amount of SF_6^-, F_2^- and F^-. TeF_6 in contrast yields predominantly TeF_6^- with some TeF_5^-, TeF_4^- and F^-. TeF_6 has a very large EA (3·34 eV), whereas the EA of SF_6 is only 0·54 eV. This result is in striking contrast with the very large electron attachment rate of SF_6 and the almost zero attachment rate of TeF_6.

Finally, results for Cs and organic molecules containing 'bent' CO_2 as a basic unit have been published by Cooper and Compton (1972), together with the results of electron attachment experiments (see also Compton *et al.*, 1974). Metastable CO_2^-* ions are produced with lifetimes of the order of 65 μs, suggesting a negative electron affinity for CO_2. This was confirmed by recent results for the system Cs/CO_2 (Compton, 1974). Further results on organic molecules have been announced (Compton, private communication). Differential ionization cross sections have recently been reported by Wexler (1973), but these have already been discussed.

4. Non-Alkali Charge Transfer Processes

Differential charge transfer cross sections between other than alkali-atoms M (such as Ta, Fe, U, C, etc.) and molecules XY have been observed by Cohen *et al.* (1973) and Fite and Irving (1972). The most interesting aspect is, that in a few cases, besides simple or dissociative electron transfer, associative ionization

$$M + XY \rightarrow MXY^+ + e$$

and reactive ionization

$$M + XY \rightarrow MX^+ + Y^-$$

have also been observed. The last process was observed earlier by Neynaber *et al.* (1969) in their merging beam study of Na/O_2 collisions.

Associative ionization allows a very simple kinematical analysis of the differential cross section, even if unselected beams have been used, because the product ion carries away essentially all CM-kinetic energy. Thus the angular ion distribution is determined by the energy distribution of primary and secondary beams and the angular beam width. If these factors are known, a complete analysis is possible. From the maximum angle at which the process is observed the threshold energy can be determined.

Angular distributions of the positive ions produced by reactive ionization were found to be much broader in accord with the expectation that rather close collisions are necessary for this process. As an example, Fig. 27 shows Cohen's results on $Ba + O_2$.

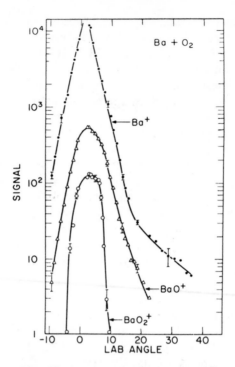

Fig. 27. Angular distribution of Ba^+, BaO^+ and BaO_2^+ formed in collisions of sputtered Ba atoms (not velocity selected) with O_2 molecules. (Cohen *et al.*, 1973.)

A peculiar example of simple electron transfer is offered by $C + O_2$ (Wexler, 1973). Here both processes are observed:

$$C + O_2 \rightarrow C^+ + O_2^-$$
$$\rightarrow C^- + O_2^+$$

due to the almost equal endoergicity (10.8 eV). The angular distributions of the positive ions from both channels exhibit a remarkable but not well understood difference. Other chemiionization channels were found recently by Können *et al.* (1975):

$$C + O_2 \rightarrow CO^+ + O + e^-$$
$$\rightarrow CO^+ + O^-$$
$$\rightarrow CO^- + O^+$$

The experimental information is still rather limited. Rather than discussing all details we refer to Wexler's recent review paper (Wexler, 1973).

Recently Können *et al.* (1974a) observed ion production from collisions between sputtered halogen atoms and organic molecules such as aniline. No reactive ionization was found. The positive organic fragment ion spectra resemble those obtained by electron and photon impact.

Earlier experiments on reactive ionization, simple charge transfer or electron ejection between molecules or atoms were performed by Utterback *et al.* and the Denver group. All processes, studied by them, are characterized by large endoergicities and very small cross sections below 20 eV. A detailed discussion is outside the scope of this paper, so we refer to Utterback (1969) and Wexler (1973).

5. Thresholds and Electron Affinities

From the threshold energy E_{thr} of the electron transfer reaction an upper limit of the electron affinity $A(XY)$ of the target molecule XY can be derived, using the relation (55):

$$A(XY) \leq I(M) - E_{thr} \qquad (59)$$

in which $I(M)$ is the ionization potential of the projectile atom M. This raises two problems, a technical and a physical one. In the first place an accurate determination of E_{thr} is desirable, which requires the knowledge of the real cross section $Q(E)$ near threshold. The apparent total cross section $Q_{app}(E_0)$ at a nominal beam energy E_0 is given by the convolution of $Q(E)$ with the energy distributions of the target particles and the primary beam. A detailed discussion can be found in Chantry (1971) and Baede (1972b). The problem of extracting $Q(E)$ from the apparent cross section has been attacked along different lines.

A solution, which requires an advanced technology, is to improve experimental conditions so that the energy distributions approximate δ-functions. Leffert *et al.* (1972) describe a high resolution apparatus with a time-of-flight selected primary beam and a narrow perpendicular crossed beam. The first results show sharp onsets at threshold, indicating a small energy spread. The NO_2 electron affinity has been determined by this technique (Leffert *et al.*, 1973).

The mathematical solution of the convolution equation is not an easy task in view of the fact that Q_{app} is in general known only at discrete, not necessarily equidistant, points in a finite energy range and affected with noise. A review of the many earlier attempts to solve this problem for other related applications is outside the scope of this paper. Nalley *et al.* (1974) have devised and applied an iterative deconvolution procedure. It is well known however that such procedure often suffer from instability due to noise on the data, which makes smoothing necessary. In the previous section we briefly discussed the deconvoluted Cs/NO_2 results of Nalley, which show considerable structure.

Finally the most widespread solution of the problem is the reversal of the deconvolution by choosing a parametrized model cross section $Q(E)$, integrating the convolution integral and fitting the calculated apparent cross section with the measured one. The problem of course is the choice of an adequate model cross section. Step functions, exponentials, (broken) linear functions and functions based upon theoretical considerations (Section I.B.6) have been applied. Application of this method to processes, the thresholds of which were accurately known, revealed its usefulness. The larger the energy range along which the calculated Q_{app} fits with the measured one, the larger the confidence that the choice of the model cross section and the numerical values of its parameters were the right ones. Chantry (1971) formulated a criterion to decide if a fit is acceptable from this point of view.

The next, physical, problem is to decide if the upper limit in (59) corresponds with the EA and, if so in the case of a molecule, with which EA. Physical arguments have to be used to support such a decision. In previous sections we met different examples of this problem. In general it can be made plausible that adiabatic EA's are found. If both potential well minima are situated at an almost equal internuclear distance, this seems trivial, but even in the case of the halogen molecules we have seen that there are arguments in favour of the adiabatic EA. In the case of N_2O and the methylhalides, the vertical EA is most likely, which in the latter case corresponds even to a transition from a higher vibrational state of the molecule.

The collisional ionization technique discussed in this paper has contributed considerably to our knowledge of many molecular EA's. A comparison with results from other techniques sometimes reveals good agreement (O_2, the halogens), in other cases however serious discrepancies are found. A notorious example is the NO_2 electron affinity, reported values of which range from $1 \cdot 6$ to 4 eV. Fortunately recent values from collisional ionization are all confined within $2 \cdot 5 \pm 0 \cdot 1$ eV, whereas negative ion charge transfer values are systematically lower, but not inconsistent (Hughes et al., 1973).

Electron affinities from collisional ionization experiments have been tabulated in Tables IV and VI. For a discussion and a comparison of these values with those obtained by other techniques we refer to a number of publications (e.g. Berkowitz et al., 1971; Nalley et al., 1971; Baede, 1972; Leffert et al., 1973; Wexler, 1973; Hughes et al., 1973; Lifshitz et al., 1973).

Acknowledgements

The author wrote this article in remembrance of the pleasant time which he spent with his colleagues of the F.O.M.-Institute in Amsterdam. Particular thanks are due to Prof. Dr. J. Los and Dr. A. E. de Vries for their critical reading of the manuscript, to Dr. M. M. Hubers for his assistance and to all those cooperators who contributed to the technical realization of this article.

References

R. W. Anderson (1968). *Thesis*, Harvard University.

R. W. Anderson, V. Aquilanti and D. R. Herschbach (1969). *Chem. Phys. Lett.*, **4**, 5.

R. W. Anderson and D. R. Herschbach (1974). *J. Chem. Phys.*, to be published.

E. A. Andreev (1972). *High Temperature*, **10**, 637.

E. A. Andreev and A. I. Voronin (1969). *Chem. Phys. Lett.*, **3**, 488.

D. J. Auerbach, M. M. Hubers, A. P. M. Baede and J. Los (1973). *Chem. Phys.*, **2**, 107.

A. P. M. Baede, A. M. C. Moutinho, A. E. De Vries and J. Los (1969). *Chem. Phys. Lett.*, **3**. 530.

A. P. M. Baede and J. Los (1971). *Physica*, **52**, 422.

A. P. M. Baede (1972a). *Thesis*, University of Amsterdam.

A. P. M. Baede (1972b). *Physica*, **59**, 541.

A. P. M. Baede, D. J. Auerbach and J. Los (1973). *Physica*, **64**, 134.

G. G. Balint-Kurti (1973). *Mol. Phys.*, **25**, 393.

A. D. Bandrauk (1969). *Mol. Phys.*, **17**, 523.

A. D. Bandrauk, (1972). *Mol. Phys.*, **24**, 661.

J. M. Bardsley and F. Mandl (1968). *Rep. Progr. Phys. XXXI*, 471.

D. R. Bates and B. L. Moiseiwitsch (1954). *Proc. Phys. Soc.*, **A67**, 805.

D. R. Bates (1966). *Proc. Roy. Soc.*, **A257**, 22.

E. Bauer, E. R. Fisher and F. R. Gilmore (1969). *J. Chem. Phys.*, **51**, 4173; *Report IDA* P-471.

J. Berkowitz, W. A. Chupka and D. Gutman (1971). *J. Chem. Phys.*, **55**, 2733.

R. B. Bernstein (1966). In J. Ross (Ed.), *Advances Chemical Physics*, Vol. X, p. 75.

M. V. Berry (1966). *Proc Phys. Soc.*, **89**, 479.

R. S. Berry (1957). *J. Chem. Phys.*, **27**, 1288.

R. S. Berry (1970). In Ch. Schlier (Ed.), *Molecular Beams and Reaction Kinetics*, Academic Press, p. 193.

R. J. Beuhler and R. B. Bernstein (1969). *J. Chem. Phys.*, **51**, 5305.

A. Bjerre and E. E. Nikitin (1967). *Chem. Phys. Lett.*, **1**, 179.

N. C. Blais and D. Bunker (1963). *J. Chem. Phys.*, **39**, 315.

N. C. Blais (1968). *J. Chem. Phys.*, **49**, 9.

N. C. Blais (1969). *J. Chem. Phys.*, **51**, 856.

S. M. Bobbio, C. D. Doverspike and R. L. Champion (1973). *Phys. Rev.*, **A7**, 526.

P. R. Brooks and E. M. Jones (1966). *J. Chem. Phys.*, **45**, 3449.

P. R. Brooks (1969). *J. Chem. Phys.*, **50**, 5031.

U. Buck and H. Pauly (1968). *Z. Physik*, **208**, 390.

A. M. Bukhteev, Yu. F. Bydin and V. M. Dukelskii (1961). *Sov. Phys.—Techn. Phys.*, **6**, 496.

D. Bunker (1970). In Ch. Schlier (Ed.), *Molecular Beams and Reaction Kinetics*, Academic Press, p. 355.

Yu. F. Bydin and A. M. Bukhteev (1960), *Sov. Phys.—Techn. Phys.*, **5**, 512.

V. Bykhovski, E. E. Nikitin and M. Ya. Ovchinnikova (1965). *Sov. Phys. JETP*, **20**, 500.

P. J. Chantry (1971). *J. Chem. Phys.*, **55**, 2746.

M. S. Child (1969). *Mol. Phys.*, **16**, 313.

M. S. Child (1971). *Mol. Phys.*, **20**, 171.

M. S. Child (1973). *Faraday Disc.*, **55**, 30.

D. Coffey, D. C. Lorents and F. T. Smith (1969). *Phys. Rev.*, **187**, 201.

R. B. Cohen, C. E. Young and S. Wexler (1973). *Chem. Phys. Lett.*, **19**, 99.

R. N. Compton and C. D. Cooper (1973). *J. Chem. Phys.*, **59**, 4140.

R. N. Compton (1974). Private communication.

R. N. Compton, P. W. Reinhardt and C. D. Cooper (1974). *J. Chem. Phys.*, **60**, 2953.

C. D. Cooper and R. N. Compton (1972). *Chem. Phys. Lett.*, **14**, 29.
J. F. Cuderman (1971a). *Rev. Sci. Instr.*, **42**, 583.
J. F. Cuderman (1971b). *Surface Science*, **28**, 569.
J. F. Cuderman, (1972a). *Research Report Sandia Labs.* SC-RR-71 0756.
J. F. Cuderman (1972b). *Phys. Rev.*, **A5**, 1687.
J. B. Delos and W. R. Thorson (1972). *Phys. Rev.* A, 6, 728.
G. A. L. Delvigne and J. Los (1972). *Physica*, **59**, 61.
G. A. L. Delvigne and J. Los (1973). *Physica*, **67**, 166.
A. E. De Vries and A. Kuppermann (1971). *Proc. VIIth ICPEAC*, Amsterdam, p. 297.
G. V. Dubrovskii (1964). *Sov. Phys. JETP*, **19**, 591.
B. S. Duchart, M. A. D. Fluendy and K. P. Lawley (1972). *Chem. Phys. Lett.*, **14**, 129.
R. Düren, G. P. Raabe and Ch. Schlier (1968). *Z. Physik*, **214**, 410.
R. Düren (1973). *J. Phys. B*, **6**, 1801.
S. A. Edelstein and P. Davidovits (1971). *J. Chem. Phys.*, **55**, 5164.
B. C. Eu (1971). *J. Chem. Phys.*, **55**, 5600.
B. C. Eu (1972a). *J. Chem. Phys.*, **56**, 2507.
B. C. Eu (1972b). *J. Chem. Phys.*, **56**, 5202.
B. C. Eu and T. P. Tsien (1972). *Chem. Phys. Lett.*, **17**, 256.
J. J. Ewing, R. Milstein and R. S. Berry (1971). *J. Chem. Phys.*, **54**, 1752.
E. R. Fisher and G. K. Smith (1971). *Appl. Optics*, **10**, 1803.
W. L. Fite and P. Irving (1972). *J. Chem. Phys.*, **56**, 4227.
M. A. D. Fluendy, D. S. Horne, K. P. Lawley and A. W. Morris (1970), *Mol. Phys.*, **19**, 659.
M. A. D. Fluendy and K. P. Lawley (1973). *Chemical Applications of Molecular Beam Scattering*, Chapman and Hall, London.
K. W. Ford and J. A. Wheeler (1959). *Ann. Physics*, **7**, 259, 287.
T. D. Gaily and M. F. A. Harrison (1970). *J. Phys.*, **B3**, L25.
M. Godfrey and M. Karplus (1968). *J. Chem. Phys.*, **49**, 3602.
R. G. Gordon (1969). *J. Chem. Phys.*, **51**, 14.
R. Grice (1967). *Thesis*, Harvard University.
R. Grice and D. R. Herschbach (1974). *Mol. Phys.*, **27**, 159.
J. Grosser (1973). *Physica*, **64**, 550.
W. Hack, F. Rosenkranz and H. Gg. Wagner (1971). *Z. Naturforschung*, **26a**, 1128.
J. B. Hasted and A. Y. J. Chong (1962). *Proc. Phys. Soc.*, **80**, 441.
R. Haug, G. Rappenecker, Ch. Schlier and C. Schmidt (1973). *Proc. VIIIth ICPEAC*, Beograd, p. 603.
J. Heinrichs (1968). *Phys. Rev.*, **176**, 141; **184** (1969), 254.
R. K. B. Helbing and E. W. Rothe (1969). *J. Chem. Phys.*, **51**, 1607.
D. R. Herschbach (1966). In J. Ross (Ed.), *Advances Chemical Physics*, Vol. X, p. 319.
D. R. Herschbach (1973). In J. Lee, D. M. Hercules and M. J. Cornier (Eds), *Chemiluminescence and Bioluminescence*, Plenum Press, N.Y.
G. Herzberg (1966). *Mol. Spectra and Mol. Structure*, Vol. III.
M. M. Hubers and J. Los (1974). To be published.
B. M. Hughes, C. Lifshitz and T. O. Tiernan (1973). *J. Chem. Phys.*, **59**, 3162.
R. K. Janev and A. Salin (1972). *J. Phys.*, **B5**, 177.
R. K. Janev and A. R. Tančić (1972). *J. Phys.*, **B5**, L250.
B. R. Johnson and R. D. Levine (1972). *Chem. Phys. Lett.*, **13**, 168.
L. R. Kahn, P. J. Hay and I. Shavitt (1974). *J. Chem. Phys.* To be published.
P. J. Kalff (1971). *Thesis*, University of Utrecht.
K. J. Kauffmann, J. R. Lawter and J. L. Kinsey (1974a). *J. Chem. Phys.*, **60**, 4023.
K. J. Kauffmann, J. L. Kinsey, H. Palmer and A. Tewarson (1974b). *J. Chem. Phys.*, **60**, 4016.

V. Kempter (1974). This book, Article 9.
G. M. Kendall and R. Grice (1972). *Mol. Phys.*, **24**, 1373.
J. L. Kinsey (1972). In J. C. Polanyi (Ed.), *Chemical Kinetics*, Vol. 9, Chap. 7. Butterworth.
G. P. Können, J. Grosser, A. Haring and A. E. De Vries (1973). *Chem. Phys. Lett.*, **21**, 445.
G. P. Können, J. Grosser, F. Eerkens, A. Haring and A. E. De Vries (1974a). *Chem. Phys.*, **6**, 205.
G. P. Können, J. Grosser, A. Haring, A. E. De Vries and J. Kistemaker (1974b). *Rad. Eff.*, **21**, 171.
G. P. Können, A. Haring and A. E. De Vries (1975). *Chem. Phys. Lett.*, **30**, 11.
L. P. Kotova (1969). *Sov Phys. JETP*, **28**, 719.
P. J. Kuntz, M. H. Mok and J. C. Polanyi (1969). *J. Chem. Phys.*, **50**, 4623.
K. Lacmann and D. R. Herschbach (1970a). *Chem. Phys. Lett.*, **6**, 106.
K. Lacmann and D. R. Herschbach (1970b). *Abstract R6*, 23rd Annual Gaseous Electronics Conf.
L. D. Landau (1932). *Physik. Zeitschr. Sowjetunion*, **2**, 46.
L. D. Landau and E. Lifshitz (1967). *Mécanique Quantique*, Moscou.
C. B. Leffert, W. M. Jackson, E. W. Rothe and R. W. Fenstermaker (1972). *Rev. Sci. Instr.*, **43**, 917.
C. B. Leffert, W. M. Jackson and E. W. Rothe (1973). *J. Chem. Phys.*, **58**, 5801.
R. D. Levine and R. B. Bernstein (1971). *Chem. Phys. Lett.*, **11**, 552.
J. K. Lewis and J. T. Hougen (1968). *J. Chem. Phys.*, **48**, 5329.
W. Lichten (1967). *Phys. Rev.*, **164**, 131.
C. Lifshitz, T. O. Tiernan and B. M. Hughes (1973). *J. Chem. Phys.*, **59**, 3182.
J. Los (1973). In B. C. Čobić and M. V. Kurepa (Eds.), *Invited Lectures and Progress Reports of the VIIIth ICPEAC*, Beograd, p. 621.
S. M. Lin and R. Grice (1973). *Faraday Dis.*, **55**, 370.
J. L. Magee (1940). *J. Chem. Phys.*, **8**, 687.
J. L. Magee (1952). *Disc. Far. Soc.*, **12**, 33.
W. B. Maier II (1964). *J. Chem. Phys.*, **41**, 2174.
W. B. Maier II (1965). *J. Chem. Phys.*, **42**, 1790.
M. Matsuzawa (1968). *J. Phys. Soc. Japan*, **25**, 1153.
D. R. McDonald, M. A. D. Fluendy and K. P. Lawley (1973). *Proc. VIIIth ICPEAC*, Beograd, p. 56.
P. E. McNamee, K. Lacmann and D. R. Herschbach (1973). *Chem. Soc. Faraday Disc.*, **55**, 318.
J. E. Mentall, H. F. Krause and W. L. Fite (1967). *Disc. Faraday Soc.*, **44**, 157.
W. H. Miller and Th. F. George (1972). *J. Chem. Phys.*, **56**, 5637.
M. H. Mittleman and L. Wilets (1967). *Phys. Rev.*, **154**, 12.
J. Moseley, W. Aberth and J. R. Peterson (1970). *Phys. Rev. Lett.*, **24**, 435.
J. Moseley, W. Aberth and J. R. Peterson (1971). *Proc. VIIth ICPEAC*, Amsterdam, p. 295.
N. F. Mott and H. S. W. Massey (1965). *The Theory of Atomic Collisions*, Clarendon Press, Oxford, p. 369.
A. M. C. Moutinho, A. P. M. Baede and J. Los (1970). *Physica*, **51**, 432.
A. M. C. Moutinho (1971). *Thesis*, University of Leyden.
A. M. C. Moutinho, J. A. Aten and J. Los (1971). *Physica*, **53**, 471.
A. M. C. Moutinho, J. A. Aten and J. Los (1974). *Chem. Phys.*, **5**, 84.
J. J. C. Mulder, B. J. Botter and J. A. Kooter (1974). To be published.
S. J. Nalley and R. N. Compton (1971). *Chem. Phys. Lett.*, **9**, 529.

S. J. Nalley, R. N. Compton, H. C. Schweinler and V. E. Anderson (1973). *J. Chem. Phys.*, **59**, 4125.

R. H. Neynaber, B. F. Myers and S. M. Trujillo (1969). *Phys. Rev.*, **180**, 139.

E. E. Nikitin (1968). In H. Hartmann (Ed.), *Chemische Elementarprozesse*, Springer-Verlag, Heidelberg, p. 43.

E. E. Nikitin (1970). *Comments on Atomic and Mol. Physics*, **1**, 166.

E. E. Nikitin and A. I. Reznikov (1972). *Phys. Rev.*, **A6**, 552.

C. Nyeland and J. Ross (1971). *J. Chem. Phys.*, **54**, 1665.

R. E. Olson, J. R. Peterson and J. Moseley (1970). *J. Chem. Phys.*, **53**, 3391.

R. E. Olson (1970). *Phys. Rev.*, **A2**, 121.

R. E. Olson and F. T. Smith (1971). *Phys. Rev.*, **A3**, 1607.

R. E. Olson, F. T. Smith and E. Bauer (1971). *Appl. Optics*, **10**, 1848.

Th. F. O'Malley (1971). In D. R. Bates (Ed.), *Adv. At. Mol. Phys.*, Vol. 7, p. 223.

M. Oppenheimer and R. S. Berry (1971). *J. Chem. Phys.*, **54**, 5058.

M. Oppenheimer (1972). *J. Chem. Phys.*, **57**, 3899.

E. K. Parks, A. Wagner and S. Wexler (1973). *J. Chem. Phys.*, **58**, 5502.

J. Perel and H. L. Daley (1971). *Phys. Rev.*, **A4**, 162 and references cited there.

W. B. Person (1963). *J. Chem. Phys.*, **38**, 109.

J. R. Peterson, W. H. Aberth, J. T. Moseley and J. R. Sheridan (1971). *Phys. Rev.*, **A3**, 1651.

L. G. Piper, L. Hellemans, J. Sloan and J. Ross (1972). *J. Chem. Phys.*, **57**, 4742.

M. Polanyi (1932). *Atomic Reactions*, London, Williams and Norgate Ltd.

J. Politiek and J. Los (1969). *Rev. Sci. Instr.*, **10**, 1576.

G.-P. Raabe (1971). *Thesis*, University of Göttingen.

G.-P. Raabe (1973). *Z. Naturforschung*, **28a**, 1642.

D. Rapp and W. E. Francis (1962). *J. Chem. Phys.*, **37**, 2631.

K. Razi Naqvi (1972). *Chem. Phys. Lett.*, **15**, 634.

K. Razi Naqvi and W. Byers Brown (1972). *Int. J. Q. Chem. VI*, 271.

J. F. Reddington, M. A. D. Fluendy and K. P. Lawley (1973). *4th Symp. Int. Jets Moleculaires*, Cannes.

R. D. Rundell, K. C. Aitken and M. F. A. Harrison (1969). *J. Phys.*, **B2**, 934.

A. Russek (1971). *Phys. Rev.*, **A4**, 1918.

B. M. Smirnov (1965a). *Sov. Phys. Dokl.*, **10**, 218.

B. M. Smirnov (1965b). *Dokl. Akad. Nauk SSSR*, **161**, 92.

B. M. Smirnov (1967). *Sov. Phys. Dokl.*, **12**, 242.

F. J. Smith (1966). *Phys. Lett.*, **20**, 271.

F. T. Smith (1969a). In S. Geltmann (Ed.), *Topics in Atomic Collision Theory*, Academic Press, New York, p. 95.

F. T. Smith (1969b). *Phys. Rev.*, **179**, 111.

E. C. G. Stueckelberg (1932). *Helv. Phys. Acta*, **5**, 370.

E. Teller (1937). *J. Phys. Chem.*, **41**, 109.

W. R. Thorson and S. A. Boorstein (1965). *Proc. IVth ICPEAC*, Québec.

W. R. Thorson, J. B. Delos and S. A. Boorstein (1971). *Phys. Rev.*, **A4**, 1052.

J. P. Toennies (1968). In H. Hartmann (Ed.), *Chemische Elementarprozesse*, Springer-Verlag, Heidelberg, p. 157.

J. P. Toennies (1974). In W. Jost (Ed.), *Physical Chemistry, An Advanced Treatise*, Vol. VIa: Kinetics of Gas Phase Reactions, Academic Press, New York, Chap. 5.

F. P. Tully, Y. T. Lee and R. S. Berry (1971). *Chem. Phys. Lett.*, **9**, 80.

F. P. Tully, H. Haberland and Y. T. Lee (1973). *Proc. VIIIth ICPEAC*, Beograd, p. 101.

J. C. Tully and R. K. Preston (1971). *J. Chem. Phys.*, **55**, 56; see also: J. Krenos *et al.* (1971), *Chem. Phys. Lett.*, **10**, 17; R. K. Preston and J. C. Tully (1971), *J. Chem. Phys.*, **54**, 4297.

N. G. Utterback (1969). In L. Trilling and H. Y. Wachmann (Eds), *Rarefied Gas Dynamics*, Vol. II, Academic Press, p. 1361.

J. Van den Bos (1970). *J. Chem. Phys.*, **52**, 3254.

J. Van den Bos (1971), *Proc. VIIth ICPEAC*, Amsterdam, p. 299.

A. Van der Meulen, A. M. Rulis and A. E. De Vries (1973). Private communication.

J. Von Neumann and E. P. Wigner (1929). *Z. Physik*, **30**, 467.

R. B. Vora, J. E. Turner and R. N. Compton (1973). *Report ORNL-TM-4329: Phys. Rev. A.*,**9**, 2532 (1974).

F. T. Wall, L. A. Hiller and J. Mazur (1958). *J. Chem. Phys.*, **29**, 255.

J. Weiner, W. B. Peatman and R. S. Berry (1970). *Phys. Rev. Lett.*, **25**, 79.

J. Weiner, W. B. Peatman and R. S. Berry (1971). *Phys. Rev.*, **A4**, 1824.

S. Wexler (1973). *Ber. Bunsengesellschaft*, **77**, 606.

E. Wigner (1948). *Phys. Rev.*, **73**, 1002.

B. W. Woodward (1970). *Thesis*, University of Colorado; JILA-report, No. 102.

A. M. Woolley (1971). *Mol. Phys.*, **22**, 607.

C. E. Young, R. J. Beuhler and S. Wexler (1974). *J. Chem. Phys.*, **61**, 174.

A. A. Zembekov (1971). *Chem. Phys. Lett.*, **11**, 415.

A. A. Zembekov and E. E. Nikitin (1972). *Dokl. Akad. Nauk. SSSR*, **205**, 1392 (in Russian).

A. A. Zembekov (1973). *Teor. i Eksper. Khimiya*, **9**, 366 (in Russian)

C. Zener (1932). *Proc. Roy. Soc.*, **A137**, 696.

SUBJECT INDEX

Abelian integral 338, 353
Action angle variables 86, 91, 110, 129
Action classical 87, 321
Action integral 105
Active vibrations 304
Adiabatic approximation 20
  vibrational 31
Adiabatic states 305, 465, 469, 512
Alkali dimers, reactions 259, 280, 430
Analytic continuation 114, 337
Angular momentum coupling 372
Angular coupling 470
Anisotropy,
  in potential 229, 332, 337, 375, 380,
    393, 407, 410
  in cross section 380, 397
Associative electron detachment 277, 527
Atom-surface scattering 105
Autoionization, see Ionization; cross section, charge transfer
Axilrod–Teller–Muto–forces 364

Beam detectors
  differential surface ionization 248, 270
  electron bombardment 274, 281, 430
  negative surface ionization 494
Beam resonance methods 258
Beam sources
  discharge 284, 287, 289, 292, 318
  dissociative 285, 289
  excited states 318
  hyperthermal 464
  ion neutralization 427, 446
  nozzle 248, 259, 401
  seeded 256, 295, 427, 430
  sputtering 429, 494
  unstable species 318, 373
  vibrationally hot 514
Bolometers 319, 367
Born–Oppenheimer approximation 465
Boundary value method 15
Branching ratio 39
Broken path method 26

Cd, collision induced emission, 445
Canonical transformation, 79, 81, 89, 113
Charge exchange, resonant 479

Chemical Reaction, see also scattering,
    reactive; cross section, reaction
  orientation dependence 390
Chemiionization 263, 269, 464, 528
Chemiluminescence 9, 263, 269, 276,
    418, 454, 505
  infra red 288, 294
Collisions, ion-molecule, see also Potential parameters; scattering
  elastic 231-242
  inelastic 218–227
  reactive 9, 193–242
Collisions, neutral–neutral
  collinear 10, 13, 90, 92
  diatomic-atom 394
  dissociative 493
  elastic Chapter 7
  electronic excitation 59, 193, 218, 445,
    446, 447, 449
  inelastic 218–227, 489
  ionizing 443, 449, 453, 458, Chapter 10
Collision complex 68, 102, 105, 216, 302
  osculating 284
  reactive Chapter 2, 297 et seq.
  statistical 37, 47, 491
Complex time 118, 130
Coordinates
  atom transfer 24, 55
  curvilinear 11, 33
  natural bifurcation 35
  reaction path 13
Correlation diagrams 204, 272, 293
  alkali-rare gas 438
  Ar–Ar 438
  Ar–K 438
  Br_2–Cl_2 300
  He–He 452
  K–Hg 441
  K–K 440
  K–Na 441
  Ne–Ne 453
  O–N_2 306
Correspondence relations 81, 85
Corresponding states 357, 367
Coupled channel method 52
Coupling
  matrix elements 176, 426, 468, 482,
    499, 512, 519

radial 423, 458
rotational 423, 458, 468
symmetry effects 487
Cross section, differential
elastic 458, Chapter 7
emission 422
excitation 422, 457
inelastic 229, 456
ionization 458, 499
orientation dependence 397
oscillations 327, 333, 344, 378, 382
reaction Chapter 2, 252 et seq.
Cross section, see also Scattering, Reactive
absorption 48
charge transfer 458, 477, 495, 528
alkali–halogen, tabulated 509
alkali–non halogen, tabulated 522
emission 419, 431, 450
inelastic 122
ionization 524
orientation dependence 397, 512
reaction 38
total 315, 419, 425, 431, 458
undulations 17, 91, 93, 329, 351, 361, 379, 426, 436
Curve crossing
avoided 59, 347, 423, 454, 456, 469, 476
conical 485
covalent–ionic 304, 306
pseudo 464

Deflection function 321, 474
with curve crossing 476, 521
Delay time 19
Diabatic states 372, 424, 438, 464, 466
Diatomics in molecules method 139, 489
Diffraction 131, 327
Direct interaction model 61
Distorted wave approximation 27, 54, 67, 402, 481

Elastic scattering, see Cross section; Scattering
Electron affinity
adiabatic 306, 513, 529
vertical 304, 512, 529
Electronic excitation, see Collisions; Cross section; Scattering
Energy loss spectra 222, 456, 457
Energy release 287, 291, 292, 295, 304

Exchange oscillations 329, 479
Exponential method 14

Faddeev–Watson equations 11, 28, 62, 68, 167
Fast beams 426
Fine structure components 433, 444
Fission model 41
Fitting procedure 334
Flux lines 26
Focusing, molecular
alternating gradient 396
four pole field 395
six pole field 396
Four centre reactions 300
Frank–Condon factor 2, 60, 82, 84, 277, 306, 490
Frank–Condon principle 423, 445

Gallium beam, orientation 390
Generating function 80
Generator, canonical transformation 84, 113
Glory effect, see also Cross section, total
322, 328, 344, 379
with angle dependent potential 375, 402, 408
Green's function 115

H, collisional excitation 454
Harpooning, see Reaction mechanism
High energy approximation 332, 353
Hund's rules 372, 423, 433, 468

Identical particle scattering 68, 329, 346, 361, 374, 379
Impulsive model 60, 66
Inert gases, catalogue of total cross sections 419
Interference structure, see also Cross section, Glory effect
in differential cross section 234, 327, 378, 382, 476
in total cross section 329, 351, 369
Intermolecular forces, see also Potential energy surfaces
ab initio calculations 163
anisotropic, long range 331, 333, 337, 390, 407
anisotropic, short range 393, 408–414
effective 51
three body 365

Inversion procedures 336, 354
Ion beams, generation of 189
  internal energy 191
Ionization, *see also* Cross section; Scattering
  autoionization 5, 419, 448, 449, 455, 458
  Penning 418, 448
Isotopic substitution 15, 61

K, collisional excitation 420, 432, 439, 441

Landau–Zener model 424, 432, 449, 453, 477
Landau–Zener–Stuekelberg model 60, 469
Laser excitation 318
Laser induced fluorescence 277
Lennard-Jones potential 334, 345, 356, 381
Lippmann–Schwinger equation 57, 67
Lyman radiation 454

Mass spectrometer
  conventional 186
  scattering 189
  tandem 186
Microscopic reversibility 42, 67, 418, 432
Molecular orbitals 217, 291, 451
Monte Carlo methods 103, 112, 113
Multiple collision expansion 63
Mutual neutralization 495

Na, collisional excitation 421
Ne, collisional excitation 447
Negative surface ionization 494
Non-adiabatic transitions 4, 59, 79, 418, 466
Nuclear symmetry 65, 423

O, autoionizing states 455
Opacity function 48
Optical model 11, 51, 67, 257, 268, 326, 333, 337
Orbiting 305, 322, 328, 339, 369
Orientation, in beams 258, 389

Phase integral 100
Phase shift 88, 100, 330, 347, 472

Phase space theory, 16, 38, 66
Photodissociation 85, 304
Polarization of emission 425, 431, 441
Population inversion 28
Potential energy surfaces 8, Chapter 4
  two body (potential curves)
    Br–Br 254
    CCl_3–Cl 254
    CH_3–I 254
    Cs–Hg 358
    H_2–He 391
    He–He 363, 366, 373
    HgCl–Cl 254
    I–Br 511
    K–Br 306
    K–Hg 358
    Li–Hg 358
    N–H 150
    $N–NO^+$ 206, 207
    Na–Hg 358
    Ti^{4+}–F 162
  many dimensional
    H_3, H_3^+, H_4, H_n, HeH_2^+, LiH_2^+, Li_2H, Li_3 Li_3^+, Li_4, Li_2H^+, BeH_2, ArH_2^+, H_2He^+ 165
    FH_2 167
    LiF_2 176, 307, 489, 507
    Li_2F 176
    KCl_2 490, 507
    HNO, H_2O, MgF_2, LiO_2, H_2NO, CH_2, CaF_2, HeH_2, $BH_2CH_2^+$, BeF_2, CO_2, LiH_2^+ 145
    H_2F^-, LiHF, $HeBeH_2$, HeH_2^+, CH_5^+, CH_4, FCH_3F^-, $CNCH_3F$, NaH_2O^+, FH_2O^-, ClH_2, CaF_2Cl, $ClLi_2$, $ClCH_3NC$, H_3, H_2O, NO_2 146
    HeH_2, H_3, H_3^+, H_4, N_2O^+, O_3, O_3^-, C_3, CO_2, KrF_2, NH_3, HeH_3^+, CH, FH_2, HF_2, HO_2, CH_2 150
    H_2S_2, Cl_2S_2, HCHO, LiH_2, Li_3, BH_2 162
    H_3, H_4 166
    Li_2F, Na_2Cl, LiNaCl, NaKCl, KNaCl, K_2Cl, Li_2Cl, LiKCl 168
    $ClCH_3Br$, KCl_2, CH_4, BrH_2, HBr_2 169
    NaKBrCl 172
    H_3^+, H_3, H_2F 174
    K_2XY 264
    $NaCs_2$, $NaRb_2$, NaK_2, KRb_2 274

Potential energy surfaces, calculation of
  ab initio   139, 140, 371
  atoms-in-molecules   158
  closed shell interaction   162
  configuration interaction   148
  diatomics-in-molecules   163
  empirical   170, 334, 345, 351, 356, 381
  geminal method   156
  generalized valence bond   147, 154
  Hartree–Fock   141, 143
  iterative natural orbital   174
  LEPS   22, 171
  maximally paired HF   147
  perturbation methods   152
  Porter–Karplus   25, 54
  pseudopotential   139, 160, 162
  Rittner   168, 172
  SCF   143
  separated pairs   157
  spin optimized SCF   146
Potential energy surfaces, fitting   334, 355
Potential parameters
  atom/atom
    H-rare gases   371
    Hg–alkali metals   357
    rare gas–rare gas   361, 368
  atom/molecule
    H–H_2 379
    rare gas–H_2, N_2, O_2, NO, CO, CH_4, CO_2 378
  ion/molecule   236, 239, 513
  molecule/molecule
    H_2–H_2 379
    H_2–N_2, O_2, CO, CH_4, NO, CO_2, NH_3, SF_6 382

Quenching, collisional   375, 380, 418, 432, 442, 490

RRKM theory   45, 46, 215, 298, 300, 303
Radiative recombination   276
Rainbow effect   323, 324, 327, 333, 344
  double   376
Reactive scattering, see Scattering, reactive; Reaction mechanism; Collisions
Reaction mechanism
  collision complex   37, 66, 253, 265, 297, 302
  electron jump   60, 67, 249, 257, 260, 263, 266, 301, 507

four centre   260, 262, 300
impulsive   60, 66
rebound   248, 252
stripping   60, 64, 201, 248, 250, 275, 284
Reaction probability   12, 34, 61, 94, 121
Reaction threshold   94
Rearrangement operator   57
Regge pole   347
Resonance
  compound state   21, 30, 66
  Feschbach   104
  shape   21, 30, 66, 99, 330, 369
Resonance lines, in emission   437, 438
  polarization   425, 431, 437
Rosenthal effect   436
Rotational transitions   331, 332, 337
Rotational excitation   227, 258, 377
Rotational polarization analysis   258

SO_2, collisional excitation   422
S matrix   38, 40, 93, 99, 347, 481
  classical   20, 59, 78, 86, 90, 93, 94
  for forbidden processes   35, 114, 117, 118
Scattering, see also Cross sections; Collisions
  ion/molecule
    Ar^+–D_2 199, 224
    C^+–D_2 209
    $C_2H_2^+$–C_2H_4 212
    $C_2H_4^+$–C_2H_4 212
    D^+–HD   195
    H^+–Ar   234, 237
    H^+–CO   220
    H^+–H_2 194, 224
    H_2^+–H_2 210, 225
    H^+–N_2 238
    I_2^+–C_2H_4 215
    Kr^+–D_2 202
    Li^+–H_2 226
    N^+–H_2 203
    N^+–O_2 203
    N_2^+–H_2 211
    O^+–D_2 62, 208, 240
    O^+–HD   241
    O^+–N_2 204
    O_2^+–D_2 210
  charge transfer   463
  elastic, see Potential parameters
  inelastic   78, 223
  reactive

alkali atoms, superthermal 256
alkali atoms, thermal 49, 250
alkali dimers, alkali atoms 272
  halogen atoms 269
  halogen molecules 260
  hydrogen atoms 270
  polyhalides 265
alkaline earth atoms 274
halogen atoms–halogen molecules 283
halogen atoms–hydrogen 294
hydrogen atoms–halogen molecules 285
hydrogen atoms–H_2 296
hydrogen atoms–unsaturated hydrocarbons 296
methyl radicals–halogen molecules 289
oxygen atoms–halogen molecules 292
Second virial coefficient 359, 365
Seeded beams 430
Selection rules 58
  optical 221
  spin 221
Semi-classical approximation 77, 78, 320, 341, 353
Shuler's rule 58
Spectator stripping, see Reaction mechanism
Spectrometers single beam 185
  double beam 188
Spectroscopy absorption 359
Spin–orbit coupling 176
Stark effect 389, 395
  linear 397
  negative 396
  quadratic 396
Stationary phase approximation 32, 58, 81, 82, 114, 117, 321, 474
Statistical models 37
Steepest descent, method of 115
Stern–Gerlach magnet 395
Stueckelberg oscillations 426
Sudden approximation 78, 405
Supernumerary rainbow 327, 342
Superposition principle 99, 103, 131
Surface hopping 59, 194
Surprisal analysis 42, 307

Symmetric top molecule, scattering form 377, 397
Symmetry oscillations, see Identical particle scattering; Interference

Temperature 43
Termolecular reaction rates 41, 48
Threshold behaviour
  energy 422, 432
  for excitation 17, 419, 450, 451, 454
  for ionization 492, 529
  region of 121
Time of flight analysis 285, 319, 456, 457
  spectra 457
Trajectory
  classical 77, 78, 87, 97, 201, 480
  complex 103, 132
  Monte Carlo 10, 489
  surface hopping 59, 175, 194
  theories of 10
Transition, see also Collisions
  non-adiabatic 59, 132, 423, 465
  probability 91, 93, 126
  vibrational 49, 59, 92, 122
Transition state theory 16, 45
Transitional approximation 323
Transvibronic reaction 418
Triple collisions 41
Tritium, beams of 296
Tunnelling 31, 99, 118, 453, 471
Two state approximation 468

Uniform approximation 96, 322, 326
Unimolecular reactions 8

Valence bond method 139
Variation method 23, 27, 53
Velocity analysis 251
Velocity selection 250, 284, 319
Vibration–rotation excitation 223–227
Vibrational excitation 333

Wigner's rule 446
Wigner–Witmer rule 58
WKB approximation 82, 88, 102, 129, 321, 338, 377, 471, 480
Woodward–Hoffman rules 58

Zeeman effect 395
Zn, collisional excitation 445